NEW ENGLAND INSTITUTE
OF TECHNOLOGY
LIBRARY

ELECTRONIC ASSEMBLY FABRICATION

Electronic Packaging and Interconnection Series
Charles M. Harper, Series Advisor

Alvino PLASTICS FOR ELECTRONICS

Classon SURFACE MOUNT TECHNOLOGY FOR CONCURRENT ENGINEERING AND MANUFACTURING

Ginsberg and Schnoor MULTICHIP MODULE AND RELATED TECHNOLOGIES

Harper ELECTRONIC PACKAGING AND INTERCONNECTION HANDBOOK

Harper and Miller ELECTRONIC PACKAGING, MICROELECTRONICS, AND INTERCONNECTION DICTIONARY

Harper and Sampson ELECTRONIC MATERIALS AND PROCESSES HANDBOOK, 2/e

Hwang MODERN SOLDER TECHNOLOGY FOR COMPETITIVE ELECTRONICS MANUFACTURING

Lau BALL GRID ARRAY TECHNOLOGY

Lau FLIP CHIP TECHNOLOGIES

Licari MULTICHIP MODULE DESIGN, FABRICATION, AND TESTING

Related Books of Interest

Boswell SUBCONTRACTING ELECTRONICS

Boswell and Wickam SURFACE MOUNT GUIDELINES FOR PROCESS CONTROL, QUALITY, AND RELIABILITY

Byers PRINTED CIRCUIT BOARD DESIGN WITH MICROCOMPUTERS

Capillo SURFACE MOUNT TECHNOLOGY

Chen COMPUTER ENGINEERING HANDBOOK

Coombs ELECTRONIC INSTRUMENT HANDBOOK, 2/e

Coombs PRINTED CIRCUITS HANDBOOK, 4/e

Di Giacomo DIGITAL BUS HANDBOOK

Di Giacomo VLSI HANDBOOK

Fink and Christiansen ELECTRONICS ENGINEERS' HANDBOOK, 3/e

Ginsberg PRINTED CIRCUITS DESIGN

Juran and Gryna JURAN'S QUALITY CONTROL HANDBOOK

Jurgen AUTOMOTIVE ELECTRONICS HANDBOOK

Manko SOLDERS AND SOLDERING, 3/e

Rao MULTILEVEL INTERCONNECT TECHNOLOGY

Sze VLSI TECHNOLOGY

Van Zant MICROCHIP FABRICATION

To order or receive additional information on these or any other McGraw-Hill titles, please call 1-800-822-8158 in the United States. In other countries, contact your local McGraw-Hill representative.
WM16XXA

ELECTRONIC ASSEMBLY FABRICATION

Chips, Circuit Boards, Packages, and Components

Charles A. Harper Editor-in-Chief
Technology Seminars, Inc., Lutherville, Maryland

First Edition

NEW ENGLAND INSTITUTE
OF TECHNOLOGY
LIBRARY

McGRAW-HILL
New York   Chicago   San Francisco
Lisbon   London   Madrid   Mexico City   Milan
Montreal   New Delhi   San Juan   Seoul   Singapore
Sydney   Toronto

Cataloging-in-Publication Data is on File with the Library of Congress.

McGraw-Hill
A Division of The McGraw Hill Companies

Copyright © 2002 by McGraw-Hill, Inc. All rights reserved. Printed in the United States of America. Except as permitted under the United States Copyright Act of 1976, no part of this publication may be reproduced or distributed in any form or by any means, or stored in a data base or retrieval system, without the prior written permission of the publisher.

1 2 3 4 5 6 7 8 9 0 DOC/DOC 0 9 8 7 6 5 3 2

ISBN 0-07-137882-0

The sponsoring editor for this book was Stephen S. Chapman and the production supervisor was Sherri Souffrance. It was set in Century Schoolbook by J. K. Eckert & Company, Inc.

Printed and bound by R. R. Donnelley & Sons Company.

> Information contained in this work has been obtained by McGraw-Hill, Inc., from sources believed to be reliable. However, neither McGraw-Hill nor its authors guarantee the accuracy or completeness of any information published herein and neither McGraw-Hill nor its authors shall be responsible for any errors, omissions, or damages arising out of use of this information. This work is published with the understanding that McGraw-Hill and its authors are supplying information but are not attempting to render engineering or other professional services. If such services are required, the assistance of an appropriate professional should be sought.

This book is printed on acid-free paper.

CONTENTS

Preface xi
Contributors xiii
Dedication xv

Chapter 1 Printed Circuit History and Overview 1
Ken Gilleo
1.1   Introduction 1
1.2   Circuit Technology 4
1.3   Circuit Materials 21
1.4   Circuit Processes 26
1.5   Circuit Construction and Types 32
1.6   Circuitry for Electronic Packages 40
1.7   Printed Circuit Trends 45
1.8   The Printed Circuit Business and Economics 46
1.9   Long-Term Expectations 46
1.10 Acknowledgments 48
1.11 References 48
1.12 Bibliography 49

**Chapter 2 Development and
Fabrication of IC Chips** 51
Charles Cohn
2.1   Introduction 51
2.2   Atomic Structure 52
2.3   Vacuum Tubes 55
2.4   Semiconductor Theory 58
2.5   Fundamentals of Integrated Circuits 70
2.6   IC Chip Fabrication 78
2.7   References 103

Chapter 3 Packaging of IC Chips 105
Charles Cohn
3.1   Introduction 105
3.2   The IC Package 106
3.3   Package Families 108
3.4   Package Technologies 117
3.5   Comparison of Package Technologies 153
3.6   IC Assembly Processes 156
3.7   Summary and Future Trends 186
3.8   References 188

Chapter 4 Laminates and Prepregs as Circuit Board Base Materials 191
Doug Sober
4.1 Introduction 191
4.2 Paper-Based Materials 192
4.3 FR-4 Materials 194
4.4 Composite Materials 199
4.5 High-Performance Materials 203
4.6 Microvia Materials 205
4.7 High-Speed/High-Frequency Materials 207
4.8 Conclusion 208

Chapter 5 Printed Circuit Board Fabrication 211
Joseph Fjelstad
5.1 Introduction 211
5.2 Background and History 212
5.3 Materials of Construction 216
5.4 Laminate Material Preparation 223
5.5 Lamination Methods 223
5.6 Laminate Forms for PCBs 225
5.7 Laminate Selection 228
5.8 Solder Masks 228
5.9 Generic Processes Overview for a Plated-Through-Hole Printed Circuit 231
5.10 Additive and Subtractive Processing 241
5.11 Single-Sided Circuit Process Examples 242
5.12 Double-Sided Circuit Process Examples 245
5.13 Standard Multilayer Circuit Process Example 247
5.14 Mass Lamination 251
5.15 Metal-Core Printed Circuit Boards 252
5.16 Flexible Circuits 254
5.17 High-Density Interconnection (HDI) Structures 257
5.18 Future Directions 265
5.19 Summary 265
5.20 References 265

Chapter 6 Package and Component Attachment and Interconnection 267
Charles G. Woychik
6.1 Introduction 267
6.2 Levels of Interconnections 268
6.3 Types of First-Level Interconnections 269
6.4 Types of Second-Level Interconnections 272
6.5 Ceramic Packages 279

6.6	Organic Packages.	281
6.7	Module-Level Assembly.	283
6.8	Ceramic Flip Chip Modules.	284
6.9	Organic Flip Chip Modules	286
6.10	Organic Wire Bond Modules	289
6.11	Second-Level Assembly	292
6.12	Migration to Pb-Free Solders and Soldering Processes	301
6.13	References	303

Chapter 7 Solder Materials and Processes for Electronic Assembly Fabrication 305
Jennie S. Hwang, Ph.D.

7.1	General Trends	305
7.2	Solder Materials.	309
7.3	Physical Properties	313
7.4	Metallurgical Properties	315
7.5	Mechanical Properties.	316
7.6	Solder Alloy Selection—General Criteria.	317
7.7	Lead-Free	317
7.8	Reflow Soldering	328
7.9	Inert and Reducing Atmosphere Soldering	340
7.10	Printing	344
7.11	PCB Surface Finish	347
7.12	"Green" Manufacturing	357
7.13	References	358
7.14	Suggested Readings.	359

Chapter 8 Printed Wiring Board Cleaning 363
William G. Kenyon

8.1	Introduction	363
8.2	Board Fabrication	368
8.3	Solder Mask over Bare Copper	379
8.4	Solder Mask over Tin-Lead	397
8.5	Environmental Controls and Considerations.	402
8.6	Regulatory (OSHA and EPA) Considerations	415
8.7	Applicable Documents.	424
8.8	Source and Reference Materials	426
8.9	Acknowledgements and Thanks	427

Chapter 9 Board Coating Materials and Processes 429
John Waryold, Edward B. Mines

9.1	Introduction	429
9.2	Purpose and Function of Conformal Coatings	430
9.3	Specifications that Address Conformal Coatings.	430

9.4	Generic Types of Coating Materials	430
9.5	Criteria for the Selection of a Generic Coating Type	433
9.6	Engineering or Performance Aspects	433
9.7	Processing or Application Aspects	435
9.8	Basic Prerequisites for Optimal Coating Performance	437
9.9	Conformal Coating Application Methods	441
9.10	Multiple Coats	445
9.11	Health and Safety Considerations Government Regulations	449
9.12	References	450

Chapter 10 Flexible and Rigid Flexible Fabrication Process ... 453
Martin W. Jawitz, Michael J. Jawitz

10.1	Introduction	453
10.2	Classification	454
10.3	Flexible Materials	456
10.4	Fabrication Processes	459
10.5	References	482

Chapter 11 Fabrication and Properties of Electronic Ceramics and Composites ... 483
Jerry E. Sergent

11.1	Introduction	483
11.2	Surface Properties of Ceramics	486
11.3	Thermal Properties of Ceramic Materials	490
11.4	Mechanical Properties of Ceramic Substrates	494
11.5	Electrical Properties of Ceramics	500
11.6	Ceramic Fabrication	503
11.7	Ceramic Materials	506
11.8	Composite Materials	515
11.9	Forming Ceramics and Composites to Shape	524
11.10	References	526

Chapter 12 Hybrid Microelectronics and Multichip Module Technologies ... 529
Jerry E. Sergent

12.1	Introduction	529
12.2	Thick Film Technology	530
12.3	Screen Printing	538
12.4	Drying	548
12.5	Firing	549
12.6	Cermet Thick Film Conductor Materials	555
12.7	Thick Film Resistor Materials	559

12.8	Thick Film Dielectric Materials	567
12.9	Overglaze Materials	570
12.10	Thick Film—Conclusions	571
12.11	Thin Film Technology	571
12.12	Thin Film Materials	577
12.13	Comparison of Thick and Thin Film	580
12.14	Copper Metallization Technologies	581
12.15	Summary of Substrate Metallization Technologies	590
12.16	Resistor Trimming	590
12.17	Assembly of Hybrid Circuits	601
12.18	Packaging	612
12.19	Design Of Hybrid Circuits	621
12.20	Multichip Modules	625
12.21	References	629

Chapter 13 Environmental Considerations in Electronic Assembly Fabrication 631

John W. Lott

13.1	Introduction	631
13.2	Material Considerations	636
13.3	Processes for Environmentally Conscious Manufacturing	640
13.4	Regulatory and Nonregulatory Considerations	648
13.5	Summary	654
13.6	References	654

Index   659
About the Editor   672

PREFACE

This modern era of high-technology electronics has a major impact on everyone in one way or another. Yet, the complexities of modern electronics assemblies, and the many stages required to fabricate a final high-technology electronic product, make it very difficult for most except high-technology engineering experts to understand. Adding to this problem of understanding high-technology electronics are, first, the myriad of types of electronic assemblies and products, and, second, the many different materials and fabrication processes required to create a final product. In fact, the broad range of materials and processes required are so varied that even the technical specialists in one area of materials and fabrication processes often do not understand the other materials and fabrication processes required to create the final electronic assembly product. With this being the case, it is understandable that all of the many nontechnical people whose interests, or even livelihood, are impacted by these technologies would not understand them. The bottom line of all of the above discussion is that there is an urgent need for a book to explain these multiple electronic assembly fabrication materials and processes in easy to understand presentations. This book, *Electronic Assembly Fabrication*, fully meets this need.

Electronic Assembly Fabrication is organized to present the fabrication materials and processes in the sequence required to create an electronic assembly—from fabrication of silicon semiconductor devices through fabrication of circuit boards and assemblies, to fabrication with various substrates and component attachment technologies, to environmental considerations, which are becoming increasingly important. After an initial chapter explaining the very interesting history of electronic assembly fabrication, which has led us to our modern technology era, there are two chapters dealing with semiconductors. The first of these explains the fabrication of semiconductor devices, such as integrated circuits, from raw sand and refined silicon metal. The second explains the fabrication of packaged semiconductor devices, ready for attachment to some substrate, such as circuit boards or ceramic substrates. After this comes a series of chapters explaining the fabrication of base laminates used in circuit boards, the fabrication of the various types of single-layer and multilayer circuit boards, and the myriad of fabrication technologies required to attach components and devices to circuit boards. Included in this series of chapters is a chapter on the always critical, and often controversial, subject of soldering materials and fabrication processes. The also highly critical and often controversial electronics assembly fabrication areas of electronic assembly cleaning and circuit board coating are also covered in two separate chapters. Then comes a chapter that covers electronic assembly fabrication and interconnection using flexible circuits, another rapidly growing technology. Finally, two chapters are included to explain the other major substrate and electronic assembly fabrication technologies that utilize ceramic substrates. The first of these covers the materials and processes associated with fabrication and use of the various types of ceramic substrates. The second covers the sensitive and often complex array of materials and fabrication processes that are used to manufacture ceramic based electronic assemblies. Last, and also increasingly important, is a chapter explaining the critical environmental considerations that must be dealt with to keep electronic assembly fabrication operations within environmentally safe bounds, both from safety and governmental regulation viewpoints.

Needless to say, a book of this breadth, covering so many disciplines, could not have been written without an outstanding author team. I have been fortunate to have worked with a group of the most highly regarded industry leaders in the preparation of this book. I would like to publicly express my deep appreciation to all of them. Their participation in this book will benefit many groups of readers, from students to business and industry readers to users of the myriad types of electronic equipment. Again, on behalf of all of these readers, I thank them.

Charles A. Harper
Technology Seminars, Inc.
Lutherville, Maryland

CONTRIBUTORS

Charles Cohn, Agere Technologies, Allentown, Pennsylvania (Chaps. 2, 3)

Joe Fjelstad, Pacific Consultants LLC, Mountain View, California (Chap. 5)

Ken Gilleo, Cookson Electronics, Cranston, Rhode Island (Chap. 1)

Jennie S. Hwang, H-Technologies Group, Inc., Cleveland, Ohio (Chap. 7)

Martin W. Jawitz, Consultant, Las Vegas, Nevada (Chap. 10)

William G. Kenyon, Global Center for Process Change, Inc., Montchanin, Delaware (Chap. 8)

John W. Lott, Du Pont Electronics, Research Triangle, North Carolina (Chap. 13)

Edward B. Mines, Humiseal Company, Woodside, New York (Chap. 9)

Jerry E. Sergent, Consultant, Corbin, Kentucky (Chaps. 11, 12)

Douglas Sober, Polyclad Laminates, Franklin, New Hampshire (Chap. 4)

John Waryold, Humiseal Company, Woodside, New York (Chap. 9)

Charles G. Woychik, IBM Corporation, Endicott, New York (Chap. 6)

Dedicated to my Little Buddy T.J.

Chapter 1

Printed Circuit History and Overview

Ken Gilleo
Cookson Electronics
Cranston, Rhode Island

1.1 Introduction

The printed circuit, sometimes called *printed wiring,* has been with us for about a century. Modern electronics is completely dependent on this technology, which replaced hand wiring long ago. Hand wiring, or point-to-point wiring, cannot even come close to the density requirements of today. Even if we could somehow make discrete wiring compact and dense, the economics would be prohibitive compared with highly automated circuit production that creates all the "wires" simultaneously. This chapter will provide a short history, some background, and an overview of the fascinating world of printed circuits.

1.1.1 Circuitry Terms

The simplest definition of a printed circuit is *a patterned array of electrical conductors supported on a nonconductive platform.* The conductor pattern is designed to interconnect components and to provide a path to a subsystem or to the main system (sometimes called the *backplane*). The two main constituents of circuits are thus electrical conductors and insulators, called *dielectric.* Such a product that only has conductors and dielectric would appear to be a simple system, but the types of materials and their different constructions create an almost endless variety of types and categories. The number of different processes is also very extensive and on the increase. Later, we will describe and discuss the many materials, processes, and classes of printed circuits. You may be surprised to learn that only a few types of circuits are actually printed and that some of the methods used almost a century ago have come back as the "new" methods. You will encounter two abbreviations, PCB (printed circuit board) and PWB (printed wiring board). PWB is an older term that was repopularized when polychlorinated biphenyls (PCBs) made the headlines as some of the nastiest chemicals to harm humans and the environment. However, PCB is a more accurate term, since the fabricated wires are really part of the circuit. The conductor pattern for many products, especially high-frequency devices, is adjusted and tuned. The printed circuit is much more than just mass-produced wires, and we will proudly use the term PCB.

1.1.2 The Purpose and Value of Printed Circuits

The printed circuit has basically been a replacement for wires. Early electronics used metal wires, rods, and strips to connect coils, capacitors, resistors, batteries, and vacuum tubes of that day. The most complex device at the dawn of electronics was the vacuum tube triode amplifier, followed by the pentode. These original amplifiers required less than half a dozen connections. Radio receivers and transmitters, such as the Marconi wireless sets, were built by hand, and devices were connected with copper wires and strips that were crimped or bolted first and soldered later. While it would seem that discrete wiring could handle the job during the early years of electronics, this was not really the case. The first telecom revolution of the early 1900s required high density. While the telegraph and wireless radio could be built with wires, the telephone exchange switches had already hit the density barrier. Surprisingly, the first printed circuits attempted to solve the density problems of the phone exchange central office. A denser switchboard would allow an operator to connect more customers faster.

Analyzing the printed circuit from the wire replacement perspective of the early 1900s, we would see several advantages. First, density could be increased by at least an order of magnitude. Today, that figure is many orders. Manufacturing costs could be cut drastically, since wiring would occur simultaneously. One printed circuit technician could produce as many wire structures as a roomful of others doing hand wiring. Another important benefit was error reduction, still one of the most valuable attributes of printed circuits. Anyone who has ever wired a home electronics product, such as a Heathkit, knows how easy it is to miss-wire. Correctly designed PCBs preclude this possibility, although it is still possible to attach the wrong component. Other attributes are shorter run lengths for higher performance, a platform for components, miniaturization, weight reduction, and many more. The PCB has also facilitated standardization, and its value to the entire industry should not be minimized. The huge contract assembly industry would not be practical without PCB standardization. Next, we will begin to look at how the simple concept of a patterned conductor array can be deployed into different configurations for various applications.

1.1.3 Classes of Circuits

The simplest circuit comprises only a pattern of conductors on one side, which we will refer to as the top. This class is referred to as a single-sided circuit and sometimes abbreviated as SS-circuit. It is the simplest configuration and the easiest to manufacture, and this level satisfies a good portion of the industry.

The next level of density is called back-bared or double-access and is found in the flexible circuit industry where thinness of dielectric enables this constriction. Dielectric is selectively removed so that conductors can be accessed from the back or bottom of the circuit. This may seem like a trivial feature, but it has wide ramifications, especially for packaging. The ability to connect from the bottom without adding a second layer of conductors makes low-cost area array packaging possible. Bumps can be formed on the bottom of the circuit that connect to the single top circuit layer that makes the connection to the chip. Double-access circuits are used for many of the flex-based packages that include many of the ball grid arrays (BGAs). We will go into more detail in later sections.

The next class is the double-sided circuit, which is very popular and only moderately difficult to build. However, manufacturing complexity increases more than two-fold, since vertical connections, typically plated-through holes, must be also constructed using many additional steps. There is a subset of double-sided circuits without any connections between the two layers, but this represents a special product that will be ignored for simplicity. Although the plated-through hole vertical interconnection method has been practiced for decades, technologists have worked just as long to replace it. Finally, this environmentally unfriendly and density-limiting process is beginning to see replacement. The method of vertical interconnection is another way of defining the circuit class and will be discussed later.

We will mention one more density-increasing process before moving on to mutlilayers circuits: *jumpers* or *crossovers*. Here, a conductor is created on the circuit layer side, and it "jumps" over one or more conductors without making contact, which would result in a short. Many methods have been developed with varying degrees of success. A simple process, but one with potential reliability problems, is the printed jumper. Dielectric is first applied over the existing conductors that are to be "jumped." Then, conductive ink is applied so that it connects two conductors without making electrical contact with those being jumped. Screen printing and needle deposition are used. Problems can arise if there are defects in the dielectric coating such as voids produced in the printing process or from solvent evaporation when the dielectric is hardened (if solvent borne). A second problem can be the interface between the jumper and the conductors. Unless the metal conductor is finished with a nonoxidizing material (e.g., gold), oxide formation over time can cause a loss continuity with the printed ink. Jumpers can be used for all kinds of circuits but have been more common with flex circuits, especially polymer thick film (PTF), which will be covered later.

The final and ultimate class is the multilayer circuit. The construction requires at least three separate conductor layers or planes, although layer pairs that produce even numbers of layers are more common. The original processes involved laminating a stack of double-sided circuits together, drilling holes through the entire lay-up, and then adding vertical connections. Pins, rivets, and bolts were used early on, but the plated-through hole method was a major breakthrough, since all connections formed simultaneously. Over the years, alternative methods have been developed with varying levels of success. Some use a build-up strategy wherein one layer at a time is constructor. Other methods mate finished double-sided pairs with conductive adhesive films called *interposers*. Both the old and new multiplayer processes will be covered later.

1.2 Circuit Technology

1.2.1 The Dawn of Electronics

The printed circuit was invented at the turn of the previous century during the *first telecom revolution* that was gathering strength and popularity. Let's turn back to this exciting time that has so many parallels with the *second telecom revolution* of our new millennium. We are now in the second century of electronics, and telecom is now the most important driver for business, technology, and printed circuit innovation.

The time is Dec. 31, 1899. The Morse telegraph key had celebrated its 50th birthday, and the Western Union Company has been in business for half a century. Alexander Graham Bell had already proven telephonic communication, and the telephone and telegraph are essential communication links for countries throughout the world. Marconi had already transmitted messages across the Atlantic, but it would still be a few years before Fleming would perfect the vacuum tube and the first diode (the Fleming valve), and Lee De Forest would build the first triode amplifier (Audion) to fully enable wireless broadcasting. Marconi would later share the Nobel prize for his wireless contributions, the invention that the Titanic used to send its fateful "mayday." We have already entered the *information age*. Figure 1.1 shows an early Marconi wireless key radio key operated by the author (amateur license W1PHD).

The embryonic electronics industry, comprising the telegraph, telephone, and radio, was off to a strong start as we entered the new twentieth century. All of these communications industries were rapidly expanding, and that was creating an immense need for circuitry—mass-produced circuitry! The telephone systems, with their hundreds of phone exchange lines, required manual switching units, or PBX consoles, that would allow operators to make manual line connections. The increasingly complex radio circuits needed an alternative to tedious and error-prone hand wiring if that technology was to become as widespread and far-reaching as that industry envisioned. The electronics industry sought circuit technology that could enable mass production and automation.

Figure 1.1 Marconi wireless.

1.2.2 The Pioneers (1900–1925)

The year was 1903 when Albert Hanson, of Berlin, filed his "printed" wire patent in England. His invention was aimed at solving the telephone exchange needs. The Hanson process, although not a true "printed circuit" method, did produce a patterned array of electrical conductors on dielectric substrate. Metal foil was first cut or stamped out into conductor patterns. The copper or brass traces were then adhesively bonded to paraffin-coated paper and the like. This appears to be the first documented circuit invention. Figure 1.2 shows the basic concept.

But Hanson added some additional innovations that can still be considered "modern" circuit principles. This circuit pioneer had already realized that high density would be of great importance, and he therefore designed his circuits with conductors on both sides, as can be seen in Fig. 1.3. Also recognizing that interlayer connections were critical, he added access holes to permit the top and bottom conductors to be selectively connected together. Although the connections were crude crimping and twisting, his 1903 invention clearly describes *double-sided through-hole circuitry*. Hanson also stated that conductors could be formed *in situ* by electrodeposition or by applying metal powder in a suitable medium (conductive ink). We find it remarkable that the first circuit patent touched on so many concepts that are considered to be of modern origin.

Thomas Edison, who had recently commercialized the first incandescent electric light bulb, also tackled the printed wire problem. When asked by close friend Frank Sprague, founder of Sprague Elec-

Figure 1.2 Hanson patent figure.

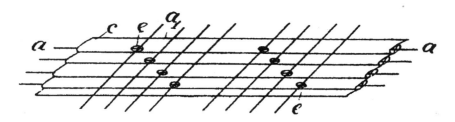

Figure 1.3 Hanson double-sided construction.

tric Company, how to "draw" conductive traces on paper, Edison offered several ideas in a written response. The approaches included (1) selectively applying glue (polymer adhesive) and dusting the wet "ink" with conductive graphite or bronze powder, (2) patterning a dielectric with silver nitrate solution and reducing the salt to metal, and (3) applying gold foil to the patterned adhesive. While Edison, in his short note, did not mention printing, the first two methods could be easily adapted to several printing processes. In fact, the concept (1) of using polymer-based adhesive with conductive particles, is the basis for today's polymer thick film technology, which continues to gain importance because of its low cost and intrinsically clean attributes. Figure 1.4 shows the Edison PTF concept that will be described later. Concept (2) is actually electroless plating, a method used today. Perhaps if Edison had worked the problem, he would have included copper plating

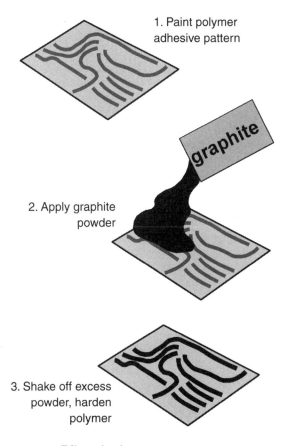

Figure 1.4 Edison circuit concept.

and vacuum deposition, since America's most prolific inventor had already patented these processes.[3] Incidentally, Edison invented the electrolytic copper foil—the process that is used to manufacture circuit laminates. We should note that Edison's approaches are *additive*.

Several other ideas emerged over the next several decades as electronics continued to expand at an exciting pace. Radio soon became the most important driver for printed circuitry. Wireless was capturing the attention of the world. The first public radio station, KQW (San Jose, CA) had gone on the air in 1912. By the end of the second decade of the twentieth century, radio had been introduced to most of the major countries. Ships now carried the Marconi radio system, and the wireless was beginning to save lives. Perhaps all lives would have been lost if the Titanic had not been equipped with a Marconi radio. Soon, there could be a radio in every household, as predicted by David Sarnoff, who headed RCA and NBC. Electronics pioneers could see the immense market for mass-produced circuits, and inventors were strongly motivated to answer the challenge of electronic interconnect.

1.2.2.1 Subtractive or additive? The earliest circuit principles were based on additive, or build-up methods. Conductors were deposited onto dielectric. This, of course, is the most straightforward approach and the most obvious process. However, the graphics printing industry had long used subtractive methods for making plates. Initially, wood had been carved away to yield raised letters and graphics. Later, the industry switched to metal. Metal type could be cut or cast and then set. Finally, printers moved to the printing plate. Metal was etched away to produce raised letters as background was removed. It was only a matter of time before the electronics industry would discover what the printers had known for decades.

In 1913, Arthur Berry filed for a patent[4] that claimed a method of making circuits where metal was etched away. He described a process of coating metal with a resist, prior to etching, as in improvement over die cutting, which left stress concentrator sharp corners. Berry appears to be the first to describe etched, or subtractive circuitry. Figure 1.5 shows his etched heater circuit. Later, Littledale[5] described a similar method.

Photolithography was well known during the early days of circuit development while the subtractive process was mostly being ignored. Bassist[6] gave specific details on photoengraving, including the use of photosensitive chromium salts. Although his patent dealt with print plate making, the process could easily be adapted for circuitry, since Bassist described preparing compliant plates by electrodepositing copper foil onto dielectric laminate. We can see that all of the basic circuit

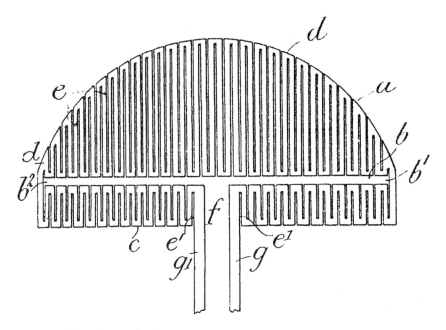

Figure 1.5 Berry heater circuit.

concepts and the process principles were know during the first quarter century of electronics.

1.2.3 A Real Industry Develops (1915–1939)

1.2.3.1 Early commercial circuit processes. One successful inventor, Max Schoop, commercialized a metal flame spraying process that was used for many years. We need to keep in mind that early electronics were "power hungry." Vacuum tubes required heated filaments and high voltages. The practical circuit of the vacuum tube era would need to carry substantial currents. The Schoop process, which could deposit thick patterns of metal by flame spraying through a mask, produced the hefty and robust circuits that were required. Figure 1.6 shows the 1918 Schoop process.[7]

The Schoop flame spray approach had problems of cost and wasted metal and, although subsequent inventors added improvements, a true printed circuit process was still needed. The next circuit inventor of note was Charles Ducas,[8] who described both etching and plated-up conductors. One version involved electroplating a copper, silver, or gold pattern onto a low-temperature metal alloy through a contact mask. Heating allowed the conductor (typically a coil) to be separated

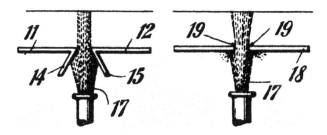

Figure 1.6 Schoop metal spray process.

from the fusible bus plate and mask. Another Ducas process involves forming grooves in dielectric such as wax and filling them with conductive paste (composition not disclosed). The paste is then electroplated. Conductive paste could also be printed, or stenciled, onto dielectric and then electroplated. Both sides of the dielectric layer could be used for circuitry. In fact, Ducas goes on to describe multilayer circuits and a means of interconnection layers: "Two or more panels may be positioned adjacent to each other...." Figure 1.7 shows a method that may be employed for interconnecting the circuits of different planes. "The electroplated metal forming conductor 62 extends through an aperture in panel... which in turn contacts with conductor 60 on panel 58." Croot, of Paragon Rubber, also described filling grooves and electroplating.[9]

A year later, Frenchman Cesar Parolini came up with improvements in the field of additive plating.[10] He patented the printing of patterns with adhesive onto dielectric followed by application copper powder to the wet ink. The excess copper particles were shaken off, and the ink hardened with heat. This is the basic Edison concept and one of the Ducas methods, but Parolini reduced it to practice and added one more concept—jumpers or crossovers. "U"-shaped metal wires were imbedded into the wet conductive ink, which was then hardened. Electroplating formed continuous metal over the copper ink

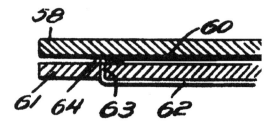

Figure 1.7 Ducas multilayer.

and also plated the jumpers to the conductors. Figure 1.8 shows the Parolini process, complete with a jumper.

Other inventors of that era also employed print and plate methods for circuitry. Seymour[11] used printed graphite paste to make the platable patterns. But this inventor was already moving into the flexible circuit niche. Seymour described "...pathways along or across which impulses may be conveyed, transmitted or regulated..." on "...flexible, relatively thin, pliable body...capable of being made to assume various forms...." He used waxed paper and gutta percha as the dielectric with graphite, lead, and copper conductive pastes. Copper plating was the final step. Figure 1.9 shows this 1923 flexible circuit used as a radio tuner. This is the first dynamic flexible circuit, since the end user flexed the circuit to tune the radio.

A parade of other inventors quickly followed, but most proposed variations on the themes that had been disclosed by predecessors, as is typically the case today. For example, Franz,[12] in 1933, added conductive particles to the polymer ink. He made a printing paste with carbon filler that could be screen printed or stenciled onto cellophane or similar lamina. The hardened ink, although stable, had high resistance as compared with metal. Franz, perhaps aware of the Parolini process that was disclosed seven years earlier, added a copper electroplating step. Once again, modern circuit makers have used this concept. Figure 1.10 shows a clever "accordion" circuit that Franz described in his patent. This is clearly a 3-D flex circuit principle. Surprisingly, the flex circuit industry has not yet fully utilized 3-D flex concepts. And what was purpose of the accordion circuit? The Franz folding circuit was a replacement for windings in transformers, as can be seen in Figure 1.10. Since the Franz invention, several modern circuit practitioners have attempted to reinvent the printed winding concept. While not all circuit inventions have yet been made, a search of early patents can be a humbling experience for the would-be inventor.

1.2.4 World War II Boosts Technology

1.2.4.1 The war effort and hybrid circuits. World War II brought on circuit developments that took a different twist. The need for extremely

Figure 1.8 Parolini circuit with jumper.

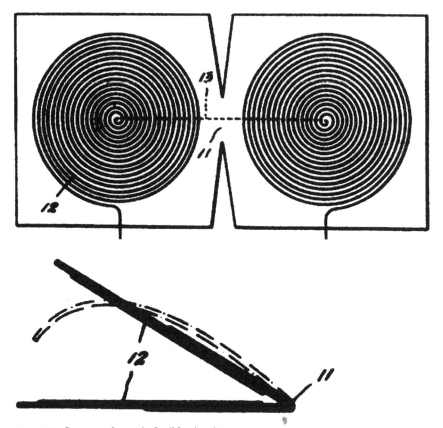

Figure 1.9 Seymour dynamic flexible circuit.

robust microelectronics for military ordnance spurred development of ceramics. Secret projects developed highly reliable ceramic substrate and conductive inks, call *cermets;* ceramic-metal. This process, now widely practiced in the ceramic hybrid industry, involved screen printing and stenciling circuit inks followed by high-temperature firing. The process was used to produce tens of thousands of various ordnance electronic fuses and is discussed in great detail by Cadenhead and DeCoursey.[2] The war efforts resulted in both the development and optimization of high-volume thick film printed circuit manufacturing. Note that the hybrid process and most of the earlier inventions are additive methods. Figure 1.11 shows an early hybrid printed circuit from the National Bureau of Standards achieves.

After the war, the U.S. government, under the auspices of the National Bureau of Standards (NBS), disseminated printed circuit technology. Conferences were held, and publications resulted that de-

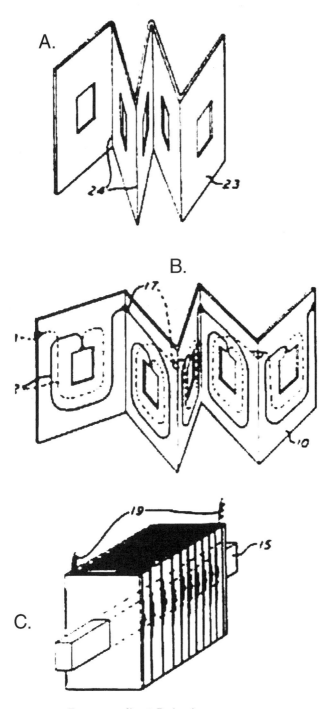

Figure 1.10 Franz accordion 3-D circuit.

Figure 1.11 Hybrid circuit.

scribed virtually all of the circuit-making concepts, including subtractive etching.[13]

1.2.4.2 The U.S. government defines circuit processes. Now that the war was over, it was time to move printed circuit technology into the commercial sector. A Circuit Symposium, sponsored by the U.S. Aeronautical Board and the National Bureau of Standards, was held in Washington, DC, in October 1947. Dozens of speakers and hundreds of attendees interacted in this unusually well attended conference. The more than two dozen processes where distilled down to six methods with a seventh included as an alternate step in some of these. The government settled on the following categories:

1. *Painting (really, printing).* Metal-filled inks are applied and cured or fired. This method includes ceramic thick film (CTF) and polymer thick film (PTF), and it remains a key method today. Application methods included, brushing, coating, pen writing, stenciling, printing, and dipping. CTF and PTF utilized screen printing almost exclusively, although conductive adhesives for component assembly are often stenciled. Printing is expected to remain an important process in the thick film circuit area for some time to come.

2. *Spraying.* Molten metal or composite conductor material is sprayed through a mask or onto a stencil. The mask can be a resist applied to the substrate. Schoop[7] first described metal flame spray-

ing in 1918, and later it was depicted by several others. The method achieved only limited commercial success because of high cost and waste. One alternative was to spray the entire substrate and then selectively etch away metal to form the circuit pattern. Spraying is an obsolete circuit process today.

3. *Chemical deposition.* Electroless and electrolytic plating are included here. Dozens of early patents described electroless, electrolytic, and combination plating. Chemical deposition remains an important process in many circuit-making schemes. Electroless deposition of copper was patented[14] over a century ago, however. Thomas Edison's circuit concept of 1904 was to selectively apply electroless silver.[2] Chemical deposition methods can be expected to increase as additive, semiadditive, and build-up processes become more prevalent.

4. *Vacuum deposition.* Sputtering and evaporation through a mask are the key processes mentioned. Thin film circuits are made by vacuum depositing copper, gold, and other metals. Modern thin film processes involve applying the mask, or photoresist, directly to the substrate and removing it after coating with metal. Although sputtering is gaining popularity today for producing adhesiveless flexible circuit substrate, Thomas Edison patented a metal vacuum deposition apparatus and process over 100 years ago.[3]

5. *Die stamping.* Many of the early patents claimed cutting and die stamping as the process for patterning conductors. More modern methods simultaneously bonded the weakly adhered metal foil to the substrate during the die cutting process. This was accomplished by using B-staged adhesive and a heated die bed. The method, although low in cost and environmentally friendly, is diminishing in use as tolerances become tighter and density demands increase. The major market for die stamping has been automotive instrument cluster circuits. General Motor's Packard Electric Division used die stamping to produce hundreds of millions of automotive circuits. The process was abandoned in the 1980s as high tool costs and long changeovers, combined with higher density requirements, diminished the value of the process. The die stamping circuit process is still used today for very high-volume runs of low-density circuits.

6. *Dusting (conductive powder over tacky ink).* Application of graphite or metal powder over wet ink or adhesive is one of the earliest processes reported. Thomas Edison was apparently the first to suggest dusting graphite over pattern adhesive. As mentioned earlier, a dozen or more patents used the basic dusting concept, but with

various improvements like electroplating. The dusting idea still seems to intrigue the modern circuit technologist, as evidenced by recent patents. Some of the later patents apply solder to the dusted conductors.[15] However, Parolini, in 1927, patented all the important elements of the process including overplating with copper.

X. We should note that the conference to not give subtractive chemical etching its own category. The majority of processes covered in the conference and in much of the early patent literature dealt only with additive circuit processes. The idea of applying metal to the substrate only to remove most of it in a later step apparently did not appeal to the pioneers. The conference considered etching as an obvious auxiliary process that could be used, if really needed, with some other processes.

One more method mentioned, but not separated out, is "groove and fill," or channel circuitry. This involves forming grooves or recesses in the substrate and filling them with conductive material. The two basic subdivisions are (1) doctor blading conductive paste into the grooves and (2) using a fusible alloy (solder) that will flow off the board surface into the grooves. The groove-and-fill concept, mentioned by early inventors, still seems to attract modern technologists.[16,17] The method has never been an important circuit manufacturing technique. However, over the past several years, circuit interposers (multilayer interconnects) have been made by doctoring conductive materials into holes in dielectric. Figure 1.12 shows some of the modern printed circuits and products.

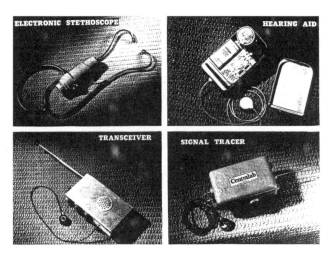

Figure 1.12 Modern pre-1950 circuits and products.[13]

1.2.5 Post-War Boom (1948–1975)

The NBS conferences sparked much interest among technologists and businessmen. Companies were started and a new industry began to develop. But, as is often the case, legal battles ensued, and the U.S. courts were asked to decide who invented the printed circuit. Paul Eisler claimed to be the inventor of the practical printed circuit, and many books refer to him as the "father of the printed circuit." Eisler's company, which essentially pitted him against the entire industry, eventually brought a lawsuit. The case would have major ramifications on the entire industry, and so it fought back. The Eisler vs. Bendix case was pivotal and one that would uncover prior art that the author has relied on in this chapter.

1.2.5.1 The rise and fall of the father of the printed circuit. Paul Eisler began developing circuit manufacturing processes during the war era and perhaps as early as 1936, when he moved to England, although documentation is lacking on his earliest claim. The goal of this Austrian was to develop a low-cost, mass production circuit process. Eisler eventually obtained several British patents[18,19] dealing with etching processes. The etching method ultimately became the well known photo-etching process that remains the most popular method in use. The process steps are shown in Figure 1.13.

The printing industry had used etching for print plate making well before there was even a need for circuitry. Printing technologists had already perfected the copper etching process, which initially used an etch resist that was mechanically patterned by scraping away with a sharp tool. During the 1800s, photosensitive coatings were discovered and perfected, which enabled the widespread use of photo engraving. Eisler appears to have borrowed this well established technology and imported it to the electronics industry, but was he really the first, as he claimed? The long list of inventions previously presented indicates that Eisler was not first by any measure.

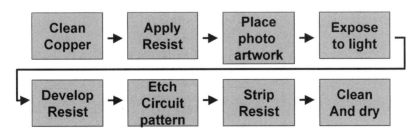

Figure 1.13 Photo-etching process.

The primary difference in the photo engraving process and Eisler's circuit-making method is the substrate construction. The printers used relatively thick copper plates, whereas Eisler used copper foil laminated to dielectric—although some printing plates were made of copper laminate.[6] The thick copper plates were engraved by the etchant to a depth of several mils, and copper typically remained at the base. Eisler's thinner copper was etched all the way through so that the patterns were isolated from one another. The Eisler patent referenced the print plate making technology and did not actually describe the etching chemistry. The phrase "as used in the printing industry" is found throughout the Eisler circuitry patents.

Several years later, Eisler received several U.S. patents[20–23] by referencing his British patents. There were a number of irregularities in the proceedings, and the U.S. patents were held up for about four years. Initially, the U.S. Patent Office rejected all of the Eisler claims because of prior art, much of which was covered in the earlier sections. After meetings and appeals, most of the patent claims were allowed, but there is no explanation by the Patent Office. A patent's "file wrapper" normally contains all of the written communication as well as summaries of meetings. The patent examiner simply allowed the patent without explaining what had transpired in the meeting and why he had decided to ignore the substantial prior art that would seemed to invalidate Eisler's claims.

Now armed with U.S. and British patents, Eisler commercialized circuit-making under Technographic Printed Circuits, Ltd. All appears to have gone well for the company until it sought to limit competition by lawsuit. Technograph, through its U.S. counterpart, Technograph Printed Electronics, Inc., sued the Bendix Corporation, which was producing printed circuits by an etching process in the U.S. The weary judge pointed out that the case came to trial in a lengthy process that essentially reviewed the entire history of printed circuitry *ad nauseum*. Eisler was not able to substantiate claims of earlier work and could not produce his "book of circuit samples." His prime exhibit, a three-tube radio receiver, never worked throughout the long months of trail.

Bendix, the defendant, countered with an overwhelming amount of prior patent art and asked that the patents be declared invalid. They pointed out that many, including Spaulding[24] and Miller[24B] as well as Stevens and Dallas,[25] had used photoetching to make circuits before 1940. Others, such as O'Connell,[26] had used print and etch. But a key point argued in the trial was that Eisler had maintained claims in the U.S. patents that had already been rejected in the earlier British patent filings. In other words, he was attempted to get U.S. coverage by referencing nonexistent documentation. Another important defense was that Eisler had simply patented well understood photolithography that had long been used by the printing industry. Certainly, Eisler's own statements in his patents supported this accusation.

On May 27, 1963, the case was decided: "Action by Technograph Printed Circuits, Ltd. and Technograph Printed Electronics, Inc. against the Bendix Corporation for patent infringement. Complaint dismissed."[27] Eisler had been defeated and dethroned as the father of printed circuitry. Until the day he died, Eisler felt that he had been wronged by the system, but it is clear from the vast amount of prior art that he was not first. The printed circuit was not invented by a single person but by many inventors who contributed to the total concept.

1.2.6 Summary of Modern Circuitry

1.2.6.1 The industry today. Subtractive photolithography, borrowed from the ancient printing industry, still remains the workhorse circuit-making method. While one may argue that today's circuit making is akin to using the eraser end the pencil instead of the point, the process has survived because it works and has reasonably good economics. The backward process has been assaulted for decades by matter-of-fact technologists who want to set circuit-making right—make it additive. However, fully additive processes have not delivered enough on their promises and have garnered only a small niche. The etching process has also weathered many storms brewed by environmentalists as better waste recovery methods evolved. However, subtractive etching is now under the most serious attack yet.

The demands of *faster-smaller-cheaper* and the responses have propelled the electronics industry into a revolution. Perpetual progress by the semiconductor industry has brought more power and more problems. Chips with millions of transistors needed 1000-plus connections to the outside world. Consumers demanded portability for converging technologies. Something had to give, and it would be the interface between the IC and circuit board. The IC could no longer be housed in a body that was more than an order of magnitude larger than the brain. The electronic package was under severe pressure to shed girth.

The ever-downsizing package was quickly running out of real estate with its peripheral wiring scheme, and micro-pitch leads were becoming an assembler's nightmare. We were already feeling the stranglehold of perimeter paralysis packaging. The salvation was the "packaging revolution" that was launched when area array designs, like the ball grid array (BGA), became a major focus. Chip on board (COB), flip chip on organic circuits, and new chip scale packages (CSPs) also become part of the shrinking landscape. But the packaging revolution puts extraordinary demands on the circuit industry for higher and higher density. And that pulls us right into the next stage of the domino affect, the "printed circuit revolution."

The two areas for circuit densification are conductor linewidths/spaces and vertical interconnect. Both areas are receiving considerable attention, with impressive government and industrial funding. ARPA has funded development in both areas, and significant progress has resulted that will likely change the circuit industry. One important trend is emphasis on semiadditive processing. Let's take a closer look.

The flexible circuit industry is noted for very fine-pitch, high-density products. Flex has had a long association with products the require intense miniaturization. This includes packaging like tape automated bonding (TAB, a flex-based package), disk drives, watches, and hearing aids. More recently, flex has become the choice for many new packages, including the elegant µBGA. So it is no mystery that the flex circuit industry has been striving for high density and is the recipient of much of the ARPA money.

Density attributes begin with flex materials that lend themselves to fine pitch and microvias. The substrate is very thin and smooth, with no reinforcement that could interfere with hole fabrication. But flex materials have been undergoing dramatic changes that can enable new circuit processes. Flex substrate producers have been steadily replacing foil lamination with direct metal deposit processes. Vacuum deposition will likely become the main process for producing adhesiveless laminate (actually, the correct term is *clad*, since no adhesive is used). Although the newer materials are superior, because the weak link adhesive has been eliminated, a more important feature is the thinness of the copper. Flex material is now available in very thin configurations, all the way down to a few microns of copper. This has important ramifications for the circuit industry.

The semiadditive circuit processes require a thin conductor layer or "seed" metal that can serve as a removable plating bus. A typical process involves applying a photoresist over the thin metal followed by imaging, developing, electroplating, stripping resist, and etching away the thin seed layer. Although etching is still involved, the process does not define the conductor shape and very little metal is removed. The semiadditive process is starting to gain momentum, since it can produce very fine lines (<25 microns), build straight-wall conductors (no etch factor), and produce minimal waste. The flex-based package industry has already adopted semiadditive circuit making to produce the needed conductor width of down to 1 mil. TAB and µBGAs are now produced by plating-up conductors.

Vertical connection processes are also undergoing changes, and this may be the most important area today. The original plated-through hole process, launched in 1953 by Motorola (Placir process), is being replaced by more efficient structures and processes. Lasing and plasma etching are replacing drilling. The plated-through hole is replaced by direct plating through microvias. Build-up multilayer methods have become commercial where holes are photodefined. Interposer methods, where anisotropic conductive adhesives mate circuit pairs together, are also enjoying success.[28–30] Future high-density circuits will likely be plated-up instead of etched.

Now that we know how the industry developed, the next sections will provide an overview of the printed circuit technology in more detail, starting with materials. The purpose is to give the unfamiliar reader the background to better understand the forthcoming chapters and perhaps add new information for the circuitry experts.

1.3 Circuit Materials

1.3.1 Board Materials

The dielectric substrate is commonly referred to as the *board*, even when flexible material is used, a practice that irritates many involved with *flex*. A better term is *substrate,* and it will be used for the remainder. A substrate can be virtually any nonconductor, even paper. Circuits have even been built on metal, but with a dielectric coating to prevent shorting. The primary requirement is to provide a platform that can support conductors (and components) without shorting them together.

1.3.1.1 Ceramic. Ceramic substrate, as we saw earlier, came into use during WWII and was used for making electronic bomb fuses and other ordnance electronics. One popular material is aluminum oxide, or alumina (Al_2O_3), which is still used today. The ceramic-based circuits are still often referred to as *hybrid,* because the early circuits used a mix of technologies and types of components. Today, ceramic substrate is used for high-performance military products as well as high-frequency modules. While ceramic is the most expensive of the three classes, it has extremely good high-temperature performance and excellent thermal management characteristics. Some of the ceramics, while excellent dielectrics, are better thermal conductors than metals. Ceramic also has unparalleled dimensional stability. This feature, coupled with high strength, makes it a top choice for many types of package platforms. Many computer chips are mounted on ceramic chip carriers that are configured as area array pages such as pin grid arrays (PGAs), ball grid arrays (BGAs), or column grid arrays (CGAs).

The circuit as an electronic package chip carrier is covered in a later section.

1.3.1.2 Rigid organic. The organic class of substrate is by far the most popular because of low cost, ease of manufacturing, and high versatility. Polymers, especially epoxies, are used as the binder or continuous phase. Glass fiber is commonly used to provide higher strength, dimensional stability, and lower thermomechanical expansion. The most common class of materials fall into the broad category called FR-4.

FR-4 consists of aromatic epoxy resins, hardener, flame retardant, glass weave, and a variety of additives. While epoxy systems typically have a coefficient of thermal expansion (CTE) range of 70 to 90 ppm/°C, the glass fibers restrain movement. A desirable CTE value is near that of the copper conductors or about 18 ppm/°C. A matched conductor-substrate pair gives minimum stress. When unmatched, the metal-substrate construction, called *laminate*, tends to warp when heated—like a bimetallic strip. Low-expansion FR-4 reduces the warping problem and also minimizes stress on assembled components. But we should note that the glass weave only constrains movement in the x-y plane. The epoxy is more or less free to expand in the vertical, or z-direction. In fact, z-axis expansion may actually increase, since the expanding epoxy has no other place to go but up. Z-axis CTE values can exceed 100 ppm/°C.

Many subclasses of FR-4 are available and can be categorized by thermal characteristics. The glass transition temperature (T_g) of the system is a common criterion. The T_g is the temperature at which a phase change occurs in the polymer and may not always be a precise value when several resins are blended together. A polymer that is below its glass transition temperature is in a more rigid, *glassy* state in which strength is higher and the CTE is lower. Raising the temperature above the T_g transforms the material into its rubbery phase, where there is more molecular movement (heat is molecular motion), and the polymer becomes softer but expands at a much higher rate. Other changes can occur, but it is the change in CTE that is of most concern. The CTE below the T_g (called α_1) can be 15–18, an ideal range. Above the T_g, the CTE value, called α_2, can more than triple, and this excessive expansion can produce warpage. The situation is exacerbated by the fact that the polymer is softer and more easily distorted. Since a high CTE value and warpage are undesirable, laminate suppliers continue to work on improving properties and raising the T_g value. Higher grades of laminates typically have higher T_g values. The CTE value is raised when more polymer cross-linking occurs by adding more functional monomers. You may hear terms like *multi-*

functional, polyfunctional, and *high functionality* in reference to higher-T_g circuit materials.

Although FR-4 is the workhorse substrate, many others are available, and more are coming on stream. There are two key driving forces for new materials. One is performance, where higher temperature stability and the ability to make higher-density circuits are paramount.

A possible transition to higher-temperature, lead-free alloys could boost thermal performance requirements. Although defining high-temperature characteristics is straightforward, the high-density factor is less clear. Attributes that facilitate higher-density circuit fabrication include smoother surfaces and easier and more controllable hole formation. Laser drilling has become a popular method of achieving smaller holes, called *microvias*, and substrates may need to improve laser-drilling characteristics in the future. Both the polymer properties and the reinforcement type and structure make this a more difficult to define area. Some of the newer substrates include, polyimides (PIs), bismaleimide triazine (BT), cyanate ester (CE), and epoxies blended with other polymers, including CEs and PIs. The so-called advanced laminates have T_g values that exceed 200°C and even approach 300°C. Epoxy systems have remained dominant, even though they may be modified with other polymers. But this may be changing.

One more factor that could greatly alter the laminating chemistry is the halogen-free movement that appears to be coming from Japan and spreading to Europe. Some, with no scientific support, are proposing the ban on halogens in printed circuits, because dioxins have been detected in incinerator gases (apparently operated below correct combustion temperatures). The true source of chlorine-containing dioxins, the prevalent pollutants, cannot be bromines from FR-4. Bromine simple cannot transmute to chlorine. The likely source is the millions of pounds of PVC products that are disposed of by consumers each year. However, brominated epoxies could nevertheless be banned, opening the door for intrinsically flame-retardant polymers like higher-priced polyimide.

1.3.1.3 Flexible substrate. Flexible circuitry is the original, and still the most versatile, type of substrate. The pioneers sought to replace flexible wire with flexible printed wire. Flex substrate is much more than a thin version of rigid board. Flex is designed to remain flexible throughout a life that can exceed many millions of flexural cycles, as is the case for disk drive circuits. The first difference, compared with rigid, is that there is no reinforcement—at least not in the most common products. Glass weave, so widely used in rigid, would interfere with flexing, and the glass would eventual fracture and break. Absence of low-expansion reinforcement requires that the substrate polymer have an intrinsically low CTE value. The most common dielectric for flex, polyimide, such as Dupont's Kapton®, the first commercial PI film. The PIs can easily match copper's 18 ppm/°C value. In fact, polyimides without filler or reinforcement can be made with CTE values low enough to match bare silicon die. The polyimides, because they contain nitrogen atoms, have intrinsic flame retardancy and will not run afoul of any pending legislation that targets halogen flame retardants. In many ways, the thin (25 to 125 µ) flex dielectrics are the best printed circuit materials in the world. The absence of glass fiber and their thinness allow very small microvias to be formed using lasers, plasma, and even chemical etching. The polyimides are unique among all laminates in that they are extremely tough and heat-resistant but are etched (dissolved) by special caustic solutions that do not attack copper conductors. The two factors that limit the widespread use of polyimides are higher moisture absorption (requiring prebake for assembly) and higher cost, the more significant consideration.

Other polymers are used for flex, and the second most popular one is polyester such as Dupont's Mylar®, the original polyester film. The polyesters are used only where some level of flammability is acceptable (e.g., for computer keyboards) and solder will not be used. The polyesters begin to shrink above 150°C and can be soldered only by using special techniques in which heat is localized. Polyester remains popular because of its excellent electrical properties and very low cost. Higher-temperature thermoplastics, such as PEN, are used to a smaller extent, but they can be soldered only with highly controlled processes. One more class of flex substrate is represent by the polyamides such as Dupont's Nomex®, the same polymer used for bullet-resistant vests and flame-protective clothing. One Dupont inventor claims that the name Nomex came from *No More Ex*cuses.

1.3.1.4 Conductors. Copper is king in both the world of wiring and the printed circuitry. Readily available copper was used on the earliest circuits, and continuing use has made it the *de facto* standard of electronics as the integrated circuit (IC) continues to move from aluminum to copper to boost performance. The circuit chemistry of copper is well established, and processes have been optimized for etching, plating, and applying other finishes.

Conductive inks, both ceramic and polymeric, are also used in different areas of the printed circuit industry. Ceramic inks, called *cermets*, are used with ceramic substrate. The ceramic conductive inks can be made with copper, but silver and its alloys have been used since the beginning of this technology, starting in the 1940s. Copper-based inks can require special reducing gas atmospheres during the ink hardening stage. Palladium may be alloyed with silver to retard potential silver migration.

Polymer-based inks of the polymer thick film (PTF) class are also used but for applications where their lower conductivity and density are acceptable. The PTF conductive inks are almost exclusively based on silver filler, since this metal has stable conductivity due to its unusually conductive oxide. Copper PTF ink would quickly become highly resistive and eventually nonconductive as oxide formed around the copper particles. PTF inks are used alone and also with copper circuitry. Work continues to make stable conductive copper ink, but success has been marginal.

1.3.2 Masks and Coatings

The conductor traces are often protected with dielectric coatings or masks. The soldering process typically requires a non-wetting stop, called a *solder mask,* to prevent the molten solder from flowing too far beyond the component assembly zone. Both surface mount assembly and wave soldering generally require a solder mask. The mask must be selectively applied so that the conductors in the component assembly and connection areas are open. The two common masking methods are printing and photomaging that will be described in the process section. Conformal coatings that essentially seal the entire assembly from the environment are also used, but usually for harsh-environment applications like military products.

1.3.3 Conductor Finishes

While copper conductors provide very good electrical paths that are easy to fabricate, the surface of this metal is active and will readily react with many materials. Copper will tarnish and corrode if not protected. Copper quickly reacts with oxygen from the air to form an oxide layer, and also with just about any other active gas, including CO_2 and sulfur compounds. Soluble copper salts (e.g., copper chloride) that form when chloride contamination is present often cause fatal electromigration problems in the field. The simple remedy is to apply a protective finish or treatment over the bare copper.

Copper can be readily electroplated with nickel, gold, silver, platinum, and other metals. Since gold is very conductive but very inert, a gold finish is highly desirable. Wire bonding, the most common method for connecting ICs to the outside world, generally requires a gold finish. But there is an issue with applying gold to copper, and it relates to high compatibility. Gold atoms will alloy with copper and eventually diffuse into the copper conductor so that the gold protective layer slowly disappears. The simplest solution is to first plate nickel over the copper and then plate gold onto the nickel. This finish, called gold-over-nickel, remains popular and can be used for component assembly areas, wire bond pads, and contacts. Note that a different thickness of gold (and alloys) is used, depending on the application. Also note that gold is readily dissolved, or leached, by molten solder and that an excessive amount of gold will embrittle the solder. However, only a very thin layer of gold is used as a finish, primarily to keep costs low, and this level is not a problem for solder. Other inert metals, including platinum, palladium, and alloys can also be used, but gold remains the standard.

Organic finishes, called organic solderability preservatives (OSPs), are also used, especially when the primary objective is to keep the printed circuit ready for soldering during the interval during which it waits for assembly. Certain nitrogen-ring compounds will readily react with clean copper, and most OSPs are based on this chemistry. The most common product line of OSPs is Entek,® from Polyclad/Entek. The resulting copper complex can protect the circuit for more than a year, depending on storage conditions. The copper complex readily decomposes, leaving bare copper, when subjected to the heat of soldering. Since the OSPs are present as only a molecular layer and decompose during soldering, there is no problem with residues or contamination. Application involves a simple dipping after cleaning.

1.4 Circuit Processes

The circuit processes are so varied that completely opposite methods are used to create conductors. As described in the historical sections, conductors can be "added" by painting, printing, or hand application. This category is appropriately called *additive circuitry*. Later, a removal method similar to the one used to make printing plates came into use, and this method is called *subtractive*. A purely additive process is difficult at best, and this has led to a hybrid scheme called "semiadditive. There is one more possible process that we could call *conversion*, but not much headway has been made here. A conversion process would selectively convert a dielectric surface to conductors. The photographic process, where a silver salt is converted to silver metal where struck by light, would be the analogy. Some limited success was obtained in the 1970s but not pursued. This is a worthwhile area for future research.

1.4.1 Substrate Manufacturing

Laminates typified by FR-4 are made by bonding copper foil to the dielectric using a large, heated press. Woven glass and epoxy resin mix are combined and partial cured or polymerized (B-staged) to a point where the sheet, called *prepreg*, can be handled. Copper foil is then applied to one (for single-sided) or both sides (double-sided) of the prepreg sheet. Manufacturing efficiency is obtained by stacking many lay-ups in the same press. The press is closed to force the copper against the prepreg and exclude air bubbles. The press plates are heated (an oven or external heat can also be used), and the stack is allowed to polymerize to a permanent, nonmelting state. The laminate sheets are removed and may be cut to size, and the edges may be trimmed.

Flexible circuit substrate manufacturing may be very different, depending on the type. However, the older laminating method is somewhat analogous. The dielectric film is first coated with adhesive and is dried (if solvent borne), copper foil is applied to the adhesive surface, and the material is either placed in a flat press, run through a special moving sheet press, or run through heated rollers. The final product can be a roll of laminate or sheets cut to size.

A newer method avoids the use of adhesive altogether, and the product is either called *adhesiveless* or *clad*. While some still call this copper-dielectric substrate *laminate*, it is actually a clad, since there is no adhesive layer. Today's most common method of making adhesiveless clad is to vacuum deposit copper onto the dielectric surface using a roll-to-roll process. The film is unwound inside a vacuum chamber while copper is deposited by sputtering or by thermal evaporation. The copper adheres, and the coated film is wound on the take-up reel. The chamber is opened after all the film is coated and is now ready for additional processing such as building up thickness by electroplating copper. Lower adhesion was once an issue but is no longer a concern, as products have improved. An extremely thin adhesion promoter, or tie coat of metal such as chromium, is often first applied to the dielectric before applying copper. This process step improves adhesion and thermal stability of the thermal interface. A secondary etch may be needed to remove the nearly invisible chrome deposit.

The vacuum-coating process is economical when only a few microns thickness of copper is applied, and it is customary to build up the thickness by electroplating more copper in a continuous roll step. A key factor for this adhesiveless process is that, unlike the foil lamination, nearly any copper thickness can be produced. Foil lamination cannot use ultra-thin copper, since production and handling would be very difficult. But adhesiveless flex can be made with exactly the right thickness of copper for a specific circuit process. The thinner the copper layer, the finer the etching. The result is higher-density circuitry than for rigid. But very thin copper enables another process that gives even higher density circuitry with much better uniformity. The high-density process is called semiadditive, and it requires thinner copper than can be achieved in the rigid substrate area by any practical method. The circuit patterning processes are covered in the next section.

1.4.2 Conductor Patterning

The conductor patterning process defines the quality and density limits of the printed circuit and is the most important step in manufacturing, if we also include the vertically conductive conduits.

1.4.2.1 Subtractive. Historically, the subtractive or etching process was second, but it remains the first in general use at this time. The substrate must include the metal that will become that conductor pattern when the undesirable part is subtracted away as waste. A mask or resist is applied selectively by printing or by photolithography onto all exposed copper to protect the desired conductor pattern from attack. The resist-coated laminate or clad is then placed in an etching machine that typically sprays heated etchant onto surfaces. The etchant chemically dissolves the exposed copper by conversion to a soluble compound. The process continues until no copper remains in the exposed regions. The resist is then removed with stripper that chemically debonds the film leaving the copper pattern. The copper conductors, when viewed as a cross section, have a somewhat trapezoidal profile, since copper is etched downward and sideways, even though the design of the spray etcher attempts to maximize downward etching. The resulting conductor with slanted walls is undesirable but usable. Other patterning processes can produce straight walls, and they are covered next.

1.4.2.2 Additive. There are two widely used, fully additive methods, and they involve screen printing conductors onto a dielectric. The substrate is therefore just dielectric and not a copper laminate. The ceramic thick film (CTF) and polymer thick film processes use screen printing almost exclusively and are fully additive methods. In fact, CTF and PTF are among the few methods that really produce a printed circuit. The CTF method, originated during WWII, prints cermet ink onto ceramic substrate that is fired and fused into extremely durable conductors. PTF technology also screen prints conductive ink onto dielectric. The PTF ink is usually hardened by rapidly drying the circuit at 120 to 150°C, since these materials are typically made up of polymer dissolved in solvent with silver added for electrical conductivity. Cermet inks for ceramic circuits are composed of metal and glass frit that must be fused at high temperatures approaching 1000°C.

Continuous metal can be applied by plating, however. The copper additive process generally involves sensitizing the substrate so that electroless copper plating will occur. A plating resist is applied over the sensitized surface so that conductors will plate up within the channels and have straight sidewalls. A core issue is that electroless plating is slow, and the process can take an entire day. Work continues, and progress is being made. However, semiadditive methods have solved the rate issue and are becoming more popular, especially for flexible circuits.

1.4.2.3 Semiadditive. This process starts with a thin conductive "seed" layer that can conduct, or bus, current for plating. The thin copper adhesiveless flex materials are ideal. There only must be enough copper to permit electrolytic plating to occur without damaging the thin copper layer. A plating mask is applied over the copper surface, and the sheet or roll is placed in a copper-plating bath. At least one chip carrier product uses a gold bath, since gold conductors are required for bonding and flexural fatigue resistance. Copper is plated into the open channels in the resist and nowhere else. The copper fills these channels perfectly, atom by atom, so that the conductor sidewalls flawlessly replicate the walls of the resist. If the resist has a perfect channel, the resulting plated conductor will have straight and smooth walls. Since the plating resist is photoimaged with straight walls, the conductors likewise have straight walls, unlike the trapezoidal-shape of etched circuits. Conductors can also be placed much close together, and spacing can be very narrow. The semiadditive process produces conductors with widths under 25 µ (1 mil) and spacings of less than 50 µ (2 mils), making it extremely valuable for high-density circuits and packages. Some are already producing traces of less than 0.5 mil (12 µ) in width.

The next step is to strip off the plating mask and then remove the seed layer of copper. So far, the process has been only additive. The copper is removed by immersing the circuit into a mild etchant that dissolves away the thin copper between the conductors. This also removes a tiny amount of copper from the conductors, but the amount is so small that it is more like a polishing step. Sharp edges and imperfections tend to etch faster, and the final result is a micropolishing of conductors while removing seed copper.

1.4.3 Hole and Via Formation

Holes are necessary for classes of circuits beyond the single-sided type. Access holes, now called *vias*, can be formed at various stages in the circuit-making processes. In fact, one adhesiveless flex circuit process forms vias as the very first step so that the vacuum-deposited copper will coat their walls to allow electroplating to occur while copper is electroplated to build up thickness on the surface. Vias have been traditionally formed in rigid circuits by drilling, and, in flex, by punching. Laser drilling continues to gain popularity as smaller vias, called *microvias*, are required for higher density. There are probably at least 300 laser drills now used in the circuit industry, with more coming on stream. The flex circuit industry welcomed lasers even earlier, because the thin, unreinforced flex is the ideal laser-compatible substrate. UV lasers, like the eximer, depolymerize polyimide instead of "burning" it like the heat lasers. Photoablation with UV energy enables extremely fine holes to be formed with straighter walls.

The rigid circuit industry, working with equipment manufacturers, has achieved good results with laser drilling, and this method may eventually become dominant. But the presence of somewhat random levels of glass fabric makes lasing a more difficult process. Greater thickness is also more problematic. The laser does not work well for a stack of circuit layers for mutlilayers, so these must be mechanically drilled.

1.4.4 Vertical Connections

Vertical conduits must interconnect the various conductor patterns residing on separate layers. The most common method is the plated-through hole (PTH) process that has been used for many decades. Prior to the PTH breakthrough, rivets, pins, and various mechanical methods were used, but with low efficiency and density. The plated-through hole process may start with a cleaning step if there is any adhesive smear from drilling through a multilayer stack. After desmearing, the hole may be chemically roughened for better adhesion, unless this was accomplished by the desmearing chemistry. The hole is then sensitized for plating in a two-step process that binds catalytic palladium to the hole walls. The circuit board is then placed in an electroless plating solution, and copper is deposited onto the catalyzed walls but not on other parts of the dielectric. Electroless copper is autocatalytic, and plating continues on the walls and also onto any other copper surfaces. Since the electroless plating is slow, only enough is deposited to enable fast electrolytic copper plating to occur. The circuit is then plated-up with copper until the hole walls are thick enough. Any exposed copper on the surface will also be plated-up when either of two processes is used. The added copper can be plated onto the bare copper foil in the panel plating process. Alternatively, a plating resist can be applied over the foil so that copper is added to the vias and onto the copper foil as the circuit pattern in a process aptly called *pattern plating*. In both cases, undesired copper must be removed by etching, and this can require masking the plated vias.

Newer processes have been developed over the years that attempt to eliminate the electroless copper step. Most methods place a platable material on the via walls. Conductive carbon inks and intrinsically conductive polymers have had some success. However, vacuum depositing copper onto the via walls has achieved commercial status and should increase in popularity, especially in the flex circuit industry where copper metallization is widely used. We can expect to see continuing developments in the vertical connection area.

1.4.5 Solder Masks and Other Protection

Solder masks can be applied by screen printing if the feature sizes are not too small. However, the minute sizes and higher precision required by surface mount technology have forced circuit makers to the more costly but much higher-resolution photoimaging systems. The photoimagable solder mask can be screen printed, sprayed, roll coated, or applied as a film. The mask is exposed to selective actinic light, typically intense UV, and then developed to remove the undesired mask material. Today's products are aqueous systems that are developed in water solutions that contain a small amount of alkali.

Cover coats are similar to solder masks in that they protect dielectrics, but their purpose is to isolate the circuit in use rather than during the soldering process. They are commonly used for flex circuitry—especially PTF products that are not soldered. In some cases, a protective film, called a *cover layer*, is applied to the flex instead.

1.4.6 Singulation/Depanelization

Circuits are typically made in arrays that may contain just a few or up to many hundreds of individual units to increase manufacturing efficiency and achieve some level of standardization. The individual circuits must be separated at some point in the process. The general rule is to keep the circuits in an array or roll (for flex) as long as possible. Some vertically integrated companies don't separate, or singulate, until after assembly, but singulation after test is more common. Several separation methods have evolved, and the process will often depend on the type of substrate. The terms used within different parts of the industry for the separation process include *singulation* (borrowed from semiconductors) and *depanelization,* since the array contains "panels."

Ceramic circuits have long been cut with lasers such as the CO_2 type. This is a reasonably clean and highly versatile process that can also be used to form mechanical mounting holes and slots. The array can be precut so that the individual circuit areas are only joined by a small amount of material that allows manual separation. After the circuit is completed, they are snapped out of the array. The precut substrate has been referred to as "snapstrate." Ceramic circuits can also be sawn with a diamond blade as is done with silicon wafers in the semiconductor industry.

Rigid organic substrate can also be precut, but this method is reserved for simpler, low-end products built where labor is cheap and equipment is scarce. FR-4 and other board materials can also be sawn, and this is a common method. A water-jet cutting process was introduced a few decades ago and was used by IBM. Laser cutting, while feasible, has limitations, especially with thicker boards, but the increasing amount of lasers within the PCB industry may increase such use.

Flexible circuitry lends itself to a larger range of processes. High-volume flex circuit runs are typically singulated by blanking out the circuits with automatic presses. This method remains popular today and is supported by dedicated equipment. Circuits with components can also be blanked out from either sheet or roll, adding to the value of this method. Lasers, long used for drilling, are being used for singulation, and we can expect this process to increase. Figure 1.14 shows a laser system designed specifically for singulating flex circuits.

1.5 Circuit Construction and Types

Circuit construction is one of the most interesting areas of the industry, and it has seen considerable innovation. This section will describe standard constructions and those under development. In some cases, especially for multilayer constructions, new designs have been described but only developed to the prototype stage. Many of these concepts seem to have merit and will hopefully receive more attention in the future. Figure 1.15 shows the breakdown by type for world production of PCBs.

1.5.1 Single-Layer Circuits

Single-sided circuitry is obviously the simplest construction, and this construction would seem to be reserved for low-density, simpler prod-

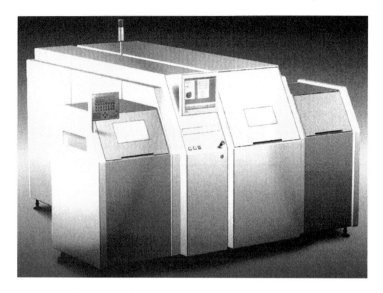

Figure 1.14 Singulation laser (courtesy of Siemens).

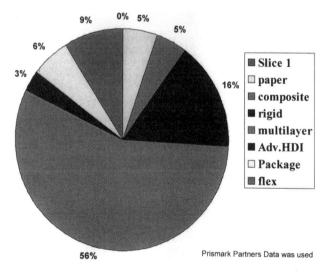

Figure 1.15 Pie chart of circuit types.

ucts. This is not always the case, since some very high-density packages use a single metal layer construction. Conductors are applied or formed on only one side of the substrate, as shown in Figure 1.16. Additive, semiadditive, and subtractive processes are all used to make single-side circuits.

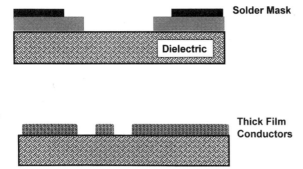

Figure 1.16 Single-side circuit diagram.

1.5.2 Double-Access Circuits

The double-access circuit construction was once only a small niche type of product within the flexible circuit arena. Today, it is one of the most important designs for advanced area array packaging. The original purpose was to allow access to a single layer of conductors from the opposite side at one end. This was a valuable feature for flex connector cables. Later, the access windows were moved to other parts of the circuit for various purposes, including grounding to a metal chassis.

One of the earliest methods for creating the access holes was to first coat the flex dielectric film with adhesive and then dry it. Openings were next punched at the desired locations. Copper foil was then laminated to the precut dielectric using pressure with heat to activate the adhesive. The result was a single-sided laminate with windows in selected locations. The standard circuit patterning processes could now be used, but the copper exposed from the back had to be covered with resist (an easy task).

The advent of affordable precision lasers about a decade ago allowed the windows and holes to be formed after the circuit pattern was completed. The eximer laser was ideal here, since it could easily remove polyimide dielectric but stop at the copper layer. IBM pioneered this photoablation process, but Tessera, founded by former IBM personnel, continued to advance the method for packaging and high-density circuits made with polyimide. Today, automatic laser drills are available for the purpose of opening vias and windows in flex circuit material. While larger openings are still referred to as *double-access*, or *back-bared*, small holes are called *blind vias*, since the hole does not go all the way through. The blind via array is ideal for forming an interconnect structure that permits the tiny circuit to be connected to a larger circuit board. The vias can be plated-up with solid metal, like copper or nickel, to form posts. Continued plating will result in a mushroom shape once the metal being plated is beyond the constraints of the via walls. Careful control of the plating will produce an array of bumps that can be used to connect the circuit, properly called a *chip carrier*, to a larger board. While this method was used originally, the plated bumps were difficult to control, and solder had to be added for assembly.

The next evolution in the process was to terminate plating at the post level and not allow metal to form above the hole. Now, solder spheres could be attached to create the ball grid array (BGA). Alternatively, solder paste could be printed on the posts and reflowed into balls. The final construction, one now widely used, simply forms blind vias that are large enough to accommodate micro-solder spheres. The solder spheres are automatically placed into the opens and reflowed in an oven so that they directly attach to the copper layer that is on the opposite side. Millions of packages, including the popular chip scale packages (CSPs), are made this way. But the package chip carrier is still a circuit, and the construction is still the double-access type. More on this subject will be presented later.

1.5.3 Jumpers/Crossovers

Taking density increase one step at a time, we next move to single-sided circuits with conductors added to the same surface. While a second layer of conductors is added, this is not really a double-sided circuit, since there are no conductors on the bottom. This method was originally developed over 75 years ago, when an innovator added small U-shaped wires that jumped over a conductor to allow adjacent ones to be connected without shorting the one that was bridged. This method is still used today for low-cost circuits. A U-shaped wire is soldered into place. A cleaver alternative has been to use a "0"-Ω resistor to jump over one or more conductors so that the assembly process actually becomes part of circuit making. Figure 1.17a shows the first jumper.

PTF ink, as mentioned under in Sec. 1.4, "Circuit Processes," has been used to build jumpers more efficiently. The PTF industry tends to use the term *crossovers*. Figure 1.17b shows the construction.

1.5.4 Double-Sided Circuits

Double-sided circuits are quite popular, and their manufacture is well established. Ironically, the first circuit patent, filed in 1903, described a double-sided construction. While circuit processes have improved

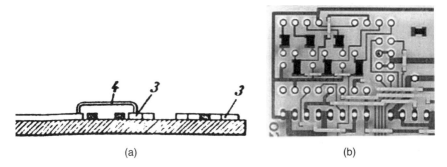

Figure 1.17 (a) Parolini jumper and (b) PTF jumper or "crossover."

since then, the major changes have been in the area of the vertical connection, as mentioned earlier. The double-sided circuit usually begins with double-sided laminate that is etched from both sides simultaneously. The resulting structure is shown in Fig. 1.18. We can note that this is a small subset of double-sided circuits with no vertical interconnect, as shown in Fig. 1.18a. The majority, however, have plated-through holes, solid vias, or some other conductors to selectively connect the top and bottom layers (Fig. 1.18b).

The same conductor patterning process is commonly used on both sides, but there are alternatives. Patterning for double-sided circuits is usually done concurrently.

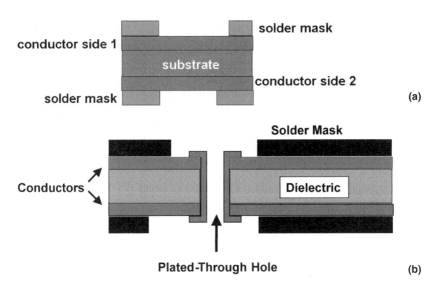

Figure 1.18 Double-sided circuit diagram.

1.5.5 Multilayer (M-L)

Each substrate tends to use different multilayer processes so that different constructions result. In some cases, the constructions are restricted to a single substrate and cannot be used in the others.

1.5.5.1 Rigid M-L. A number of constructions are possible, but all must have at least three conductor layers that can be interconnected. The most common construction is shown in Fig. 1.19. The laminated stack of double-sided circuits is mechanically drilled so that holes go all the way through. The holes are plated with copper so that a long copper "barrel" connects to all layers. While this construction has been used for decades, it has limits that have encouraged development of different constructions. The first issue is that of density. The holes go all the way through. To connect just the top and bottom layers of a 50-layer circuit, we would need to form holes in the other 48 layers, thus robbing them of density. The other problem is that the holes must be large enough to allow plating solution to travel through the entire length. Very high layer counts tend to have larger-than-desirable holes that rob density.

1.5.5.2 Ceramic circuits. Ceramic M-L constructions can be manufactured by using a build-up process wherein one layer at a time is cre-

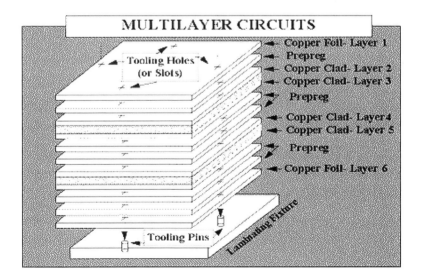

Figure 1.19 M-L diagram.

ated. The build-up process, while often touted as "new," goes back to the 1940s. A single-layer circuit is overprinted with dielectric, but openings are produced at every point where a connection is desired. The dielectric is hardened, and then a second conductive layer is printed and fired. The process is repeated until the desired number of layers has been produced. Figure 1.20 shows this build-up construction. A nearly identical process is used in the PTF industry that results in the same construction but with different materials.

The build-up process has been replaced to some degree by a co-lamination method whereby double-sided pairs of "green" ceramic circuits are fired and fused together. The double-sided layers have through holes that are produced by pulling conductive ink through the opens while the circuit pattern is being printed. The vertical conduits are completed when the second pattern on the backside of the circuit is printed. The green circuits can be handled, but they have not been fired to a final hard ceramic. The layers are stacked together and fired to fuse the connections and complete the ceramic formation. The resulting construction allows blind and buried (no access to the outside) vias to form, and this construction is shown in Fig. 1.21.

1.5.5.3 Flexible circuitry. Multilayer flex circuits can be made using a construction similar to the tradition rigid M-L that was shown in Fig. 1.19. However, the thinness and compliancy of flex has prompted this innovative industry to consider other constructions. One, called Z-Link, laminated double-side circuits together with aniostropic conduc-

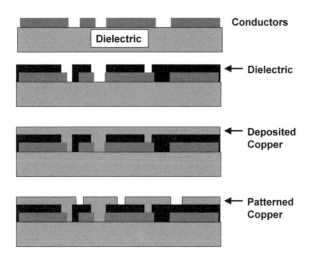

Figure 1.20 Build-up M-L.

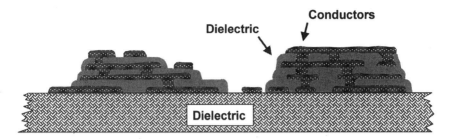

Figure 1.21 Cofired ceramic M-L diagram.

tive adhesives (ACA). The Z-Link process allows blind and buried vias to be used and avoids the full stack, drilling and plating processes. The Z-Link construction is shown in Fig. 1.22.

This is a relatively simple process that uses the common laminating press to bond layers and create vertical connections, but density is limited by the random nature of the adhesive. Several developers have attempted to replace the random ACA (RACA) with a patterned material (PACA). Such a material, consisting of an array of conductive columns in a dielectric adhesive, can be called an *interposer*. A stack of alternating double-sided circuits and interposers could be stacked and

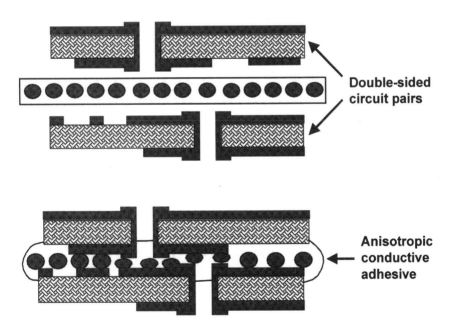

Figure 1.22 Z-Link M-L.

laminated to produce the M-L construction with high density. This construction seems to offer the ultimate in density, since vertical connections are placed only where needed. However, the development is incomplete, and we can hope that the industry will meet the challenge and finalize the process. Figure 1.23 shows the PACA interposer concept.

1.5.5.4 Other. The previous discussion has covered all of the disclosed constructions of any notoriety, but certain others are possible. However, many of these methods have been combined. For example, the traditional drill-and-plate M-L process can be used for inner layers. The top layer or both outer layers can be completed with a high-density build-up step.

1.6 Circuitry for Electronic Packages

While most ICs are connected to a freestanding metal construction, called a *lead frame,* higher-density packages developed during the last decade changed this. Many area array packages, including the BGA

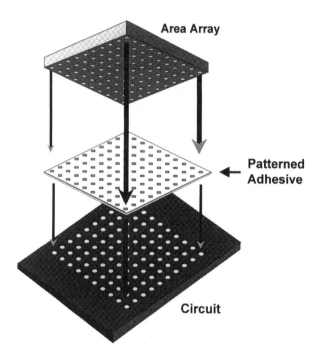

Figure 1.23 PACA concept.

and PGA, use printed circuits. A look at a typical BGA construction will show that a printed circuit is mostly hidden within, but the familiar structure can been seen by looking at the bottom. The construction may be a double-access, double-sided, or even multilayer. Millions of printed circuits now serve as chip carriers, and the market is expanding.

1.6.1 Printed Circuit Area Array Chip Carriers (BGAs, Etc.)

The most common area array package is the plastic ball grid array (PBGA), which consists of a dense printed circuit that serves as the platform for attaching the semiconductor device or devices; hence the name *chip carrier*. High-quality rigid circuits are commonly used that have pads on the top side for chip connection and solder balls on the bottom for assembly. The circuit and chip are generally encapsulated so that the PCB is obscured. However, when viewed from the ball side, one can usually make out the circuit pattern covered with a green solder mask. Conductors are made of copper and usually gold plated. One of the more common circuit materials is BT, since it can provide more dimensional stability and less warp during the processes, including solder assembly to the main board. It is a bit ironic that some in the IC and packaging industries have complained that the PCB has not kept pace when it is actually enabling a better package. Figure 1.24 shows PBGA printed circuit substrate before encapsulation.

1.6.2 Flex-Based Packaging

Flex circuitry has been used as a packaging medium for about 40 years, and its popularity is increasing as its ideal characteristics become known. The two most important attributes for flex packaging are the ability to fabricate into very high-density structures and superior heat management, primarily due to thinness.

1.6.2.1 TAB (tape automated bonding).
Double-access flex circuits came into use in the 1960s as a means of connecting chips to circuits boards.

Figure 1.24 PBGA circuit.

The original "spider" circuits that attempted to connect chips directly to the package were designed as unsupported metal, like today's lead frames. But is was recognized that a dielectric carrier to support the chip circuit would add many benefits. Several flex-based packaging circuit constructions were designed and tested. The first, called the *flip chip strip,* consisted of copper on dielectric film with the chip connected by direct chip attach (DCA), which is better known as the *flip chip.* The conductors fanned out so that the flex chip carrier could be bonded to a circuit board of modest density. A few years later, a window was formed in the bonding area so that the tiny chip bonding conductors were suspended over the window to form cantilevered beams. This copper-Kapton package from General Electric was called the Minimod and is usually considered to be the first TAB package. The flex circuit is made in reel form, known as a *tape,* so that chips can be automatically bonded as the bare tape is unreeled and the loaded tape is reeled up. TAB, also called tape carrier package (TCP), is still widely used today. The outer leads are bonded to a circuit board by hot bar soldering or with anisotropic conductive adhesive film. Figure 1.25

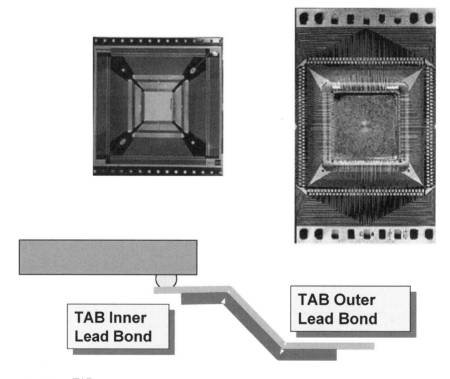

Figure 1.25 TAB.

shows the TAB configuration that is actually just a tiny, high-density circuit with double access. Figures 1.26 and 1.27 show the flip chip strip and Minimod, respectively.

1.6.2.2 Flex BGA packages. Since the inner lead chip bonding section of TAB worked quite well, but the outer lead bonding process was always a problem, innovative packaging developers sought to retain the ILB but replace the OLB. IBM appears to be the first to solve the problem. The designed a TAB-like flex circuit and added connection balls to the bottom using a welding process. The result was the tape ball grid array (TBGA). This product is widely used, and dozens of versions from many companies have been commercialized. Names vary from the IBM TBGA to Amkor's FleX-BGA®, but all are flex circuits. Figure 1.28 shows the IBM TBGA.

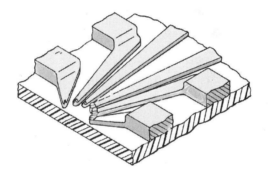

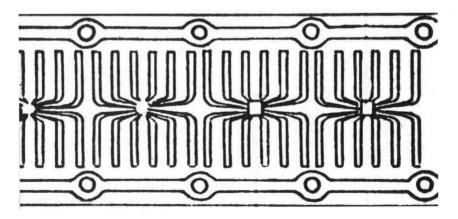

Figure 1.26 Flip chip strip circuit.

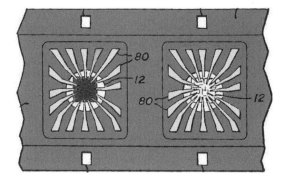

Figure 1.27 Minimod circuit.

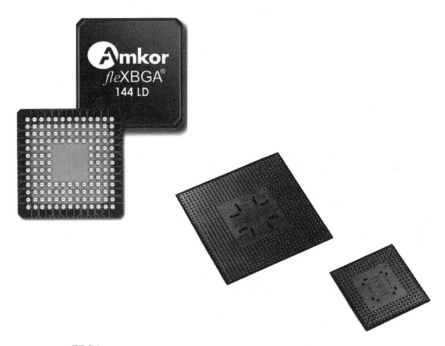

Figure 1.28 TBGA.

Many chip scale packages (CSPs) also rely on flex circuits. Some, such as the well known Tessera μBGA®, use the TAB-like ILB and micro-balls on the bottom akin to the IBM concept, but with fanned-in leads to achieve small size. Others, like Texas Instruments' MicroStar®, bond the chip with wire bonds. All flex-based packages, however, rely on the printed circuit. Figure 1.29 shows the μBGA that is rapidly become the CSP of choice for memory chips.

Figure 1.29 μBGA.

1.7 Printed Circuit Trends

Ever since early 1900s, when the telephone exchanges pushed density, the industry has been seeking more and more density in the endless quest for smaller-faster-cheaper electronics. The trend to increase density continues unabated and is even accelerating. The semiconductor industry leads the charge and drives the market as more functionality and speed is added every year. The number of I/Os (input/outputs) increases as the direct result of adding more devices to the chip (Rent's rule). The package must accommodate the new chips by increasing the number of connections. Simultaneously, attempts are made to shrink the package size. The success of area array packages has made it practical to now build packages with more than 2,000 connections and this number will approach 10,0000 over the next several years as super-super computers, like IBM's Blue Gene, are built to help sort out the mass of genome DNA data.

    The printed circuit must follow the density curve and accommodate the newest compact packages. Direct chip attach (DCA), or flip chips, are being bonded directly to the circuit, avoiding the package completely. Flip chips place tremendous challenges before the PCB industry that are only partially being met, and only by a small part of the industry. We have finally reached many limits using the traditional circuit processes and must move on. Both the subtractive etch process and mechanical hole drilling are under attack, as well they should be. The flex industry, the often neglected stepchild of circuitry, has led the way with newer processes for at least a decade. Semiadditive conductor fabrication is now producing traces under 1 mil (25 μ) and laser machine is crafting microvias of 2 mils (50 μ) and less. Small research lines have cut these numbers in half, and we can expect these developments to move to the commercial world quickly.

The rigid circuit industry has adopted some of these methods, although some are more difficult to implement in this segment where vacuum deposition is uncommon. We can expect to see laser drilling continue to gain share as more high-density interconnect (HDI) circuits are required for packaging and electronic products. The rigid PCB industry will also add more vacuum coating to enable high-density semiadditive conductor formation. Finally, the multilayer process will continue to evolve, and build-up methods will gain share. We may also see epoxy systems lose market share as better polymers are used for laminates. The move away from epoxies could be accelerated if brominated epoxy flame retardant materials are banned. Again, we note that flex circuits have solved many of the density problems, they can handle the higher temperatures of the lead-free alloys coming into use and the flex dielectrics contain no bromine or other elements that are on environmental "hit lists."

1.8 The Printed Circuit Business and Economics

The printed circuit industry surpassed the $35 billion mark as we entered the new millennium. Persistent growth continues, and we can expect dollars to increase at about six percent even though board area may drop slightly because of miniaturization. PCB manufacturing is divided into four major manufacturing sectors of the world, as shown in Fig. 1.30. We will expect small shifts that may favor Asia, but this is speculative, since so many unpredictable events influence the world. Electronics will continue to grow as it has in the past, but the rate may increase as we joyfully embrace the communications age and the world of all-pervasive technology.

1.9 Long-Term Expectations

Once the printed circuit industry moves to energy drilling (lasers) and additive conductor formation, where will it go from there? Both circuitry and semiconductors have been able to create fairly dense connective structures by adding and subtracting material. While our

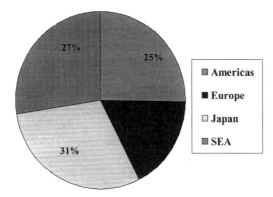

Figure 1.30 PCB manufacturing by region.

modern processes may seem to be fine-scale, they are crude when we use a molecular "yardstick." In a sense, the best processes still hack and chisel (etching) to remove and bombard with massive clusters of atoms (vacuum and electroplating) to add material. The pattern-defining processes are still masks that are conceptually the same as 1000-year-old stencil printing methods. But "information processing engines," produced by the millions in nature, are using molecular-level processes that appear to be guided by intelligence within the molecules (or directed by the Creator). Self-assembly, with no intervention from machines, produces structures that are orders of magnitude more complex than our best attempts to build thinking machines.

DNA, the blueprint of life, contains the equivalent of 1 million pages of information, yet a strand is too tiny to see. At the same time, a single DNA molecule, if unwound and stretched, would be 9 ft long, but still too thin to be seen. DNA also automatically replicates through processes that are only partially understood, even though the double-helix structure was discerned nearly 50 years ago.

We have now entered the *genome age* and hope to *read* the blueprint and eventually fathom the design rules for the most complex machine—the human being. The brain uses electrons and chemical reactions to carry out tasks that are beyond today's best computers. The circuitry is three-dimensional to a level at which a term like *multilayer* is meaningless. More than 20 billion neurons are interconnected, but in a web-like array that makes it effectively 1 million times more sophisticated than our 20 million transistor chips, which are even simpler than the numbers suggest, because they are only digital. So what has this all to do with the printed circuit?

Will circuits and computing machines of the future finally move to the molecular level? There really is no other place to go but down. We can continue to hack away at atoms and molecules or throw relatively large clusters through crude masks for only so long. We have spent more than a century using *mechanical chisels* (using chemical reactions), and a point of diminishing returns is ultimately reached at which smaller tools don't make sense. The architecture and processes are the problems, and genuine breakthroughs in future will require abandoning old tools, techniques, and models. So we should expect to eventually move down to the molecular level following the clues from molecular biology. Will the circuit of the future be grown? Perhaps, but maybe not in the sense of a living and breathing organism. Circuitry, packaging, and devices will likely merge, especially as we move to the molecular level. And when will we have molecular-level products? The sciences have been moving in that direction from the beginning but accelerating as we reach knowledge convergence in other fields. We can expect to see developments in the field of molecular circuits within 20 years. But they won't be printed!

1.10 Acknowledgments

The author is indebted to Joe Fjelstad, who has provided considerable information over the years and located the rare NBS reports.

Much thanks to Jerry Murray who co-authored the monthly series on the "History of the Printed Circuit" in *PC Fab* magazine during 1999 and 2000.

Others who have helped include associates from Sheldahl, Inc., Poly-Flex Circuits, Inc., and Tessera, Inc. The author gained experience and friendship while working at these companies.

1.11 References

1. Hanson, A., British patent 4,681, 1903.
2. Cadenhead R. and DeCoursey D., "The History of Microelectronics—Part One," *International Journal of Microelectronics,* vol. 8, no. 3, Sept., pp. 14–30, 1985.
3. Edison, T., U.S. patent 395, 963, Jan. 1889.
4. Berry, A., British patent 14,699, 1913.
5. Littledale, British patent 327,356, April 1930.
6. Bassist, E., "Halftone Plate Process and Process of Producing Same," U.S. patent 1,525,531, Feb. 1925.
7. Schoop, M., U.S. patent 1,256,599, 1918.
8. Ducas, C., "Electrical Apparatus and Method of Manufacturing the Same," U.S. patent 1,563,731, Dec. 1925.
9. Croot, W., "Improvements in Wireless and Other Electrical Apparatus," British patent 267,172, Mar. 1927.
10. Parolini, C., British patent 269,729, 1926.
11. Seymour, F., "Variable Pathways," U.S. patent 1,647,474, Nov. 1927.
12. Franz, E., "Article," U.S. patent 2,014,524, Sept. 1935.
13. *Proceedings of the First Technical Symposium on Printed Circuits,* Washington, DC, Oct., 1947 (NBS Publication # 192, 1948).

14. Von Liebig, J., "Improvements in Electroplating with Copper and Other Metals the Silvered Surfaces of Mirrors and Other Articles," U.S. patent 33,721, Nov. 1891.
15. DesMarais Jr., R., U.S. patent 4,421,944, Dec. 1983.
16. Stepan, W., U.S. patent 4,508, 753, Apr. 1985.
17. Elarde, V., U.S. patent 4,532, 152.
18. Eisler, P., British patent 639,178, June 1950.
19. Eisler, P., British patent 639,179, June 1950.
20. Eisler, P., U.S. patent 2,441,960, May 1948.
21. Eisler, P., U.S. patent 2,587,568, Feb. 1952
22. Eisler, P., U.S. patent reissue 24,165 June 1956.
23. Eisler, P., U.S. patent 2,706,697, Apr. 1955.
24A. Spalding, A., "Halftone Photomechanical Printing Plate and Method for Producing the Same," U.S. patent 1,709,327, April 1929.
25B. Miller, F., "Process for Reproducing Designs in Metal," U.S. patent 1,804,021, May 1931.
26. Stevens and Dallas, U.S. patent 2,219,494, Oct. 1940.
27. O'Connell, "Method of Making Electrostatic Shields," U.S. patent 2,288,735, July 1942.
28. "Technograph Printed Circuits, Ltd. et al. v. The Bendix Corporation," 137 USPQ p. 725, no. 11421, May 1963.
29. Barnwell, P. G., "Alternative Interconnection Technologies—A Personal View," *5th European Hybrid Microelectronics Conference Proceedings,* Stresa, Italy, May 22–24, 1985.
30. Gilleo, K., "A Simplified Version of the Multilayer Process," *Electronic Packaging & Production,* pp. 134–137, Feb. 1989.
31. Gilleo, K., "A New Multilayer Concept Based on Anisotropicity," *Proceedings, NEPCON WEST,* Anaheim.

1.12 Bibliography

Gilleo, K., et al., *Handbook of Flexible Circuits;* being reprinted in 2001 by Kluwer Publishing U.S.A.

Gilleo, K., *Polymer Thick Film,* Kluwer Publishing U.S.A.

Chapter 2

Development and Fabrication of IC Chips

Charles Cohn
Agere Systems
Allentown, Pennsylvania

2.1 Introduction

At the end of the nineteenth century, the consumer products of that time included simple electrical circuits for lighting, heating, telephones, and telegraph. But the invention of radios and the need for electrical components that could rectify and amplify signals spurred the development of vacuum tubes. Vacuum tubes were found in products such as radios, televisions, communication equipment, and in early computers. Their use lasted until the late 1960s, when the development of semiconductor devices ushered in a new era in electronics. The semiconductor, containing an array of complex transistors and other components on a single IC chip, provided improved reliability and reduced power, size, and weight, and it made possible today's sophisticated electronic products.

This chapter, which is subdivided into five sections, presents a simplified approach to the understanding of the fundamentals of semiconductors, IC development, and IC chip fabrication. The topics cover

- Atomic structure
- Vacuum tubes
- Semiconductor theory

- Fundamentals of integrated circuits
- IC chip fabrication

2.2 Atomic Structure

All matter, whether solid, liquid, or gas, is composed of one or more of the 109 presently recognized elements referenced in the periodic table (Fig. 2.1). Of these, 91 elements occur naturally, and the rest are either man-made or are by-products of other elements. An element is composed of molecules, which are divisible into even smaller particles called atoms. The atomic structure for each element is unique and defines the element's properties.

Materials can be categorized according to the way they conduct electricity when a voltage is applied across them. Insulators, as the name implies, do not conduct electricity, whereas conductors allow a large flow of current, depending on the voltage applied and the conductance properties of the material. Semiconductors have properties in between those of resistors and conductors, having limited current flow capabilities that depend on their atomic structure, the purity of the material, and temperature.

The structure of an atom, as was first proposed by Neils Bohr in 1913 and later supported by extensive experimental evidence, consists of negatively charged electrons rotating in somewhat defined orbits, or energy levels, about a highly dense nucleus consisting of protons and neutrons (Fig. 2.2). The protons are positively charged, and the neutrons have no charge, or are electrically neutral. Each atom has an equal number of (+) protons and (–) electrons, but the number of neutrons may vary.

Each element in the periodic table is assigned an atomic number, which is equal to the number of protons, and therefore electrons, contained in its atom. The atomic number is shown in the upper part of the box representing the element (Fig. 2.1).

The actual weight of an atom is extremely small, which makes it very difficult to work with. As a result, a weight scale was devised that assigns weights to atoms that show their weights relative to one another. The weights assigned are based on the densest part of the atom; namely, the sum of the number of protons and neutrons in the nucleus.

The positively charged protons exert an inward force on the negatively charged electrons, which is balanced by an outward centrifugal force created by the electrons spinning in their orbits around the nucleus. Thus, the two opposing forces provide a balanced structure for the atom.

	IA	IIA	IIIB	IVB	VB	VIB	VIIB	VIIIB	VIIIB	VIIIB	IB	IIB	IIIA	IVA	VA	VIA	IA	IIA
1	1 H																1 H	2 He
2	3 Li	4 Be											5 B	6 C	7 N	8 O	9 F	10 Ne
3	11 Na	12 Mg											13 Al	14 Si	15 P	16 S	17 Cl	18 Ar
4	19 K	20 Ca	21 Sc	22 Ti	23 V	24 Cr	25 Mn	26 Fe	27 Co	28 Ni	29 Cu	30 Zn	31 Ga	32 Ge	33 As	34 Se	35 Br	36 Kr
5	37 Rb	38 Sr	39 Y	40 Zr	41 Nb	42 Mo	43 Tc	44 Ru	45 Rh	46 Pd	47 Ag	48 Cd	49 In	50 Sn	51 Sb	52 Te	53 I	54 Xe
6	55 Cs	56 Ba	57 La	72 Hf	73 Ta	74 W	75 Re	76 Os	77 Ir	78 Pt	79 Au	80 Hg	81 Tl	82 Pb	83 Bi	84 Po	85 At	86 Rn
7	87 Fr	88 Ra	89 Ac	104 Rf	105 Db	106 Sg	107 Bh	108 Hs	109 Mt	110	111	112		114		116		118

58 Ce	59 Pr	60 Nd	61 Pm	62 Sm	63 Eu	64 Gd	65 Tb	66 Dy	67 Ho	68 Er	69 Tm	70 Yb	71 Lu
90 Th	91 Pa	92 U	93 Np	94 Pu	95 Am	96 Cm	97 Bk	98 Cf	99 Es	100 Fm	101 Md	102 No	103 Lr

Figure 2.1 Abbreviated periodic table of the elements.

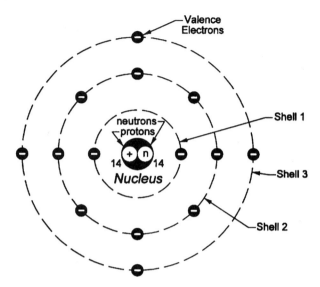

Figure 2.2 Bohr model of silicon atom.

The maximum number of electrons that a given orbit or shell can support is governed by the $2n^2$ rule, where "n" is the shell number.[6] That is, shell #1 (closest to the nucleus) can hold a maximum of two electrons, shell #2 can have a maximum of 8 electrons, and so on. If the number of electrons for a given shell exceeds the maximum indicated by the $2n^2$ rule, then the extra electrons are being forced into the next higher shell. An atom is chemically stable if its outer shell is either completely filled with electrons, based on the $2n^2$ rule, or has eight electrons in it. The electrons in the outer shell are called valence electrons and, if their number is less than eight, the atom will have a tendency to interact with other atoms either by losing, acquiring, or merging its electrons with other atoms.

In the periodic table (Fig. 2.1), elements with the same number of valence electrons have similar properties and are placed in the same group. For example, elements in Group I have atoms with one electron in their outer shell. Group II shows elements that have atoms with two electrons in their outer shell, and so on. Elements on the left side of the periodic table have a tendency to lose their valence electrons to other atoms, thus becoming electropositive. The elements on the right side of the periodic table show a tendency to acquire electrons from other atoms and become electronegative.

The type of interaction occurring between atoms, as they are brought together, depends largely on the properties of the atoms themselves. The interaction may form bonds that can be classified as ionic, covalent, molecular, hydrogen bonded, or metallic. Since this chapter is concerned with semiconductors, which tend to form covalent bonds with other elements and with themselves, the emphasis will be on covalent bonding.

Covalent bonds occur when two or more atoms jointly share each other's valence electrons. If the outer shell is partially filled with electrons, the atom will be attracted to other atoms also having a deficiency of electrons, so sharing each other's valence electrons will result in a more stable condition. As an example, two chlorine atoms will attract and share each other's single electron to form a stable covalent bond with eight electrons in each shell (Fig. 2.3).

2.3 Vacuum Tubes

Modern electronics can trace its roots to the first electronic devices called vacuum tubes. Although, today, solid state devices have totally replaced the vacuum tube, the fundamental principle as to its usage remains relatively unchanged. For more than 40 years, until the late 1960s, the most important part in a consumer electronics product was

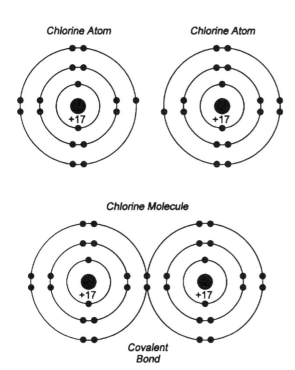

Figure 2.3 A chlorine molecule forms a covalent bond.

the vacuum tube. It is with this historical perspective in mind that this section is presented so that readers will not lose sight of where it all started.

The vacuum tube got its start in 1883, when Edison was developing the incandescent lamp. To correct the premature burnout of the red-hot filament in light bulbs, Edison tried a number of experiments, one of which was to place a metal plate sealed inside a bulb and connect it to a battery and ammeter, as shown in Fig. 2.4. Edison observed that, when the filament was hot and the plate was positively (+) charged by the battery, the ammeter indicated a current flow through the vacuum, across the gap between the filament and the plate. When the charge on the plate was reversed to negative (−), the current flow stopped. As interesting as this phenomena was, it did not improve the life of Edison's lamps and, as a result, he lost interest in this experiment and went on to other bulb modifications that proved more successful. For about 20 years, Edison's vacuum tube experiment remained a scientific curiosity. In 1903, as radios were coming into

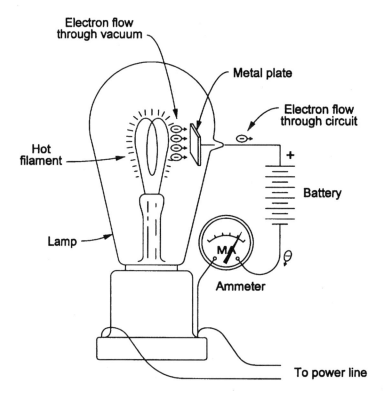

Figure 2.4 Edison's vacuum tube.

use, J. A. Fleming, in England, found just what he needed to rectify alternating radio signals into DC signals required to operate headphones. By hooking up Edison's vacuum tube to a receiving antenna, the tube worked like a diode. When the signal voltage increased in one direction, it made the plate positive (+), and the signal got through. When the signal voltage increased in the other direction of the AC cycle, applying a negative (–) charge to the plate, the signal stopped.

The vacuum tube, also called the electron tube, required a source of electrons to function. In Edison's original electron tube, the electron source, called the *cathode,* was the filament that, when heated red-hot, emitted electrons that flew off into the vacuum toward the positively charged plate, called the *anode*. The effect of heating the cathode to activate the electrons was called *thermionic*. Other electron tubes used high voltage to pull the electrons out of a cold cathode. Electronic emission also occurred by applying light energy to a photosensitive cathode. Tubes using this effect were called *photoelectronic* vacuum tubes. Although a variety of methods existed to remove electrons from the cathode, the thermionic vacuum tubes were the most widely used. The cathode was either heated by resistors within or used a separate source of power for heating. The vacuum tube consisted of a glass or metal enclosure with electrode leads brought out through the glass to metal pins molded into a plastic base (Fig. 2.5).

When the electron tube contains two electrodes (anode and cathode), the circuit is called a *diode*. In 1906, Lee DeForest, an American inventor, introduced a grid (a fine wire mesh) in between the cathode and the anode. The addition of a third electrode expanded the application of electron tubes to other electronic functions. The grid provided a way of controlling the flow of electrons from the cathode to the plate (anode). Even though the grid had a weak positive or negative charge, its proximity to the cathode had a strong effect on the flow of electrons from cathode to plate. The open weave in the grid allowed most of the electrons to pass through and land on the stronger positively charged anode. When the grid was negatively charged, it repelled the electrons from the cathode, stopping the current flow (Fig. 2.6).

Thus, with the three electrodes (i.e., cathode, anode, and grid), it was possible to both rectify and amplify weak radio signals using one tube. The three-electrode vacuum tube was called a *triode*. Additional electrodes, such as a suppressor grid and screen grid, were also enclosed in electron tubes, making it possible to expand the functions of electron tubes.

Vacuum tubes, although widely used in the industry for a half a century, had a number of disadvantages, among them that they were bulky, generated a lot of heat, and were subject to frequent replace-

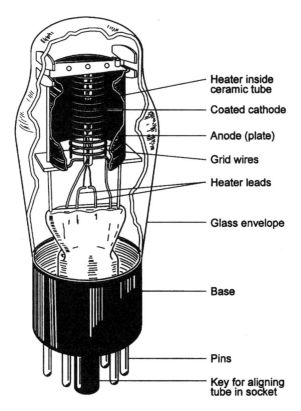

Figure 2.5 The construction of a triode vacuum tube.

ment because they would burn out. With the advent of solid state devices, which had none of the disadvantages of vacuum tubes, vacuum tubes started to fade from use in electronic products.

2.4 Semiconductor Theory

Semiconductor materials have physical characteristics that are totally different from those of metals. Whereas metals conduct electricity at all temperatures, semiconductors conduct well at some temperatures and poorly at others.

In the preceding section, it was shown that semiconductors are covalent solids. That is, the atoms form covalent bonds with themselves, the most important being silicon and germanium in Group IV of the periodic table (Fig. 2.1). Others may form semiconductor compounds where two or more elements form covalent bonds, such as gallium (Group III) and arsenic (Group V), which combine to form gallium arsenide.

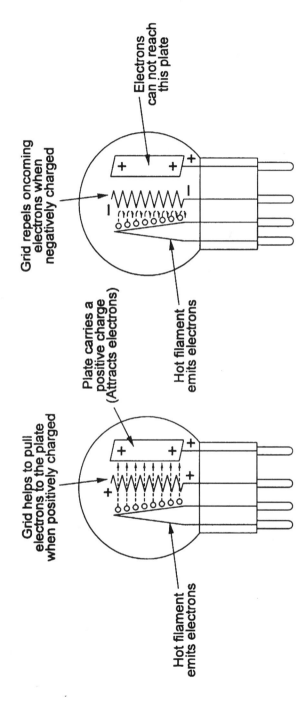

Figure 2.6　Grid controls the flow of electrons to the plate of a triode.

Typical semiconductor materials used in the fabrication of IC chips are

- Elemental semiconductors
  - Silicon
  - Germanium
  - Selenium
- Semiconducting compounds
  - Gallium arsenide (GaAs)
  - Gallium arsenide–phosphide (FaAsP)
  - Indium phosphide (InP)

Germanium is an elemental semiconductor that was used to fabricate the first transistors and solid state devices. But, because it is difficult to process and inhibits device performance, it is rarely used now.

The other elemental semiconductor, silicon, is used in approximately 90 percent of the chips fabricated. Silicon's popularity can be attributed to its abundance in nature and retention of good electrical properties, even at high temperatures. In addition, its silicon dioxide (SiO_2) has many properties ideally suited to IC manufacturing.

Gallium arsenide is classified as a semiconducting compound. Some of its properties, such as faster operating frequencies (two to three times faster than silicon), low heat dissipation, resistance to radiation, and minimal leakage between adjacent components, makes GaAs an important semiconductor for use in high-performance applications. Its drawbacks are the difficulty of growing the ingots and fabricating the ICs.

An elemental or compound semiconductor that was not contaminated by the introduction of impurities is called an *intrinsic semiconductor*. At an absolute zero temperature, intrinsic semiconductors form stable covalent bonds that have valence shells completely filled with electrons. These covalent bonds are very strong, so that each electron is held very strongly to the atoms sharing it. Thus, there are no free electrons available, and no electrical conduction is possible. As the temperature is raised to relatively high temperatures, the valence bonds sometimes break, and electrons are released. The free electrons behave in the same way as free electrons in a metal; therefore, electrical conduction is now possible when an electric field is applied.

If an impurity, such as phosphorus or boron, is introduced into the crystal structure of an intrinsic semiconductor, its chemical state is altered to where the semiconductor will have an excess or deficiency of electrons, depending on the impurity type used. The process of adding a small quantity of impurities to an intrinsic semiconductor is called *doping*. As an example, consider an intrinsic silicon crystal structure with its covalent bonds, shown as a two-dimensional sketch in Fig. 2.7. Each atom is surrounded by four other atoms, with which it shares one pair of electrons, to form four covalent bonds. If the silicon crystal (Group IV) is doped with a controlled quantity of an impurity (dopant), such as phosphorus (Group V), the newly formed covalent bonds (Fig. 2.8) have an excess of electrons that are free to move from atom to atom when a voltage is applied across the semiconductor. The material thus altered is called an n-type (n for negative) semiconductor. Another semiconductor type, called p-type (p for positive), can be formed by doping the silicon crystal with a dopant from Group III, such as boron. The resultant combination (Fig. 2.9) has a deficiency of electrons and thus creates "holes," or electron vacancies, in the positively charged atoms. A single semiconductor crystal structure can be selectively doped with two different kinds of impurities that will form adjacent p-type and n-type semiconductors (Fig. 2.10). The transition between the two types of semiconductors is the p-n junction and is where electrons and holes recombine. As the electrons enter the p-type region, filling the holes, the atoms become negatively charged while the atoms left behind, with fewer electrons, and new holes, become positively charged (Fig. 2.11). The process can be considered as a flow of holes or a current flow of positively charged vacancies, which is opposite to the electron flow. Since there is a depletion of electrons and

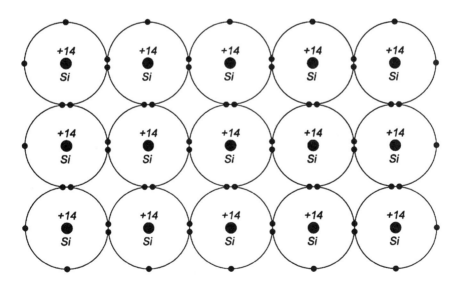

Figure 2.7 Two-dimensional representation of an intrinsic silicon crystal (only valence electrons are shown).

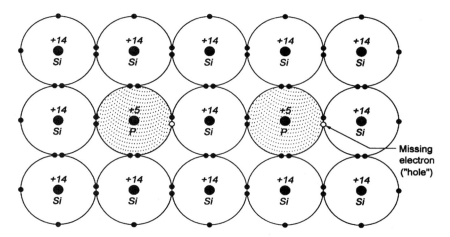

Figure 2.8 Two-dimensional representation of silicon crystal doped with phosphorus to create a p-type semiconductor (only valence electrons are shown).

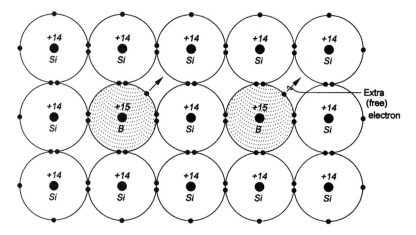

Figure 2.9 Two-dimensional representation of silicon crystal doped with boron to create an n-type semiconductor (only valence electrons are shown).

holes in the contact region, the p-n junction is referred to as the *depletion region*. The double layer of charged atoms sets up an electric field across the contact that prevents further intermixing of electrons and holes in the region, creating a barrier.[1]

2.4.1 The Diode

When an external battery is placed across the p-n junction, with the positive (+) terminal of the battery connected to the n-type side of the

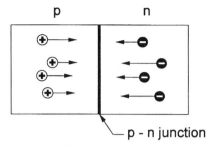

Figure 2.10 P-type/n-type semiconductor junction (after Tedeschi[1]).

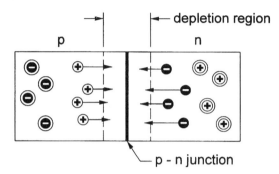

Figure 2.11 P-type/n-type semiconductor junction with depletion region (after Tedeschi[1]).

semiconductor and the negative (−) terminal connected to the p-type side, a so-called reverse bias condition is created across the junction. As the electrons are attracted to the positive terminal of the battery, and the holes are attracted to the negative side, the electrons and holes move away from the junction, thus increasing the depletion region and preventing current flow (Fig. 2.12).

If the battery terminals are reversed (Fig. 2.13), the electrons in the n-material and the holes in the p-material are repelled by their respective negative and positive potentials of the battery and move toward the junction. This reduces the barrier junction, allowing electrons and holes to cross the junction and continue to recombine. As the electrons and holes recombine, new electrons from the (−) terminal of the battery enter the n-region to replace the electrons that crossed into the p-region. Similarly, the electrons in the p-region are attracted by the (+) terminal, leaving new holes behind, which are filled by electrons coming from the n-region. The continuous recombining process creates a forward current flow across the p-n region, which is referred to as *forward biased*. Thus, a p-n junction acts as a diode (rectifier); i.e., when the junction is forward biased, it conducts current, and when the bias is reversed, the current stops.

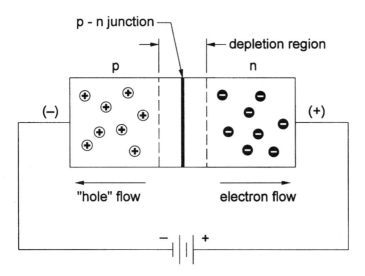

Figure 2.12 Reverse-biased p-n junction (after Tedeschi[1]).

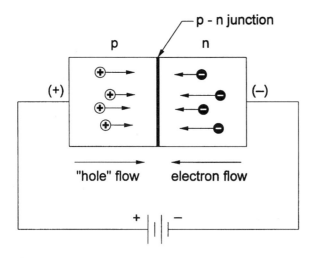

Figure 2.13 Forward-biased p-n junction (after Tedeschi[1]).

2.4.2 The Junction-Type Bipolar Transistor

Combining two or more p-n junction arrangements (p-n-p, n-p-n, etc.) into one device resulted in the development of the transistor. The transistor is a device capable of amplifying a signal or switching a current *on* and *off* billions of times per second. Its development dawned a new age in electronics.

Since its inception in 1948 by W. Shockley, J. Bardeen, and W. Brattain of Bell Laboratories, the transistor has evolved into many forms. The original device (Fig. 2.14) used point contacts to penetrate the body of a germanium semiconductor. Subsequent transistors were of the junction (bipolar) type with germanium as the semiconductor. The semiconductor material was later replaced with silicon.

To illustrate how a bipolar transistor works, an n-p-n semiconductor configuration (Fig. 2.15) is used as an example. In this structure, a very thin, lightly doped p-region, called the base (B), is sandwiched between two thicker outer n-regions, called the emitter (E) and collector (C). The emitter generates electrons, the collector absorbs the electrons, and an input signal applied at the base controls the electron flow from emitter to collector.

Figure 2.16 shows a typical circuit of a bipolar transistor functioning as a digital switch. A supply voltage V_{CE} is applied across the emitter and collector terminals, with the (+) positive terminal of the voltage source connected through a load resistor R_L to the collector terminal. Applying a positive voltage between the base and emitter

Figure 2.14 The original point-contact transistor (courtesy of Bell Laboratories).

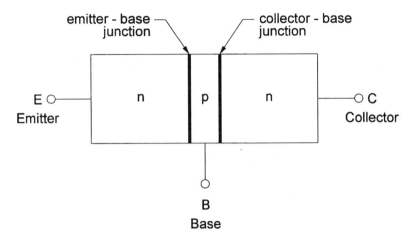

Figure 2.15 Typical n-p-n transistor.

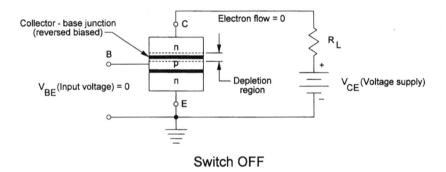

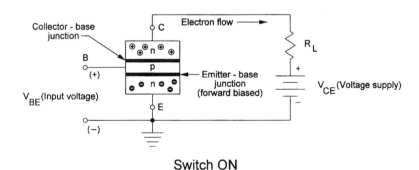

Figure 2.16 Bipolar transistor functioning as a digital switch (after Levine[2]).

terminals, $V_{BE} > 0.5$ V, turns the transistor *on*. Since the emitter-base junction is forward biased, the electrons in the emitter region will cross the junction and enter the base region where a few of the electrons will recombine with holes in the lightly doped base. Because the base region is very thin, and the free electrons are close to the collector, the electrons are pulled across the collector-base junction by the positive potential of the collector and continue to flow through the external circuit. Decreasing the input voltage to zero no longer sustains a flow of electrons across the emitter-base junction and the transistor is turned *off*.

When the bipolar transistor is used as an amplifier, the strength of the emitter-to-collector current flow follows the variations in strength of the input voltage, but at a magnified level. That is, increasing the strength of the input voltage at the base causes proportionally more electrons to cross the emitter-base junction, thus increasing the current flow between the emitter and collector. Decreasing the input voltage causes the electrons to reduce their speed of crossing the emitter-base junction, and the current flow decreases. Since the bipolar transistor can equally amplify both current and voltage, the transistor can also be considered a power amplifier.

The characteristic of the bipolar transistor is its high-frequency response capability, which equates with high switching speed. But to achieve high switching speeds, the transistor must operate at high emitter-to-collector current flow, causing increased power losses.[2]

2.4.3 The Field-Effect Transistor (FET)

The FET transistor operates on a different principle from that of the bipolar transistor. The input voltage creates an electric field that changes the resistance of the output region, thus controlling the current flow. Its unique characteristic of having a very high input resistance will prevent a preceding device in the circuit from being loaded down, which could degrade its performance. The working principle of the FET transistor was known long before the bipolar transistor was developed, but, because of production difficulties, it was abandoned in favor of the bipolar transistor. The 1960s saw a revival of interest in FET transistors after the earlier production issues were resolved. The FET transistor has three semiconductor regions, similar to the bipolar transistor, but, because its principle of operation is different, the FET regions are called the *source,* the *drain,* and the *gate.* These regions are equivalent to the emitter, collector, and base of the bipolar transistor. If we again consider an n-p-n structure, the source and the drain regions are n-type semiconductors, and the gate region is a p-type material.

2.4.4 The Junction Field-Effect Transistor (JFET)

In a junction field-effect transistor (JFET), the electrons do not cross the p-n junction but, rather, flow from the source to the drain along a so-called n-channel, which is formed between two p-type materials (Fig. 2.17). The n-channel is considered the output section of the transistor, and the gate-to-source p-n junction is the input section. In a typical JFET circuit (Fig. 2.18), where the transistor functions as a digital switch, the voltage supply V_{SD} is applied across the (−) source and the (+) drain terminals, through a load resistor R_L. The input voltage V_{GS} is connected between the gate and source terminals with the

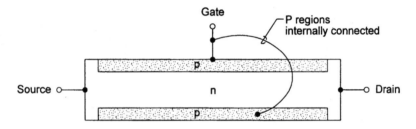

Figure 2.17 Junction field-effect transistor (JFET) construction (after Levine[2]).

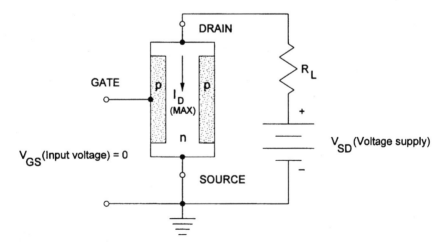

Figure 2.18 JFET functioning as an "on" switch, p-n junction forward biased (after Levine[2]).

negative polarity on the gate. With a reversed bias input voltage, the effect of the electric field creates depletion areas around the two p-n junctions, which are characteristically devoid of electrons. As the input voltage increases, the depletion areas penetrate deeper toward the center of the channel, restricting the electron flow between the source and the drain (Fig. 2.19). If the input voltage is large enough, the depletion areas will totally fill the n-channel, choking off the flow of electrons. Reducing the input voltage V_{GS} to zero, the depletion areas disappear, and the n-p channel is wide open, with very low resistance; thus, the electron flow rate will be at its maximum. When the JFET transistor is used as a linear amplifier, the input voltage variation will have an equivalent effect on the current flow in the n-channel and cause an output voltage gain across the source and drain terminals.[2]

2.4.5 The Metal-Oxide Semiconductor Field-Effect Transistor (MOSFET)

Another type of FET transistor is the metal-oxide semiconductor field-effect transistor (MOSFET). It operates on the same principle as the JFET transistor but uses the input voltage, applied across a built-in capacitor, to control the source-to-drain electron flow.

A MOSFET typically consists of a source and drain (n-type regions) embedded in a p-type material (Fig. 2.20). The gate terminal is connected to a metal (aluminum) layer that is separated from the p-type material by a silicon dioxide (SiO_2) insulator. This combination of metal, silicon dioxide (insulation), and p-type semiconductor layers forms a decoupling capacitor. The gate region is located between the

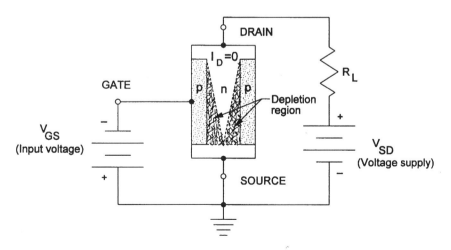

Figure 2.19 JFET functioning as an "off" switch, p-n junction reverse biased (after Levine[2]).

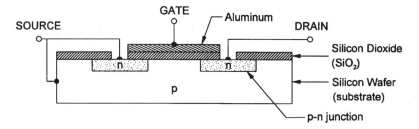

Figure 2.20 Typical construction of a MOSFET (metal-oxide semiconductor field-effect transistor) (after Levine[2]).

source and drain regions, with a fourth region located under the gate, called the *substrate*. The substrate is either internally connected to the source or is used as an external terminal.

The flow of electrons from the source to the drain is controlled by whether the gate has a positive or negative voltage. If the input voltage applied to the gate is positive, free electrons will be attracted from the n-regions and the p-region to the underside of the silicon dioxide layer, at the gate region. The abundance of electrons under the gate forms an n-channel between the two n-regions, thus providing a conductive path for the current to flow from the source to the drain (Fig. 2.21). In this case, the MOSFET is said to be *on*. If the input voltage at the gate is negative, the electrons in the p-region under the gate are repelled, and no n-channel is formed. Since the resistance in the p-region between the two n-regions is infinite, no current will flow, thus turning the MOSFET *off*. Although the MOSFET used in the above description was of an n-p-n type, a p-n-p type MOSFET can also be constructed, but its voltage polarities are reversed.[2]

2.4.6 The CMOSFET Transistor

When two MOSFET transistors, one an n-p-n type and the other a p-n-p type, are connected, the combination (Fig. 2.22) is called a complementary MOSFET or CMOSFET. The advantages of a CMOSFET transistor are simplified circuitry (no load resistors required), very low power dissipation, and the capability to generate an output signal, which is the reverse of the input signal. For example, a positive input will have a zero output, or a zero input will create a positive output.

2.5 Fundamentals of Integrated Circuits

An integrated circuit (IC) chip is a collection of components connected to form a complete electronic circuit that is manufactured on a single

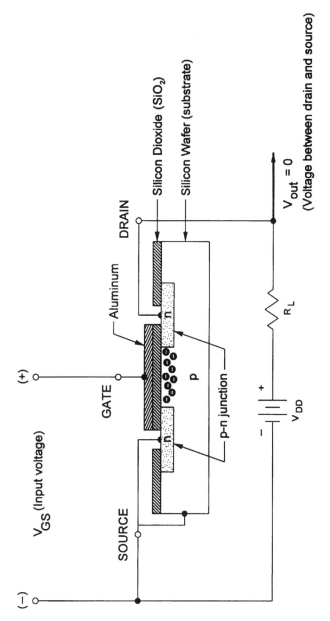

Figure 2.21 MOSFET functioning as an "on" switch (after Levine[2]).

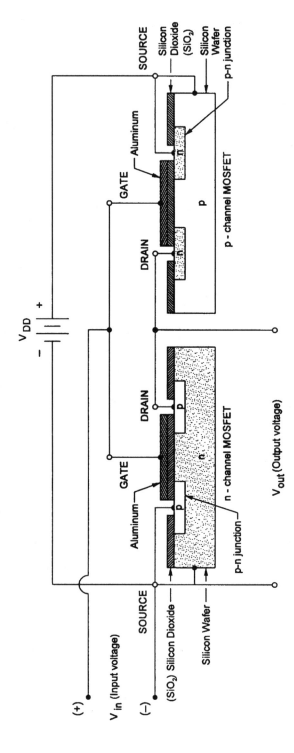

Figure 2.22 CMOSFET (n-p-n MOSFET connected to a p-n-p MOSFET to form a switch).

piece of semiconductor material (Fig. 2.23). As described, the function of most solid state components is dependent on the properties of one or more p-n junctions incorporated into their structures. Figure 2.24 illustrates the combination of various electrical components on an IC, showing their p-n junction structures.

Although the development of ICs was the result of contributions made by many people, Jack Kilby of Texas Instruments is credited with conceiving and constructing the first IC in 1958. In the Kilby IC, the various semiconductor components (transistors, diodes, resistors,

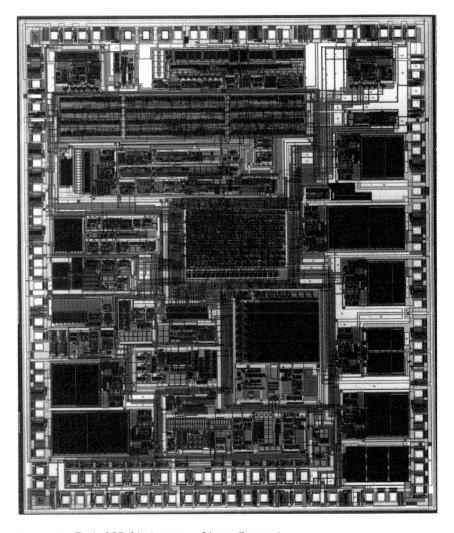

Figure 2.23 Typical IC chip (courtesy of Agere Systems).

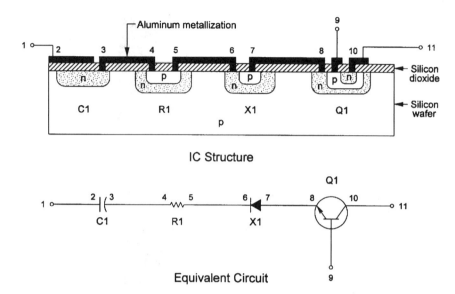

Figure 2.24 Typical silicon structure of electrical components.

capacitors, etc.) were interconnected with so-called "flying wires" (Fig. 2.25). In 1959, Robert Noyce of Fairchild was first to apply the idea of an IC in which the semiconductor components are interconnected within the chip using a planar fabrication process, thus eliminating the flying wires[4] (Fig. 2.26).

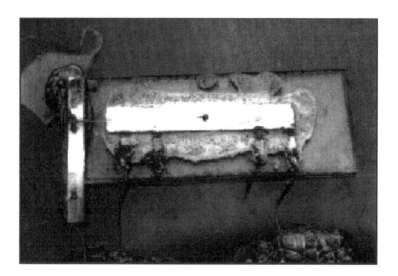

Figure 2.25 Jack Kilby's first integrated circuit.

Development and Fabrication of IC Chips 75

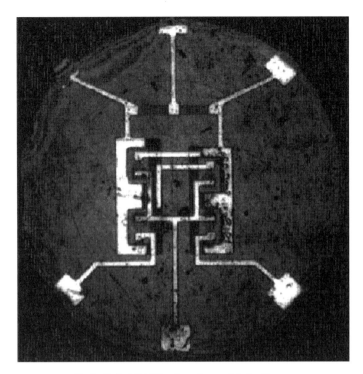

Figure 2.26 Early Fairchild IC using planar fabrication process.

Over the last four decades, the electronics industry has grown very rapidly, with increases of over an order of magnitude in sales of ICs. In the 1960s, bipolar transistors dominated the IC market but, by 1975, digital metal-oxide semiconductor (MOS) devices emerged as the predominant IC group. Because of MOS's advantage in device miniaturization, low power dissipation, and high yields, its dominance in market share has continued to this day.

IC complexity has also advanced from small-scale integration (SSI) in the 1960s, to medium-scale (MSI), to large-scale integration (LSI), and finally to very large-scale integration (VLSI), which characterizes devices containing 10^5 or more components per chip. This rate of growth[3] is exponential in nature (Fig. 2.27) and, at the current rate of growth, the complexity is expected to reach about 5×10^9 devices per chip by the year 2005.

Continued reduction of the minimum IC feature dimensions[3] (Fig. 2.28) is a major factor in achieving the complexity levels mentioned. The feature size has recently been shrinking at an approximate annual rate of 11 percent. Thus, by the year 2006, it is expected to reach a minimum feature size of 10^2 nm (0.10 µm).

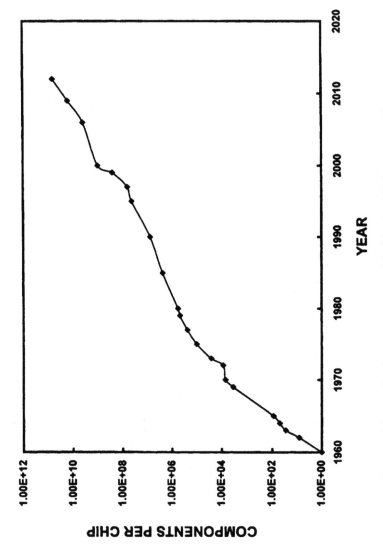

Figure 2.27 Exponential growth of components per IC ship for MOS memory (after Harper[3]).

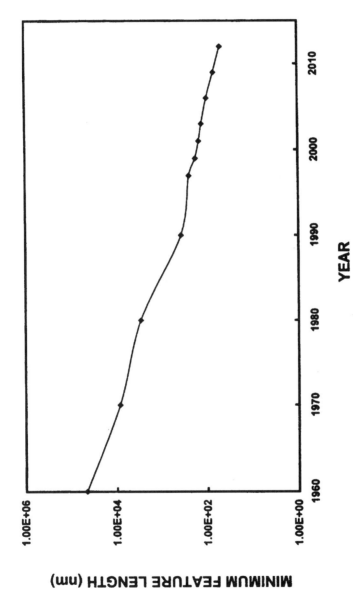

Figure 2.28 Exponential decrease of minimum device dimensions (after Harper[3]).

Device miniaturization has further improved the circuit-level performance, one improvement being the reduction of power consumption at the per-gate level. Figure 2.29 illustrates the exponentially decreasing trend in the power per gate for five major IC application groups: automotive, high-performance, cost-performance, hand-held, and memory. Figure 2.30, on the other hand, shows that the power dissipation per chip actually increased over the same period of time for the high-performance and cost-performance groups, whereas, for the automotive, hand-held, and memory groups, the power dissipation remained relatively constant. This is explained by the fact that, while the power per gate scales linearly with feature size, the power dissipation per chip, P, is largely influenced by the inverse square of the feature-size, as shown below.

$$P = f(Freq, C, V^2, Gate\ Count)$$

where $Freq$ = clock frequency
$\quad\quad\quad\ C$ = capacitance
$\quad\quad\quad\ V$ = voltage
$Gate\ Count$ = chip area / (feature size)^2

While the clock frequency and gate count have been increasing exponentially over the years (Figs. 2.27 and 2.31), the capacitance and voltage have been decreasing. Therefore, the increase in chip power dissipation is primarily due to the greater number of gates on a chip made possible by the decrease in the feature size.

Device miniaturization has resulted in significant improvements in on-chip switching speeds. Off-chip driver rise-time trends for ECL, CMOS, and GaAs are shown in Fig. 2.32. MOS circuits are known to be more sensitive to loading conditions due to their relatively high output impedance. Hence, interconnect density is more important in MOS systems than for bipolar designs. As the applications for these devices tend toward the nanosecond and subnanosecond signal rise times, more attention will be directed to the electrical design consideration of packages and interconnections.

Reduced unit cost per function is a direct result of miniaturization. The cost per bit of memory chips was cut in half every two years for successive generations of DRAMs. By the year 2005, the cost per bit is projected to be between 0.1 and 0.2 microcents for a 1-Gb memory chip. Similar cost reductions are projected for logic ICs.

2.6 IC Chip Fabrication

This section describes wafer preparation and the processes involved in fabricating the solid state components (ICs). The IC chips, which are

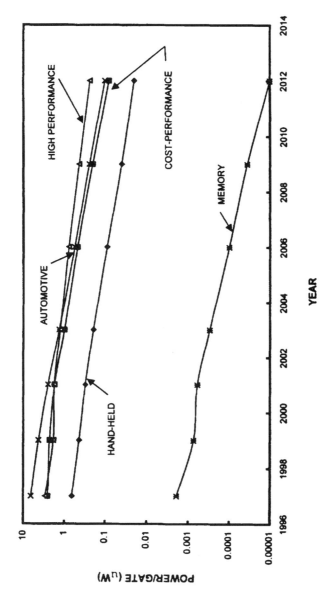

Figure 2.29 Trends in circuit power dissipation per gate (after Harper[3]).

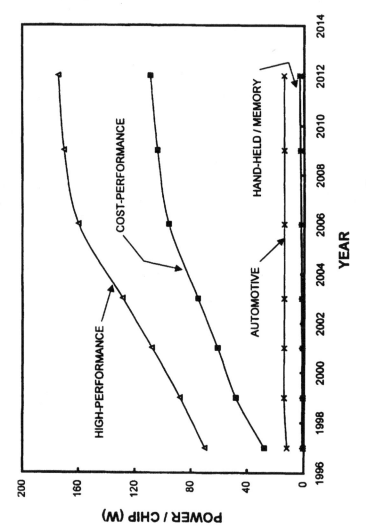

Figure 2.30 Trends in circuit power dissipation per chip (after Harper[3]).

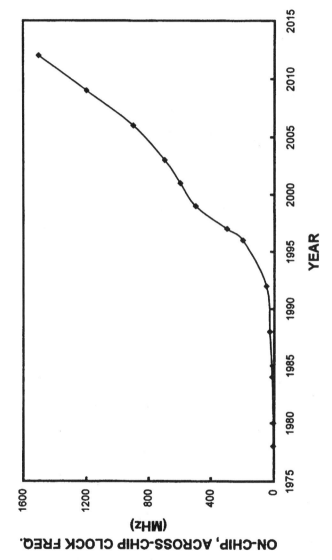

Figure 2.31 Frequency trends of high-performance ASIC chips (after Harper[3]).

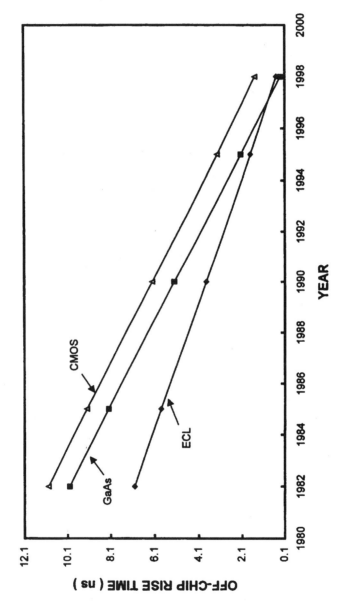

Figure 2.32 Off-chip rise times (typical loading) (after Harper[3]).

configured on the wafer in a step-and-repeat pattern, are formed in a batch process. The pitch of the chip array pattern is dependent on the IC chip size and the width of the "saw street" separating the chips from each other. The width of separation is equal to the thickness of the saw used in singulation. The economics of chip fabrication dictate that as many chips as possible be processed at the same time on a given wafer. Thus, reducing the size of the chips by decreasing their feature dimensions and using larger-diameter wafers are the most cost-effective ways of fabricating ICs. Figure 2.33 shows a typical wafer with chips covering the entire wafer surface.

Of all the semiconductor materials described in Sec. 2.4, silicon is used the most, because it is found in abundance in nature and its silicon dioxide (SiO_2) has many properties ideal for IC fabrication. As a result, this section will use silicon as the exemplary material to describe IC fabrication.

IC fabrication comprises many physical and chemical process steps (Fig. 2.34) that involve state-of-the-art equipment in ultra-clean environments. The following are the step-by-step processes used to fabricate ICs.

2.6.1 Ingot Growth and Wafer Preparation

Before starting on the fabrication of ICs, the silicon wafer, defined as the semiconductor substrate upon which ICs are formed, must be fabricated.

The first step in producing a silicon wafer is to refine raw silicon, which is obtained from either beach sand or quartz mined from agatized rock formations. The sand or quartz is heated along with reacting gases at approximately 1700°C to separate and remove the impurities. The remaining material is chemically purified silicon (nuggets), which has a polycrystalline structure that lacks uniformity in the orientation of its cells. The polycrystalline silicon cannot be used to fabricate wafers but has to be further processed to convert it into a monocrystalline structure containing a single-crystal silicon with uniform cell structures. The silicon nuggets are placed in a quartz crucible (Fig. 2.35) and heated to 1415°C (the melting point of silicon). From the molten silicon, a single-crystal ingot is grown and then sliced into wafers upon which ICs are fabricated.

There are several methods used to grow silicon ingot, but the Czochralski (CZ) method is the most popular. A single silicon crystal seed is placed at the end of a rotating shaft and lowered into the heated crucible until the seed touches the surface of the molten silicon (Fig. 2.35). By continually rotating the shaft and crucible in opposite directions and simultaneously pulling the seed away from the molten sili-

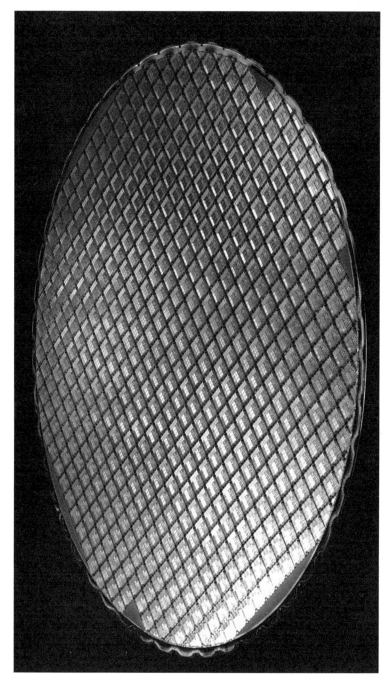

Figure 2.33 Typical wafer with an array of chips (courtesy of Agere Systems).

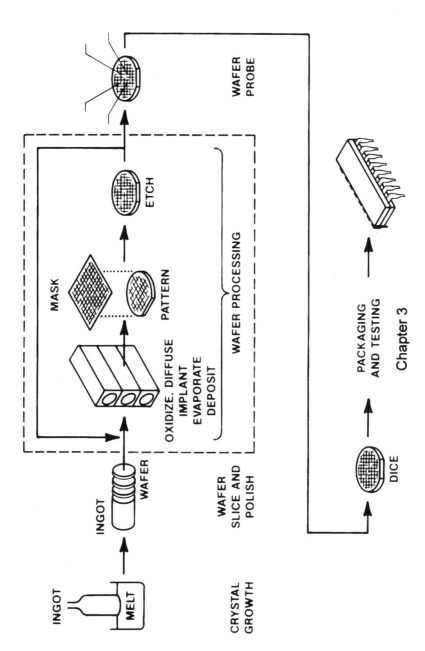

Figure 2.34 Typical IC fabrication processes.

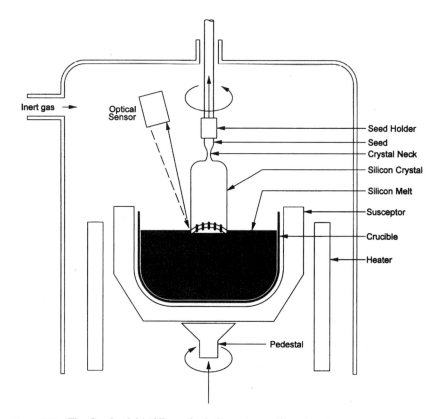

Figure 2.35 The Czochralski (CZ) method of growing a silicon ingot.

con, a silicon crystal is formed at the seed/melt interface with an identical crystal structure as the seed. The monocrystalline silicon ingot continues to be formed as the seed is slowly withdrawn from the crucible and the supply of molten silicon is replenished. To grow an n- or p-type crystal structure, small amounts of impurities (dopants) are introduced to the melt. For example, a phosphorus dopant, when mixed with the pure silicon melt, will produce an n-type crystal, whereas a boron dopant will produce a p-type.

The shape of the ingot consists of a thin circular neck formed at the seed end [approx. 0.12 in (3.0 mm) dia.], followed by the main cylindrical body, and ending with a blunt tail. The length and diameter of the ingot is dependent on the shaft rotation, withdrawal rate of the seed, and the purity and temperature of the silicon melt. Ingot sizes vary from 3 in (75 mm) to 12 in (300 mm) dia. and have a maximum length of approx. 79 in (2 m) (Fig. 2.36). The ingots are grown at a rate of about 2.5 to 3.0 in/hr (63.5 to 76.2 mm/hr).

Development and Fabrication of IC Chips

Figure 2.36 Typical silicon ingots (courtesy of Agere Systems).

The following are typical processing steps to prepare a silicon wafer for IC fabrication (Figs. 2.37 and 2.38):

1. The ingot is cut to a uniform diameter and then checked for crystal orientation, conductivity type (n- or p-type), and resistivity (amount of dopant used).

2. A flat is ground along the axis to be used as reference for crystal orientation, wafer imaging alignment, and electrical probing of the wafer. Sometimes, a secondary, smaller flat is also ground, whose position with respect to the major flat signifies the orientation and type of conductivity (p- or n-type) the crystal has. Larger-diameter ingots may use a notch for this purpose.

3. The ingot is now ready to be sliced into thin disks, called wafers, which may vary in thickness from 0.020 in (0.50 mm) to 0.030 in (0.75 mm), depending on the wafer diameter. Wafers are sliced with either an inner diameter saw blade or a wire saw. The saw blade slicing technique consists of a 0.006-in (0.152-mm) thick stainless steel blade with an inside diameter cutting edge that is coated with diamonds. The cutting edge, being on the inner diameter of a large hole cut out of a thin circular blade, is fairly rigid. The slicing process is sequential; that is, one wafer is cut at a time, which takes approximately nine minutes.

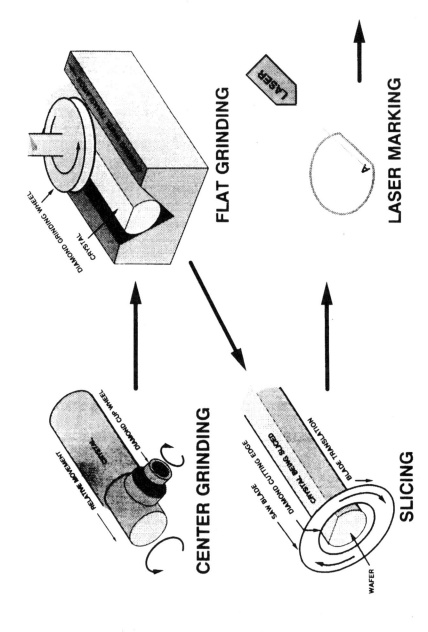

Figure 2.37 Cutting silicon ingot into wafers.

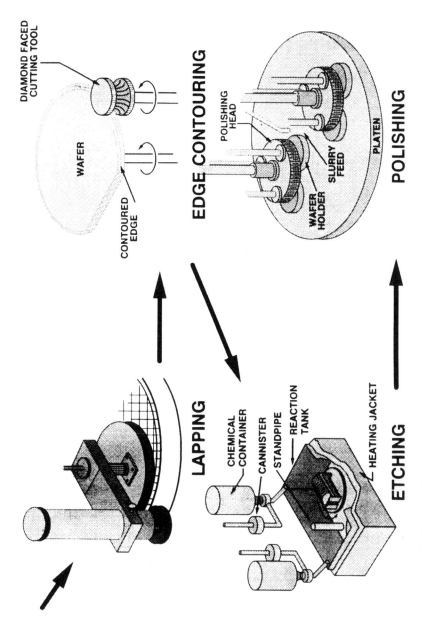

Figure 2.38 Wafer processing.

The wire saw, on the other hand, slices the wafers in a batch process, cutting all the wafers at once in a 16-in (410-mm) length of ingot. The process consists of a wire-winding mechanism, which positions the wires parallel to each other at a pitch equal to the wafer thickness to be cut. The wires are 0.007 in (0.170 mm) dia. and are made of stainless steel coated with brass. The slicing equipment includes a wire guiding unit and a tensioning and wire feed-rate mechanism. The wires continually travel in a closed loop by winding up on one spool and unwinding from another. A silicon carbide slurry, which acts as an abrasive, coats the wires prior to cutting through the silicon ingot. The wires travel about 10 m/s, and it takes approximately 5.5 hr to cut through all the wafers at once.

4. The wafers are laser marked for identification.

5. The sliced wafers are lapped, to remove any imperfections caused by sawing, and then deburred and polished on the top side, to a mirror-like finish. This provides a flat surface for subsequent IC fabrication processes.

2.6.2 Cleanliness

The processes explained so far involved preparation of the wafers for the next phase of IC fabrication, i.e., forming the circuitry. Before proceeding to describe new processes, we must first examine a critical aspect of IC fabrication that affects the yield at every step, namely the cleanliness of the environment where ICs are being produced. Contamination control in the fabrication area is of great concern, because lower yields, caused by unwanted particles, chemicals, or metallic ions in the atmosphere, increase IC costs.

To control the environment, all IC fabrication processes are housed in clean rooms that are classified by how many particles, 0.5 µm in diameter, are allowed in one cubic foot of air. In general, clean rooms range in classification from Class 1 to Class 100,000, with particle size distributions as shown in Figure 2.39. For example, a Class 1000 clean room can have 1000, 0.5-µm size particles in one cubic foot. For IC fabrication, clean rooms range from Class 1 to Class 1,000, depending on the needs of the process.

2.6.3 IC Fabrication

Having explained the importance of cleanliness on IC fabrication, let us resume with the processes involved in forming the circuitry in and on the wafers. The following ten basic IC fabrication processes are described:

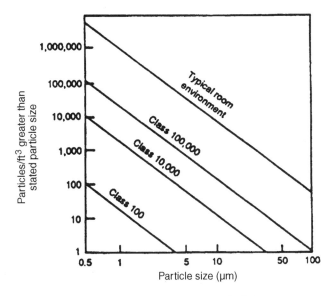

Figure 2.39 Particle size distribution in typical clean room atmosphere and in three classes of clean environments (after Harper[3]).

- Oxidation
- Photolithography
- Diffusion
- Epitaxial deposition
- Metallization
- Passivation
- Backside grinding
- Backside metallization
- Electrical probing
- Die separation

2.6.3.1 Oxidation. Oxidation is the process of forming a silicon dioxide (SiO_2) layer on the surface of the silicon wafer. The silicon dioxide is an effective dielectric that is used to construct IC components, such as capacitors and MOS transistors. Because it acts as a barrier to doping and can easily be removed with a chemical solvent, the silicon dioxide is also an ideal template when used in the doping process. Silicon dioxide is formed by heating the wafer in an atmosphere of pure oxygen at a temperature between 900 and 1200°C, depending on the oxidation rate required. The oxidation can be speeded up if water vapor is introduced into the oxygen. The silicon dioxide growth on the silicon wafer, as the oxygen in contact with the wafer surface diffuses through the oxide layer to combine with the silicon atoms. As the oxide layer grows, it takes longer for the oxygen to reach the silicon, and the rate of growth slows. The growth of a 0.20-μm thick layer of silicon dioxide, at 1200°C and in dry oxygen, takes approximately 6 min, whereas, to double the oxide layer thickness to 0.40 μm takes 220 min or 36 times as long.

The parameters that affect the silicon dioxide growth rate are

- Use of dry oxygen or in combination with a water vapor
- Ambient pressure within the furnace
- Temperature in the furnace
- Crystal orientation
- Time

The silicon dioxide layers vary in thickness from 0.015 to 0.05 μm for MOS gate dielectrics or 0.2 to 0.5 μm thick when used for masking oxides or surface passivation.

2.6.3.2 Photolithography. Photolithography is a patterning process whereby the elements representing the IC circuit are transferred onto the wafer by photomasking and etching. Photolithography has similarities to photographic processes. The images of the various semiconductor element layers are formed on reticles or photomasks made of glass, which are then transferred to a photoresist material on the surface of the silicon wafer. The resist may be of a type that changes its structure and properties to either UV light or laser. If it is a negative-acting photoresist, the areas that are exposed to UV light polymerize (harden) and thus are insoluble during development, whereas the unexposed areas are washed away. This results in a negative image of the photomask being formed in the photoresist. An alternative to the negative image forming photoresist is a positive-reacting photoresist, where the material behaves in the opposite way. Areas exposed to UV light become unpolymerized, or soluble when immersed in chemical solvents.

Until the advent of VLSI circuits in the mid 1980s, the negative-reacting photoresist, because of its superior developing characteristics, was the resist most commonly used in the industry. However, due to its poor resolution capability, the negative photoresist could no longer provide the requirements demanded by the high-density features of VLSI circuits. As a result, the semiconductor industry has transitioned to the positive-reacting photoresist because of its superior resolution capability. The transition was difficult, because not only was the photomask or reticle changed to a positive image, but the industry had to overcome the resist's lower adhesion capability and reduced solubility differences between polymerized and unpolymerized areas. Photomasking is used for patterning both the silicon dioxide and the metallization layers.

The increasing need for ICs to be smaller and operate at higher speeds has forced the industry to develop ICs with ever smaller features (see Sec. 2.5 for feature size trends). As feature sizes decrease, the patterning technology has to advance to where the requirements of high resolution, tight pattern registration (alignment), and highly accurate dimensional control are met. Photomasking is the most critical element of the IC fabrication process in that alignment of the different photomask overlays and mask contamination have an overwhelming effect on fabrication yield. The following photomasking methods are used for patterning:

- Optical exposure
  - Contact printing
  - Proximity printing
  - Scanning projection printing
  - Direct wafer stepping
- Non-optical exposure
  - Electron beam
  - X-ray lithography

The characteristics of each patterning method are described in Table 2.1.

A typical photolithography process for patterning the silicon dioxide layer consists of the following steps (Fig. 2.40):

1. The silicon wafer undergoes an oxidation process (Sec. 2.6.3.1) where a silicon dioxide (SiO_2) layer is grown over its entire surface.

2. A drop of positive photoresist is applied to the SiO_2, and the wafer is spin coated uniformly across the surface.

3. If contact printing is used, a photomask, containing transparent and opaque areas that define the pattern to be created, is placed directly over the photoresist. In areas where the photoresist is ex-

Table 2.1 Characteristics of Patterning Methods

Patterning methods	Description	Advantages	Disadvantages
Contact printing	Mask is placed directly on resist with UV light shining through the mask onto the wafer.	Good resolution High throughput	Causes defects such as scratching of mask and resist Adherence of dirt to the mask may block the light
Proximity printing	Small separation between mask and resist with UV light shining through the mask onto the wafer.	Less damage to mask and resist High throughput	Poor resolution due to some light scattering Not used for VLSI photomasking
Scanning projection printing	The UV light source is a slit projected through the mask. By use of optics, the pattern image of one slit width at a time is projected onto the wafer, exposing the resist	Good resolution High throughput	Alignment problems Possible image distortion from dust and glass damage
Direct wafer stepping	Based on refractive optics, the image of one or several chip sites is projected onto the wafer exposing the resist. The process is step repeated until the whole wafer is patterned.	Good resolution with fewer defects Better alignment Less vulnerable to dust and dirt Most used for VLSI fabrication Medium throughput	Tight maintenance requirement of humidity and temperature control
Electron beam	An electron beam produces a small diameter spot that is directed in an x-y direction, onto the wafer. The electron beam is capable of being turned ON and OFF to expose the resist as needed to form the pattern.	No mask is used Excellent resolution	High cost Low throughput
X-ray lithography	The process resembles the UV light system of the proximity printing method, but high energy x-rays are used instead.	Produces smaller pattern then light sources Excellent resolution	Requires masks made from gold or other refractory materials capable of blocking x-rays Low throughput

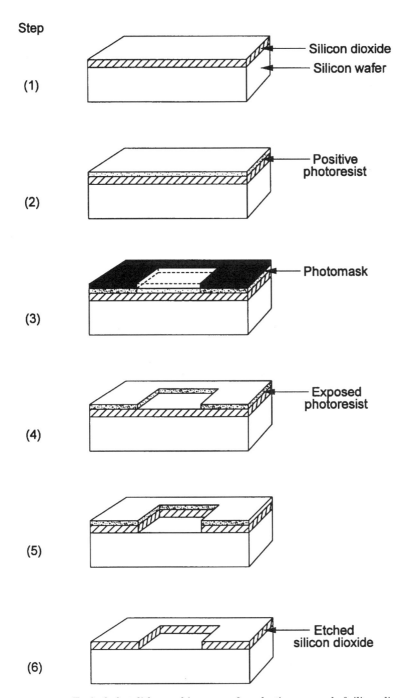

Figure 2.40 Typical photolithographic process for selective removal of silicon dioxide.

posed to UV light, projected through the mask, it becomes unpolymerized (does not harden), and where the UV light is blocked, the material polymerizes (hardens).

4. The photomask is removed, and the resist is developed to dissolve the unpolymerized areas, exposing the silicon oxide below.

5. The wafer is then wet or dry etched to remove the exposed oxide, resulting in a pattern identical to the photomask. Wet etching typically consists of immersing the wafer in a diluted solution of hydrofluoric acid for a specified time that will result in complete etching. The wafers are then rinsed and dried. Wet etching is primarily used for wafers with IC feature sizes greater than 3 µm. For high-density etching, the dry etching technique is used, because it's more precise. Dry etching can be accomplished by three different etching techniques: plasma, ion beam milling, and a reactive ion etch. All three techniques use gases as the etching medium.

6. The remaining photoresist is removed with a chemical solvent.

This process is repeated a number of times to create the desired semiconductor elements on the wafer surface.

2.6.3.3 Diffusion. As was discussed in Sec. 2.4, when forming solid state components, silicon is not used in its natural or intrinsic state but is converted to either an n-type or p-type semiconductor. The n- or p-type materials, by themselves, are of little value unless they are joined to form a p-n junction. Diffusion or doping is the process of implanting impure atoms in a single crystal of pure silicon so as to convert it into n-type or p-type material. Depending on the dopant element used, antimony, arsenic, and phosphorus will produce an n-type material, whereas boron will produce a p-type structure. The basic dopant elements are available either in solid, liquid, or gaseous states as shown in Table 2.2. Type of dopant, dopant concentration, time of exposure, and temperature affect the diffusion process.

2.6.3.4 Epitaxial Deposition. This is a process whereby a thin layer of silicon (approximately 25 µm thick) is deposited upon the surface of an existing silicon wafer and doped using the same dopant types and delivery systems used in the diffusion process. Thus, this is another technique for fabricating p-n junctions. Although there are several deposition methods available, chemical vapor deposition (CVD) is the

Table 2.2 Common Dopant Sources (after Zant[6])

Type	Element	Compound name	State
n-type	Antimony	Antimony trioxide	Solid
	Arsenic	Arsenic trioxide	Solid
		Arsine	Gas
	Phosphorus	Phosphorus oxychloride	Liquid
		Phosphorus pentoxide	Solid
		Phosphine	Gas
p-type	Boron	Boron tribromide	Liquid
		Boron trioxide	Solid
		Diborane	Gas
		Boron trichloride	Gas
		Boron nitride	Solid

most commonly used technique. The basic CVD process consists of the following:

1. Silicon wafers are placed in a reaction chamber with an inert gas and heated to a temperature that depends on the reaction and parameters of the deposition method used and layer thickness required.

2. Reactant gases are introduced into the reaction chamber at a specified flow rate, where they come in contact with the wafer surface.

3. As the reactants are absorbed by the silicon wafer, the chemical reaction forms the deposition layer. The surface reaction rate is dependent on the temperature; increasing the temperature increases the reaction rate.

4. To dope the deposition layer, dopant gases are introduced into the reaction chamber where they combine with the deposited layer to form an n- or p-type material.

5. The gaseous by-products are flushed from the reaction chamber.

6. The wafers are removed from the chamber, and the deposited layer is checked for thickness, coverage, purity, cleanliness, and n- or p-type composition.

Variations in the CVD techniques, involving changes in vapor pressure and temperature in the chamber, have resulted in process enhancements. There are three different CVD techniques used in the industry:

- Atmospheric pressure CVD (APCVD)
- Low-pressure CVD (LPCVD)
- Plasma-enhanced CVD (PECVD)

The characteristics of the above techniques are shown in Table 2.3.

Table 2.3 CVD Techniques (after Wolfe[5])

Process	Advantages	Disadvantages	Application	Temp. range
APCVD	Low chemical reaction temperature Simple horizontal tube furnace Fast deposition	Poor coverage Particle contamination	Low temperature oxides, both doped and undoped	300–500°C
LPCVD	Good coverage and uniformity Vertical loading of wafers for increased productivity	High temperature Low deposition rate	High temperature oxides, both doped and undoped	580–900°C
PECVD	Lower chemical reaction temperature Good composition, coverage and throughput	High equipment cost Particulate contamination	Low temperature insulators over metals or passivation	200–500 °C

2.6.3.5 Metallization. The deposition of a conductive material, to form the interconnection leads between the circuit component parts and the bonding pads on the surface of the chip, is referred to as the *metallization* process. As chip density increases, interconnection can no longer be accomplished on a single level of metal but requires multi-level metallization with contact holes or vias interconnecting the various levels.

Materials such as aluminum, aluminum alloys, platinum, titanium, tungsten, molybdenum, and gold are used for the various metallization processes. Of these, aluminum is the most commonly used metallization material, because it adheres well to both silicon and silicon dioxide (low contact resistance), it can be easily vacuum deposited (it has a low boiling point), it has a relatively high conductivity, and it patterns easily with photoresist processes. In addition to pure aluminum, alloys of aluminum are also used for different performance related reasons; i.e., small amounts of Cu are added to the aluminum to reduce the potential for electromigration effects. Electromigration may occur during circuit operation, when high currents are carried by the long aluminum conductors, inducing mass transport of metal between the conductors. Sometimes small amounts of silicon or titanium are added to the aluminum to reduce the formation of metal "spikes," that occur over contact holes.

The aluminum metallization process consists of depositing aluminum on the wafer surface and again using the photoresist process to etch away the unwanted metallization. One of the techniques used to apply the aluminum is the vacuum deposition process wherein the aluminum is evaporated in a high-vacuum system and redeposited over the wafer surface. This process has the disadvantage of nonuniform metal coverage. Sputtering is another method for depositing aluminum metallization. Because it offers better control of the metallization quality than the vacuum deposition method, it's currently being used in the majority of IC metallization processes. Sputtering is a physical (nonchemical) method of deposition, which is performed by ionizing inert gas (Argon) particles in an electric field and then directing them toward an aluminum target. There, the energy of the incoming particles dislodge or "sputters off" atoms of the aluminum target, which are then deposited onto the wafer.

One of the problems encountered when pure aluminum is in contact with silicon, while being heated, is the formation of an eutectic aluminum-silicon alloy. The alloy formation penetrates into the wafer, where it can reach shallow junctions, causing leakages or shorting. To alleviate this problem, a metal barrier such as titanium tungsten (TiW) or titanium nitrate (TiN) is placed between the aluminum and the silicon. Adding silicon (1 to 1.5 percent by weight) to the aluminum is another way of preventing the formation of aluminum-silicon alloy, although this is less effective. Some alloying with the silicon wafer still occurs, but to a lesser extent.

The electrical performance of any given type of metallization is dependent on its resistivity, contact resistance, and the length and thickness of the conductor. To improve electrical performance in MOS circuits, the resistivity and contact resistance of the conductors are reduced through the use of barriers made of refractory metals such as titanium, tungsten, platinum, and molybdenum, in combination with silicon, to form silicides of $TiSi_2$, WSi_2, $PtSi_2$, and $MoSi_2$, respectively. The silicides can also be used as conductors or via plugs.

As more and more chips are required to operate at higher frequencies, the current aluminum metallization can no longer meet the lower resistances needed to prevent data processing delays. As a result, copper has started to replace aluminum because of its lower resistance and reduced electromigration problems.

2.6.3.6 Passivation. The passivation layer is deposited at the end of the chip metallization process and is used to protect the aluminum interconnecting circuitry from moisture and contamination. An insulating or passivation layer of silicon dioxide or silicon nitride is vapor deposited over the chip circuitry (Fig. 2.41), with bond pads remaining exposed for wire bonding or flip chip interconnection (see Chapter 3).

2.6.3.7 Backside grinding. At the end of the IC fabrication process, after the passivation layer is applied, wafers are sometimes thinned to fit the overall package height requirements. The thinning process consists of back grinding the wafer, similar to the procedure used in lapping the wafer, to remove any imperfections caused by sawing (Fig. 2.38)

2.6.3.8 Backside metallization. In cases in which the chip is to be eutectically bonded to a ceramic package (see Chapter 3), or where the back of the chip has to make electrical contact with the die attach area, it is necessary for the chip to have a gold film backing. The gold film is deposited by vacuum evaporation or sputtering and is done after backgrinding.

2.6.3.9 Electrical probing. The last step in wafer processing is to test the die. A test probe makes contact with the bonding pads on the sur-

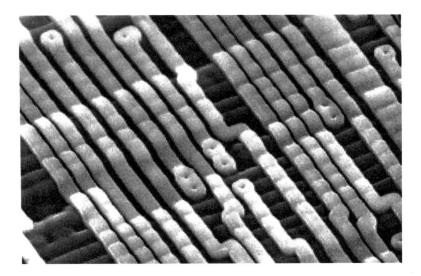

Figure 2.41 IC circuitry covered by a passivation layer (courtesy of Agere Systems).

face of the wafer, and the chips are electrically tested against predetermined specifications. Chips thought to be faulty are inked, or an electronic map is developed indicating the bad chips.

2.6.3.10 Die separation. After the chips have been electrically tested, the chips are separated by two different methods:

1. *For chips thinner than 0.010 in (0.25 mm):* The chips are separated by first scribing shallow, fine, diamond-cut lines between the chips and then mounting the wafer onto a release tape affixed to a steel ring. Pressure from a roller is then exerted on the wafer, breaking it up into individual chips. The individual chips that tested good are removed by pushing the chip up (with a pin) from the underside of the tape and then picking them up with a vacuum tool called the *collet*. The chips are placed in a tray or are automatically transferred to the die attach process for IC packaging (see Chap. 3). This type of separation method may cause rough and cracked edges on the chip.

2. *For chips greater than 0.010 in (0.25 mm) thick:* As above, the wafer is mounted onto a release tape affixed to a steel ring and then cut between the chips, through the silicon thickness, using a diamond-impregnated round saw. The method for removing the good chips from the tape is similar to that for thinner chips. Unlike the break-up method, this separation process leaves smooth edges on the chip.

2.6.3.11 Typical construction of a p-n-p bipolar transistor (Fig. 2.42)

1. A silicon dioxide (SiO_2) layer is grown on a p-doped silicon wafer (Sec. 2.6.3.1).

2. A positive photoresist layer is applied to the SiO_2 (Sec. 2.6.3.2).

3. A photomask is created with opaque and clear areas, patterning the clear areas in locations where windows in the SiO_2 are to be formed. The photomask image is transferred onto the positive photoresist, which becomes polymerized in the areas where it is not exposed to the UV light (opaque areas in the photomask).

4. The resist is developed, and the unpolymerized areas dissolve, forming a window that exposes the SiO_2.

5. The silicon dioxide is etched away in the photoresist windows, exposing the silicon wafer.

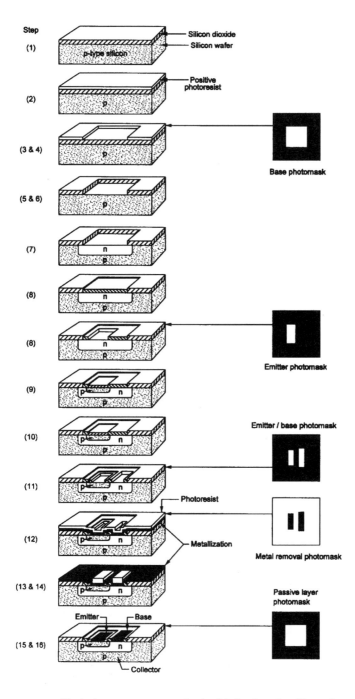

Figure 2.42 Typical process sequence in the fabrication of a silicon planar bipolar transistor.

6. The photoresist is removed.

7. Using phosphorus as the dopant, an n-type region in the p-type silicon base is created by diffusion (Sec. 2.6.3.3).

8. A new layer of silicon dioxide is grown on the surface of the n-region, and steps (2) through (6) are repeated to create a new window in the SiO_2.

9. A second diffusion creates the p-type region in the n-type base by using boron as the dopant.

10. Silicon dioxide (SiO_2) is again grown over the exposed silicon wafer, and the photoresist is applied over the SiO_2.

11. The photomask, containing the two clearances for the emitter and base, is placed over the positive photoresist, and steps (2) through (6) are repeated.

12. The structure is now ready for metallization. An aluminum film is deposited over the entire surface, followed by a coating of positive photoresist.

13. The photomask, with the emitter and base areas opaque, is placed over the photoresist and exposed to UV light.

14. The photoresist is developed, leaving the resist over the emitter and base areas.

15. The exposed metallization is etched away, followed by the removal of the resist over the emitter and base areas.

16. A passivation layer of silicon nitride is applied to the circuitry, leaving the bond pads exposed.

17. The silicon planar bipolar transistor is now complete.

2.7 References

1. F. P. Tedeschi and M. R. Taber, *Solid-State Electronics,* Van Nostrand Reinhold, 1976.
2. S. Levine, *Discrete Semiconductors and Optoelectronics,* vol. 2, Electro-Horizons Publications, 1987.
3. C. Harper, ed., *Electronic Packaging and Interconnection Handbook,* 3rd ed., McGraw-Hill, 2000, Chap. 7.
4. T. R. Reid, *The Chip,* Simon and Schuster, 1984.
5. S. Wolf and R. N. Tauber, *Silicon Processing for the VLSI Era,* vol. 1, Lattice Press, 1986.
6. P. V. Zant, *Microchip Fabrication,* 4th ed., McGraw-Hill, 2000.

Chapter

3

Packaging of IC Chips

Charles Cohn
Agere Systems
Allentown, Pennsylvania

3.1 Introduction

Integrated circuit (IC) chips are at the heart of electronic system controls, and since they are typically sensitive to electrical, mechanical, physical, and chemical influences, they require special considerations by the packaging engineer. IC chip packaging is the middle link of the process that produces these systems. Hence, it must respond to demands from both ends: wafer fabrication and device trends upstream (circuit level), and circuit board assembly and system performance trends downstream (system level). Today's circuit- and system-level requirements of high performance, high reliability, and low cost have placed greater demands on the packaging engineer to have a better understanding of the existing and emerging IC packaging technologies. Many electronic system performance problems result from a lack of knowledge and understanding of the interaction of the many materials and processes that are a part of this first level of packaging. Packaging engineers must be aware of the latest industry demands and be able to trade off among the various packaging technologies on what works and what is economically feasible to satisfy these demands. This chapter deals with the first level of packaging (Fig. 3.1), where the IC chip is assembled into a package, such as a quad flat pack (QFP), pin grid array (PGA), or ball grid array (BGA), using wire bonding (WB), tape automated bonding (TAB), or flip chip (FC) bumping interconnection techniques.

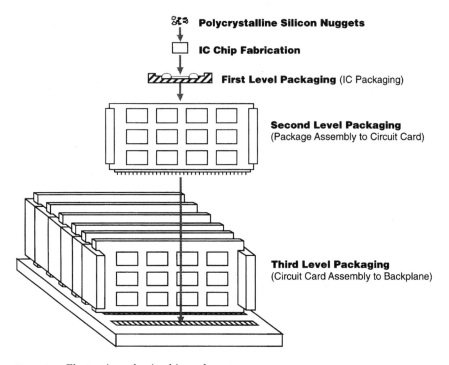

Figure 3.1 Electronic packaging hierarchy.

3.2 The IC Package

An IC package is the housing, which assures protection to the IC chip from mechanical stress, environmental stress (such as humidity and pollution), and electrostatic discharge during handling. In addition, the package provides mechanical interfacing for testing, burn-in, and electrical interconnection to the next-level of packaging. The package must meet all device performance requirements, such as electrical (inductance, capacitance, crosstalk), thermal (power dissipation, junction temperature), quality, reliability, and cost objectives. Until recently, the packaging and interconnecting technologies have not been the limiting factors in the performance of most silicon devices in high-volume production. But, as circuit and system demands are increasing, more attention is needed in selecting package technologies, materials, and designs that will meet these challenges. Some package designs may need to provide special design enhancements to accommodate such demands as higher power dissipation, signal terminations, power and ground distribution, and matched impedances.

Within a given package, there is a distribution of traces that connect the chip to the external leads. Large packages, in particular those with rectangular shapes and side brazed leads, such as DIPs, require very long traces to the corner leads of the package. Array-type packages, typically square, also contain a distribution of traces, with the longest trace going to the outermost rows of leads in the corner of the package. This distribution of trace lengths within the package, in series with similar distributions of trace lengths on the PWB, could result in undesirable signal delays, missed line driver requirements, or crosstalk between critical signal leads. Hence, the demands on packaging are for smaller physical-size packages with higher-density circuitry, thus minimizing the variation of trace lengths within the respective distributions and reducing their undesirable effects on performance. In addition, lead spacing or pitch, and the method of attachment to the next level of interconnection, also play an important factor in the performance of the device. Figure 3.2 compares the package areas for some typical surface mount (SM) and through-hole (TH) mount packages.

Improved quality, reliability, and cost are other system demands receiving much attention today. Systems manufacturers are driving down their costs and consequently are putting great pressure on the quality and reliability of the components they purchase. Terms such as *zero defect* are used to define quality and reliability objectives and *just-in-time (JIT)* manufacturing to describe the strategy for reducing

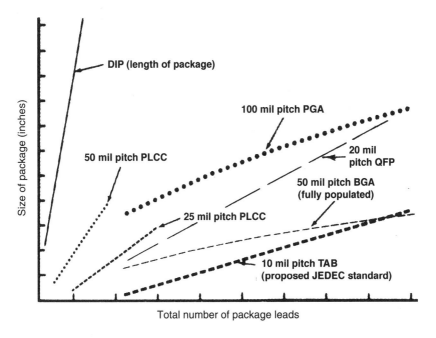

Figure 3.2 Sizes of some common TH and SM packages as a function of number of package leads and pitch.

cost associated with having to carry large in-plant inventories. JIT manufacturing requires that the assembly lines be running defect-free, which demands very high-quality components and assembly processes. Today, the quality of the product received by the system user is expected to have incoming defect levels in the parts per million (ppm) or even parts per billion (ppb) range.

Reliability requirements are also becoming more stringent. System users require that packaged devices pass moisture preconditioning followed by several reflow simulations, temperature-humidity-bias (THB) test, and an accelerated temperature-humidity test such as the pressure-cooker test (PCT). Figure 3.3 illustrates typical test requirements for a PBGA device qualification.

No longer can the packaging engineer specify a generic package for a system and expect it to meet all system requirements without considering all possible conditions that the component will encounter in the field.

3.3 Package Families

IC packages come in a variety of lead arrangements and mounting types (Figure 3.4). The packages are grouped into families defined by the method of mounting to the circuit board [through-hole (TH) or surface-mount (SM)] and by the physical arrangement of the leads on the package (in-line, perimeter, or array). These package families are governed by standard package outlines that control all dimensions, tolerances, and all other information necessary to define, manufacture, and assemble the package. In the United States, the Electronic Industries Alliance (EIA, formerly Electronic Industries Association)/Joint Electron Device Engineering Council (JEDEC) JC-11 committee is the dominant body developing new package outlines. Over the past several years, JEDEC has been working closely with the Electronic Industries Association of Japan (EIAJ) to reduce proliferation of outlines and to focus similar outlines from each organization into one worldwide standard outline.

Table 3.1 lists the packages available in each of the families, grouped by the mounting method to second-level board assembly (TH vs. SM) and by package technologies, as described in Sec. 3.4. Figures 3.5 and 3.6 illustrate some of the more popular package types used today in the TH and SM families.

3.3.1 Through-Hole Mounted Packages

For assembly onto a PWB, through-hole (TH) type packages require one plated-through hole (PTH) in the PWB for each package pin or

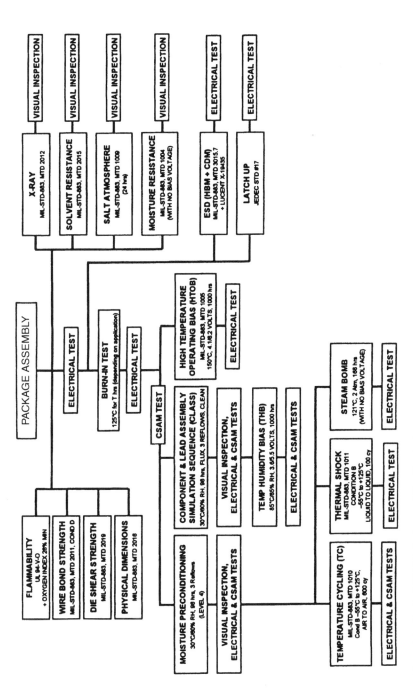

Figure 3.3 Typical PBGA device qualification tests.

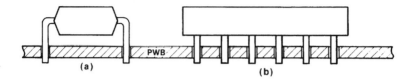

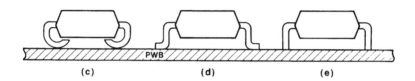

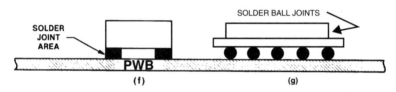

Figure 3.4 Examples of packages, package lead geometries, and mounting methods in current use: *(a, b)* typical TH packages, *(c–g)* typical SM packages. Lead geometry options are *(a)* DIP, *(b)* GPA, *(c)* SOJ and PLCC, *(d)* SOIC, SOG, QFP, PQFP, and fine-pitch LDCC, *(e)* butt-leaded packages, *(f)* LLCC, and *(g)* BGA.

lead, typically 0.035-in. (0.90 mm) in diameter on a 0.100-in. (2.54 mm) grid. Pin or lead lengths are not critical with TH components, since the assembly process usually calls for the leads to be cut off after insertion into the PWB and then crimped to provide mechanical backup.

The DIP family, introduced in the early 1970s, with dual in-line leads on 0.100-in. (2.54 mm) centers (Figure 3.6), is generally available in three package technologies: molded plastic, pressed ceramic, and laminated ceramic. DIPs can accommodate devices with as few as 8 and up to 64 I/Os. The family is defined by width, or lead row spacing, ranging from 0.300-in. (7.62 mm) for the narrowest (referred to in the industry as *skinny* DIPs) to 0.600-in. (15.2 mm) for the widest family. Another variation, known as *shrink* DIPs (SDIPs), have the same lead row spacing as the standard family but with the lead pitch reduced from 0.100 to 0.070-in. (2.54 to 1.78 mm) to provide higher pin-count capability for the same size package. Similar DIP families exist in the molded plastic technology, pressed ceramic technology (CER-

Table 3.1 Package Families by Packaging Technology

	Package technology			
	Molded plastic	Laminated plastic	Laminated ceramic	Pressed ceramic
Sealing method	Molded	Molded/ encapsulated	Metal lid seal	Glass lid seal
Surface mount	SOIC SOJ PLCC BQFP MQFP TQFP TSOP TAB TapePak*	PBGA	CQFP LLCC LDCC SMPGA CBGA	CERQUAD CQFP
Through-hole mount	DIP SIP ZIP QUIP	PPGA LLCC	DIP SIP PGA	DIP CERDIP

*Registered trademark of National Semiconductor Corporation.

DIP), and in the laminated ceramic technology. In addition, laminated ceramic DIPs come in a wide variety of configurations identified by the method of lead attachment to the ceramic body, which can be side-brazed, top-brazed, or bottom-brazed, as shown in Figure 3.7. Side-brazed DIPs are footprint compatible with the plastic and pressed ceramic DIP families. Top- and bottom-brazed DIPs are generally supplied by a vendor as a flat pack and the leads are then formed to the specification of the end user, which may or may not be footprint compatible with the side-brazed variety.

DIPs are limited in pin count to 64 leads by the physical size of the package, which approaches 3.0 in. (76.2 mm) in length for 64 I/Os. At this level, the IC package becomes very expensive, difficult to handle, and, in general, the electrical performance of the device begins to degrade—not to mention that it uses up a lot of expensive PWB real estate.

Area array packages, for example PGAs, were developed to extend the pin-count capability for TH-mount-type packages with a significant saving in PWB area. Figure 3.2 compares the physical size of a PGA package with pins on an area array of 0.100-in. pitch to the DIP at an in-line 0.100-in. (2.54-mm) pitch. The 0.100-in. (2.54-mm) pitch PGA family is available in both laminated plastic and ceramic technologies and can provide pin-count capability above 1000 pins. Ceramic

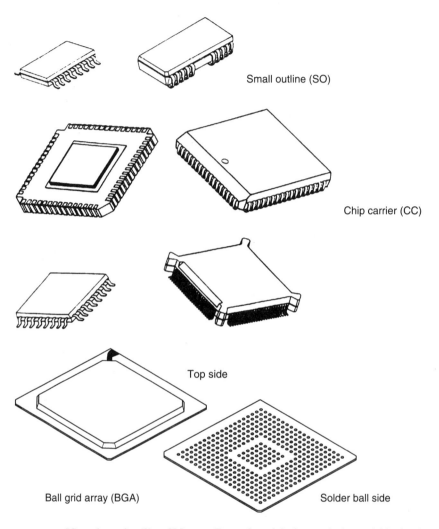

Figure 3.5 IC package families, SM type. Examples of design variations within families are lead geometry (gull wing, J, leadless, or solder balls).

PGA pin arrays range from 9 by 9 (81 max. pin count) to 20 by 20 array (400 max. pin count), in both cavity-up and cavity-down configurations. The largest registered ceramic PGA outline, at 1300 pin count, has an overall body size of 2.66 in. (67.6 mm) with a staggered pin arrangement.

Other in-line families are the single-in-line package (SIP) family, the zigzag-in-line package (ZIP) family, and the quad in-line package (QUIP) family. All were introduced to meet special product needs; for example, SIPs were needed for memory chips and memory modules.

Packaging of IC Chips   113

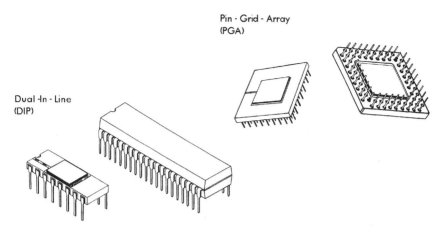

Figure 3.6 IC packaging families, through-hole (TH) type. Design variations exist within families, such as the way leads are attached to or exit the body, as illustrated in the DIP family, and in the direction the cavity faces with respect to the PWB (up or down), as shown in the PGA family.

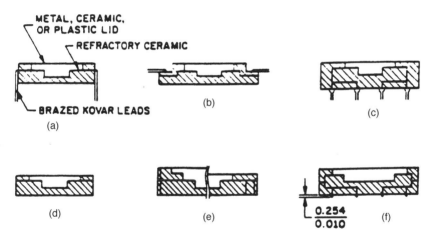

Figure 3.7 Cross-sectional sketches of several package types. (a) Side-brazed, (b) top-brazed, (c) pin grid array, (d) leadless with edge metallization, (e) leadless with via holes, (f) leadless array package with stand-offs. (After Kyocera[6])

These package families have pins typically on 0.100-in. (2.54-mm) pitch and are available up to 40 pin counts for ZIPs and to 64 for QUIPs. A QUIP has four in-line rows of leads, with the rows generally spaced 0.100 in. (2.54 mm) apart, and the leads in each row staggered with respect to each other to maximize hole-to-hole spacing on the PWB. This package option effectively reduces the package length by one-half over the equivalent DIP of the same pin count.

3.3.2 Surface Mounted Packages

The first SM packages were introduced in the early 1980s to achieve higher board density and better electrical performance. Surface-mount technology (SMT) eliminated the need for through holes. Initially, industry acceptance of SM did not advance as rapidly as expected. However, the acceptance rate in the 1990s accelerated, and the demand for SM packages has exceeded TH packages. Today, the availability of new SM package types has expanded greatly into a very formidable list (Table 3.1).

High-density packaging cannot be achieved with TH-type packages, such as DIPs and PGAs, because they are limited to 0.100-in. (2.54-mm) lead pitch, by current IC package and PWB design rules and process capabilities. Either perimeter or array-type SM packages with finer lead-pitch capability are needed to satisfy this demand. Perimeter leads for SM packages with a pitch as fine as 0.016 in. (0.40 mm) have been achieved, but handling and assembly at these levels has been difficult. Some packages now require over 1000 I/Os, an unachievable task for perimeter packages. But, given an array type package like the BGA, with a 0.040-in. (1.0-mm) solder ball pitch, the feasibility of meeting the over 1000 I/Os requirement becomes realistic.

The demand for SM packages has placed a great deal of pressure on the IC package technology. Processes such as vapor-phase (VP) and infrared (IR) solder reflow have placed stringent requirements on the physical design of SM packages that were not required for the TH types. Some of the most important aspects are lead coplanarity, lead finish, lead geometry, lead true position, and thermal effects. To ensure a reliable and high-yield (cost-effective) SM attachment, the perimeter leads on the SM package must be coplanar to within 0.002 to 0.004 in. (0.05 to 0.10 mm). Similarly, the solder balls on a BGA package must be coplanar to within 0.006 to 0.008 in (0.15 to 0.20 mm). The solder joint is another critical item for SM packages. Since the solder joint provides both electrical connectivity and mechanical attachment of the packaged device, its integrity must be preserved through all the stress encountered in subsequent assembly and test, over the life of the system in its operating environment. Solder embrittlement must be avoided; hence, lead-finish materials on perimeter type packages and the gold plating thickness on BGA solder pads must be specified with care.

Mature SM families today are those introduced in the early 1980s, such as the small-outline (SO) and PLCC types in molded plastic, CERQUAD in pressed ceramic, and leadless (LLCC) and leaded (LDCC) chip carriers in laminated ceramic. These packages have leads on 0.050-in. (1.27-mm) pitch instead of the 0.100-in. (2.54-mm) pitch used in the TH families. The SO package is an in-line type defined by body width, or lead row spacing, similar to the way the DIP family is defined. Both narrow-body and wide-bodied SOs are gull wing (GW) families. The SO package with J leads (SOJs) has the same body width as the SO; that is, it uses the same plastic molding die but has a J-formed lead for surface mounting.

PLCCs, CERQUADs, LLCCs, and LDCCs are perimeter-type packages with the leads on all four sides of the package. The lead geometries vary with the technology used. PLCCs use J-shaped leads designed to provide the highest possible density on the PWB. CERQUAD packages generally come from the vendor as a leaded flat pack, and the device manufacturer, or end user, does the final lead forming (GW or J) prior to SM. The LLCC packages are generally restricted to surface mounting on substrates with similar thermal expansion coefficients. This limitation is overcome by converting an LLCC to an LDCC by means of one of the commercially available soldered clip-lead techniques. Today, LDCCs can be purchased either as leaded flat packs with the leads formed to the appropriate shape (GW or J) by the device manufacturer or end user, or as finished packages with the leads in final form ready for SM. Examples of SO, PLCC, and CC packages are shown in Fig. 3.5.

The drive for higher pin counts and higher density has been very evident in the development of fine-pitch SM packages. Since the pin count of a perimeter package is a function of package size and lead pitch, as illustrated in Fig. 3.2, to achieve a high pin count with PLCCs at a 0.050-in.(0.013-mm) lead pitch requires very large packages. Difficulties in molding large plastic packages and maintaining coplanarity of all leads necessary to achieve high board assembly yields limit the practical PLCC to a lead-to-lead overall size of 1.699 in. (43.2 mm), equivalent to 124 I/O.

A new package family evolved to fill the needs for high pin counts, namely the quad flat pack (QFP) with gull wing leads on all four sides (see Figs. 3.8 and 3.9). The QFP has a lead pitch that varies from 0.40 to 0.16 in. (1.0 to 0.40 mm) and a maximum pin count of 376 leads. Two QFP families exist; the metric EIAJ MQFP version, shown in Fig. 3.8, and the bumped JEDEC BQFP, shown in Fig. 3.9. The MQFP family is characterized as fixed body size, variable pitch. That is, a variety of pin counts can be manufactured from a single body size using leadframes of various external lead pitches from 0.040 in. (1.0 mm) down to 0.016 in. (0.40 mm). The BQFP family is characterized as having a variable pitch, consisting of 0.020-in. (0.50-mm), 0.025-in. (0.60-mm), and 0.500-in. (1.27-mm) pitches, and a variable body size, with a variety of pin counts. Major advantages of the JEDEC over the EIAJ version are that the bumpers provide for safe handling in tubes, trays,

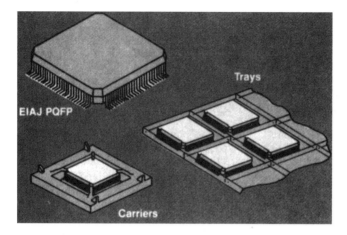

Figure 3.8 EIAJ MQFP packaging and handling options.

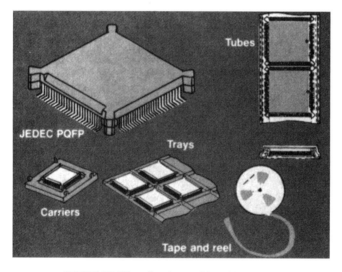

Figure 3.9 JEDEC BQFP packaging and handling options.

carriers, and tape and reel. The generous stand-off also provides for easy removal of flux from under the package after board assembly. The major advantage of the EIAJ over the JEDEC is the multiple lead counts available per package, compared with only one pin count per package size for the JEDEC. EIAJ packages are more difficult to handle without bending the leads, and the low stand-off on higher pin counts causes high stress on solder joints due to board warp. EIAJ's head start by several years provided valuable experience in high-volume production, which enabled EIAJ to become the industry *de facto* standard before the JEDEC version left the ground. Both versions are in use today, but the EIAJ version has gained in popularity. The EIAJ MQFP family provides pin counts up to 376 I/Os, while the JEDEC BQFP has leveled off at 244 I/Os.

Another family of SM packages was developed that are shorter, smaller, thinner, and lighter, called TQFPs. Whereas the standard MQFP body thickness varies from 2.0 to 3.8 mm, depending on the package outer dimensions, the TQFP body thickness are 1.0 and 1.4 mm. The TQFP packages vary in lead count, from the smallest at 32 I/Os with a body size of $5.0 \times 5.0 \times 1.0$ mm and an 0.80-mm lead pitch, to the largest with a maximum lead count of 256 I/Os in a body size of $28.0 \times 28.0 \times 1.40$ mm and a 0.40-mm pitch.

As the demand for packages with high I/O counts and high performance increases, the industry has turned to BGAs as a favorable alternative to the fine-pitch QFPs, which are difficult to handle. The BGA arrays range from 4×4 to 49×49, in sizes of 7.0×7.0 to 50.0×50.0 mm.

Reliability of SM packaged devices could be affected directly by the stresses imposed during VP or IR solder reflow. This is particularly true if the packaged device is inadvertently exposed to high levels of moisture during the period between device assembly and board assembly. Immersion of the package into the hot vapor of VP soldering, or direct heating by IR, occurs so rapidly that even small amounts of moisture (approximately 0.20 percent by weight) could subject the package to severe internal stresses and initiate cracks from the device surface to the outside. Most of these fine cracks are not readily detected by visual inspection after board assembly, because initially they are usually non-fatal to the device and are hidden from view at the package surface. But, if left undetected, the fine cracks may provide a relatively easy path for the ingress of contaminants from fluxes and solvents used in board assembly. Moisture from subsequent operations such as testing, shipping, storage, or from the ambient may also enter the package through the fine cracks. These contaminants may result in conditions, at the IC-to-package interface, that could lead to premature reliability failures.

3.4 Package Technologies

IC package technologies fall into three basic categories: molded plastic, pressed ceramic, laminated ceramic, and laminated plastic. In the molded plastic technology, the package is constructed around the IC chip assembled on a metal leadframe, in strip form. The IC chip is first mechanically bonded to the die attach paddle and then electrically interconnected by fine wires from chip bond pads to the corresponding package wire bond fingers. If TAB bonding is used, then the IC chip pads are directly connected to the bumped leadframe fingers. The final package configuration is formed by plastic molding around the leadframe subassembly. The portion of the leadframe that is external to the package body is subsequently trimmed and formed into specific geometries suitable for either TH or SM attachment.

The pressed ceramic package contains a leadframe embedded and glass sealed into a pressed ceramic base. The IC chip is attached and wire bonded to the leadframe and hermetically sealed using a ceramic cap or lid. The leads external to the package body are trimmed and formed to the geometries needed for TH or SM attachment.

A laminated ceramic or plastic package consists of a substrate with an integral metallized fanout circuitry, and external terminals which are either leadless, leaded or with solder balls for SM, or leaded for TH attachment. The substrate body may come with either a cavity to accept the chip or an in-plane configuration where the chip attach area and the bond pads are on the same level. The substrate material may be either ceramic or plastic and of single or multilayer construction, depending on the functional requirements. Both the ceramic and plastic substrates can equally provide high performance, but the ceramic has hermetic sealing capability whereas the plastic substrate is nonhermetic and requires some form of epoxy encapsulation for environmental protection of the chip. The electrical interconnections are done using assembly techniques similar to those used with the plastic molded package types.

A comparison of cost, performance, and reliability among the various package technologies shows the multilayer ceramic package to be best suited when the requirements call for high performance, hermeticity, and high reliability, but the trade-off is high cost. A multilayer plastic package, with electrical and thermal enhancements, can provide equal or better performance at a lower cost, trading off hermeticity. For the lower end of consumer electronic products, the package requirements, in order of importance, are low cost, reliability, and performance. These attributes can be satisfied with plastic molded packages and the proper selection of materials and rigid process controls. Sections 3.4.1–3.4.4 discuss packaging technologies that are generally applicable to a variety of package types and families. Section 3.5 presents package technology comparisons and trade-offs.

3.4.1 Molded Plastic Technology

The molded plastic package technology is widely accepted in the industry as reliable and low in cost, but the packages are nonhermetic and can have only a single metallization fanout layer. Package family outlines have been standardized in the Electronic Industries Alliance (EIA), the Joint Electron Device Engineering Council (JEDEC), and the Electronic Industries Association of Japan (EIAJ).

3.4.1.1 Post-molded plastic package. A metal leadframe, usually copper, copper alloy, or Alloy-42 (42 percent Ni, balance Fe) provides the mounting surface for the chip; that is, the chip support paddle. The leadframe also provides the electrical fanout path from the bonding fingers to the outside leads, which vary in pitch from 0.040 in. (1.0 mm) to 0.016 in. (0.40 mm) as shown in Fig. 3.10. The leadframe is usually spot plated with silver or gold on the paddle and at the tip of the wire-bond fingers to provide reliable chip attachment and wire bonds. Silver spot plating is preferred for lowest cost and better adhesion to the molding compound. Wire-bond interconnects are gold wires, typically 0.00100 or 0.00125 in. (0.025 or 0.032 mm) in diameter, with current maximum wire span lengths, for the respective wire diameters, ranging from 0.20–0.25 in. (5.1–6.3 mm). Thermosetting (cross-linking) epoxy resin is molded around the leadframe–chip subassembly after the chip has been wire-bonded to the leadframe. After molding, the external portion of the leadframe must be processed to the final form and lead finish for assembly to the next level. This assembly sequence is shown in Fig. 3.11 and applies to all plastic molded package types. Details, such as leadframe design and format, workholders for chip and wire bonders, molding dies, trim and form tools, and handling hardware will vary according to package types. Figure 7.14 illustrates the construction of a QFP in the post-molded plastic package technology.

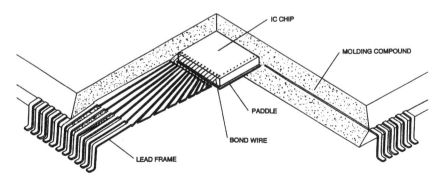

Figure 3.10 Sectional view of a typical QFP illustrating molded plastic (post-molded) technology.

120   Chapter 3

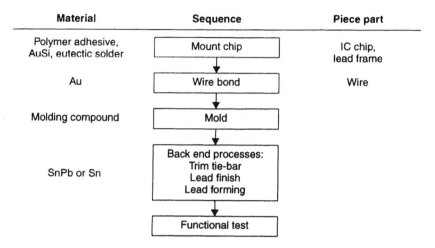

Figure 3.11 Assembly sequence for post-molded plastic technology packages.

3.4.1.2 Pre-molded plastic package. A pre-molded package concept, shown in Fig. 3.12, provides a more benign environment for packaging very sensitive IC devices requiring low-cost assembly. Delicate wire spans and strain-sensitive features on the chip surface are decoupled from the molding process, thus avoiding the stresses associated with post-molded packages. The assembly sequence for a pre-molded plastic technology package is shown in Fig. 3.13.

The package is fabricated using a preplated leadframe, in strip form with multiple sites per strip, and then either transfer molded,

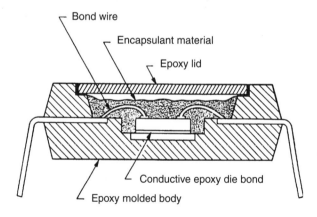

Figure 3.12 Cross-sectional view of a completed pre-molded plastic DIP, including IC chip.

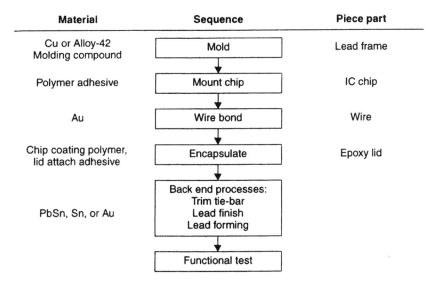

Figure 3.13 Assembly sequence for pre-molded plastic technology packages. Package is fabricated using preplated leadframe then either transfer or injection molded using a thermoset epoxy B or a thermoplastic polymer molding compound.

using a thermoset epoxy B molding compound, or injection molded with a thermoset or a thermoplastic polymer, such as polyphenylene sulfide. The leadframe, usually made of a copper alloy for better thermal performance, is spot plated with either silver or gold in the die attach paddle area and wire bond fingers. The external leads may be selectively plated with thin gold to retain solderability for final lead finishing. After molding, the package may need to be deflashed to remove any molding compound flash deposited on the chip-bonding areas and then cleaned to remove any residual particulate matter. Chip attachment is done with a polymer adhesive, either conductive (silver-filled) or nonconductive epoxy, or polyimide paste. Wire bonding is either thermosonic or thermocompression ball-wedge bonding using 0.001 or 0.00125-in. (0.025 or 0.032-mm) dia. 99.99 percent pure gold wire. This is followed by chip encapsulation with a die coating or, preferably, a flow coating to fill the entire cavity, thus environmentally protecting the chip and the wedge bond areas in the cavity. Candidate materials include room-temperature vulcanized (RTV) silicone rubbers and combinations of silicon gels and cover coats. In some cases, a plastic lid is used instead of encapsulation. Back-end processes include trim and form to the final package configuration, for a DIP, or a plastic leaded chip carrier (PLCC). Final lead finish is then applied by solder dipping. Another option is to use the original thin gold plate as the final finish, making sure that solderability is not degraded in the

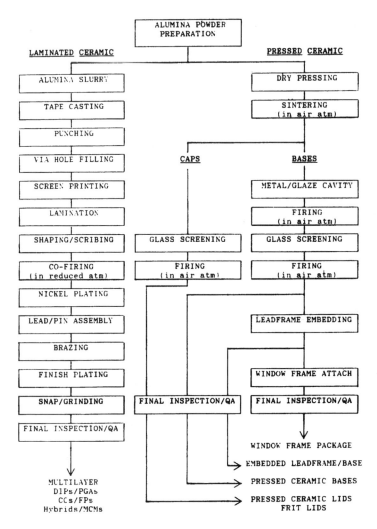

Figure 3.14 Process flowchart for piece parts and prefabricated packages for pressed ceramic and laminated ceramic package technologies. *(After Kyocera[5])*

assembly process. A gross leak tight-lid seal will preclude the subsequent ingress of cleaning solvent (used in PWB assembly) into the chip cavity, where it could interact with the flow coat to cause swelling and stress the delicate wire bonds or degrade the interface integrity between chip and encapsulant. Of the two plastic package technologies, the post-molded package technology is the more cost-effective of the two and by far the predominant package technology in use today.

3.4.2 Pressed Ceramic (Glass-Sealed Refractory) Technology

Pressed ceramic technology packages are used mainly for economically encapsulating IC chips requiring hermetic sealing. Hermeticity means that the package, with the assembled IC chip, must pass both gross and fine leak tests and also exclude environmental contaminants and moisture for a long period of time. Furthermore, any contaminants present in the package before lid sealing must be removed to an acceptable level during the sealing process. Glass is an effective material for achieving a hermetic seal for high-reliability applications. Leak rate, as defined by MIL-STD-883, is "that quantity of dry air at 25°C in atmospheric cubic centimeters per second flowing through a leak or multiple leak paths from a high pressure side of one atmosphere (760 mm Hg absolute) to the low pressure side ≤ 1 mm Hg absolute." Standard leak rates are expressed in units of atmosphere cubic centimeters per second (atm $\cdot$ cm^3/s). Procedures for detecting gross and fine leaks in hermetic packages are detailed in MIL-STD-883. Gross-leak tests involve die penetrants and bubble tests, whereas fine-leak testing procedures use helium and radioactive tracer techniques.

The process flowchart for fabricating the basic piece parts used in the pressed ceramic packaging technology is shown in Fig. 3.14. Also shown alongside is the fabrication process flow sequence for the laminated ceramic package technology, which is described in Sec. 3.4.3. Pressed ceramic piece parts require fewer and simpler process steps, resulting in the lowest-cost piece parts for a hermetic package technology. This technology, when implemented with a leadframe in strip format (that is, multiple sites per strip) can be automated and therefore can be competitive with the plastic technology for low-cost packaging. The IC package designer has a choice of three pressed ceramic options, as shown in Fig. 3.14; that is, basic piece parts (bases and caps), partially assembled subassemblies (embedded leadframe and base), and a prefabricated package known also as a *window-frame* package. The IC assembly sequence is described in Fig. 3.15.

In the first option, the package is constructed around the IC chip using basic piece parts, i.e., ceramic base, leadframe, and ceramic cap. When done in-house, this is the lowest-cost option, because the IC manufacturer controls all of the assembly process steps. Option 2 is a minor variation of Option 1, where the IC manufacturer purchases the leadframe already embedded into the base ceramic and ready for IC chip mounting. This option is attractive to manufacturers who do not have, nor wish to invest in, the equipment needed for embedding the leadframe. These options are known in the industry as the CERDIP or CERQUAD packaging technology. The assembly structure is shown in Fig. 3.16*a*.

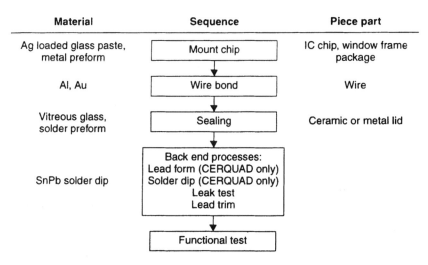

Figure 3.15 IC assembly sequence for pressed ceramic package technology.

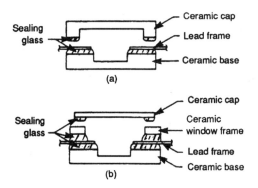

Figure 3.16 Structures of *(a)* CERDIP and *(b)* window frame. *(Courtesy of Kyocera America, Inc.)*

In both options, the IC chip is attached to the leadframe using either AuSi eutectic or silver-loaded glass adhesive technologies. Hermetic sealing is done with a glassed ceramic cap or lid. Due to the relatively high temperature used for glass sealing, typically greater than 400°C, an all-aluminum monometallic system is used. This consists of the aluminum bond pads on the chip, ultrasonic aluminum wedge-to-wedge wire bonding, and leadframe fingers that have aluminum deposited or cladded surface finishes in the bond areas. The gold-aluminum intermetallic interface, which has caused many reliability problems in the past, is thus avoided. Lid sealing is done in a controlled nitrogen atmosphere conveyor-type furnace, in which the heat is transferred by conduction and convection or by infrared radiation. Back-end processes include leak testing, lead finishing, and lead trim and form. External leads are typically plated with matte tin, which provides a finish that is compatible with the subsequent lead trim and form operation and final assembly to the next-level packaging. CERDIP leadframes are procured already formed and require only to be finished and trimmed prior to testing. Leadframes for SM types, such as the CERQUAD, usually come planar and are processed planar through lead finishing, then trimmed and formed into the desired shape for the final package configuration.

The third pressed ceramic option, known in the industry as a *window-frame package,* is the structure shown in Fig. 3.16b. The assembly sequence is given in Fig. 3.15. This design concept provides the option of decoupling the IC chip from the high-temperature glass-sealing procedure by providing a low-temperature solder seal capability for the final hermetic seal. This option is used for temperature-sensitive chips that cannot go into a glass-sealed package without degrading their performance or reliability. In addition, this option also allows the use of thermosonic or thermocompression gold ball-wedge wire-bonding equipment, which extends this package technology to those manufacturers who have this wire-bonding capability but not the ultrasonic aluminum wedge-to-wedge bonders needed to achieve the desired all-aluminum monometallic structure for glass seals. Back-end processes are similar to those used with the other options.

Seal glass is an important constituent of the pressed ceramic technology, because it influences both electrical performance and reliability of the packaged device. The dielectric constant of the glass used to embed the leadframe can affect the capacitance between the leads; that is, the speed of the I/O signals through the leadframe. Seal glass is exposed to the cavity; hence, alpha-particle emission from the glass fillers may affect the soft error reliability of some sensitive devices. The hermeticity of the finished package is influenced by the adhesion of the seal glass to the ceramic, the seal glass to the leadframe, the mechanical strength of the glass, and the dissolution of fillers in the mother glass. Any residual stress developed in the mother glass is caused by the mismatch in the thermal expansion between ceramic, Alloy-42 leadframe, and the glass. The particle-size distribution of the filler in the glass also influences the mechanical strength or the ability to provide reliable hermetic seals.

The raw material for the pressed ceramic technology is alumina ceramic (90 percent minimum Al_2O_3) in black-brown characteristic color. Package vendors are also providing ceramic materials such as aluminum nitride (AlN), silicon carbide (SiC), and low-thermal-expansion sealing glasses (4.7 ppm/°C), for improving hermeticity yields of pressed ceramic.

3.4.3 Cofired Laminated Ceramic Technology

The cofired laminated ceramic technology is one of the most reliable IC packaging technologies currently available. In addition, the technology is capable of meeting today's demands for higher performance at the device and system levels. This section addresses the technology for single-chip packages. The cofired laminated ceramic technology is also being used for such applications as hybrids and MCMs.

3.4.3.1 High-temperature cofired ceramic technology. The process flowchart for the cofired laminated ceramic package technology is shown in Figs. 3.14 and 3.17. A dispersion or slurry of alumina ceramic powder (90 to 92 percent Al_2O_3) in a liquid vehicle (solvent and plasticized resin binder) is first prepared, then cast into thin sheets by passing a leveling or doctor blade over the slurry. After drying, the sheets are in the green-tape stage, ready for cutting to size, punching of via holes (through the dielectric layers for future interconnection), and forming the cavities. Custom conductor paths are then screened onto the surface (usually a slurry of tungsten powder), and via holes are filled with metal. Several of these sheets are press-laminated together in a precisely aligned fixture, and the entire structure is fired at 1600°C, in a reducing atmosphere, to form a monolithic sintered body. The cofired laminated refractory ceramic technology is a complex process requiring careful control throughout.

After the laminate has been sintered, it is ready for the finishing operations of lead/pin attachment and metallization plating. For packages that require either leads or pins, nickel is first plated over the exposed tungsten in preparation for lead/pin brazing. The lead/pin material is either Kovar (a Fe-Ni-Co alloy) or Alloy-42, and the brazing material is a Ag-Cu eutectic alloy. After lead/pin brazing, all exposed metal surfaces are electroplated or electroless plated (usually gold over nickel) for bondability and environmental protection. A typical single-chip package could contain up to seven tape layers and four screened dielectric layers. The thickness of a taped layer, with tungsten-filled vias, may range from 0.002–0.025 in. (0.05–0.64 mm). Screened dielectric layers are typically 0.001–0.003 in (0.025–0.076 mm) in thickness.

The cofired laminated ceramic package technology is very effective for constructing complex packages with signal, ground, power, bond-

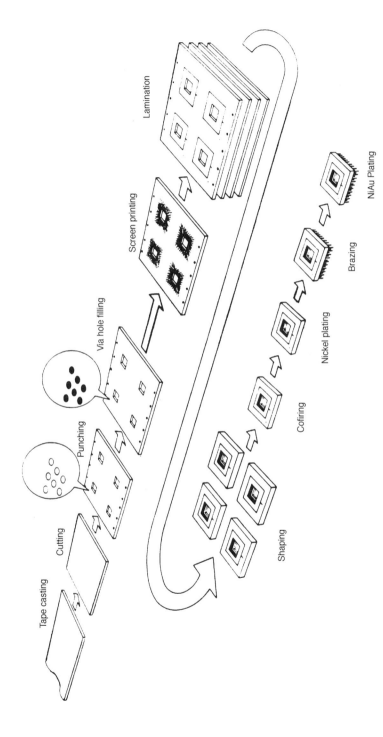

Figure 3.17 Process flowchart of laminated ceramic package technology. *(After Kyocera[6])*

ing, and sealing layers. Electrical characteristics, such as line capacitance, propagation delay, impedance, and inductance, can be customized for a particular application by design layout and specification of layer thicknesses.

This technology, however, has three technical areas of weakness.

- Hard-to-control tolerances caused by high shrinkage in processing
- A high dielectric constant of 9.5, which affects signal-line capacitive loading
- A modest thermal conductivity of Al_2O_3

The use of beryllia (BeO), AlN, or SiC, instead of the 90 to 92 percent Al_2O_3, will result in a greatly superior thermal performance and a significantly lower dielectric constant.

An assembly sequence flowchart for cofired laminated ceramic IC packages, with wire bond interconnection, is shown in Fig. 3.18. Chip bonding is done with either an AuSi eutectic, AuSn eutectic, or polyimide system to achieve a high yield and highly reliable mechanical and electrical interconnection. AuSi and AuSn preforms vary from 99.99 percent pure gold, with dopants to provide a reliable electrical back contact, to alloys of a variety of constituents such as 80Au/20Sn for lower eutectic temperature bonding. Polymer die attachment, in hermetic packages, has also been used. Filled and unfilled polyimides

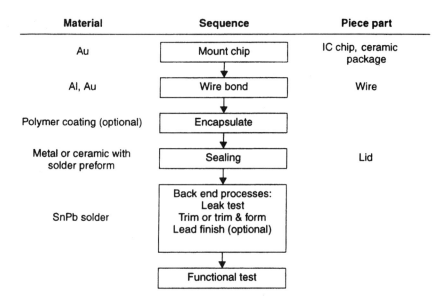

Figure 3.18 IC assembly sequence for cofired laminated ceramic packages.

were found to withstand the hermetic assembly processing temperatures. Electrical interconnections are generally done using thermosonic ball-wedge gold wire bonding with wire diameters typically at 0.001 in. (0.025 mm). Wire spans are usually limited to a maximum of 0.150 in. (3.8 mm) to meet shock, vibration, and acceleration requirements. Excessively long wire spans are avoided by going to a two-tiered, or in some cases three-tiered, wire-bond construction, as shown in Fig. 3.19. Final hermetic sealing is done with an 80Au/20Sn braze, in a dry nitrogen ambient, either in a belt furnace with a temperature profile peaking at about 350°C or in a controlled-ambient glove box containing a parallel-gap seam sealer. The seam seal brazing operation confines the heat locally to the seal ring area; thus, the package and IC are not significantly heated by the operation. The environment in the sealed package is predominantly nitrogen, and the moisture content is typically at or below 1000 ppm H_2O by volume, which meets the military standard requirement for high-reliability applications.

Figure 3.19 Tiered ball-wedge wire bond configuration. *(Courtesy of Agere Systems)*

Back-end processes are similar to those for pressed ceramic, that is, leak test, lead trim or lead trim and form, and sometimes lead finish for SM packages. The most popular lead finishes available from package vendors range from electroplated gold for soldering and socketing to tin plating for soldering applications only. In between, there are several other combinations involving proprietary alloys of Au-Pd-Ni, Au-Ni, and electroless gold and nickel. Some of these finishes are not compatible with one or more of the assembly processes and may degrade solderability to the point where the SM solder reflow quality may be jeopardized. Leads with thick gold or alloys with significant amounts of gold will cause brittle solder joints that cannot be tolerated in SM. Fine-pitch leaded ceramic chip carriers (LDCCs) may require a solder dip coating to achieve high yields and reliability.

Ceramic packages come with brazed leads/pins for TH or SM, and leadless perimeter or array pads for SM (Fig. 3.7). The package is delivered to the IC assembly manufacturing line with leads/pins, cavity, and bond sites finish-plated to customer specification and ready for all subsequent level 1 and 2 assembly steps.

Dual in-line package (DIP). A typical DIP construction (Fig. 3.20a) consists of three cofired laminated ceramic layers, with a chip cavity formed by the middle layer and a wire-bond ledge formed by the top layer. There are two buried layers and one top-layer metallization in the construction. The bottom buried layer provides the metallization for the chip cavity base, while the top buried layer provides the signal fanout path and bond pad from the wire-bond ledge to the external pins. The different levels can be interconnected by vias, as shown in Fig. 3.20b. Top-layer metallization provides the seal ring for hermetic sealing. The seal ring may be tied to a ground plane through a via, which may be located near the notch, as illustrated in Fig. 3.20a. The same construction applies to a wide variety of DIP configurations, such as the package with two cavities where the IC chips are electrically interconnected to each other (Fig. 3.21).

Ceramic pin grid array package (CPGA). As demand for high I/O pin counts increased, the DIP, due to its size limitations, was no longer able to fulfill IC packaging needs. As a result, a TH-mount ceramic pin grid array (CPGA) package was developed to accommodate the increase in I/Os at an effective pin count to area package configuration.

The CPGA is a ceramic cofired multilayer package (Fig. 3.22) with a cavity facing up or down, to accommodate the chip, and pins attached to the underside in an array pattern. The base of the cavity is normally gold plated for eutectic die bonding. Within the cavity are wedge-bond pads, located on single or dual ledges, for wire bonding the chip to the package. The CPGA is manufactured of alumina (90 to

Packaging of IC Chips

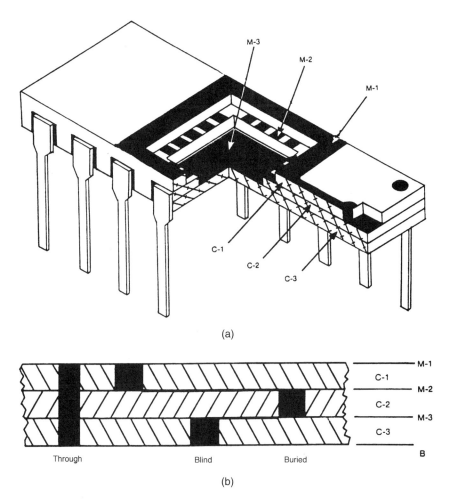

Figure 3.20 (a) Isometric sectional view of cofired laminated ceramic technology in DIP configuration and (b) cross section illustrating blind, buried, and through-hole vias in structure of part a.

92 percent Al_2O_3) or aluminum nitride (AlN) ceramics and follows the same processing steps (Figs. 3.14 and 3.17) as for any other cofired laminated ceramic multilayer package, like the DIP.

Kovar or Alloy-42 (a Fe-Ni-Co alloy), 0.018-in. (0.46-mm) dia. pins are brazed onto the underside of the package, in an array configuration, at a 0.100-in. (2.54-mm) pitch or staggered rows, which results in an effective pitch of 0.050 in (1.27 mm) between rows. The four corner pins have small tabs brazed on at approximately 0.050 in (1.27 mm) from the package base so as to provide a stand-off height when the CPGA is TH mounted to the PWB. The stand-off height provides pin

Figure 3.21 Ceramic dual in-line package (DIP) with two electrically interconnected cavities. *(Courtesy of Agere Systems)*

compliance and facilitates cleaning under the package. The finish on the pins is either gold or solder plating. A metallized seal ring around the cavity provides for either seam sealing or eutectic bonding a metal lid that hermetically seals the chip cavity.

Ceramic ball grid array (CBGA) packages. With the advent of high-density, high-pin-count, and high-performance SM package requirements, the ceramic ball grid array (CBGA) was developed to meet related demands. The CBGA is constructed in a similar fashion as other laminated ceramic packages (Fig. 3.7) but, instead of perimeter leads and area array pins, the CBGA contains area array pads for solder ball attachment. The substrate consists of a single or several cofired laminated ceramic layers, with molybdenum metallization, and with vias interconnecting the circuitry on the top side to the solder pads on the bottom. The IC chip is interconnected to the ceramic substrate by wire bonding or flip chip techniques, as discussed in Secs. 3.6.1.2 and 3.6.1.4. The substrate is lidded or capped to protect the chip, and solder interconnects are attached to the underside to complete the assembly.

Figure 3.22 Ceramic pin grid array (CPGA) package. *(Courtesy of Agere Systems)*

The CBGA-to-board interconnects consist of high-temperature 90Pb/10Sn solder balls, which are attached to the substrate with eutectic 63Sn/37Pb solder. The solder balls are 0.035 in. (0.89 mm) in dia. for grid arrays on 0.050-in. (1.27-mm) pitch, and 0.025-in. (0.64 mm) in dia. for a 0.040-in. (1.0-mm) pitch.

To overcome the thermally induced fatigue strain on the solder balls, caused by the CTE mismatch between the ceramic substrate and the glass-epoxy motherboard, the solder balls are replaced with more compliant solder columns; thus, the CBGA becomes a CCGA. The solder columns, which are 0.020 in. (0.50 mm) in dia., and 0.050 or 0.087 in. (1.27 or 2.2 mm) high, can be either directly attached to the underside of the alumina substrate, or a 63Sn/37Pb solder can be used as an interface for attachment. The different types of solder interconnects are shown in Fig. 3.23.

The CBGA and CCGA packages offer high reliability and high performance with pin counts exceeding 1000 at 0.039-in. (1.00-mm) pitch. For packages with a fewer number of pins, the area array pitch can be as small as 0.020 in. (0.5 mm).

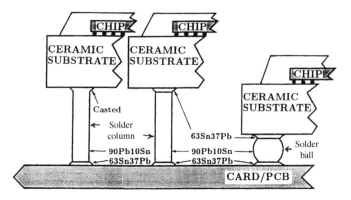

Figure 3.23 IBM's ceramic ball grid array (CBGA), showing different types of solder interconnects and attachments. *(After Lau³)*

3.4.3.2 Low-temperature cofired ceramic technology. MCP and MCM packages with functional substrates are gaining popularity for RF applications. The most common use of the low-temperature cofired ceramic (LTCC) technology is to provide embedded inductors, capacitors, and resistors for functional substrates. The manufacturing processes of the LTCC substrate are exactly the same as those of high-temperature cofired ceramics, except the firing for the LTCC substrate can range from 850 to 1000°C. Most of the Japanese manufacturers use their own proprietary material and green tape for LTCC manufacturing. Several other manufacturers in the U.S. and Europe use green tapes provided by DuPont. The most common metallizations in use for vias and internal conductors are Cu or Ag. For the external conductors, Cu, Pt/Ag, or Pd/Ag can be used. There are several technical issues and challenges, such as cost, shrinkage factors, quality and accuracy of the embedded component, that need to be improved.

3.4.4 Laminated Plastic Technology

The laminated plastic packaging technology is a spin-off of the traditional PWB technology where bare chips are mounted directly to the substrate in a technique known as *chip on board (COB)*. Early vintage laminated plastic packages used FR-4 epoxy substrates, which exhibited high contamination levels due to the flame-retardant fillers. FR-4 epoxy substrates also suffered from a low glass-transition temperature (140°C), making them incompatible with many standard IC package assembly processes. Substrate materials such as high-temperature epoxies, polyimides, and triazines are being used today to overcome these deficiencies. These materials exhibit glass-transition temperatures in the range of 180–240°C, thus overcoming a major deficiency of the FR-4 material. In addition, these materials also have a much lower level of contamination than the FR-4 boards.

The substrate may have single-sided, double-sided, or multilayer circuitry, depending on the application needs, and can be provided with or without a cavity to house the chip. The exposed copper metallization is generally electro or electroless plated with nickel and gold to prevent oxidation and provide wire-bondable pads. The most popular package types using this technology are the plastic pin grid array (PPGA) for TH mounting and the plastic ball grid array (PBGA) for SM.

3.4.4.1 Plastic pin grid array package (PPGA). The PPGA, illustrated in Fig. 3.24, is a nonhermetic prefabricated package that houses the IC chip, providing electrical interconnection to the motherboard and protection from hostile environments. The package consists of a square plastic PWB body with round pins press-fitted and solder reflowed to the underside. The pins can be configured in any desired footprint as long as the pin field is in a straight or staggered array of 0.100-in.

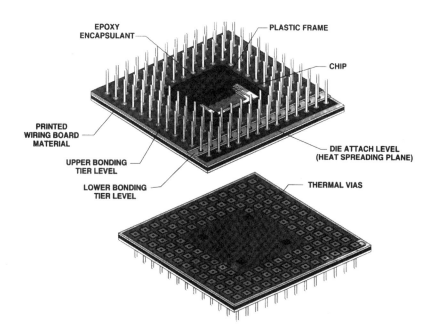

Figure 3.24 Multilayer PPGA with chip cavity facing down. *(After Cohn et al.[7])*

(2.54-mm) pitch and is restricted to the outside of the die attach and wire-bond pad areas. The pin array configuration results in the highest density of pins per package area and highest pin count of any other TH type package. PPGAs come in various forms (Fig. 3.25); chip down facing the motherboard or chip up mounted to the top of the package, double-sided or multilayer construction, and with or without a chip cavity. In addition, thermally enhanced PPGAs may also contain plated THs as "thermal vias" under the chip or a copper slug to spread the heat dissipated by the chip.

The PPGA package is constructed by first fabricating the substrate from a rigid double-sided or multilayer PWB, by either the additive, semi-additive or subtractive process, using copper conductors. The substrate contains plated-through holes and, depending on design complexity, may also have buried and semi-buried vias for electrical interconnection.

Pins are made of phosphor bronze, typically 0.018 in. (0.46 mm) in dia., instead of the more expensive Kovar or Alloy-42 used with ceramic packages. The solder-plated (60Sn/40Pb or 90Pb/10Sn) pins are inserted through gold or solder-plated-through holes in the substrate, and the connection is made by a press fit and solder reflow. A star or knurl is incorporated into the shank of the pin to facilitate a more reliable interconnection.

Solder reflow is accomplished by either dipping the entire length of the pin into a solder bath or laser spot heating the base of the pin. During solder dipping, the solder wicks up into the interface between the plated hole and the shank of the pin, and at the same time forms a fillet at the base of the pin. This provides for a strong reliable contact but results in a variable solder thickness along the length of the pin, which may affect solderability when tested per MIL-STD-883.

Another solder reflow process uses a laser beam to spot heat the base of individual pins at a rate of approximately 25 pins per second, causing the solder in the hole-to-pin interface to reflow. Sufficient solder must be available at the hole-to-pin interface so that, when reflowed, it will produce a strong interconnection. Since laser reflow heating affects only the base of the pin, the original solder-plated thickness along the length of the pin remains unchanged.

The chip is assembled to the PPGA using most of the same assembly processes and techniques used for the plastic molded technology. The assembly flowchart is shown in Fig. 3.26. The chip is first attached using one of the polymer adhesive systems, such as silver-filled epoxies or polyimides, then wire bonded by either the thermosonic gold ball-wedge or the ultrasonic aluminum wedge-wedge process. The chip is then polymer-coated for environmental protection using a "glob-top" liquid epoxy, flow-coated silicone gel, or RTV rubber. A resin dam is de-

**Double Sided,
Chip Down, W/O Cavity**

**4 Metallization Layers
Chip Down W/O Cavity**

**4 Metallization Layers
Chip Down, W/Cavity**

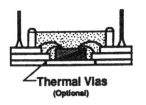

**6 Metallization Layers
Chip Down, W/Cavity**

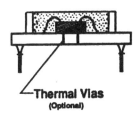

**Double Sided
Chip Up, W/O Cavity**

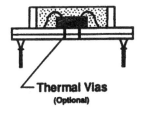

**4 Metallization Layers
Chip Up, W/O Cavity**

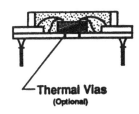

**4 Metallization Layers
Chip Up, W/Cavity**

**6 Metallization Layers
Chip Up, W/Cavity**

Figure 3.25 PPGA package configurations. *(After Cohn et al.[7])*

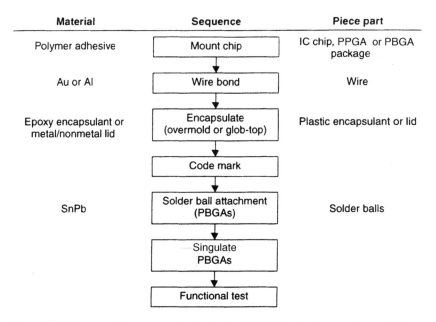

Figure 3.26 IC assembly sequence for laminated plastic technology packages (PPGAs and PBGAs).

sirable, when encapsulating a chip that is not in a cavity, to confine the materials to the IC and wire-bond areas. A metal or nonmetal lid, attached with a polymer adhesive, is optional and is provided for a chip-up configuration, mainly for physical protection when gels or RTV rubber systems are used.

The PPGA comes in either a low-cost or an enhanced, high-performance configuration, depending on the application and customers' design objective. The low-cost version, shown in Figure 3.27, consists of a conventional 0.060-in. (1.5-mm) thick copper-clad laminate processed as a single- or double-sided substrate with no cavity. Pins are press fitted and solder reflowed. The chip is encapsulated with either a "glob-top" liquid epoxy or covered with a silicone gel or RTV and lidded.

A performance-driven device will require a multilayer substrate with power and ground planes and multiple bonding tiers to yield the appropriate interconnection density within the allowable wire-bond design rules. Liquid epoxy or a lid is used to encapsulate the chip. Pins may be inserted to provide either a cavity-up or a cavity-down configuration. A typical cross section through the chip cavity of such a laminated PPGA, showing the multilayer construction, two-tier bond pads, and thermal vias, is illustrated in Fig. 3.28. For IC chips dissipating approximately ≥3 W, and depending on the environmental conditions and the maximum IC junction temperature (T_j) allowed, it may be

Packaging of IC Chips   139

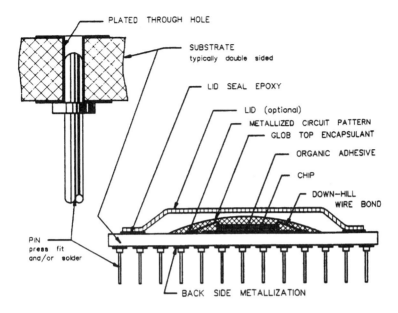

Figure 3.27 Cross section of a low-cost PPGA in laminated plastic technology. Substrate is either single- or double-sided PWB with no cavity for IC. Chip is environmentally protected by glob-top epoxy encapsulant. Metal or plastic lid is optional for physical protection.

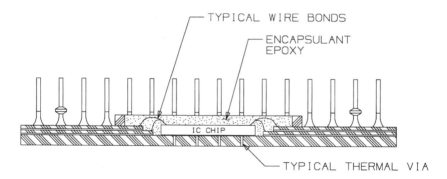

Figure 3.28 Cross section of a multilayer PPGA with chip cavity facing down.

necessary to mount the chip directly to a heat-spreading copper slug of a cavity-down package configuration, as shown in Fig. 3.29.

The low dielectric constant of PPGA substrate materials (in the range of 4 to 5) and the ability to provide thermal vias under the chip, or a heat spreading copper slug, result in equivalent or better electrical and thermal performance when compared with cofired laminated ceramic CPGAs using Al_2O_3.

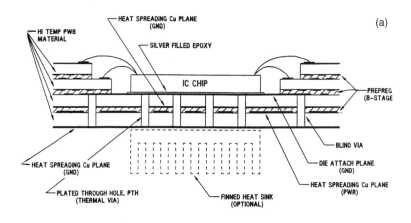
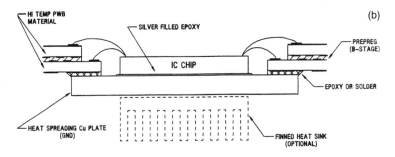

Figure 3.29 (a) PPGA package thermal enhancements using heat-spreading Cu planes and thermal vias, and (b) PPGA package thermal enhancements using a heat-spreading Cu plate. (*After Cohn et al.*[7]

3.4.4.2 Plastic ball grid array (PBGA) packages. Another IC package that utilizes the laminated plastic technology is the plastic ball grid array (PBGA) package. The PBGA (Fig. 3.30) leverages on the PPGA technology but replaces the pins with solder balls for surface mounting (SM).

PBGAs were first introduced by Motorola in 1989 for low-I/O devices and were then called *over molded pad array carriers (OMPACs)*. The PBGA has since emerged as the SM package of choice, because it offers many attractive features such as the following:

- High I/O-to-package-area ratio
- Multilayer substrate capability
- Improved thermal/electrical performance (Fig. 3.31)
- Increased interconnect density

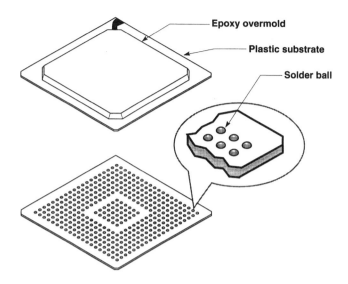

Figure 3.30 Plastic ball grid array (PBGA) package (overmolded type).

- Ball coplanarity less critical during SM, ≈0.006 in. (0.15 mm) to 0.008 in. (0.20 mm)
- Self-centering during SM solder reflow
- Elimination of fine pitch solder paste printing
- Pad-to-pad shorting diminished
- Low profile
- Expected solder joint defect level < 5 PPMJ

A PBGA package consists of a chip mounted and interconnected to a square double-sided or multilayer PWB substrate. Vias interconnect the top surface signal traces to the respective solder pads under the substrate. After chip attachment and wire bonding, the assembly is overmolded using a transfer or injection molding process. Another encapsulation method is to "glob top" the chip and its bond wires with a liquid-dispensed epoxy, which is confined within a resin dam or a chip cavity. Attaching solder balls to the underside of the package then completes the assembly.

The PBGA substrate provides structural support for the chip, encapsulant, and solder balls. It also facilitates electrical interconnection between the chip and motherboard and helps dissipate the power generated by the chip. Some of the issues encountered with substrates are: electrical interference to chip's functionality (electrical noise, cur-

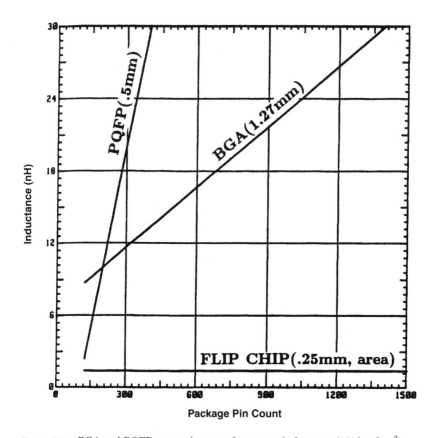

Figure 3.31 BGA and PQFP comparison: performance (inductance.) *(After Lau[3])*

rent leakage, high trace resistance), affinity for excessive water absorption that may lead to internal delamination (popcorning), warpage affecting coplanarity, and CTE mismatch that may affect solder joint reliability. Proper design and material selection can minimize these issues.

The PBGA substrates are usually fabricated by a PWB vendor in singulated or strip (multi-site) form (Fig. 3.32), according to a customer's design specifications, and then shipped to the semiconductor manufacturer for assembly. When the substrates are fabricated in strip form, with multiple sites, the vendor identifies by marking (x-out) those sites that are mechanically or electrically defective. During assembly, the chip pick-and-place machine avoids placing chips on an "x-out" site. The substrates may be either selected from standard open-tooled artwork configurations, available from vendor stock, or are custom designed to fit special applications. General-purpose sub-

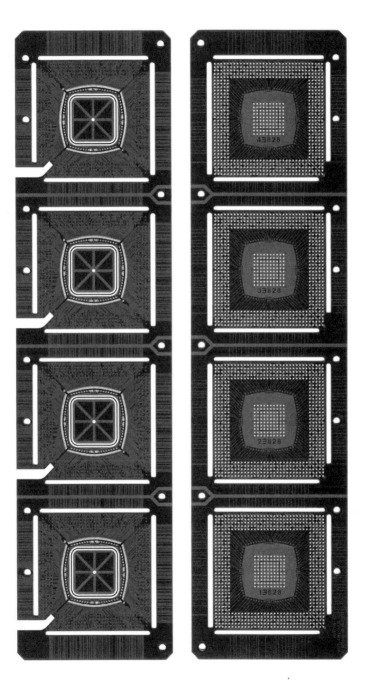

Figure 3.32 Laminate strip with slots outlining the four BGA substrates. *(Courtesy of Agere Systems)*

strate designs help reduce assembly turn-around time and inventory logistics problems. In addition, open tool designs eliminate tooling charges and take advantage of possible volume discounts.

The most common PBGA substrate material used in the industry today is bismaleimide triazine (BT), although other materials are being used as well, but to a lesser extent. The pads are sized according to the solder ball diameter used and second-level solder joint reliability effects. The pads are arranged in a standard JEDEC area array on a pitch of either 0.020, 0.030, 0.040, 0.050, or 0.060 in. (0.5, 0.8, 1.0, 1.27, or 1.50 mm). The exposed Cu metallization is either electroplated or electroless plated with nickel and gold to prevent oxidation and facilitate wire bonding to the bond pads and solder ball attachment to the solder pads.

The composition of the solder balls is either 63Sn/37Pb or 62Sn/36Pb/2Ag. The solder ball diameter varies from 0.012 to 0.030 in. (0.30 to 0.75 mm) and is dependent on the ball pitch used for the given application. Table 3.2 shows the industry's commonly used sizes for solder ball/pad diameter vs. ball pitch.

Table 3.2 Typical PBGA Solder Ball Pitch vs. Ball/Pad Diameters

Solder ball pitch (mm)	Solder ball dia.* mm (mils)	SMD		NSMD
		Cu pad dia. (mm)	Solder mask opening dia. (mm)	Cu pad dia. (mm)
0.5	0.30 + 0.10/–0.05 (12)	0.40	0.30 ± 0.05	0.30
0.8	0.40 + 0.10/–0.05 (16) or 0.50 ± 0.10 (20)	0.45 0.60	0.32 ± 0.05 0.45 ± 0.05	0.32 0.45
1.0	0.50 ± 0.10 (20) or 0.63 ± 0.10 (25)	0.60 0.75	0.45 ± 0.05 0.60 ± 0.05	0.45 0.60
1.27	0.75 ± 0.15 (30)	0.83	0.63 ± 0.05	0.63
1.50	0.75 ± 0.15 (30)	0.83	0.63 ± 0.05	0.63

*Before reflow.

Where device heat dissipation is critical, the PBGA design may contain plated THs as "thermal vias" under the chip to provide a direct path for the heat to flow to the solder balls and dissipate into the mother board. PBGAs contain many materials with various coefficients of expansion, thermal conductivities, and elastic behavior. Typical PBGA materials are shown in Table 3.3. Maintaining adequate thermal-mechanical performance, mechanical tolerances (e.g., flatness, coplanarity), moisture sensitivity, and overall device reliability with these materials is a major challenge to the industry.

Table 3.3 Typical PBGA Materials

Item	Specification
Substrate material	High glass transition temperature, $T_g > 160°C$ Epoxy/glass laminate, e.g., bismaleimide triazine (BT)
Metallization	0.5 oz min. Cu on outside layers and 1 oz min. Cu on internal layers
Plating	20 μin min. gold over 50 μin min. nickel
Die attachment	Silver-filled epoxy
Bonding wire	1.0 mil dia. gold wire
Encapsulant	Epoxy overmold or dispensed epoxy
Balls	Eutectic solder, e.g., 62Sn/36Pb/2Ag or 63 Sn/37Pb

Source: after Cohn et al.[4]

Figure 3.33 shows the cross section of a typical low-cost, double-sided PBGA substrate, with the chip mounted to the top surface, wire bonded, overmolded, and solder balls attached to the underside. For enhanced power dissipation, the substrate contains "thermal vias" under the chip and a center array of thermal solder balls that are normally grounded. In a high-performance PBGA configuration, the substrate is multilayered with internal power and ground planes for improved electrical inductance and heat spreading capability (Fig. 3.34). The internal Cu planes are normally 1 oz (0.035 mm) thick, but better heat spreading and lower junction to air thermal resistance (ap-

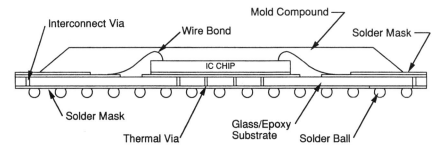

Figure 3.33 Cross section through a PBGA with a double-sided metallization substrate. (After Cohn et al.[4])

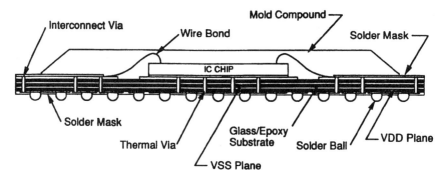

Figure 3.34 Cross section through a PBGA with a multilayer metallization substrate. *(After Cohn et al.[4])*

prox. 10 percent improvement, depending on substrate size) can be achieved if a 2-oz (0.070-mm) Cu thickness is used. The substrate for a high-power-dissipating device typically contains a chip-down cavity with single or double bonding tiers and a heat-spreading copper slug to which the chip is directly attached (Fig. 3.35).

Wire bonding is the leading chip interconnection process used today for PBGAs. Flip chip bonding, although not widely used, is expected to become a leading contender in the near future (see Sec. 3.6.1.4).

Each substrate size (27 × 27 mm, 35 × 35 mm, etc.) has a given plastic overmold configuration that does not change with different artwork or pin count. This enables the package designer to take advantage of existing mold cavities.

Figure 3.26 illustrates a typical PBGA assembly sequence, processed in strip form with wire bond interconnection. The assembly consists of the following:

- Silver-filled epoxy is transfer printed onto the die attach area, the chip is placed over the epoxy, and the assembly is cured.
- The chip is thermosonically ball-wedge gold wire bonded and visually inspected.

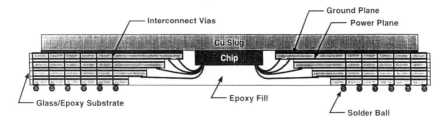

Figure 3.35 Typical cross section through a thermally enhanced PBGA.

- The strip sites are overmolded or encapsulated with a liquid dispensed epoxy resin system and cured.
- After overmolding, the sites are code marked, and solder balls are attached to the underside of the individual sites.

The individual PBGA devices are separated from the strip by routing or punching and inspected, cleaned, electrically tested, burned-in when required, electrically retested, placed in bakeable shipping trays, baked/bagged, and shipped to customer.

Prior to committing a PBGA device to production, the device is usually qualified by conducting a series of tests according to the customer's requirements or to the semiconductor manufacturer's own standard specifications. Requalification of an already qualified device may be required whenever materials change, different substrate or assembly vendors are used, the chip technology is changed, or the design rules have drastically changed. Figure 3.3 illustrates typical test requirements for PBGA device qualification.

The PBGA has a number of drawbacks that may be of concern to the end user, such as the following:

1. *Moisture sensitivity.* Many of the materials that make up the present PBGA configurations are susceptible to moisture absorption. As a result, the PBGAs, while being stored in a factory environment awaiting assembly to the PWB, are prone to absorb moisture that may cause internal delamination (popcorning) and cracking during solder reflow.

2. *Solder joint reliability.* The number of solder joints on a circuit board has increased dramatically over the last few years. Thus, it is critical that the effects on solder joint integrity are understood and solutions are implemented to improve their reliability. This is particularly important when dealing with PBGAs, which have solder joints far less compliant than leads. Cycle-to-failure tests and finite element modeling have been used to predict the life cycle of PBGA solder joints in a given application. Most of the solder joint failures in PBGAs have been attributed to thermal fatigue induced by temperature cycling and caused by the CTE mismatch between the PBGA structure and its PWB attachment surface. Other failure mechanisms may be creep rupture and solder joint/pad interface cracking due to intermetallics. Solder joint integrity is affected by factors such as joint location relative to the chip, solder joint configuration (as influenced by the solder ball/pad size), intermetallic structure, substrate thickness, number of temperature cycles and range, stress concentration from the sharp edges of the solder

mask defined pads (SMD), etc. The PBGA and its board interconnection are to be designed for reliability at the outset so as to achieve the required service life for a given application.

3. *Solder joint inspection/repairability.* Other PBGA technology issues of concern to end users are that (a) inspection of solder joints requires expensive x-ray equipment and (b) repairability is complicated.

3.4.4.3 Tape BGA (TBGA). A tape BGA is defined as a package having a copper/polyimide flex tape for a substrate. The tape contains one or two metallization layers interconnected by plated-through holes. Tape substrates have been utilized in EMBGAs, thermally enhanced cavity type packages, low I/O flip chip BGAs, etc.

In a cavity-type TBGA (Fig. 3.36), one side of the tape contains the wedge bond pads and the circuitry to fan out to the solder ball pads. The metallization layer on the other side of the tape can be utilized as a ground plane.

Another example of a TBGA package (Fig. 3.37), developed by IBM, utilizes flip chip (FC) bonding to a copper/polyimide tape containing two metallization layers. The top circuitry layer fans out the I/Os from the chip to solder ball pads on the bottom layer, via plated-through holes. The remaining space between the BGA pads is utilized as a functional ground plane. The circuitry is formed by electroplating Cu onto the polyimide tape, as defined by the resist mask. A stiffener ring is attached to the tape, using an adhesive, to provide rigidity and solder ball coplanarity to the package.

The interconnection between the chip and the tape is accomplished by a modified solder bumped flip chip process developed by IBM, which consists of a chip with 97Sn/3Pb solder bumps in an area array placed on matching footprints on the tape. Bonding to the tape is achieved by a partial reflow of the solder bumps without flux while applying a slight pressure to the bumps.

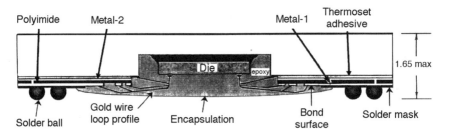

Figure 3.36 Cross section through a thermally enhanced tape BGA. *(After ASAT[11])*

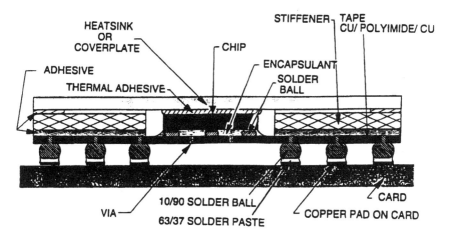

Figure 3.37 Cross section through a flip chip tape BGA. *(After IBM[10])*

An epoxy underfill is applied between the chip and the substrate tape. The underfill provides moisture protection to the chip surface and reduces the thermal stresses induced in the solder joints due to coefficient of expansion mismatches of the bonded materials.

On the BGA side, 0.025-in. (0.635-mm) dia., 10Sn/90Pb solder balls on a 0.050-in. (1.27-mm) pitch are attached to the pads on the tape, using a resistance heating process. At the center of each BGA pad is a 0.008-in. (0.200-mm) plated-through hole that is used for electrical interconnection and to anchor the solder ball. To enhance the thermal characteristics of the TBGA, a Cu plate is attached to the stiffener, and to the back of the chip, to provide heat spreading.

As a result of the relatively short conductor lengths, and close proximity of the ground plane, the FC type TBGA has low signal-line inductances as well as low power and ground inductances.

The capability to route higher-density circuitry on polyimide tape is an advantage that tape has over laminate substrates. This advantage is slowly eroding because of the advances in laminate technology, which resulted in laminate design rules to approach that of tape. In addition, the polyimide tape is more expensive than a laminate substrate of the same size, except at very high volumes.

3.4.4.4 Chip-scale package (CSP). A new class of IC packages has emerged that are only slightly larger than the chip itself (max. 1.2 × the size of the chip). The new technology, called *chip scale packaging (CSP),* comes closest to direct chip attachment, but without the responsibilities usually imposed on the board assembler for handling, attachment, wire bonding, and encapsulation of known good die (KGD). Using CSPs, the above tasks are shifted back to the semiconductor manufacturer in that the device comes fully tested, burned in, and ready for SM assembly to the PWB by conventional means, like any other IC package. Several chip-scale package configurations were developed to provide a lower-cost, high-I/O, high-density, compact package. One such package was developed by Tessera Inc., called the micro-BGA (μBGA®). It combines a unique chip interconnection with BGA technology that results in a highly miniaturized package (Fig. 3.38). The μBGA consists of a copper/polyimide flex-tape substrate with an array of pads interconnected to Cu ribbon leads extending from the substrate edge and joined at their tips to a common bus bar. Both the pads and ribbon leads are Ni/Au plated. The ribbon leads are thermosonically bonded to the perimeter pads on the chip and separated from the bus bar at the same time. A silicone elastomer layer between the flex-tape and the chip surface cushions any impact to the chip during socketing or board assembly. The compliant layer also reduces any stresses caused by the CTE mismatch between the silicon and the solder ball joints when surface mounted to a PWB. A molding compound encapsulates the ribbon leads and supports them during package handling.

3.4.4.5 Edge molded BGA (EMBGA). The plastic edge molded BGA (EMBGA) (Fig. 3.39), also called fpBGA™ by ASAT and ChipArray® BGA by Amkor, is a conventional PBGA with the overmold extending to the edge of the substrate. The thickness of the overmold, being uniform across the substrate, provides protection to the bond wires, even at the edge. The substrate is designed with the wedge bond pads

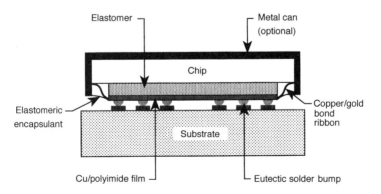

Figure 3.38 Typical cross section through a μBGA®. *(After Tessera[9])*

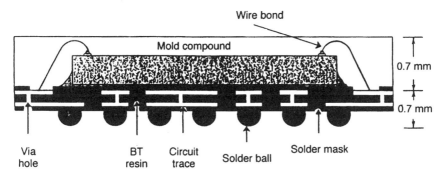

Figure 3.39 Typical cross section through an edge molded BGA. *(After ASAT[11])*

placed close to the edge and interconnected to the solder ball pads on the BGA side through fan-in circuitry and plated-through holes, mostly located under the chip. A solder mask covers the entire surface of both sides of the substrate, with the exception of clearances to expose the wedge bond pads and solder ball pads.

Typically, EMBGA substrates range in size from 4.0 × 4.0 mm to 17.0 × 17.0 mm, with ball counts from 24 I/O to 256 I/O, in an area array of 0.5, 0.8, or 1.0 mm pitch. The organic substrate is usually made of bismaleimide triazine (BT), or equivalent, with two metallization layers, although additional inner layers of power and ground planes can be incorporated for electrical and thermal enhancement. The substrate consists of multiple sites in an array configuration with tie bars connected to common buss bars to facilitate Ni/Au electroplating. Since the tie bars cannot be cut after substrate fabrication to isolate each site for electrical testing of opens and shorts, the sites are usually checked, prior to applying the solder mask, by automatic optical inspection equipment (AOI). The defective sites are identified with markings that will prevent placement of a chip on the affected site during assembly.

Assembly follows the same process flow as for conventional PBGAs (Fig. 3.26), with the exception of overmolding and singulation. The overmold covers the entire array of sites instead of individual sites. After code marking and ball attachment, the sites are singulated using a wafer saw that cuts through the overmold and substrate.

The overall thickness of an EMBGA, including solder balls, is 1.40 mm (Fig. 3.39). The demand for thinner packages has forced the assembly vendors to reduce the overall thickness to 1.25 mm by reducing the thickness of the substrate from 0.36 to 0.25 mm.

The EMBGA package has an interconnection density (substrate area to solder ball count ratio) that comes close to that of the chip scale package (CSP).

3.4.4.6 Flip chip BGA (FCBGA).

Combining a BGA package configuration with flip chip (FC) bonding results in a most effective high-density interconnection (Fig. 3.40). The solder bumped chip is placed face down onto matching wettable bonding pads on a multilayer organic BGA substrate, and the assembly is reflowed. The bonding pads on the substrate may initially be bare of any solder or solder bumped and coined prior to chip assembly. After flip chip attachment, the chip is underfilled with an epoxy resin formulated to relieve the stresses induced by the thermal mismatch between the chip and the substrate, and to prevent moisture from getting to the chip surface. The flip chip can be capped or topped with a liquid epoxy, which is retained within the confines of a resin dam. For low cost, the back of the flip chip is left exposed, but this leaves the edges of the chip susceptible to chipping during handling. Where the chip power dissipation is of concern, or stiffening the substrate to improve solder ball coplanarity is required, a Cu stiffener ring slightly thinner than the chip stand-off is attached to the substrate prior to chip assembly. After chip assembly, an adhesive is applied to the top surface of the stiffener ring, and the back of the chip is covered with a pliable conductive adhesive. A Cu plate is placed over the assembly, making contact first with the adhesive on the back of the chip. The plate is pressed on the chip adhesive until a predetermined adhesive thickness is reached. At the same time, the Cu plate also becomes attached to the adhesive on the stiffener ring, completing the lidding process. The purpose of having a slightly higher chip stand-off than the stiffener ring thickness is to maintain a constant adhesive thickness on the back of the chip regardless of the tolerances in thickness that either the chip or stiffener may have. The heat dissipated by the chip is conducted through the adhesive to the heat-spreading Cu plate. A heat sink can be attached to the plate to further lower the case-to-ambient air thermal resistance. After chip encapsulation, solder balls are attached to the underside of the substrate to complete the BGA.

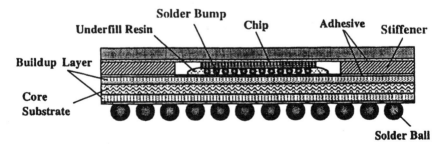

Figure 3.40 Typical cross section through a flip chip PBGA.

The high-density interconnection emanating from the chip requires equal high-density circuitry on the substrate. To achieve these circuit densities, the substrates have to be fabricated by the more expensive additive or build-up process. Substrate vendors utilizing the build-up process have reported current production capabilities of 30/30 µm lines/spaces and 55/100 µm via/land dia. The exposed metal surface finishes on the substrate may be electrolytic or electroless plated with Ni/Au or coated with an organic solderability preservative (OSP).

Although the infrastructure for flip chip bonding is currently not industry wide, it is expected that FC will be a leading interconnection technology in the near future, particularly for high I/Os and where high electrical performance is required.

3.5 Comparison of Package Technologies

A comparison of the attributes and trade-offs of the four contending IC packaging technologies, described in Sec. 3.4, is shown in Table 3.4.

Pressed and laminated ceramic technologies have hermetic sealing capabilities that provide the best environmental protection for the IC; that is, the technologies provide a leak-tight seal, a cavity free of contaminants, and moisture that is controlled to an acceptable level. Nonhermetic molded and laminated plastic technologies can provide equal reliability performance for most applications.

The electrical, thermal, and mechanical performance of a given package technology is greatly influenced by the properties of the materials inherent in that technology. For example, the capacitance between adjacent leads is a property that is influenced by the dielectric constant of the material in direct contact with the leads. Table 3.5 lists the materials and dielectric constants for each package technology. Plastic technologies have the potential for best electrical performance due to their lower dielectric constants. The high dielectric constant inherent in the laminated alumina ceramic technology can be avoided by the use of thin-film-metal-polymer multilayer structures as used in MCMs. MCM structures make laminated ceramic competitive with laminated plastic. Electrical performance is further optimized by the use of power and ground planes and high-density routing within the package. Both laminated plastic and ceramic technologies have this as a standard design capability.

Thermal performance is influenced by the thermal conductivity of the basic package materials (Table 3.5) and the use of thermal design enhancements (Table 3.4). The thermal conductivity of alumina ceramic materials is an order of magnitude better than that of plastic materials. Materials such as AlN and SiC, with higher thermal conductivity and heat-spreading copper-tungsten slugs, have further im-

Table 3.4 Package Attributes and Trade-Offs vs. Package Technologies

Attributes and trade-offs	Molded plastic	Pressed ceramic	Laminated ceramic	Laminated plastic
Hermeticity	No	Yes	Yes	No
Power/ground distribution capability	Doable custom design	Not feasible	Standard	Standard
High-density routing capability	Custom TAB design	Not feasible	Standard	Standard
Power dissipation capability	Fair	Fair	Excellent	Excellent
Thermal design enhancements	Heat spreaders	New materials	New materials	Thermal vias and heat spreaders
Capability for handling bond-pad-limited chips	Yes	Yes	Yes	Yes
Automated assembly	Excellent	Feasible	Good	Excellent
Max. I/O capability	375	196	1000+	1000+
Cost rank order (lowest = 1)	1	2	3	2+

proved the thermal performance of ceramic packages. Design enhancements in laminated plastic packages (e.g., thermal vias under the chip, center array of thermal solder balls, heat spreading internal planes or copper slugs, or heat spreaders in molded plastic packages) have closed the gap and are competing in thermal performance with laminated ceramics.

Mechanical performance is related to internal stresses developed within the package as a result of the mismatches of the coefficients of thermal expansion (CTE) of the different materials used in the package construction. CTEs of the silicon IC and of the substrate material to which the chip is mounted, for each technology, are listed in Table 3.5. The best match is between silicon ICs and ceramic-type packages. Further complications are encountered in technologies where the chip is not decoupled from other materials, such as molding compounds in molded plastic and "glob-top," or flow coats, in laminated plastic technologies. Chips in the pressed ceramic and cofired laminated ceramic

Table 3.5 Physical Properties vs. Package Technologies

Physical properties	Molded plastic	Pressed ceramic	Laminated ceramic	Laminated plastic
Dielectric constant at 1 MHz				
Plastic overmold	5.0			5.0
Ceramic body		9.0	9.0	
BT resin lamina				3.9
Thermal conductivity, W/m K				
Plastic overmold	0.80			0.80
Ceramic body		20.9	20.9	
BT resin lamina				0.26
CTE, ppm/°C				
Chip mount surfaces				
Copper LF	17.2			
Alloy-42 LF	8.0			
Ceramic body		6.0–7.7	6.0–7.7	
BT resin lamina				15.0
Plastic overmold	13.0			13.0
IC chip (silicon)	2.7–3.5	2.7–3.5	2.7–3.5	2.7–3.5

technologies are usually decoupled from other materials; however, their assembly processes, which include high-temperature exposure during chip mounting and hermetic sealing, can present problems, particularly for very large chips, greater than 0.500 in. (12.7 mm). Other material properties, such as Young's modulus and Poisson's ratio, also affect the magnitude of stresses in packages where the chips are not decoupled. The finite-element analysis is a useful CAD tool for modeling stress conditions in new package designs before committing them to development and manufacturing stages.

The cost of packaging is best addressed by using a rank-order system rather than an absolute cost. There are many factors and assumptions that enter into a cost analysis of packaging, and these are beyond the scope of this chapter. Molded plastic technology packages, such as plastic quad flat packs (QFPs), which are assembled on leadframes, and plastic ball grid arrays (PBGAs), which are assembled on single-layer, multi-site strip substrates, rank low. In contrast, the multilayer laminated technology packages such as CPGAs or PPGAs, which are singulated and need to be assembled in "boats," rank higher.

The ability of a package to accommodate bond-pad-limited chips is an important factor influencing the cost of the chip. For example, the ability to accommodate bond-pad-limited chips in molded plastic and pressed ceramic packages is sometimes limited by the fabrication process for the leadframe. The leadframe feature size is dictated by its thickness. The use of thinner leadframes and TAB tape, such as in fine-pitch high-I/O packages, has extended the capability for the plastic molded technology. The laminated ceramic and plastic technologies solve the wire bonded-pad-limited chip problem, for high-I/O devices, by using tiered wedge bond sites or a tighter wedge bond pad pitch, thus bringing the wire bonds closer to the chip and minimizing the wire span length. The advent of flip chip interconnection in laminated ceramic and plastic technologies has made it possible to further reduce the size of the chip, but it requires a more advanced substrate technology. Thus, the cost savings realized by shrinking the chip size are sometimes reduced by the higher packaging cost. Each application must be evaluated thoroughly to determine the bottom-line cost for both chip and package.

Table 3.6 compares most of the attributes discussed in this section by rank order, providing a simple format to evaluate trade-offs between the competing package technologies.

Table 3.6 Attribute Rank Order vs. Package Technologies

Physical properties	Molded plastic	Pressed ceramic	Laminated ceramic	Laminated plastic
Moisture sensitivity	2*	1	1	2
Lead capacitance	1	3	2	1
Power dissipation	3	2	1	†
CTE mismatch, chip-to-chip-mount surface	‡	1	1	2
Automated assembly capability	1	1	2	**
Capability of accommodating bond-pad-limited chips	1	1	1	1
Cost	1	2	3	2+

*Rank order: best = 1.
†Rank order depends on thermal design of substrate; 1+ = design with thermal vias and heat-spreading planes, 2+ = design without thermal vias.
‡ Rank order depends on leadframe material used: 1– = Alloy-42; 3 = copper alloy.
**Rank order depends on package type; 1– = PBGs (overmolded), 2 = PPGAs.

3.6 IC Assembly Processes

This section covers the first-level assembly processes used with each of the four package technologies. A detailed listing of all major package families, showing the preferred method of chip attachment and first-level interconnection, is presented in Table 3.7. Also shown are lead form, lead arrangement, and type of mounting used at second-level assembly.

Table 3.7 Interconnection Levels and Major Process Options

Package	Chip attachment	Level 1 connection	Level 2 connection
Molded plastic			
PDIP	Ag–loaded organic	WB, Au	TH, IL
	None	TAB	
SOIC/J	Ag–loaded organic	WB, Au	SM, IL, GW/J
PLCC	Ag–loaded organic	WB, Au	SM, P, J
PQFP	Ag–loaded organic	WB, Au	SM, P, GW
	None	TAB	
TapePak	Nonconductive organic	TAB	SM, P, GW
Pressed ceramic			
CERDIP	Ag–loaded inorganic	WB, A1	TH, IL
CERQUAD/CQFP	Ag–loaded inorganic	WB, A1	SM, P, GW
Laminated ceramic			
DIP–STB/TB/BB	Metal, AuSi	WB, Au	TH, IL
PGA	Metal, AuSi	WB, Au	TH, AA
LLCC	Metal, AuSi	WB, Au	SM, P
LDCC	Metal, AuSi	WB, Au	SM, P, CL, BZL
FP–LDCC/QFP	Metal, AuSi	WB, Au	SM, P, GW/BZL
MCM–Si on C	Ag–loaded organic	WB, Au	
	None	TAB	
	None	FC	
BGA	Metal, AuSI	WB, Au	SM, AA
	None	FC, SB	
Laminated plastic			
LLCC	Ag–loaded organic	WB, Au	SM, P
PGA	Ag–loaded organic	WB, Au	TH, AA
SMBA/LGA	Ag–loaded organic	WB, Au	SM, AA
BGA	Ag–loaded organic	WB, Au	SM, AA
	None	FC, SB	
Other options			
COB	Ag–loaded organic	None	WB, Au/FC, SB
	None	None	OLB, TAB
TOB	None	ILB, TAB	OLB, TAB
MCM–Si on Si	FC, SB	FC, SB	
TQFP	Ag–loaded organic	WB, Au, WBT	SM, P, GW

Note: BZL = brazed lead; J = J-formed lead; P = perimeter lead locus; SB = solder–bumped; SBZ = side–braced; AA = area array.

3.6.1 Chip-to-Package Interconnection

There are four chip-to-package interconnection options in use today, as illustrated in Fig. 3.41: wire bond (WB), flip chip (FC), beam lead (BL), and TAB. For wire-bonded packages, the chip interconnection process consists of two steps; chip attachment and wire bonding. In the first step, the back of the chip is mechanically attached to an appropriate mounting surface; for example, the leadframe paddle or the chip attach area of a laminated ceramic or plastic substrate. This attachment sometimes enables electrical connections to be made to the backside of the chip. There are three types of chip attachments in use today: metal alloy bonding, organic adhesives, and inorganic adhesives. In the second step, the bond pads on the circuit side of the chip are electrically interconnected to the package bond pads by wire bonding. Wire bonding is further split into three options: thermosonic or thermocompression ball-wedge bonding using gold wire, and ultrasonic wedge to wedge bonding using either gold or aluminum wire.

The other interconnection options (FC, BL, and TAB), as illustrated in Fig. 3.41, are done as a one-step process, where the mechanical and electrical interconnections are provided by the same feature. The most popular interconnection process in general use today is thermosonic ball-wedge bonding. Table 3.8 shows the processing temperatures used in assembling IC packages in the four technologies.

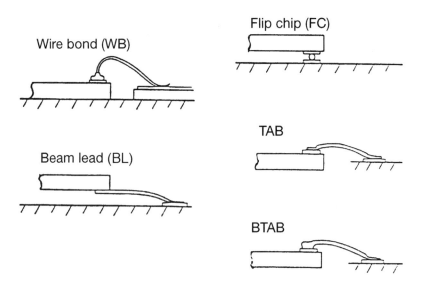

Figure 3.41 IC package interconnection options.

Table 3.8 IC Assembly Process Temperatures

		Process	Approximate process temp. (°C)
Die attachment	Eutectic bonding	Gold–silicon (98 Au/2Si)	425
		Gold–phosphorus (99.5 Au/.5P)	450
		Gold–tin (80 Au/20Sn)	280
	Soft soldering (95Pb/5Sn)		315
	Glass		450
	Silver–glass		375
	Epoxy (Ablestik #8360)		175
Die interconnection	Wire bonding	Thermocompression	300–400
		Ultrasonic	Room temp.
		Thermosonic	150
	Tape automated bonding (TAB) *(bumped chip or tape)*	Thermocompression	300–400
		Ultrasonic	Room temp.
		Thermosonic	150
	Flip chip	Soldering (63Sn/37Pb)	215
Package sealing	Hermetic packages	Soldering (95Pb/5Sn)	315
		Brazing (Au–Sn)	320–350
		Welding	1000–1500 (localized)
		Glass sealing	420
	Nonhermetic packages	Epoxy molding	175
Package interconnection	Ceramic BGA	Solder cast column (90Pb/10Sn)	315
	Plastic BGA	Solder ball (63Sn/37Pb)	215

3.6.1.1 Chip attachment. As chip sizes increase, it places more stringent demands on the processing and reliability of chip attachment. Attachment of large chips (>12 mm on a side) requires special attention to thermal and stress management. Chip size also affects wafer fabrication guidelines and assembly automation. These issues influence equipment selection and determine how easily a controllable process can be achieved.

The major choices for a chip attach process include metal alloys (AuSi eutectic, AuSn eutectic, or soft solders), organic adhesives (epoxies, polyimide paste), and inorganic adhesives (silver-filled glasses). Table 3.7 shows the preferred method of chip attachment for the various package types. Care must be exercised in the selection of a chip-attach process for a particular application, and in the implementation of stringent process controls to assure high-quality bonds.

The eutectic chip attach process, used in ceramic packages, is essentially contamination free, has excellent shear strength, provides high thermal conductivity across interface, and assures low moisture in package cavities. The major disadvantages are that the preforms are difficult to handle for high-speed automation as compared with epoxy die attach. This process may also induce high thermal stresses on the chip due to the high process temperature. Using ceramic substrate materials (such as AlN and SiC, whose coefficient of thermal expansion closely matches that of silicon) will reduce the thermal stresses, which can be detrimental to large chips.

The organic chip attach process, used in ceramic and plastic packages, utilizes an epoxy as the bonding medium, which may be electrically and thermally conductive or nonconductive. The epoxy has the advantages of being less expensive, more flexible, and easy to automate, and it requires low-temperature curing, which minimizes thermal stresses in large chips. As a result, epoxy chip-bond adhesives are preferred for attaching large chips in both ceramic and plastic packages.

Silver-filled epoxy adhesives are of major interest today as chip-bond materials. The use of silver fillers (typically flakes) makes the epoxy both electrically conductive, to provide low resistance between the back of the chip and the substrate, and thermally conductive, to allow a good thermal path between the chip and the rest of the package.

The epoxy chip-attach process can be highly automated and accurate, since the epoxy can be applied at very high rates to the die-attach area by transfer-printing, epoxy writing, or syringe dispensing, and the chip can be placed with high-speed pick-and-place tools. Accurate chip placement affects automatic wire bonding by yielding greater consistency of wire lengths and improved looping characteristics. In addition, chip placement accuracy also enhances the wire bonder's pattern-recognition performance and efficiency. The chip is removed from its film frame tape using a surface or collet pickup tool and placed over the die attach area contacting the epoxy pattern. Because epoxies are thermosetting polymers (cross-link when heated), they must be cured at elevated temperatures to complete the chip bond. Typical cure temperatures range from 125–175°C. In general, epoxy chip bonds are as good as or better than their metal counterparts, except in the most demanding applications where high temperatures, high current through the chip bond interface, and critical thermal performance are required.

A slightly modified version of the inorganic chip attach process is used in the assembly of pressed ceramic CERDIP- and CERQUAD-type packages, where the final cure is done at the same time as the high-temperature glass-sealing operation. The chip attach material is a silver-loaded glass system in an organic medium, which is applied as a paste. In the wet form, silver-loaded glass contains 66.4 percent silver, 16.6 percent glass, 1.0 percent resin, and 16.0 percent solvent by weight. After drying and organic burnout, the system is completely inorganic with 80.0 percent silver and 20.0 percent glass by weight. The diffusion of silicon from the chip backside is important in the development of the bond. Adhesion between the silver glass and the silicon chip is achieved via a glass network structure that starts at the silicon dioxide film on the silicon chip backside, while the bonding mechanism between the silver glass and the ceramic substrate appears to be mechanical in nature. These materials have excellent thermal stability, are lower in cost than the AuSi eutectic, and can provide a void-free bond for high-reliability hermetic parts.

3.6.1.2 Wire bonding. Wire bonding to silicon ICs is the most common interconnection process in use today. After attaching the IC chip to either a leadframe paddle, in a cavity or the top surface of a laminated plastic or ceramic substrate, the IC is wire bonded with either gold wire by the thermosonic, thermocompression, or ultrasonic techniques, or with aluminum wire by the ultrasonic process. Today, gold wires are mostly used for thermosonic bonding, and aluminum wires are typically used for ultrasonic bonding.

In the ball-wedge wire bonding technique (Fig. 3.42), the gold wire is ball-bonded to the chip bond pad surface (typically aluminum finish), as shown in Fig. 3.43a, and wedge-bonded to a plated leadframe or substrate pads (typically gold or silver-plated), as shown in Fig. 3.43b. The ball-wedge wire bonding process flow is illustrated in Fig. 3.44. One advantage of ball-wedge bonding comes from the symmetrical geometry of the capillary tip. The ball is formed at the tip of the capillary and then pressed against the chip (Fig. 3.44a–d). The pressure and the thermosonic action result in bonding the ball to the die. The wedge bond can then be placed anywhere on a 360° arc around the ball bond, using the capillary tip (Fig. 3.44e–i). This ability to dress the wire in any direction from the ball is the key factor that makes this process attractive for high-speed automated bonding; that is, the bonding head or package table does not have to rotate to form the wedge bond. Figure 3.45 shows a typical device that is ball-and-wedge bonded using state-of-the-art automatic wire bonders running at approximately 11 wires per second, depending on wire loop height and wire lengths. To assure reliable wire bonds, two types of in-process control tests are being performed at frequent intervals: the wire

Figure 3.42 Typical ball-wedge wire bonding, interconnecting the IC chip to the package/substrate. *(Courtesy of Agere Systems)*

pull test and the ball shear test. Current wire-bond design rules for gold wires of 0.00100 or 0.00125 in. (0.025 or 0.032 mm) in dia. are limited to maximum wire span lengths of 0.20–0.25 in. (5.1–6.3 mm), for the respective wire diameters.

Aluminum wire is typically wedge-to-wedge bonded. In a wedge-to-wedge configuration, the orientation of the wedge bond at the chip determines the direction of the wire that terminates in the package wedge bond. The complication of requiring motion in both the bonding head and the work holder results in a slower wire-bonding process. Figure 3.46 illustrates the wedge-wedge wire-bonding process flow.

The establishment of good chip design rules is absolutely essential to achieving high yields in IC package assembly. Rules must be generated for the particular package type used and must be compatible with the assembly equipment. These decisions should be made early, preferably before the chip layout is started and definitely before the layout is completed. As the I/O count grows and the active device feature size shrinks, the space required for interconnection could represent a major fraction of the chip area. To avoid this problem, the effective bonding pad sizes and pitch should be reduced. Not only must the pitch decrease, but the tolerances required to produce quality bonds must also decrease. Reliability and the ability to assemble IC chips automatically are affected directly by the chip layout. The chip-to-package

(a)

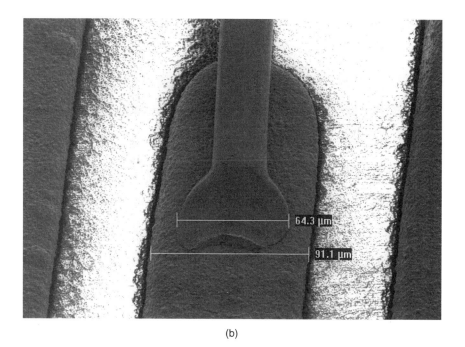

64.3 μm

91.1 μm

(b)

Figure 3.43 *(a)* Gold wire ball-bonded wedge-bonded to IC chip pad and *(b)* Gold wire wedge-bonded to Ni/Au-plated pad on substrate. *(Courtesy of Agere Systems)*

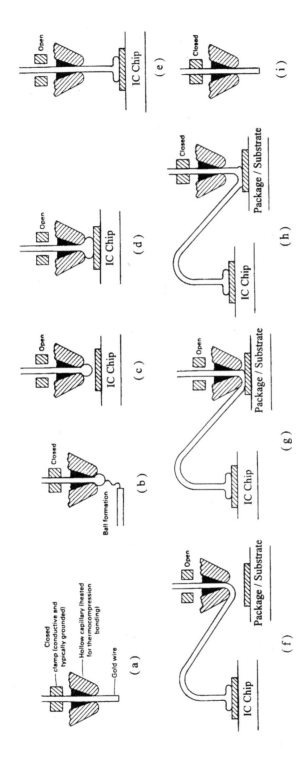

Figure 3.44 Ball-wedge thermosonic wire bonding process steps. (a) Gold wire protrudes from capillary, and wire clamp is closed. (b) An electronic spark forms the ball at the end of the wire, which is then pressed against the face of the capillary. (c) The capillary is positioned above the IC chip's bond pad. (d) Capillary descends, bringing ball in contact with the pad. Bond is created by application of heat, force, and ultrasonic vibration. (e) After ball is bonded to the chip, the clamp opens, and wire is free to feed out from the end of the capillary as it rises to loop height. (f) The wire loop is formed during the motion of positioning the capillary over the package/substrate bonding pad. (g) The capillary is lowered until the wire contacts the pad. Heat, force, and ultrasonic energy are applied at the interface, forming a stitch bond. (h) The capillary rises off the pad, leaving the stitch bond. At a preset height, the wire clamp is closed while the capillary is still rising. This prevents the wire from feeding out from the capillary, thus pulling at the stitch bond, breaking the wire at the bond. (i) The wire bonding process is complete, and the wire is now ready for a new ball to be formed for the next ball bond.

Figure 3.45 Typical thermosonic ball-wedge wire bonds (Courtesy of Agere Systems)

interface design must assure high assembly yields and avoid reliability problems.

Typical current production design rules for ball-wedge wire bonding with chip in-line and dual staggered-row bond pads are illustrated in Fig. 3.47. The wire-bonding design rules establish the wire-bond pad sizes and pitch, clearance between pads and the edge of the separated chip, wire-bond span lengths (chip to leadframe or substrate bond site), and the maximum angle of the wire from the normal direction off the chip. These rules attempt to minimize wire sweep in plastic molding, which could lead to potential edge short circuits to silicon and short circuits to adjacent conductors.

Chip-to-package interconnection, in particular wire bonding, can dictate the physical size of the silicon and thus the cost of the IC chip. The mechanical limit in gold ball bonding is the minimum pitch between adjacent ball bonds on the chip. The current production limits for in-line bond pads, using a 0.0100-in. (0.025-mm) diameter wire,

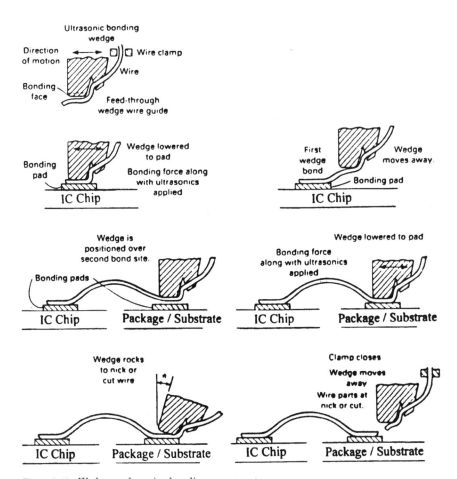

Figure 3.46 Wedge-wedge wire-bonding process steps.

are on the order of 0.0024-in. (0.060-mm) pitch but, when the bond pads are dual staggered rows, as illustrated in Fig. 3.47, the effective pitch (wire-to-wire pitch) can be reduced to approximately 0.0016 in. (0.040 mm). A triple stagger-row pad configuration is also being used as a further attempt to reduce the effective wire pitch.

The active area required on the IC chip is determined by the design rules for the particular wafer fabrication process used in manufacture. As IC features decrease (see Chap. 2), the IC designer is able to place more active circuitry, requiring more I/Os, onto each chip. A condition where, for a given number of I/Os, the chip size design is dictated by the effective bond-pad pitch on the perimeter of the chip is known as a *bond-pad-limited* chip. Since the chip bond-pad pitch has a profound effect on the chip cost, the interconnect technology vendors are driven

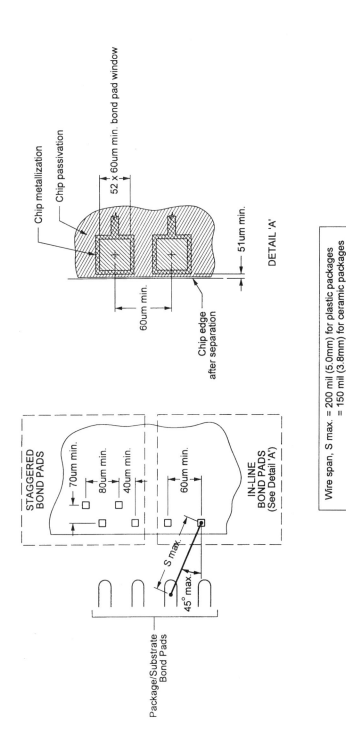

Figure 3.47 Typical ball-wedge bond design rules for chip/substrate pad sizes and placement. Chip bond pads should provide 52 × 60 μm open area and be on at least 60 μm center-to-center pitch. Maximum allowable wire span depends on package type used, with typical values noted. Wire angle, determined by chip–package interface, should not exceed 45° with respect to normal. Further restrictions on wire angle may be required. Arrangement of staggered bonding pads results in lower pitch than is attainable with single in-line pads.

to develop finer-pitch wire-bonding capability, which forces the packaging engineer to evaluate alternatives such as flip chip (FC) bonding. This is covered in Sec. 3.6.1.4.

Design rules can be integrated into chip design CAD systems. Figure 3.48 shows a typical package wire-bond CAD design template for a PBGA device and the locations of all leadframe wedge-bond targets and the optimized locations for the ball bonds on the chip. The chip designer superimposes or merges this template with the proposed chip layout and can immediately see where bonding pads should be placed to achieve a manufacturable and reliable design. Figure 3.48 also illustrates two types of CAD wire-span templates used to verify design rules and to help decide cases where bond pad placement falls outside the optimal zone. The use of a CAD system by the device designer assures the best possible design, particularly for high volume and low cost. In the case of high-performance custom designs, the same design rules apply. However, CAD tools may be too restrictive; therefore, chip and package designers must work as a team to ensure a manufacturable design.

3.6.1.3 Tape automated bonding (TAB). In the 1960s and 1970s, TAB was used for low-cost, high-volume production of products referred to

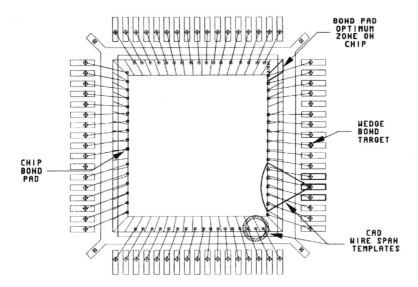

Figure 3.48 Example of CAD template for positioning bonding pads on chip in optimal location to achieve high assembly yields and high-reliability performance. Support templates check adherence to wire-span guidelines and provide design flexibility for special cases where optimal zone is too restrictive for electrical layout.

as "jelly bean" devices. The driving force was strictly low cost, which was achieved by using automated gang bonding of the inner and outer leads in place of the then very slow manual wire bonding. The process was faster and cheaper than the other options illustrated in Fig. 3.41 and resulted in higher bond strength (reliability).

In the 1980s, TAB was pursued because of its many attributes, such as lower inductance, controlled lead geometry, higher bond strength (reliability), and controlled impedance capability, all resulting in better electrical performance than wire bonding. In addition, the characteristic smaller bond pad and pitch that TAB had at the time over wire bonding significantly reduced the chip size, resulting in better wafer yields and lower chip cost. The physical size of TAB devices also resulted in a significant reduction of real estate at the PWB level.

TAB technology uses a photo-imaging/etching process to produce conductors on a dielectric conductor tape in a "movie-film" format. Carrier tape is stored on reels in widths from 1.4 to 2.8 in. (35 to 70 mm). The final tape design has a window with "beam-type" leads that extend over the windows of the tape. Conductors are fanned out from the high-density beams on 0.002- to 0.004-in. (0.05- to 0.10-mm) pitch to the external outer bond locations on 0.006- to 0.02-in. (0.15- to 0.50-mm) pitch and then beyond to test pads placed on a 0.050-in. (1.27-mm) pitch, compatible with conventional test probes.

In inner-lead bonding (ILB), the beams are bonded to an IC chip that is precisely located under the window, as shown in Fig. 3.49. An example of a 300-I/O tape with inner beams on a 0.004-in. (0.10-mm) pitch is shown in Fig. 3.50. Figure 3.51 is a *scanning electron microscope (SEM)* view of an ILB chip. Once the chips are bonded to the tape, the leads are isolated electrically by a punching operation, and then the chip-in-tape is electrically tested. A burn-in operation is also done, if required. Finally, the outer-lead bonding (OLB) operation transfers the chip, with its leads, to the next-level package. This could be the plastic molding of a QFP with more than 164 leads, an MCM, or direct bonding to a PWB or other substrate.

There are basically two methods of TAB interconnection: the bumped chip technique, as shown in Figs. 3.52 and 3.53, and the bumped tape technique, as illustrated in Figs. 3.54 and 3.55. The TAB tapes may consist of a single-layer, two-layer, or three-layer construction. Single-layer tape is a conductive foil without any insulating layer. It is the lowest-cost TAB construction and is used in jelly-bean products where pre-electrical testing is not required. Two- and three-layer tapes are composed of conductive and insulating layers that are combined by direct deposit of copper on film (two-layer) or by the use of adhesives to hold them together (three-layer). In either tape, the in-

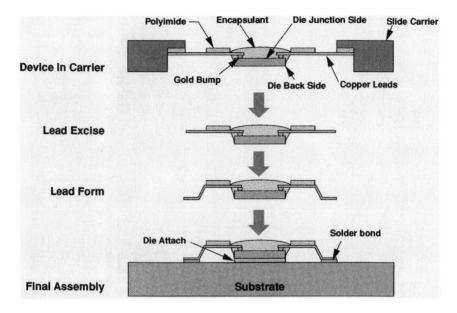

Figure 3.49 TAB process diagram with chip on tape in slide carrier.

Figure 3.50 TAB tape with inner leads on 0.004-in. pitch. *(Courtesy of Mesa Technology)*

Packaging of IC Chips   171

Figure 3.51 Inner-lead bonded chip. *(Courtesy of Aptos and Mesa Technology)*

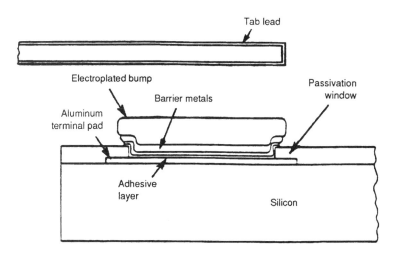

Figure 3.52 Interconnection geometry of bumped-chip TAB. TAB lead is either copper foil, ranging in thickness from 2 oz (0.0028 in./0.07 mm) to 1/2 oz, depending on feature sizes required in tape design. Lead is gold plated to facilitate TC ILB. Electroplated bump thickness is typically 0.001 in. (0.025 mm). Barrier metals are deposited in three layers, such as 600-Å Ti, 400-Å Pd, and 2000-Å Au.

Figure 3.53 SEM view of gold bumps on a chip. *(Courtesy of Aptos)*

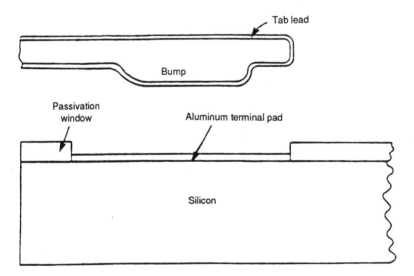

Figure 3.54 Interconnection geometry of bumped-tape TAB (BTAB). TAB lead is copper foil with etched bumps and gold plated to facilitate TC ILB.

sulator acts as a carrier and isolates the leads electrically. The various types of tape construction are shown in Fig. 3.56. Also shown is an area array tape, which contacts pads anywhere on the chip surface, and a wire-bondable tape, which allows the extension of wire bonding to chips.

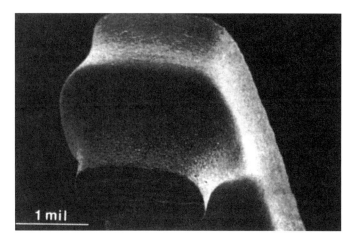

Figure 3.55 SEM view of bumped tape lead. *(Courtesy of Mesa Technology)*

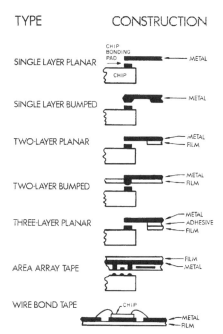

Figure 3.56 Various types of tape construction. *(Courtesy of 3M Corporation)*

With the advent of flip chip bonding and technological advances in wire bonders, TAB no longer has the fine-pitch bonding advantages it once had. IC chips on TAB tape have the advantage that they can be fully tested and burned in before committing to the next level of assembly. On the other hand, TAB requires a tape design for each chip size and pad location variation as well as special tooling and expensive bonding equipment for ILB and OLB (second-level interconnection). The ICs are typically bumped, even with bumped tape, which adds to the cost. Finally, tight tolerances are required on planarity for gang bonding. If this is not achieved, the bonding yield and device reliability will be degraded. Single-point bonding (generally modified wire bonders) is sometimes used to overcome the lack of planarity.

3.6.1.4 Flip chip bonding. Flip chip (FC) bonding is a process that facilitates direct attachment and electrical interconnection of the chip to the substrate circuitry. Whereas wire bonding requires peripheral pads on the chip, the flip chip technology utilizes area array terminals that result in the highest I/O interconnection density of any of the other bonding techniques. The terminals on the chip are placed face down (flip chip) onto matching footprints on the substrate, interconnecting through electrically conductive bumps. The bumps may be either solder or conductive epoxy, depending on the reliability requirements for a given application. Conductive epoxy bumping is a lower-cost, lower-reliability flip chip process that has been slow in gaining acceptance by the industry.

In the higher reliability solder bump interconnection process, the ICs are solder bumped at the wafer level using various bumping techniques. One bumping scheme consists of electrolytic tin/lead plating to wettable chip bonding pads, followed by a reflow to form the solder bumps, as illustrated in Fig. 3.57. The solder bumps vary in composition (97Pb/3Sn, 95Pb/5Sn, or 60Sn/40Pb), depending on the reflow temperatures that the applicable substrate can withstand. For ceramic substrates, a high-melting-temperature solder alloy (95Pb/5Sn, melting at $\approx 308°C$) is preferred to avoid remelting of the bumps when the package is reflowed to the PWB. Since organic substrates are assembly processed at lower temperatures, the solder bumps require a lower melting temperature solder such as 63Sn/37Pb which melts at $\approx 183°C$.

Flux is applied to the IC chip's solder bumps, which are then placed on matching footprints on the substrate (FC pads) and reflowed. The Cu FC pads are protected from corrosion by an electrolytic or electroless plated Ni/Au finish or with an organic solderability preservative (OSP). The substrate FC pads are sometimes presoldered and coined to increase the volume of solder at the joint, or when the solder mask defined pads (Fig. 3.58) are too small to allow the chip's solder bumps from making contact with the pad surface,. This operation can be done at the substrate manufacturer or prior to chip assembly.

After chip attachment, an epoxy encapsulant is dispensed along the edge of the chip and, by capillary action, it fills the gap between the

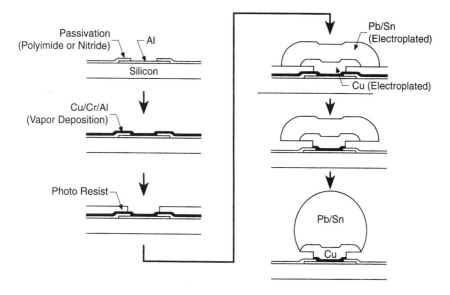

Figure 3.57 Typical solder bump-forming process. *(Courtesy of Citizen Watch Co., Ltd.; after Citizen[8])*

chip and substrate. The purpose of the epoxy underfill is to protect the chip surface from moisture and to help reduce the stresses developed in the solder joints caused by the coefficient of thermal expansion mismatches of the materials being joined.

The short and stubby solder bump interconnections between the chip and substrate provide inductances that are significantly lower than those of TAB or wire bonded chips, as shown in Table 3.9. The lower inductances, in combination with the higher interconnection density, result in improved electrical performance.

Table 3.8 Inductance of Chip Attachment Technologies*

	Cross section, in.	Length, in.	Inductance,[†] nH
Wire bond	0.001 (diameter)	0.05	0.9
TAB	0.001 × 0.003	0.05	0.7
Solder bump	0.004 × 0.004	0.002	<0.05

*Based on three-dimensional simulation of 200-I/O chip.
†25 percent of I/Os are power/ground.

The flip chip (FC) bump pitch design rules for production bonding are limited by the substrate's capability to escape route the intercon-

Non-Solder Mask (NSMD) Defined Pad

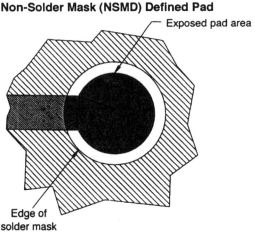

Solder Mask (SMD) Defined Pad

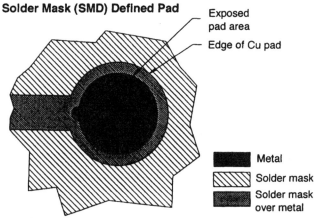

Figure 3.58 PBGA solder mask and non-solder mask defined pad designs.

nect traces from pads in the inner rows, under the chip. An FC bump array pitch of approximately 225 μm is currently in production, although a 200-mm pitch has also been bonded but to a limited degree. Further reductions in the pitch to 180 μm and even 150 μm are currently under development.

3.6.2 Other Assembly Processes

This section looks at other package assembly processes that make up the complete package assembly methodology. Among them are hermetic package lid sealing, epoxy encapsulation, cleaning, alpha particle protection, code marking, solder ball attachment to BGA packages, singulation of organic BGA packages, packing, and the assembly environment.

3.6.2.1 Hermetic package lid sealing. The major objective of package sealing is to protect the device from external contaminants during its expected lifetime. Furthermore, any contaminants present before sealing must be removed to an acceptable level before and during sealing. Packages may be hermetically sealed with glass or metals, or nonhermetically lidded with polymers. The definition of hermeticity, in this chapter, includes not only the ability to pass a fine vacuum leak test, but also to exclude environmental contaminants for long periods of time. Clearly, organic sealants are not good candidates for hermetic packages. However, in some cases, organic sealants, properly integrated into the package design, do meet the operational definition of a hermetic seal.

For almost all high-reliability applications the hermetic seal is made with glass or metal. Glass sealing was mentioned previously for pressed ceramic, and the process is essentially the same for lid sealing. Many of the metal alloys used for chip bonding are suitable for lid sealing. A leak-tight metal seal that excludes the external environment can be made without difficulty. Removing package contaminants, especially water, before sealing, has been an issue.

No package should be sealed with greater than 6000 ppm water by volume (dew point 0°C). The military specification limit is set at 5000 ppm by volume, mainly because of the technical difficulties in measuring moisture in small packages. Although 5000 ppm is the specification limit, the package assembler wants to achieve as low a moisture level as possible. A particularly effective method to achieve low moisture levels in ceramic packages makes use of the ability of atomic silicon to react with water to form SiO_2. The atomic silicon is formed during lid seal, through the melting of the chip attach material (AuSi eutectic). The reaction is:

$$Si + H_2O \rightarrow SiO_2 + 2H_2 \qquad (3.1)$$

3.6.2.2 Epoxy encapsulation. For applications where hermetic sealing is not a requirement, organic encapsulation is used. Encapsulation technologies may include either molding, potting, glob-top or cavity fill, and surface chip coatings when lid sealing. The encapsulant must possess material properties that will enable it to protect the chip from adverse environments, contaminants, package handling, storage, and second-level assembly. Mechanical strength, adhesion to silicon and substrate, CTE compatibility with silicon and substrate, temperature and moisture resistance, electrical insulation, chemical resistance, and flow characteristics are some of the traits to consider when selecting an encapsulant for a given application.

In all cases, the encapsulants are used to cover the chip surface and the fragile wire bonds. It is critical that the dispensing techniques and flow characteristics of the epoxy compound be tailored to the specific configuration so as to prevent air from being trapped or bond wire sweeps that may cause shorts.

The post-molding process is relatively harsh. Major yield and reliability problems observed in post-molded parts are wire sweep caused by the flow of viscous molding materials, and local stresses on the chip surface caused by sharp edges of filler particles. Using round-edge filler particles has reduced the IC surface microstresses. In addition, plastic packages, being nonhermetic, have a tendency of allowing moisture to ingress to the device very rapidly—on the order of days. The aluminum metallization, generally used in IC chip manufacture, is susceptible to rapid corrosion in the presence of moisture, contaminants, and electric fields. Impurities from the plastic or other materials used in the construction of the package can cause voltage threshold shifts or act as catalysts in metal corrosion. Fillers can also affect reliability and thermal performance of the plastic package. Advances have been made on the IC devices themselves in the form of improved passivation, such as silicon nitride and others, to achieve a "nearly hermetic" chip. This is desirable to protect the chip from the potentially hostile environment encountered in the plastic molding technology.

The two critical steps in producing a molded product are (1) selection of the molding material and (2) control of the molding process itself. Plastic packages, such as QFPs and BGAs, are most commonly encapsulated with a Novolac-based epoxy molding compound, typically consisting of an epoxy resin (approx. 25–35 percent by weight of the total composition), accelerator, curing agent, silica fillers, flame retardant, and a mold release agent. Novolac epoxies are generally preferred, because their higher functionality gives them improved heat resistance. The synthesis of the resins produces sodium chloride as a by-product. Both sodium and chloride ions are deleterious to device reliability. Therefore, these by-products must be washed carefully from the resins before they are synthesized into useful molding compounds. Chloride content reduction in the epoxy resin is especially effective in decreasing corrosion of the aluminum metallization in plastic packages.

Molding compound vendors are constantly upgrading the properties of their materials to achieve improved performance. Fillers, such as amorphous or crystalline silica (SiO_2) or glass fibers, and sometimes Al_2O_3, are added to the resin so that the resultant mixture is 65 to 73 percent filler by weight. Amorphous SiO_2 is used when a minimum expansion coefficient is desired, at some sacrifice of thermal conductivity, while crystalline SiO_2 improves thermal conductivity at the expense of the coefficient of thermal expansion, and Al_2O_3 has a high thermal conductivity (as a filler material) but is very abrasive to molds. Fillers greatly improve the mechanical strength of the resin and reduce its coefficient of thermal expansion, hence reduce shrinkage after molding. Silica fillers normally contain uranium and thorium, which radiate alpha particles and may cause reliability failures (soft errors) in alpha-sensitive memory devices. The introduction of low-alpha-radiation molding compounds and the development of devices with structures that are immune to alpha particles has minimized soft errors.

Small amounts of pigments, coupling agents, mold release agents, reaction accelerators, antioxidants, water getters, plasticizers, and flame retarders must also be added to complete and optimize the molding resin. Coupling agents improve the adherence of the resin to the organic fillers, metal leadframes, and silicon chips. This minimizes moisture penetration at the various interfaces and improves device reliability. Mold release agents are used to release the molded part from the cavity mold. Since they can also decrease the adherence of the resin to the leadframe, their behavior and effect on device reliability must be carefully evaluated. Reaction accelerators may decrease the volume resistivity of the molding compound. A flame retardant, normally brominated epoxy and antimony trioxide, is added to meet the industry flammability standard (UL94 V-O). Selection of all additives must be made very carefully to optimize the reliability of the molded device.

It is important to understand the rheological, chemical, and thermophysical properties of molding compounds, both for the molding process and for their interrelation with the reliability of the finished device. Moldability and the thermal expansion coefficient are two of the most important characteristics of a molding compound. The molding process is not cost-effective unless it has high yield. The most common reliability problems can be related to the mechanical quality of the molded part and its ability to withstand thermal stresses.

Thermoset molding materials are usually transfer molded in large multicavity molds (Fig. 3.59). After entering the pot, the preheated molding compound, under pressure and heat, melts and flows to fill the mold cavities containing leadframe strips or BGA substrates with attached and wire bonded IC chips. The chips are bonded with gold wires 0.001 or 0.00125 in. (0.025 or 0.032 mm) dia. and wire spans from 0.05 to 0.25 in. (1.3 to 6.3 mm) in length. To avoid damaging this

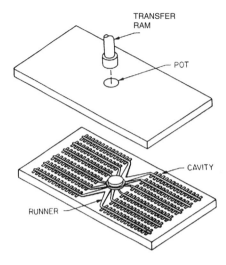

Figure 3.59 Schematic of a manual, single-plunger, multicavity transfer mold showing ram, pot, and runner system. Small packages such as DIPs or SOs use molds containing upward of 168 cavities, while larger packages such as high-I/O PLCCs use molds with fewer cavities.

fragile structure, the viscosity and velocity of the molding compound entering the mold must fall within certain ranges. Commercial molding compounds are designed to meet these requirements when molded at approximately 175°C, with pressures of about 6 MPa and a mold cycle that ranges from 1–5 min.

To control the velocity of the molten molding compound, each device cavity has a gate, or restriction, to slow the material flow. The fluid mechanics of molding is relatively complex, because the materials are non-Newtonian. In addition, partial cross-linking can occur during the molding process, affecting the material viscosity. Elaborate mold designs have been proposed to compensate for these variables. An automated multipot approach, as shown in Fig. 3.60, can also provide improved control of these molding variables.

The reliability performance criteria for post-molded plastic packages are defined by the level of stresses resulting from expansion mismatches within the package and by performance of the packaged device in a humid environment.

Potting compounds, usually consisting of a liquid two-component system, are used for hybrid modules and discrete devices. Glob-top or encapsulants used for cavity filing are thermoset epoxies or silicone resin filled with inorganic fillers. Glob-top encapsulants are used in low-cost, single-layer PPGAs and on chip-on-board assemblies, whereas cavity filled epoxies are primarily used for encapsulating cavity structured, multilayer substrates for PPGAs and PBGAs.

In cases where polymer lid sealing is utilized, surface chip coatings can be used to protect the aluminum metallurgy from atmospheric contaminants such as moisture, alpha particles, and mechanical dam-

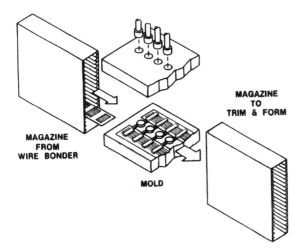

Figure 3.60 Schematic of a two-strip, automatic multiplunger molding system. Leadframes are loaded and unloaded directly from and into magazines that are compatible with assembly equipment located upstream (wire bonding) and downstream (trim and form) from molding operation. Schematic illustrates one pot for two cavities and requires fast-curing molding compounds to achieve similar economics as compared with multichip cavity molds.

age. Silicones, epoxies, and polyimides are very effective for this purpose and are used as encapsulants in premolded and laminated PPGAs in combination with a protective lid or cap.

3.6.2.3 Cleaning. In IC packaging, the most critical cleaning steps are before wire bonding, FC underfill, encapsulation, or lid sealing and before ink code marking. The cleaning process must be chemically compatible with the chip metallurgy. Aluminum has a very narrow range of pH values in which its oxide protects it from corrosion in aqueous solution. Corrosion reactions, with metal dissolution, can occur in both basic and acidic solutions. Two objectives are usually sought in cleaning. One is the removal of organic species that can affect bondability, and an organic solvent is usually required for this. The second is the removal of ionic species that can cause corrosion during the life of the device or, in an unusual instance, contribute to surface charge accumulation. Water is a very good solvent for ionic species.

Due to concerns of waste disposal and possible ozone depletion, the use of powerful solvents such as chlorofluorocarbon (CFC) and trichloroethylene (TCA) have been curtailed. Alternatives to cleaning with

CFC and TCA solvents have emerged, such as aqueous, semi-aqueous, fluorocarbon inerted organics, oxygenated organics, and other cleaning agents. But the most significant trend has been to shift to a "no-clean" IC assembly operation. The elimination of cleaning during IC assembly was made possible by (a) requiring that all incoming assembly components be in a certified clean condition and (b) minimizing contamination during assembly by conducting the operation in a Class 10,000 or better clean room environment. In addition, user-friendly cleaning processes, such as plasma cleaning with various gases (argon/oxygen, hydrogen, etc.) depending on the material to be cleaned, are being used in assembling BGAs. Plasma cleaning is typically used prior to wire bonding and prior to ink code marking the overmold and the use of aqueous cleaning agents are used to remove the water soluble flux after BGA solder ball attachment.

3.6.2.4 Alpha particle protection. Soft errors in memory circuits due to alpha particles emanating from packaging materials were first reported in 1978. Alpha particles are emitted by the decay of uranium and thorium atoms contained as impurities in packaging materials such as molding compounds and some lid-sealing materials. As the features of the chip decrease, the chip becomes more sensitive to alpha particle emission. Packaging materials (particularly ceramics) will probably not be pushed below the 0.001- to 0.01-alpha-particle/($cm^2 \cdot h$) level. Because alpha particles have low penetrating power in solids, low-alpha materials have been suggested as alpha-absorbing coatings on silicon chips. An 0.001-alpha-particle/($cm^2 \cdot h$) emission rate, the lowest level anticipated, has been reported using silicone coatings.

Molding compounds are typically filled with silica fillers to reduce the CTE, but a side-effect is alpha-particle emission that can cause soft errors in memory devices assembled in plastic packages. Applying a silicone coating to the surface of the memory chip prior to molding, reduces the alpha-particle absorption. The molding compounds used for encapsulating memory devices contain silica fillers, made by chemical vapor deposition (CVD), which do not emit alpha particles; thus, surface coating of memory chips is no longer required for alpha-particle protection.

3.6.2.5 Code marking. The packages are typically code marked after overmolding or lidding, and the mark is used for quality assurance to provide information regarding device specification, date assembled, wafer and assembly lot numbers, and where it was assembled. In addition, the marketing brand and the pin #1 location may also be shown. The methods of marking are either by ink printing or laser scribing.

Ink printing utilizes a transfer printer that applies the code mark onto the package surface, followed by curing in an oven. The ink type is selected for its permanence in withstanding the operating environment. Typically, an epoxy ink is used because of its good adhesion characteristics when applied to various package surfaces. The ink printing process has the ability of multisite printing at once but is susceptible to cosmetic defects and is unable to provide serialized part tracking.

Laser marking, which is primarily used for plastic packages, is a sequential process with excellent repeatability in the quality of the marks. The mark is permanently scribed and cannot be changed if a mistake was made. Its ability to provide serialized part numbers, coupled with its fast throughput (no curing required), has made the laser marking method the choice marking for plastic packages, in the industry today.

Regardless of the marking method used, the packages can be code marked while in strip form or, if the packages are already singulated, they are placed in boats and code marked.

3.6.2.6 Solder ball attachment to BGA packages. Ball attachment is the final process before singulation. The process includes applying flux to the BGA pads, solder ball placement, solder reflow, cleaning, and inspection.

The strips move from a magazine onto a carrier that transports each strip into a *BGA ball placement system* where fluxing of the pads and ball placement takes place. A typical ball placement process consists of screen or transfer printing flux on the exposed metallized BGA pads while automatically controlling the registration and depth of the flux. An automated ball placement mechanism places all solder balls simultaneously on all pads on the substrate, thus increasing the throughput regardless of the number of balls to be placed. There are several different ball-feed-systems available today. A typical system may consist of a plate with holes patterned to match all the solder ball pads on the strip. The plate is lowered into a reservoir filled with solder balls, which become attracted to the holes in the plate and retained in place by a vacuum emanating from the back of the plate. By conveyor belt, the strip moves under the plate and the solder balls are released onto the pads. Due to the self-alignment characteristics of the solder balls, even a misalignment of 50 percent of the pad diameter is acceptable. After solder ball placement, the strip travels on the conveyor belt into the oven where the solder balls reflow and become attached to the pads. The assembled strip continues its travel into the automatic visual inspection area where the strip is checked for missing solder balls and possible rework.

3.6.2.7 Singulation of organic BGA packages. Singulation is the process of excising the individual components from a leadframe or from an organic BGA laminate substrate in a strip or array type configuration. For leadframe-type components, singulation follows two other processes: lead trim and form. The leads are trimmed to a specified length and then formed typically straight with a 90° bend for TH insertion into a PWB, or gull/"J" bends for surface mounting. After trim and form, the components are singulated from the leadframe by punching.

Organic BGA laminate strips may have slots, prerouted by the substrate manufacturer, that surround each component, with the exception of corner tabs that hold the component attached to the strip (Fig. 3.32). The strip is fixtured, and the BGAs are singulated by either punching or routing out the corner tabs. Strips that do not have slots around the components are singulated by either punching or routing the full perimeter of the BGA.

Components such as BGA-CSPs, which are individually overmolded on the substrate in a close array configuration, require a tight excising technique such as punching with vision alignment.

Chip Array® or edge-molded-type BGAs, which are configured on a rigid organic substrate in an array pattern and are overmolded en mass over the entire array, require an alternate method of singulation. Due to the narrow saw street separation between components and the tight tolerances between the component edge and the wedge bond pads, the singulation process utilizes a die saw technology. The assembled substrate is placed, with the overmold surface face down, onto a release tape mounted to a steel ring. Fiducial marks on the BGA side, with the solder balls attached facing up, provide vision alignment for the die saw to cut through the rigid substrate and the overmold epoxy. Because of the slow throughput and the expense of replacing the release tape after each singulation, a fixturing arrangement is used instead to hold down the substrate during singulation. This increased the throughput but it requires wider saw streets to accommodate the hold-down fixturing.

3.6.2.8 Packing. After final quality inspection, the components are ready to be packaged in shipping tubes, trays, or automatic tape and reel, based on customer's requirements.

Plastic packages are prone to moisture absorption when exposed to a factory floor environment, awaiting assembly to a PWB. Given enough time, the packages laden with moisture may be susceptible to internal delamination (popcorning) or cracking during reflow soldering. Thus, to keep the packages dry, some semiconductor manufacturers ship the plastic packages prebaked and in vacuum sealed bags to be opened prior to board assembly. The plastic packages are typically baked in a dry oven at 125°C for 2–48 hr, depending on the package type. The manufacturers normally specify on the bag the maximum safe time the packages can be left outside the bag before reflow soldering. The material properties and designs of PBGAs continue to improve to where it is now possible to achieve a JEDEC moisture sensitivity Level 3 condition, defined as a floor life of 168 hr in a 30°C/60 percent RH environment before rebaking is required to solder reflow.

3.6.2.9 Assembly environment. The ever-decreasing feature sizes on chips, down into the submicrometer range, and the use of finer bond pad pitches impose stringent requirements on the control of cleanliness in the assembly environment. At a submicrometer feature size, the density of foreign particles of 1 μm or greater in a typical room environment (Fig. 3.61) are astronomical. Foreign particles not removed before molding of plastic packages, or lid sealing of ceramic packages, can degrade the reliability of the packaged device. Examples of such failures are electrical short-circuiting between adjacent interconnections by metallic foreign particles and hard failures due to metal corrosion from foreign particles with high levels of contaminants. Assembly processes such as chip attachment, wire bonding, and mold-

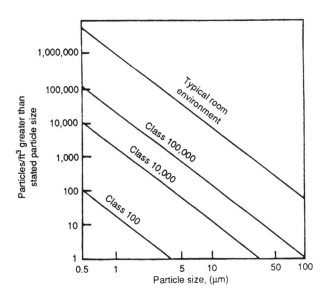

Figure 3.61 Particle size distribution in typical room atmosphere and in three classes of clean environments.

ing or lid sealing are currently being done in clean rooms having a Class 10,000 or better environment.

Another concern is the possibility of electrostatic discharge (ESD) during the assembly process. Static charges can build up on various surfaces to levels that may reach thousands of volts. These charges, if not controlled, can discharge onto the surface of an IC chip, destroying its circuitry. High-density IC structures are more sensitive to ESD than others. All IC chip assembly processing areas should have an ESD protection program whereby the operators, the equipment used in assembly, and the environment in the assembly area are all conditioned to control ESD. The operators are required to wear non-static smocks and hats, to wear grounding straps, and to handle the device parts in anti-static carriers. The equipment should also be grounded, including work benches and floor mats. Ionizers, placed in nitrogen dry boxes, air blowers, and air conditioning systems, will further reduce any potential for static buildup.

3.7 Summary and Future Trends

Four package technologies have emerged over the years, and each has enjoyed its share of success. Demand for higher pin counts will undoubtedly continue, with ever-increasing silicon capability. Laminated Al_2O_3 ceramic packages dominated the high-performance packaging technology up to several years ago, but factors such as its high dielectric constant, modest thermal conductivity, and high cost have forced the industry to consider other packaging materials, including laminated plastic packages.

Requirements for higher packaging density on the PWB level are driving package designs toward smaller perimeter lead pitches and array type packages. To illustrate the differences in mounting areas of various package technologies, a 208-I/O package was used for comparison (Fig. 3.62). The demand for BGAs has increased over the past few years because of their robust construction, ease of assembly to the PWB, and area advantage over peripheral-lead-type packages such as QFPs (Fig. 3.63). Figure 3.63 shows that, above 100 I/Os, the BGA, with solder balls on 0.050-in. (1.27-mm) pitch and fully populated, uses less area on the PWB than the QFP with peripheral leads on 0.020-in. (0.50-mm) pitch. At a peripheral lead pitch of 0.012 in. (0.30 mm), the crossover is at approximately 300 I/Os. BGAs are now challenging the fine-pitch-perimeter packages, which are difficult to handle and have a greater board assembly yield loss than the array-type packages.

Competing with traditional packaged devices will be the drive to direct chip-attach and chip-scale packaging where the ultimate in high density and performance can be achieved.

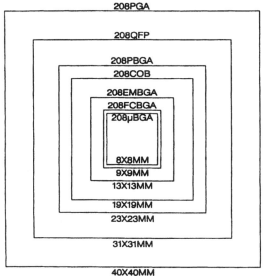

Comparisons based on the following:
208 I/O die size = 7.6 × 7.6 mm max.
208 PGA   = 40 × 40 mm, 100 mil pin pitch
208 QFP   = 31 × 31 mm, 20 mil lead pitch
208 PBGA   = 23 × 23 mm, 1.27 mm solder ball pitch
208 COB   = 19 × 19 mm, wire bonded, 0.23 mm pad pitch
208 EMBGA = 13 × 13 mm, 0.8 mm solder ball pitch
208 FCBGA = 9 × 9 mm, 0.5 mm solder ball pitch
208 µBGA   = 8 × 8 mm, 0.5 mm solder ball pitch

Figure 3.62 Mounting area comparison of IC packages.

Newer packaging materials are being developed to improve material properties to achieve better performance, better reliability, and lower costs. Designers will make more use of package enhancement features; for example, heat spreaders in molded plastic packages, and ground/power planes and heat-spreading copper slugs in laminated plastic packages.

As feature sizes of IC chips decrease, demands are placed on interconnect technology capabilities for connecting to very closely spaced bond pads. Wire bonding in particular will be challenged. If wire bonding is to survive, it will need improvements in placement accuracy to meet demands of tighter bond pad pitches. It will also be necessary to provide the capability of longer wire spans with controlled wire dressing to bridge the gap between bond pads and the second bond targets while avoiding short circuits between adjacent wires in the dense wire fan-in pattern. Wire bonding of high-I/O chips will need to be done in

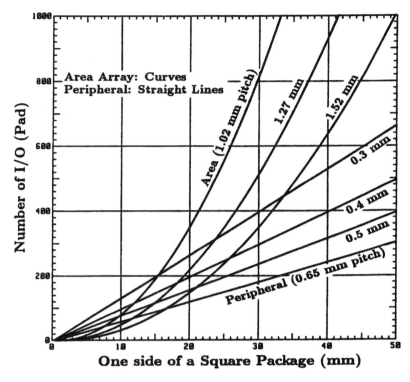

Figure 3.63 BGA and PQFP comparison: pin counts vs. package size, assumes full array for BGAs. *(After Lau[3])*

Class 10,000 or better clean rooms to optimize automatic wire-bonder performance and achieve acceptable bonding yields.

Flip chip (FC) will be giving wire bonding serious competition as the future interconnection technology for IC packaging. FC has the advantage of very low power and ground inductance connections to the package and the best capability in terms of the density of interconnection (using an area array) to the device compared to wire or TAB.

The system design will depend more and more on a systematic optimization of the entire interconnection scheme to achieve the potential benefits of improved silicon capability. This optimization may lead to completely new requirements for assembly and packaging but, more likely, it will lead to a sorting out of the various existing assembly technologies and packages.

3.8 References

1. C. Harper, ed., *Electronic Packaging and Interconnection* Handbook, 3rd. ed., New York: McGraw-Hill, 2000, Chap. 7.

2. P. V. Zant, *Microchip Fabrication,* 4th ed., New York: McGraw-Hill, 2000.
3. J. H. Lau, *Ball Grid Array Technology,* New York: McGraw-Hill, 1995.
4. C. Cohn, R. M. Richman, L. S. Saxena, and M. T. Shih, High I/O Plastic Ball Grid Array Packages--AT&T Microelectronics Experience, in *Proc. 45th Electronic Components and Technology Conf.*, May 1995, pp. 10–20.
5. *Kyocera Packaging,* Publ. A-122E, Kyocera America Inc., San Diego, Calif.
6. *Design Guidelines Multilayer Ceramic,* catalog CATD6, Kyocera America Inc., San Diego, Calif., 1995
7. C. Cohn, N. V. Gayle, L. S. Saxena, Design and Development of Plastic Pin Grid Array Packages, in *Proc. IEEE 36th Electronic Components Conf.,* May 1986, p. 100.
8. Citizen Watch Co., Ltd., data sheets on flip chip technology, 1996.
9. Tessera Inc., data sheet, The Tessera µBGA™ Package-A Low-Cost Alternative to High Pin Count Packaging, 1995.
10. Frank E. Andros and Richard B. Hammer, Area Array TAB Package Technology, *JTAP 1993 Proceedings.*
11. ASAT, 1999 Product Guide.

Chapter 4

Laminates and Prepregs as Circuit Board Base Materials

Douglas J. Sober
Essex Technologies Group, Inc.
Pembroke, New Hampshire

4.1 Introduction

The basic building blocks for the printed circuit board and electronics industries are copper-clad laminates and prepregs. A laminate can be defined as any material composed of individual layers that have been bonded together under heat and pressure to form a single, solid structure. Examples of laminates are credit cards, wooden golf clubs, surfboards, and Formica® countertops. In the case of copper-clad laminates, a thin sheet of copper foil (17 to 70 microns) is placed on one or both sides of a single ply or multiple plies of prepreg. The prepreg, when consolidated and cured, forms the dielectric space between the copper foil surfaces. During the fabrication process, conductors of the circuit board are generated by a print-and-etch process from these copper foil planes. The actual process to manufacture circuit boards will be discussed in a following chapter.

Other than the copper foil planes, the rest of the individual layers are polymer-resin impregnated reinforcements that are partially cured. These materials are commonly called prepregs, bonding sheets, or "B" stage. The process for producing these sheets is detailed in Chapter 5. An important feature of prepreg is that, although it is not tacky, there is a lot of chemical activity (polymerization) that can take place whenever the prepreg is subjected to heat. During the lamination process, the heat first melts the prepregs, which allows the resin to flow. The pressure is then applied, which forces the resin from one layer to flow into the next. Some of the resin is forced into the rough topography of the matte side of the copper foil layer providing a physical rather than chemical bond. After a period of time under heat and pressure, the resin completes the polymerization process. The resulting laminate exhibits the final product attributes such as moisture and chemical resistance, thermal stability, and insulating characteristics. In addition, the individual layers of original prepreg of the laminates cannot be separated.

A variety of reinforcements can be utilized for base materials for printed boards. The most common reinforcements are woven fiberglass fabric and cellulose paper. In addition, nonwoven fiberglass paper or felt, nonwoven aramid paper and woven aramid fabric can also be used for laminates and prepregs. Often, laminates can be composed of a combination of reinforcements so as to develop unique physical and machining properties. Reinforcements are not absolutely required, as in the case of the coated copper foil, where the dielectric space is composed only of a polymeric layer.

Most of the differences between the attributes of one laminate grade to another are the result of the resin system utilized. Most of the resins are thermosetting in nature rather than thermoplastic, which means chemistry is taking place rather than a simple melting and resolidifying. By far the most popular resin is the epoxy family, with a variety of end-result thermal and chemical performance. Other resins in wide use are phenolics, polyimides, cyanate esters (CEs), bismaleimide triazines (BTs), and polytetrafluoroethylenes (PTFEs). All fully formulated resins (also called varnishes) have significant amounts of additives and modifiers to improve overall performance or processability of the resulting prepreg or laminate. Different resins are often blended together to achieve unique properties.

4.2 Paper-Based Materials

The first laminates used for fabricating printed circuit boards were a family of substrate materials designated as paper-based materials. Paper-based, copper-clad laminates evolved from the manufacture of decorative laminates (e.g., Formica®) and industrial plate laminates. The basic manufacturing process requirements were the ability to produce phenolic-resin saturated cellulose paper prepregs and then press and cure the laminates. Horizontal treaters and large lamination presses fulfilled all the basic equipment needs. It was an easy extension of the technology of that day to replace the layer of decorative paper that looked like wood with a sheet of copper foil. The first product produced that exhibited flame-retardant character was called FR-1, a grade designator used by the National Electrical Manufacturers Association (NEMA). The original paper-based laminates were 1/16 (0.062) in. thick, and the copper foil was placed only on one side, hence the name single-sided laminates for single-sided boards.

The reinforcement used for paper-based laminates was the same as that used for decorative laminates—unbleached kraft paper. The paper had the same color and consistency as a paper bag found at a grocery store. The cellulose fibers would wet and swell when pulled through the varnish bath and finally would become totally encapsulated by the polymer. Further curing of the polymer would result in the partially reacted sheet designated as a prepreg or a "B" stage.

This use of phenolic resins was some of the first polymer technology used in a manufacturing application. Phenolic resins were produced in the simplest case by reacting formaldehyde with phenol in a condensation process, producing the final polymer and water. The base phenolic resin would then be combined with plasticizing agents, flame retardant additives, coloring agents, tung oil, and organic solvents to yield a varnish suitable for producing both decorative and copper-clad laminates. Once the curing process was complete during the lamination cycle, the resultant polymer matrix was tough and impervious to a variety of chemicals. The copper foil did not bond well directly to the phenolic polymer matrix, however, and necessitated the use of an adhesive coating to promote adequate peel strength as shown in Fig. 4.1.

In addition, the laminates were easily punchable when warmed to about 50°C. Punching produced 400 to 700 holes in one punching motion, which is an extremely economical method of generating holes. The holes provided a site for the leads of the electrical components, such as capacitors and resistors, to be placed. These leads were then soldered into the hole providing both a physical and electrical connection between the circuit board and the device.

The original paper-based laminates exhibited fairly good insulating properties under normal conditions but when, subjected to extreme levels of humidity or when immersed in water, the electrical charac-

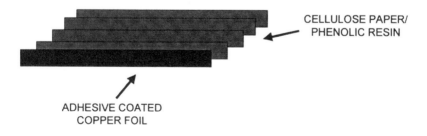

Figure 4.1 FR-1 or FR-2 1/0 0.062-in. schematic.

teristics degraded greatly. This limited the types of applications for which FR-1 laminates could be used. A variety of additives and modifiers were put into the basic resin system to improve the wet electrical properties and to lower the preheat temperature during punching. In addition, technical improvements were made to the treating process itself so as to better wet-out and encapsulate the cellulose fibers. This grade was designated by NEMA as FR-2.

Over the years, many evolutionary improvements to the phenolic resin system were made to both the FR-1 and FR-2 systems, to the point at which each name now describes a family of laminates with similar composition and properties.

The desire to generate the ultimate material in terms of wet-electrical properties led the manufacturers to develop the FR-3 laminate materials. The polymer system in FR-3 materials was composed of epoxy rather than phenolic resin. The adhesive was still required for an adequate bond between the copper foil and the dielectric material as shown in Fig. 4.2. Epoxy resins exhibited inherently improved resistance to moisture absorption over phenolic resins. In addition, epoxy resins were tough yet flexible, which was thought would improve the punching characteristics of the laminate. As in the cases of FR-1 and FR-2, FR-3 eventually became a family of materials with improved electrical properties. On the downside, FR-3 still required preheating for successful punching. Also, it could not be used as a double-sided board with plated-through holes (PTHs) routinely.

The minimum requirements for the paper-based laminates can be found in the industry document IPC-4101A, specification sheets 02 (FR-1), 03 (FR-2) and 04 (FR-3).

4.3 FR-4 Materials

Clearly, the most widely used grade of copper clad laminate and prepregs worldwide is FR-4, with over 54 percent of production represented by this single NEMA designation. About 14 percent of the total

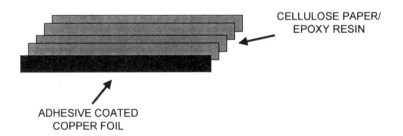

Figure 4.2 FR-3 1/0 0.62-in. schematic.

production has been destined for single- or double-sided FR-4 boards (see Fig. 4.3), while the balance (about 40 percent) has been thin FR-4 laminates (see Fig. 4.4) for multilayer applications. The reason for this historical dominance of FR-4 can be attributed to the overall thermal performance, improved moisture/chemical resistance, high flexural strength, and excellent peel strength over the paper-based laminates. Because of the moisture and chemical resistance, FR-4 was the first laminate material to be compatible with plated-through holes for double-sided applications. In addition, the cost of purchasing FR-4 materials has generated the best value in terms of performance. Over the years, the demise of FR-4 had been widely predicted, as new materials have been developed to increase packaging density. However, due to cost factors, board designers have continued to find ways of using FR-4 in high-density applications.

The reinforcement for FR-4 laminates has been woven electrical glass (E-glass). Woven E-glass has been a particularly good reinforcement for circuit board materials because of superior mechanical properties, good electrical characteristics, and resistance to heat, water,

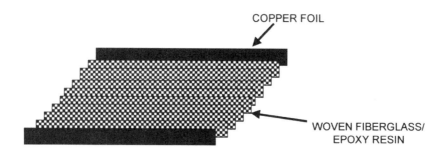

Figure 4.3 FR-4 1/1 0.059-in. schematic.

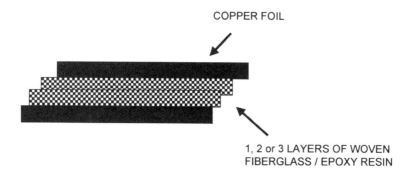

Figure 4.4 FR-4 1/1 thin laminate schematic.

and acids. All FR-4 fabrics are plain basket weaves that produce a smooth surface texture, and all are coated with a "finish," which is used to promote bonding between the glass and the resin system being applied during treating. The size of the yarn and the thread count used during the weaving process determine the basis weight and thickness of the resulting fabric. Glass styles used predominately in the rigid printed board industry range from a thickness of 0.0018 to 0.0068 in. The thickness of the pressed laminate has been achieved by careful selection of prepregs made from these glass styles. FR-4 laminates generally can be supplied from 0.002 to 0.062 in., in 0.001-in. increments, depending on the glass style and resin content percent of the prepregs used. Performance properties are highly dependent on the construction and therefore must be specified by the purchaser. For a given thickness, there are numerous possible constructions to satisfy both the nominal tolerance requirements. The variation in resin content (or resin to glass ratio as it is sometimes called) impacts the laminate properties as shown in Fig. 4.5.

Epoxy resin systems are characterized by having reactive functional groups called *epoxides*. Standard difunctional epoxy resins (resins with two reactive epoxides per polymer chain) are synthesized by combining an epoxy monomer with tetrabromobisphenol-A (TBBPA) as shown in Fig. 4.6. The length of the chain between the epoxide groups determines the toughness of the polymer and contributes to the thermal performance of the laminate. During the curing process, the epoxide groups react with the curing agent to develop a three-dimensional polymeric matrix. The TBBPA provides the flame-retardant character by incorporating aromatic bromine as part of the polymeric chain. Between 16 and 21 percent bromine by weight must be achieved to give the final laminate a V-0 flame rating according to Underwriters Laboratories UL 94 testing.

Higher glass content improves	Higher resin content improves
• Thickness control • Dimensional stability • Low z-axis expansion	• Surface smoothness • Chemical resistance • Lower DK • Measle resistance

Figure 4.5 Effect of construction on product performance.

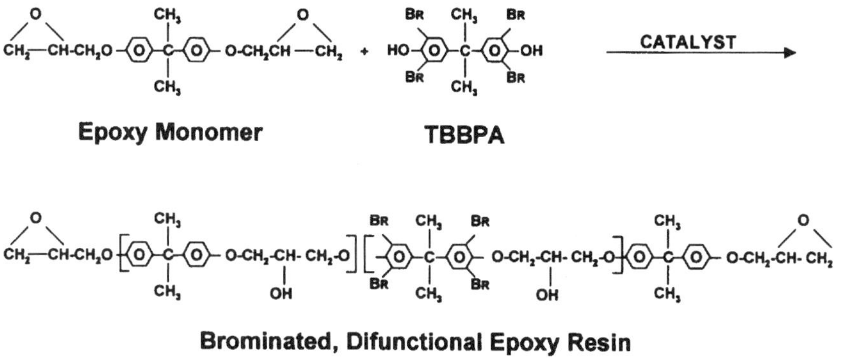

Epoxy Monomer + **TBBPA** →(CATALYST)→ **Brominated, Difunctional Epoxy Resin**

Figure 4.6 Synthesis of base epoxy resin.

Laminates made with 100 percent difunctional epoxy and dicyandiamide (dicy) as the curing agent, as shown in Fig. 4.7, exhibit a glass transition temperature (T_g) of 110 to 130°C. To drive up the T_g, some of the difunctional epoxy resin can be replaced with a multifunctional epoxy resin. The most popular multifunctional epoxy resin is tetrafunctional (see Fig. 4.8), which can be used in place of difunctional epoxy up to about 20 percent by weight, generating a T_g up to 150°C. Tetrafunctional epoxy resin also improves the optical contrast, which optimizes the ability to conduct automated optical inspection (AOI) of inner layers.

Currently, the industry has standardized on a level of multifunctional epoxy resin to yield laminates with a T_g of 140°C. The minimum requirements for this family of FR-4 laminates and prepregs can be found in the industry document IPC-4101A specification sheet 21.

In today's market, about 40 percent of all the FR-4 laminates and prepregs are considered to be high-temperature FR-4 materials. The average T_g of this family of FR-4 laminates and cured prepregs is 170°C. Multifunctional resins with still more reactive groups per weight, such as the epoxidized phenolic novolacs and the epoxided cresol novolacs, are used in the resin system to push the T_g to these levels. Novolacs also can be used as a curing agent in place of dicy to further enhance the cross-link density, which also can generate higher T_g results.

The reason why a board designer would prefer a 170°C T_g system rather than a 140°C T_g system is for PTH reliability. The z-axis expansion is reduced 30 percent for the higher T_g system, which lowers the stress on the PTH during temperature cycling. As the board gets thicker (especially over 0.100 in.) the PTH reliability has been improved when using the 170°C T_g system. Because the cost adder has not been that great (less than 15 percent, normally), many board

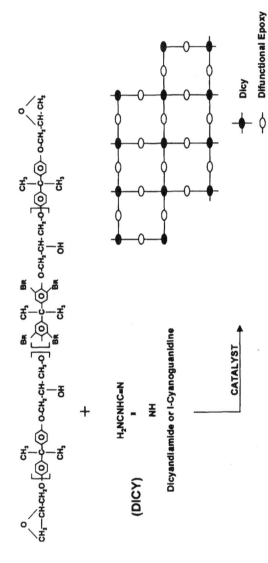

Figure 4.7 Reaction of dicy with epoxy resin.

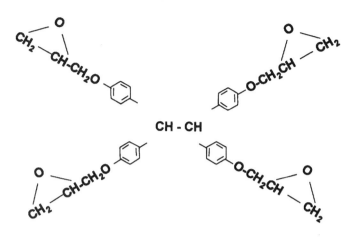

Figure 4.8 Tetrafunctional epoxy resin.

shops have standardized on the 170°C T_g system so as to carry only one inventory. The minimum requirements for this family of FR-4 laminates and prepregs can be found in the industry document IPC-4101A, specification sheet 24.

Higher glass transition temperatures than 170°C are available as well. These materials are composed of resin systems incorporating specialty additives to the normal difunctional and multifunctional epoxy resins. Other properties may be affected as well, including the dielectric constant (D_k), dissipation factor (D_f), interlaminar bond, and peel strength. Each product grade is unique to the manufacturer, and these materials tend not to be readily interchangeable with each other. The minimum requirements for this family of FR-4 laminates and prepregs can be found in the industry document IPC-4101A, specification sheets 25, 26, 28, and 29.

Although a higher T_g is desirable, especially for z-axis expansion, the improvement does not come without compromising other performance attributes. Figure 4.9 shows the performance trend as the cross-link density of the system increases. Many of the degraded properties, such as the peel strength and interlaminar bond, are due to the increased brittleness of the high T_g resin systems.

4.4 Composite Materials

Composite materials were developed for two basic reasons. The first reason was to find replacements for paper-based materials that exhibited improved wet electrical properties and punched without crazing at room temperature. The best of the paper-based laminates (NEMA

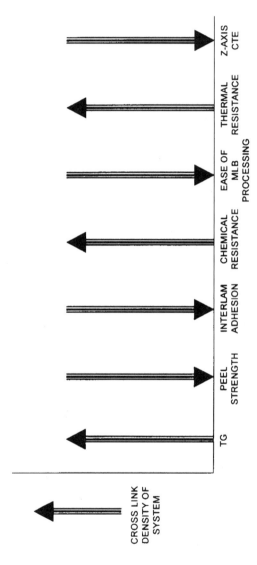

Figure 4.9 Effect of cross-link density on laminate performance.

grade FR-3) was only moderately successful in improving wet electrical properties and still required heating to 50°C for punching. In addition, the often-brittle nature, low impact strength, and weak flexural strength of the paper-based materials made them difficult to convert to the automated handling processes available in the mid 1970s to the printed board shops. The second reason for developing composite materials was to find more economical replacements for FR-4 intended for double-sided PTH applications. Although suitable for almost all printed board applications, FR-4 was overdesigned and therefore overpriced for most of the applications. The fact that it was practical only to drill FR-4 laminates made the fabrication process much more expensive than the mass production of holes using a punch press. The optimal product would then have the ability to be punched while simultaneously affording a reliable PTH for double-sided boards. Improved value would be generated through both the lower overall cost of the laminate and via the improved economics of the punching process.

The most obvious solution, then, was to combine the material makeup of both the FR-3 and FR-4 systems. The first developmental composite laminate used a ply of FR-4 prepreg composed of epoxy resin and woven fiberglass as the outer surface layers over a core of FR-3 prepregs made up of epoxy resin and cellulose paper as shown in Fig. 4.10. Although the woven fiberglass added greatly to the overall raw material cost of the laminate, the results were impressive. Not only did the composite exhibit improve wet electrical properties, the laminates punched at room temperature without cracking between holes. As an added benefit, standard copper foil could be used in place of the coated foil with no degradation of peel strength. NEMA gave this composite the grade designation of CEM-1.

CEM-1 soon became the laminate of choice in the U.S.A. for single-sided printed boards in the early 1980s. Shops fabricating boards for consumer electronics such as televisions and smoke detectors bought CEM-1 instead of FR-2 and FR-3. Although more expensive to purchase, the overall cost of the board was less; not only because of the

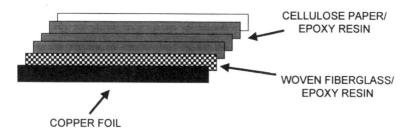

Figure 4.10 CEM-1 1/0 0.059-in. schematic.

ability to punch at room temperature but also due to compatibility with factory automation. With both hot- and room-temperature flexural strength being 2 to 3 times that of the paper-based materials, CEM-1 could be used in automatic loading, cleaning, screen printing, etching, developing, and restacking equipment. For example, RCA fabricated television boards in one semicontinuous operation, which reduced cycle time, decreased the cost of labor content, and improved inventory levels. Boards, which previously took days from start to finish, could now be completed, including electrical test, within 48 hr.

By 1984, all U.S. manufacturers of paper-based laminates had converted 100 percent to CEM-1. The same conversion was not observed in the Japanese and Asian markets. The cost of CEM-1 over FR-1 and FR-2 materials was considered to be a serious problem, and factory automation was not a priority. In these areas, paper-based laminates continued to be developed with improved wet electrical properties and punching characteristics. Boards for consumer electronics such as televisions, radios, VCRs, and automotive products to this day are still manufactured with paper-based products. These same products, when manufactured in the U.S.A., are using CEM-1. According to the IPC, CEM-1 accounts for 3 percent of all laminates manufactured worldwide.

The only drawback of the newly developed composite was PTH reliability. CEM-1 had the same issues in moisture and chemical pickup during the PTH process as FR-3. Since FR-3 laminates were composed of epoxy resin, it was then thought that the cellulose paper was the root cause of these issues. The next step was to develop a CEM-1–like product without the cellulose paper reinforcement. As a result, a new composite laminate composed of epoxy resin and fiberglass paper or felt for the core was developed, as shown in Fig. 4.11. NEMA named this product grade CEM-3 in the late 1970s.

CEM-3 satisfied all the physical requirements for a low-cost alternative version to FR-4. It could be punched rather than drilled, and

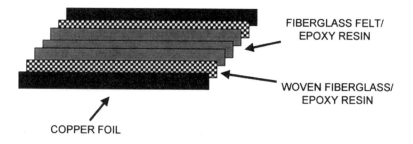

Figure 4.11 CEM-3 1/1 0.059-in. schematic.

PTHs could be successfully produced after drilling or punching. It looked and fabricated so much like FR-4 that test samples of CEM-3 at board shops were often lost when they were inadvertently used to manufacture production FR-4 boards by mistake. In terms of physical properties, however, CEM-3 exhibited lower flexural strength, worse warp, and a higher z-axis expansion than FR-4. However, for all practical purposes, CEM-3 was a viable substitute for most applications.

In the Japanese and Asian markets, buyers took advantage of the lower cost of CEM-3 to build double-sided boards. Over the years, CEM-3 took over 60 percent of the FR-4 market for double-sided applications in Japan specifically. Back in the U.S.A., CEM-3 use stalled, not because of performance or fabrication issues but due to commercial factors. The pricing of CEM-3 to the board shops was not low enough to warrant carrying two inventories. In addition, circuit board prints from the OEMs specified "FR-4," making substitution difficult. Any cost saving to be realized by the board shop would be automatically snapped up by the purchasing function at the OEM. This left no incentive to change either, so sales of CEM-3 suffered. In 2000, CEM-3 accounted for 5 percent of all laminates consumed worldwide, according to the IPC, with the greatest use being in Japan and the Far East. As a final note, thin CEM-3 laminates are currently being used as cores for multilayer applications in the Far East markets.

The minimum requirements for the composite laminates can be found in the industry document IPC-4101A, specification sheets 10 (CEM-1) and 11 (CEM-3).

4.5 High-Performance Materials

From the moment that FR-4 started to become the substrate of choice for double-sided and multilayer applications, base materials with even higher levels of performance were desired. Military board designers wanted improved thermal performance and a "guarantee" of unfailing service under less than optimal conditions. Electronics destined for the new frontier of space were to be integral parts of monitoring and control components of the capsule and therefore were critical for life support. Communications with satellites required proper detection and assimilation of even the weakest of signals. From these new requirements, novel and unique high-performance materials were soon developed.

The first of the new high-performance laminates and prepregs were the polyimide materials. The reinforcement for polyimides was the same woven E-glass fabric found in FR-4. In fact, most of the constructions and the resulting resin contents remained exactly the same as that used for FR-4, and only the resin system was different, as shown in Fig. 4.12. As base resins, polyimides were characterized by severe lot-to-lot consistency issues in terms of (1) reaction kinetics and (2) the ability to make a consistent prepreg as compared with their FR-4 counterparts. Board shops found that, in addition to prepreg inconsistencies, frequent high-temperature bakes were required to remove moisture after key manufacturing steps and to drive up the glass transition temperature. In addition, special drilling parameters were required because of the more brittle nature of the resin system. These drawbacks led to only a few suppliers of both materials and printed boards being commercially successful with polyimide materials. However, once produced, the boards exhibited all the positive attributes postulated when first commercialized.

The advantages of polyimide over FR-4 were numerous. Because of the higher glass transition temperature, the z-axis expansion was less than half that of a FR-4 board of the same thickness. MLBs could be fabricated with higher layer counts, and the overall board thickness could be increased without compromising PTH reliability. Thermal conductivity was also twice as good as FR-4, which allowed heat generated by even larger chips and devices for high-density boards to be dissipated quickly. Repair, including the "desoldering and reattachment" of devices, could be performed by military personnel in the field without concern for board failure due to pad lifting. Finally, the higher continuous operating temperature ratings allowed electronics to be employed in harsh environments. With all of these attributes, polyimides were soon the materials of choice for all space and military applications.

The disadvantages of the polyimide products as a class again are partly due to the resin inconsistencies. Only a handful of suppliers of

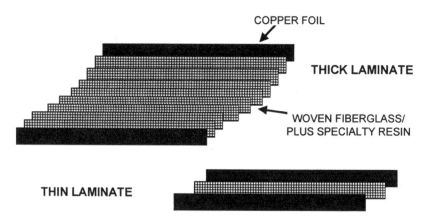

Figure 4.12 High-performance materials schematic.

polyimide-based laminates exist, and each has a highly distinctive resin system, making vendor changes difficult. In addition, high-temperature bakes for long dwell times are still required to drive off moisture and complete the polymerization process. Finally, polyimide products are more costly to purchase and fabricate, which put these materials out of reach for consumer product applications, but they are still the materials of choice for any high-reliability design.

The basic requirements for polyimide materials can be found in the industry document IPC-4101A, specification sheets 40, 41, and 42.

The second set of high-performance materials (as a class), have been around since the middle 1960s. However, as materials for circuit boards, triazines first became known for their resistance to promoting cathodic anodic filamentation (CAF) or dendritic growth. CAF resistance led designers to use triazines on boards in harsh environmental conditions, such as naval applications, where high humidity is the norm.

More recently, when combined with bismaleimide resin, the newly formed bismaleimide triazines (BTs) were found to be a suitable substrate for processor chips mounted on circuit cards. The processor cards have been found to pass the pressure cooker test for moisture resistance under the most stringent of conditions. In addition, BTs were found to be able to thermally withstand the curing temperatures for encapsulants required for ball grid arrays. Finally, BTs have been found to be compatible with MCM-L technologies. In these cases, BTs have been used for both their thermal and chemistry performance.

As in the case of polyimides, BTs utilize the same woven E-glass fabrics in their construction matrices. The blend of bismaleimide and triazine will yield a glass transition temperature of over 200°C. However, most of the BTs are mixed with epoxy to improve processing, reduce cost, and improve the flammability ratings of these materials. As a class, the BTs still exhibit brittle behavior, especially during drilling, and tend to absorb moisture rapidly when immersed in aqueous solutions. However, the electrical properties when wet are still excellent.

The basic requirements for bismaleimide triazine materials can be found in the industry document IPC-4101A, specification sheet 30.

4.6 Microvia Materials

Microvias are used to generate high-density packaging on conventional multilayer materials such as FR-4. Microvias are defined as interconnection structures incorporated into added surface layers of a multilayer board that have specific characteristics, such as a hole diameter of 0.006 in. or less, pad diameter of 0.010 in. or less, conductor widths of 0.004 in. or less, and a hole depth or thickness of 0.004 in. or less. The frequency of microvia holes in a printed board can range from less than 100 to several thousand. Although microvias appear to be fragile, based on the characteristics described above, in reality these layer-to-layer interconnects are extremely robust, making them suitable for consumer electronics such as video cameras, digital still cameras, and cell phones.

Materials found in microvia layers can be found in Table 4.1. As a class, these materials are largely unreinforced polymers in the form of semicured films or liquids. When fully cured, these materials generally exhibit a T_g of over 160°C and excellent moisture and chemical resistance properties. The z-axis expansion properties are similar to other epoxy and acrylate polymers of like glass transition temperatures. Devoid of a reinforcement layer such as woven fiberglass, microvia materials are compatible with alternative methods for the generation of holes such as laser, plasma, or the photo-development processes. Precise hole depths can be generated down to the target pad ultimately providing an electrical interconnect of 0.004 in. or less for redistribution layers.

Of the materials listed in Fig. 4.13, only the nonwoven aramid materials are reinforced. Because aramids are organic in chemical nature, as opposed to the inorganic character of fiberglass fabrics, the reinforcement can also be removed by laser or plasma techniques. In addition to being compatible with microvia production techniques, nonwoven aramids also generate improved unique properties such as a lower D_k and D_f. Nonwoven aramids also exhibit a low coefficient of thermal expansion (CTE), similar to ceramics, which makes the materials suitable for leadless chip carrier (LCC) applications. Finally, nonwoven aramids can be combined with a variety of thermosetting resin systems such as high-temperature epoxies, polyimides, and cyanate esters, which results in a variety of custom properties. The basic requirements for nonwoven aramid materials can be found in the industry document IPC-4101A, specification sheets 53, 55, and 56.

The world of microvia materials has changed rapidly over the last several years. At the time of this writing, coated copper foil was the material of choice for microvia applications. Coated copper foil benefits from already having the copper available for the surface layer as well as having a wide supplier base. Currently, coated copper foil is the

Table 4.1 Microvia Materials

Resin-coated foils
Photo-imagable liquids and films
Non-photo-imagable liquids and films
Nonwoven aramid laminates

most cost effective of the materials sold in volume. If additional properties are required in the microvia material, the nonwoven aramid materials are often used, despite the high-end cost. The front-running technique for a hole-generation method has been the laser, which can be used to cut through the foil layer as well as penetrate the dielectric material. The next 18 months will be critical to ultimately determining what materials will be used in the highest volumes for microvias as we approach the middle of the decade.

An entire industry specification has been dedicated to microvia materials. Descriptions and requirements for these products can be found in IPC-4104, "Specification for High Density Interconnect (HDI) and Microvia Materials."

4.7 High-Speed/High-Frequency Materials

As a class of high-performance base materials, the high-speed/high-frequency laminates and prepregs have been the fastest growing. A high-speed material is normally characterized as having a dielectric constant (D_k) of less than 4.0 at 1 MHz. Performance applications that require high "clock rates" call for base materials with a D_k as low as possible and with little change over temperature. A high-frequency material normally is characterized as having a dissipation factor (D_f) of less than 0.010 at 1 GHz. Performance applications operating above 800 MHz require base materials of this type. Polytetrafluroethylene (PTFE) has been the historical material of choice for these applications.

PTFE has been demonstrated over the years, in military, communications, and computer applications, to be the best material in terms of performance attributes. The D_k and D_f are stable over a variety of temperature ranges and frequencies. The moisture absorption is very low, and the thermal properties are excellent. However, PTFE, because it is a true thermoplastic and also highly inert, has been found to be very difficult to process into printed boards. In addition, PTFE has been among the most expensive to purchase. For this reason, most of the supplier base has been developing thermosetting alternatives to PTFE that, although not exhibiting the ultimate in performance properties, would be viable for high-speed/high-frequency applications. The thermosetting alternatives to PTFE exhibit a desirable blend of cost and performance and thus provide the best overall value in this market segment.

A number of resin systems can provide a reduction in D_k and D_f as outlined in Table 4.2. Most of these resin systems still are at a premium in terms of cost over conventional FR-4 epoxies, but at the same time are cost-effective alternatives for PTFE. The board designer can

Table 4.2 Electrical Properties of Resin Systems

Resin system	D_k	D_f
PTFE	2.0	0.0002
Polyester	2.4–3.0	0.001–0.020
Cyanate ester	2.6–3.0	0.001–0.006
Polyimide	2.8–3.2	0.003–0.008
Epoxy	3.5–3.8	0.018–0.024
PPO/epoxy	3.2–3.4	0.010–0.015
Alumina (filler)	10	0.0001

use the published values to choose which material best fits the performance desired. In addition, since these are thermosetting chemistries, normal printed board production techniques can be employed. Of those currently shown on the list, the polyphenylene oxide, cyanate esters, and the unsaturated polyesters are those resin systems used most actively for these applications in place of PTFE.

The D_k and D_f can also be influenced by the reinforcement used. The possible reinforcements can be found in Table 4.3. As an example, the D_k can be reduced by about 1.3 points for any given resin system simply by utilizing S-glass in place of E-glass. However, the most popular reinforcement for lowering the composite D_k from this list is the nonwoven aramid paper, which was discussed in Sec. 4.6.

Table 4.3 Electrical Properties and Cost Factors of Reinforcements

Reinforcement	D_k (@1 MHz)	D_f (@1 MHz)	Cost factor
E-glass	6.6	0.003	1
S-2 glass	5.3	0.002	4
D-glass	3.8	0.0005	10
Quartz	3.8	0.0002	30
Woven aramid	3.8	0.012	10
Aramid paper	4.0	0.012	10

An entire specification has been dedicated to high-speed/high-frequency materials. Descriptions and requirements for these products can be found in IPC-4103, "Specification for Base Materials for High-Speed/High-Frequency."

4.8 Conclusion

At one time in the United States, polyimides were used for military and high-end electronics, CEM-1 was used for single-sided applications, and FR-4 was used for everything else. In today's global marketplace, a variety of novel high-performance materials have been developed to satisfy particular printed board needs, specifically in the area of higher-density interconnects. At the same time, the old original materials have continued to evolve into higher-performing materials, as shown by the rapid conversion of FR-4 from a 125°C T_g material to a 170°C T_g material. All suppliers are working on high-speed/high-frequency substrates as alternates to PTFE as well as their own solutions to microvia dielectric materials. The competition in the paper-based and FR-4 markets is fierce.

A breakdown of the laminate grades produced on a global scale is shown in Fig. 4.13.[1] The rigid FR-4 market is the combined single- and double-sided laminate market, while the thin FR-4 segment is laminate intended for multilayer applications. All of the high-performance materials, such as polyimide, BT, and cyanate ester, fall in the high-performance category.

Where material development will go from here is uncertain, with the exceptions that thermal performance will continue to increase, D_ks and D_fs will be lower, and the low-cost laminate materials such as the FR-2 and FR-4 grades will not disappear. Fitness for use will still be the primary consideration for both existing and newly developed materials alike.

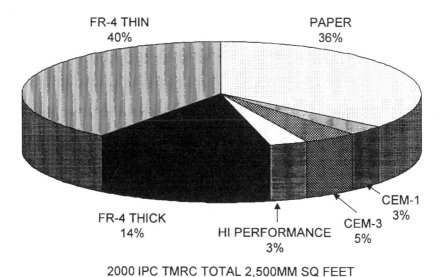

Figure 4.13 World laminate consumption by grade.

---

[1] Courtesy of IPC, 2215 Sanders Rd., Northbrook, IL 60062 (847) 509-9700.

Chapter

5

Printed Circuit Board Fabrication

Joseph Fjelstad
Pacific Consultants, L.L.C.
Mountain View, California

5.1 Introduction

Of all of the elements of an electronic system, perhaps none is more essential than the printed circuit board. At the same time, however, no other element seems to be so under appreciated. Prior to the advent of the printed circuit, electrical interconnections between components were made in a point-to-point fashion (see Fig. 5.1 for an example). This was both very time consuming and highly error prone. The printed circuit offered a way to make ordered interconnection between components radically reducing the potential for error by allowing faithful reproduction of the circuit using a combination of lithographic and etching methods. It has been that way virtually ever since.

Today, it can be easily argued that the printed circuit is the foundation of almost all electronic products and systems and is a technologi-

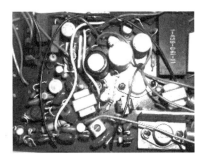

Figure 5.1 Before the printed circuit became widely available, wiring between components was carried out by soldering wires from point to point. (Photo courtesy IPC.)

cal marvel of immense importance, but it is commonly overshadowed but the more glamorous integrated circuit. Interestingly, the printed circuit is likely to have served as the inspiration for the inventors of the IC, as the concepts described by both Kilby and Noice appear to have borrowed from printed circuit manufacturing methods. Regardless, the printed circuit will assuredly remain an indispensable element of electronics for many years to come.

The purpose of this chapter is to give the reader an overview of this important electronic technology. Attention will be given to the materials and processes used in their construction and, as well, most of the many forms of printed circuits will be reviewed and described. Rigid, flexible, multilayer, metal core, and molded boards are among the types that will be covered. No attempt will be made to provide an exhaustive look at printed circuit technology. Rather, it is intended and hoped that the reader will be provided with a sufficient level of understanding of printed circuit technology to feel confident and comfortable moving through the technological forest of electronic interconnection substrates. The reader will be shown many different constructions and be provided with enough information on processes to obtain, hopefully, a good understanding of how PCBs are made, from the very simple to the very complex. Finally, it is hoped that the reader will be able to use this information to make informed choices relative to future interconnection needs.

5.2 Background and History

The origins of the printed circuit have been variously traced to either the late nineteenth century or the early twentieth century. That period marked the rise of useful inventions predicated on the use of electrons and wire. The telegraph, the telephone, and the radio are the hallmark inventions of the period. A point of entry from the older wiring perspective came in 1903 from the inventor Albert Hanson, a German living in London. Hansen conceived of a method of producing conductive metal patterns on a dielectric by cutting or stamping copper or brass foil patterns and laminating or bonding them in paraffin paper. While only two metal layers were described, it is easy to envisage that multiple layers could be produced using his concept.

The American patron saint of inventors, Thomas Edison, also took a turn at circuit processing invention at the prodding of this then assistant, Frank Sprague, later of Sprague Electric Company fame. Sprague had challenged Edison in 1904 with finding ways of patterning conductors on linen paper. Edison responded with a number of different ideas, including two that are, at least in spirit, in use today. These methods include a crude version of today's polymer thin film technology and patterning of silver salts to be reduced in situ, which, in concept, resembles today's additive processing.

Print and etch methods commonly used in production of circuits today can trace their roots to methods described by inventor Arthur Berry in 1913, in which he used the method to create resistive heaters (British patent 14,699). Another method of note was conceived of by Max Schoop, who developed a method of flame spraying of metal through a mask to create circuit patterns on a dielectric base (U.S. patent 1,256,599). Electroplating of the circuit pattern appears to have been the invention of Charles Ducas in 1927 (British patent 1,563,731). The method had the added advantage of allowing release and transfer of the circuits from the base onto which they were plated. The basic concept of transferring circuit patterns has merit and has been visited since that time by others who made improvements to the concept.

Paul Eisler is the next inventor of note in the pantheon of printed circuit technology pioneers. The self-proclaimed "father of the printed circuit," Eisler made significant contributions to printed circuit technology. Eisler's innovations spearheaded the allies' effort to make more reliable proximity fuses for munitions. The advantages offered by the PCB are credited with causing the destruction of large numbers of V2 rockets and thus ending the terror bombing of England. Examples of one of his circuit concepts are illustrated in the patent drawing shown in Fig. 5.2.

After World War II, printed circuit technology began to unfold in earnest as investigation into defining and refining methods of manu-

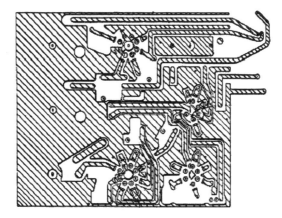

Figure 5.2 Example of an early printed circuit design layout. Note that components mainly had leads in a circular pattern consistent with vacuum tubes and the early TO packages.

facturing them began to grow and expand. The following paragraph gives and indication of how diverse the approaches were:

> "...Circuits are defined as being printed when they are produced on an insulated surface by any process. The methods of printing circuits fall in six main classifications: *Painting*—Conductor and resistor paints are applied separately by means of a brush or a stencil bearing the electronic pattern. After drying, tiny capacitors and sub-miniature tubes are added to complete the unit. *Spraying*—Molten metal or paint is sprayed on to form the circuit conductors. Resistance paints may also be sprayed. Included in this classification are an abrasive spraying process and a die casting method. *Chemical deposition*—Chemical solutions are poured on the surface originally covered with a stencil. A thin metallic film is precipitated on the surface in the form of the desired electronic circuit. For conductors the film is electroplated to increase conductance. *Vacuum processes*—Metallic conductors and resistors are distilled onto the surface through a suitable stencil. *Die stamping*—Conductors are punched out of a metal foil by either hot or cold dies and attached to an insulated panel. Resistors may also be stamped out of a specially coated plastic film. *Dusting*—Conducting powers are dusted on, either with a binder or by an electrostatic method.... Principal advantages of printed circuit are uniformity of production and the reduction of size, assembly and inspection time and cost, line rejects and purchasing and stocking problems...."

The forgoing paragraph was from a publication titled *Printed Circuit Techniques,* written by Cledo Brunetti and Roger W. Curtis and published 1947 by the U.S. government. Clearly, those involved in the early manufacture of printed circuits were very clever. Few stones, it seems, were left unturned in their efforts to find cost-effective ways of delivering patterned circuits to the nascent electronics industry.

So far in this review, most of the circuit concepts describe have been single sided; however, the drive for more functionality in an electronic assembly translated naturally to greater complexity and more wiring. Two metal layers became the next requirement. Interconnection between the two sides was accomplished initially by simple but slow methods. The z-wire interconnection served the purpose originally (see Fig. 5.3) but was supplanted by the use of eyelets (the same method as is used to reinforce lace holes in shoes), which were easier to use. However, as hole counts rose, there was a need improve productivity in making side-to-side interconnection, and the plated-through hole concept was reduced to practice.

The multilayer board was the next important milestone on the path of printed circuit innovation. Original concepts such as the one illustrated in Fig. 5.4 had layers of circuitry being laminated to prefabricated double-sided circuits, with access being provided by larger holes in the outer layers. The plated-through hole multilayer circuit fol-

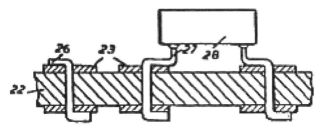

Figure 5.3 Before plated-through hole technology was developed, through hole connections were made by wires soldered to lands on both sides of the PCB, as indicated by this drawing from an early patent.

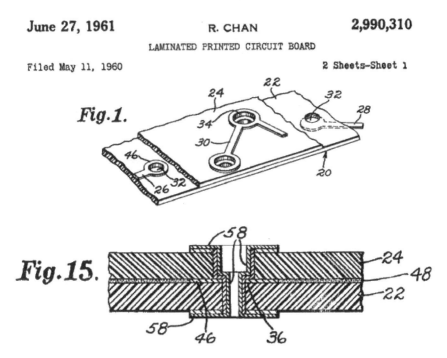

Figure 5.4 This early multilayer circuit patent used holes of different sizes to make facilitate making plated-through hole connections.

lowed as a natural extension, and the multilayer circuit was later married with the flexible circuit to create the rigid flex circuit. (See Fig. 5.26, near the end of this chapter for an example.)

In the time since the early days of printed circuit technology, there has been a steady stream of improvements to the fundamental technology, and it is safe to say that there will be many more improvements in the years ahead. Some potential directions will be discussed at the end of this chapter.

5.3 Materials of Construction

Rigid printed circuit laminates normally consist of three fundamental elements: (1) a reinforcement, such as glass cloth, paper, or other material, (2) a resin, and (3) a conductive or catalytic layer. Each of these elements serves a specific purpose. Generally, as laminate thickness increases, heavier glass cloths or reinforcements are commonly used, and the resin content in the laminate decreases. Because laminates are commonly exposed to high temperatures both in assembly and used, many laminate materials have been engineered to match the thermal expansion rate of copper. This minimizes any potential for warping in plane, which can cause problems in assembly and possibly result in reduced reliability of the plated-through hole or solder joints on the PCB.

Laminates can be made with varying resin-to-glass ratios, depending on the performance needs of the finished product. Specifically, the glass-to-resin ratio of a laminate has a direct effect on dielectric constant, with higher resin contents resulting in a lower dielectric constant being obtained. However, laminates with higher resin content tend to have higher coefficients of thermal expansion (CTEs) in the z-axis and lower dimensional stability. On the other hand, if resin content is too low, weave exposure and a phenomenon called *measling* may result. This condition has is a largely cosmetic defect, but it is a concern, because it is an indication that either the resin coating or the lamination process is not in full control.

5.3.1 Reinforcements, Laminate

Reinforcements are the conceptual foundation of a laminate and provide important mechanical properties. These are the materials that, when coated with resin, become the individual laminae or "building blocks" of the finished laminate. While it is possible to employ a wide variety of different materials as reinforcements, only a very limited number of choices are commonly used in printed circuit laminates.

The reinforcement serves several different functions. For example, it imparts important mechanical properties such as strength and rigidity. The reinforcement also provides the important attribute of dimensional stability required for accurate manufacture and assembly. Electrical properties of the laminate are also affected by reinforcement choice. The effect occurs not only from the electrical properties of reinforcement itself but also changes as the ratio of resin to reinforcement is altered. Finally, because they are normally less expensive than the resins they are combined with, reinforcements also help keep the cost of the laminate in line with customer expectations or desires.

The following is a review of some of the more common reinforcements used in the construction of printed circuit laminates.

5.3.1.1 Paper. Paper-based reinforcements have been used in the construction of laminates for a wide variety of products for many years and have proven to be valuable materials for use in the manufacture of printed circuit laminates. Paper is easily mass produced, and thus it is not surprisingly one of the lowest-cost options among reinforcements. One clear disadvantage of paper-based laminates is that they are not fireproof.

5.3.1.2 Glass fiber. Glass fibers are among the most universally employed reinforcements for resin-based laminates. Glass fibers are an excellent choice because of the good mix of electrical and mechanical properties they exhibit. Glass fibers, for example, are not only excellent insulators but also have the kind of physical and mechanical properties that are required to supply the necessary strength and dimensional stability required for manufacture, assembly, and component support in use.

There are a number of different types of glass formulations that can be used to create the glass fibers. These are given letter designators. E-type glass is the most commonly used in laminates for printed circuits; however, D-type glass is used in some applications for its lower dielectric constant, and S-type glass is used in applications demanding higher strength.

Glass fibers can be employed in one of two fashions: either as a woven cloth or as a chopped glass-fiber-based paper (see Fig. 5.5). Both forms of glass reinforcement are used but, of the two, woven glass cloths are by far the most common. While the potential number of choices of glass cloth (in terms of the different fiber diameters, yarn construction, and weave) is quite substantial, the number actually used in laminate manufacture is, fortunately, rather small. Only few different types of glass cloth see any significant use; these are characterized in Table 5.1.

5.3.1.3 Others. Many other specialty materials have been used to reinforce printed circuit laminates, including such exotic materials as quartz cloth and aramid fibers. Aramid fibers have proven especially

Figure 5.5 Glass cloths for reinforcing laminates can be either woven or a mat of chopped fibers, as seen above.

Table 5.1 Characteristics of Selected Electronic Grade Glass Cloths

Glass fabric style	Yarns per cm	Weight	Thickness
104	~24 × 20	19.7 gr/m^2	30 µm
108	~24 × 19	48.5 gr/m^2	51 µm
112	~16 × 15	71.2 gr/m^2	81 µm
116	~24 × 23	107 gr/m^2	102 µm
2112	~16 × 15	71.6 gr/m^2	76 µm
2116	~24 × 23	107 gr/m^2	102 µm
7628	~18 × 13	203 gr/m^2	173 µm
7642	~18 × 8	232 gr/m^2	279 µm

useful in applications where a low in-plane expansion rate is desired. These materials actually have a negative in-plane expansion rate, which helps offset the normally much higher expansion rate of the resin.

5.3.2 Organic Resins

Organic resins are the "other half" of a laminate. They serve as a binder to hold the reinforcements together and impart important electrical properties to the laminate. These resins are generally relatively low in their dielectric contact and loss tangent, which are key properties in electronic design especially in high performance applications.

5.3.2.1 Phenolics. Phenolics are among the longest used and best known of general-purpose thermosetting resins. Although their performance is limited, they have proven quite suitable for electronic-grade laminates. Phenolics are the lowest-cost resin used for laminates, and many consumer electronics are fabricated using these materials.

5.3.2.2 Epoxies. Epoxies are the most common resin materials that are combined with glass cloth to produce laminates. As compared with other laminate materials, epoxies offer advantages in availability and relative ease in processing. The many different types and blends of epoxies exhibit a wide range of selection for usage or soldering processes; epoxies with a T_g (glass transition temperature) from 110 to 120°C up to 180 to 190°C are available from most laminate suppliers. However, the most commonly used in the 135 to 145°C range. There are also some low-loss, low-dielectric-constant epoxy materials in development that are expected to have a T_g in the rage of 210°C.

5.3.2.3 Polyimides. Polyimides are among the highest-performance resins used in the manufacture of printed circuit laminates. These materials, with glass transitions in the range of 260°C, can withstand very high temperatures for extended periods, making them a good choice for applications where high heat is experienced or where the components used are high-wattage devices. The military also has special interest in high-temperature materials, as they make any required rework and repair easier.

5.3.2.4 Others. Several other resins are useful for creating laminates, such as cyanate esters and bismalamide triazine (BT). These products are seeing application in unique and special applications. For example, BT resins are proving very popular for the manufacture of organic laminate-based packages for integrated circuits. Liquid crystal polymers are also seeing greater levels of interest for certain types of applications because of their good combination of electrical and mechanical properties coupled with their very good dimensional stability.

5.3.3 Flexible (Unreinforced) Laminates

Flexible circuits are a special subset of printed circuit manufacture and require their own special laminates. These materials normally consist of a base film, adhesive, and copper foil; however, there is increasing interest in adhesiveless laminates. The materials for these laminates tend to be thermoplastic in nature, because they are better suited to flexing without fracturing and failing. The adhesives, when used to create flex circuit laminates, are generally formulated to be more flexible than traditional thermoset resins. There are two basic resin systems employed in flex circuit manufacture (polyester and polyimide); however, a number of materials intermediate in performance and price have seen use. One example of note is polyethylene naphthalate (PEN).

5.3.3.1 Polyester. Polyester is a low-cost resin used in a substantial number of electronic products. Common examples are keyboards and printer cables. Polyester is used both with copper and polymer thick film circuits. The major limitation for polyester is related to is thermal performance. Because of its low melting temperature, it is generally considered unsuitable for soldering; however, this can be addressed by careful engineering. Major OEMs such as Texas Instruments and Kodak have developed methods for special applications requiring soldering to polyester, and some manufacturers have specially developed tools to address the problem.

5.3.3.2 Polyimide. Polyimide is the choice of most flex circuit applications because of its excellent thermal performance coupled with its good electrical and mechanical properties. It also exhibits reasonable dimensional stability, which facilitates circuit manufacture and assembly. This material is also being used extensively in the creation of IC packages. One limitation of polyimide is that it tends to absorb moisture, which can be a concern when soldering.

5.3.4 Special materials

A wide range of niche substrate laminate products that have been developed in support of special printed circuit manufacturing methods. One example is resin-coated copper (RCC). This material is used for high-density build up technologies, which will be discussed later in this chapter.

5.3.5 Foils

Metal foils are commonly laminated to resin composites to complete the raw material need for printed circuit manufacture. Many different foils are potential candidates for the creation of a metal-clad laminate, but copper is the most common because of its excellent electrical properties and its general ease of processing and amenability to solder assembly. Thinner foils are generally used by printed board manufacturers for "fine line" designs to reduce the amount of undercutting of circuit conductors that occurs during the etch operation and to meet the requirements of high-density printed circuits for flip chip and chip scale packages.

Copper foils can be created in on of two ways: mechanical rolling and deposition. In the area of deposition, there are several potential methods: electrodeposition, electroless deposition, vapor deposition, and sputtering. The first of these methods is the one most commonly used to manufacture copper foil for printed circuit laminates. The other methods are generally used to deposit foil directly on the laminate without the aid of a lamination process. Electrodeposition and electroless deposition employ wet chemistry, whereas vapor deposition and sputtering are dry deposition methods, which are carried out in a vacuum. The following table lists the designations for the most common types of copper foil as called out by IPC-MF-150.

Copper foil type	Number	Designator	Description
Electrodeposited (E)	1	STD-Type E	Standard electrodeposited
	2	HD-Type E	High-ductility electrodeposited
	3	THE-Type E	High-temperature elongation electrodeposited
	4	ANN-Type E	Standard electrodeposited
Wrought (W)	5	AR-Type E	As-rolled wrought
	6	LCR-Type E	Light cold rolled wrought
	7	ANN-Type E	Annealed rolled wrought
	8	LTA-Type E	As-rolled wrought, low-temp annealable

5.3.5.1 Electrodeposited. Electrodeposited foil is the most commonly used for rigid printed circuit laminates. The foil is plated on to polished stainless steel of titanium drums which are negatively charged and rotated slowly through a plating bath. The thickness is controlled by either the rate of rotation, the current density or a combination of the two. Figure 5.6 illustrates the process.

222 Chapter 5

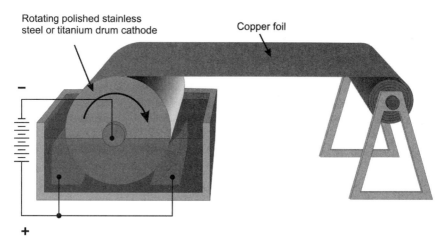

Figure 5.6 Electrodeposited copper foil is the type most often used in circuit manufacture. The foil is produced by rotating a highly polished drum of passivated stainless steel or titanium in a copper plating solution with current applied. The rate of rotation controls the copper thickness.

The surface is commonly provided with a roughening treatment to increase the surface area to improve the foil's peel strength. This can be important for surface mount components of large mass, lest they pull away from the surface from shock and vibration. The thickness of the treatment can, however, influence processing and might limit the minimum feature sizes that can be reliably etched. This will be discussed in more detail later in this chapter.

5.3.5.2 Rolled. Rolled copper foil is most commonly used in the manufacture of flexible circuits. Wrought and annealed foils have proven excellent for use in situations where the foil must be bent or formed. Figure 5.7 shows the layout of the rollers for a foil rolling mill.

5.3.6 Conductive Inks

Conductive inks are among the oldest methods of creating printed circuits and are largely responsible for giving the technology its name. The technology is used for both high- and low-end products. For example, with ceramic substrates, it is used to create hybrid circuits and is fired in an oven to create circuits of high conductivity. For lower-end products, such as keyboards and some low-cost electronic devices, the curing is done at low temperatures, using low-cost materials for substrates such as polyester film.

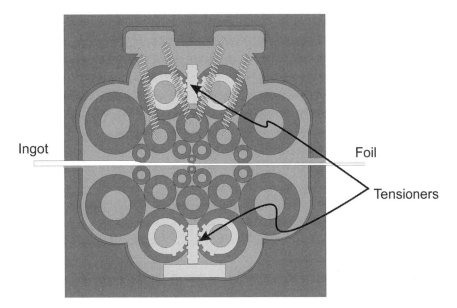

Figure 5.7 Rolled copper foil is manufactured from ingots of copper successively rolled down to the desired thickness in specially designed rolling machines such as represented in this figure.

The inks are generally combinations of resins and conductive fillers such as powered silver. The conductivity can be modified to create resistors of varying values by the addition of more resistive fillers such as carbon or graphite powders.

5.4 Laminate Material Preparation

The resin and reinforcement material are combined to create the raw material used for the creation of a laminate. The end product is commonly referred to as either *prepreg* or *B-stage*. "Prepreg" is short form for "reinforcement preimpregnated with resin," whereas "B-stage" refers to the fact that the resin is dry to the touch but is not fully cured.

This material is created by drawing the reinforcement material through a wet resin bath and then drying material and slightly advancing its state of cure normally though the addition of heat. The basic manufacturing process is illustrated in Fig. 5.8.

5.5 Lamination Methods

A lamination process is used to create laminates that can be used for the manufacture of printed circuit boards. In lamination, the prepreg

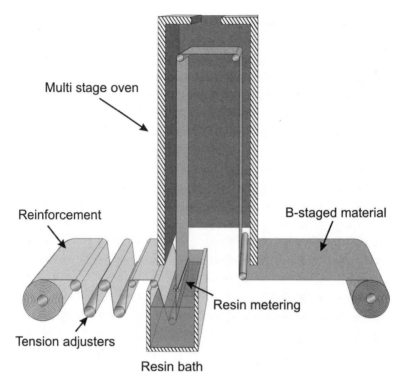

Figure 5.8 Reinforcement materials such as glass cloth are converted into prepreg materials for laminate manufacture by coating them in special resin "treaters." The material is coated with resin and then dried and partially cured in a special oven, and it emerges dry and tack free.

of B-stage material is layered most often between two sheets of treated copper foil. Following lay-up of the material, the material stack is subjected to heat and pressure, which drives the resin to a fully cross-linked or cured condition wherein it has all of the desired properties. There are several different methods of creating laminate material.

5.5.1 Batch

Batch processing is the simplest and was thus one of the first methods developed for laminate manufacture. In their simplest form, lamination presses for batch processing consist of a hydraulic ramp and a heated platen that is pressed against a parallel platen that resists the pressure of the hydraulic ram. It is obviously possible to have several openings and to have all of the platens heated. Heating can be done electrically, by steam, or by hot oil circulated through the platens. The choice is predicated on the temperature required for lamination. Figure 5.9 illustrates the basic process.

5.5.2 Continuous Lamination

Continuous lamination is a somewhat self-descriptive method wherein the laminate is produced in a continuous web. The prepreg and foil are fed into a specially designed and manufactured lamination press. There are both economical and technical advantages to such methods; however, there are only a very limited number of suppliers of such product. A conceptual example of continuous lamination is provided in Figure 5.10.

5.5.3 Vacuum-Assisted Lamination

Traditional lamination methods can be improved by carrying out the process in the presence of a vacuum. The negative pressure facilitates the removal of entrapped air and can improve the quality of the laminate. The reduced pressure created by the vacuum allows for lower lamination pressures to be used with the hydraulic ram.

5.5.4 Vacuum-Assisted Autoclave Lamination

Vacuum-assisted autoclave lamination methods were developed originally for the lamination of complex shapes. The method takes advantage of normal atmospheric pressure by simple removal of air. The effect is the same as is seen in vacuum packaged foodstuffs, as the vacuum bag conforms to shape of the materials contained within. The negative pressure is augmented by gas pressure in a pressure vessel. This creates an isostatic or uniform pressure from all directions on the laminate stack and allows for lower lamination pressures to be used.

5.6 Laminate Forms for PCBs

Laminates used in PCB manufacture, whether rigid or flexible, are created in two primary forms: single-clad laminates and double-clad laminates. They are also produced in a number of different thicknesses to facilitate the manufacture of PCBs of different thickness requirements and for the construction of multilayer PCBs. When used in the manufacture of multilayer circuits, the laminate is commonly referred to as a core material.

5.6.1 Single-Clad Laminates

Single-clad laminates have copper on one side only and are used either for the manufacture of single-sided circuits or for the manufac-

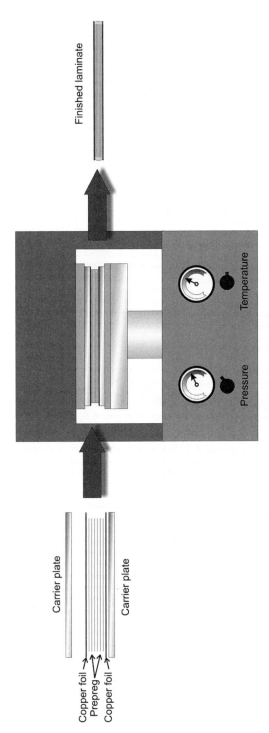

Figure 5.9 Standard laminate manufacture is accomplished by placing plies of B-stage or prepreg between sheets of copper foil and applying heat and pressure.

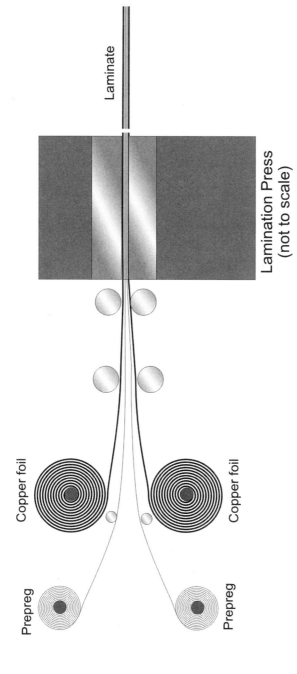

Figure 5.10 Lamination methods have been developed that allow for laminates to be manufactured in a continuous fashion. The basic concept is illustrated above.

ture of multilayer circuits as an inner layer, or more commonly as a cap laminate.

5.6.2 Double-Clad Laminates

Double-clad laminates are essentially identical to single-clad laminates in description and use except that copper is laminated to both sides of the structure.

5.7 Laminate Selection

The selection of a laminate is made based on the requirements of the product both in assembly and use. Ideally, the laminate choice should be made from standard structures to avoid delays and potentially costly qualification of new constructions. When several laminates are potential candidates the choice should be made in favor of the one which has the best balance of properties. Table 5.2 lists some of the different potential criteria for making a selection.

Table 5.2 Laminate Selection Criteria

Coefficient of thermal expansion (CTE)	Maximum continuous operating temperature
Electrical properties	Mechanical strength
Flame resistance	Overall thickness tolerances
Flexural strength	Reinforcing sheet material
Glass transition temperature (T_g)	Resin formula
Machinability	Thermal stability

As has been described, laminates come in many different resin and reinforcement combinations. The national Electronics Manufacturers Association and the U.S. military have both devised designators to describe the materials in a short hand fashion. Table 5.3 provides a list of some of the most commonly used designators for different laminate types, and Table 5.4 provides examples of the applications for different constructions.

5.8 Solder Masks

Solder masks are used to prevent solder from being deposited or flowing onto circuit features other than the areas of interest such as points of attachment. Solder masks were originally developed for PCBs that were to be assembled using wave soldering methods, and they were

Table 5.3 Laminate Designators and Descriptions

NEMA designation	Military designation (MIL-P-13949)	Basic construction of laminate
XXXP	–	Phenolic resin with paper reinforcement, heat required for punching
XXXPC	–	Phenolic resin with paper reinforcement, room temperature punching
CEM-1	–	Composite epoxy laminate, glass cloth outer with paper center
CEM-3	–	Composite epoxy laminate, glass cloth outer with mat glass center
FR-1	–	Similar to XXXP but flame retardant
FR-2	–	Similar to XXXPC but flame retardant
FR-3	PX	Similar to FR-2 but epoxy based and stronger
G-10	GE	Epoxy glass laminates without flame retardant
FR-4	GF	Epoxy glass laminate, flame retardant
FR-5	GH	High-temperature epoxy glass laminate, flame retardant
–	GI	Glass reinforced polyimide resin
–	GP, GT, GX, GY	Fluropolymer resins reinforced with various glass materials, woven or mat fiber, and having slightly different electrical properties
–	GC	Woven glass cloth with cyanate ester resin
–	GM	Woven glass cloth with bismalimide triazine resin
–	BF	Nonwoven aramid fiber with epoxy resin
–	BI	Nonwoven aramid fiber with polyimide resin
–	SC	Woven S-type glass cloth with cyanate ester resin

originally applied to only one side of the PCB—the side that was to be exposed to the solder wave. Without a solder mask, circuit traces on the bottom of the assembly were subject to shorting with solder bridges. Later, as line trace and space became finer, application of solder mask to both sides of the board was implemented to protect the traces from physical damage and inadvertent shorting. Today, with double-sided assembly in wide use, it is very common for solder mask

Table 5.4 Substrate Materials for Electronic Devices (source: IPC)

Copper-clad laminate types, general properties, and applications

Type	Composition	General Properties	Applications
CEM-1	Epoxy/glass fabric surface Epoxy/cotton paper core	Punchable at room temperature Good electricals but less than polyester Good flex and impact strength	Consumer electronics
CEM-3	Epoxy/glass fabric surface Epoxy/glass paper core	Punchable but harder than CEM-1 Good electricals Suitable for PTH applications	Computers and peripherals Keyboards
FR-4	Epoxy/glass fabric	High flex and impact strength Excellent electrical properties Commonly used for PTH applications	Computers Telecom Military
G-10	Epoxy/glass fabric Non-flame retardant	High flex and impact strength Excellent electrical properties Excellent dimensional stability	Structural applications
FR-5	Modified epoxy/glass fabric	Improved hot flex strength over FR-4 Excellent electricals	Military products
High-performance	Bismalimide triazine (BT) Polyimide PTFE (Teflon™) Cyanate ester Epoxy/PPE (Getek™)	Excellent thermal properties Lower x-y-z CTE High reliability Excellent electrical properties	Chip interposers Main frames Telecom Military
Flex	Polyimide Polyester FEP	High flexibility High ductility copper Excellent electrical properties Excellent thermal properties	Automotive Computers Military Telecom

to be applied to both sides of the PCB. Solder masks are available in several forms and chemistries.

5.8.1 Heat-Curable Resins

Heat-curable resins are a common type of solder mask. The are commonly applied using screen printing methods. In application, the solder mask resin, normally an epoxy, is screen printed onto everything except the areas and through holes that will be used for assembly.

5.8.2 UV-Curable Resins

UV-curable resins are normally applied in the same fashion as described for heat curable resins; however, they have the advantage that they can be quickly cured by exposure to UV light. The properties of these masks differ somewhat from the heat-cured types, but they are generally well suited to the task.

5.8.3 Photoimagable Resins

Photoimagable resins are similar to UV-curable resins but with the added advantage that they can be easily photo-defined or exposed and developed with an appropriate chemistry. These materials can be applied in wet form by spraying, dipping, curtain coating, or flood screening, or in dry form as a laminated film.

5.9 Generic Processes Overview for a Plated-Through-Hole Printed Circuit

The processing of printed circuits varies widely. As was indicated in the background section of this chapter, many different approaches to the manufacturing process have been tried and used since PCB technology was first developed. Still, some methods are more common than others. This is because materials, processes, and equipment for PCB manufacturing needed a focal point to create a viable infrastructure. The process description that follows is a somewhat "generic" process. The term "somewhat" is used, because there are many subtle differences that can be employed at each of the process steps described.

This particular section will describe the manufacturing steps for making a plated-through hole printed circuit board using a pattern plating process. Most of the steps described are common for both double-sided and multilayer plated-through hole PCBs. Some alternatives to this process flow will be described later in this chapter.

5.9.1 Stacking and Pinning

Stacking and pinning is the term used to describe the act of cutting sheets of copper clad laminate to a common size and the stacking them and pinning them all together to allow all the laminates to be drilled at one time. This is a common method for controlling the cost of drilling.

In the process of creating the stack, a backup material is always included on the bottom, and an entry material may be used on top. The purpose of the backup material it to prevent the drill from drilling into the drill bed. The entry material can be one of several different materials. For example, it can be a sheet of 250-µm aluminum foil or an unclad piece of a composite material of similar thickness. The purpose of the entry material is to reduce or eliminate the formation of drill burrs at the hole edge. See Figure 5.11.

5.9.2 Drilling

The drilling process is one of the more important steps in the manufacture of a printed circuit board. Today, most if not all drilling is performed on numerically controlled drill machines. The stacked and pinned laminates are placed onto the drill bed, and the pins, which extend beyond the surface of the drill stack, are placed in alignment holes located in the drill bed. The location of the holes serves as a datum for the drill program, and all holes are drilled based on their location relative to these alignment pins.

Careful control of the drilling process is required to assure the quality of the final plated-through hole. Hole quality is achieved by characterizing the materials in a drill study and by adjusting the in feed rate of the drill bit along with the speed of rotation of the drill bit. Excess time in the hole can cause resin smear, while too fast an infeed rate can break drills or result in rough holes.

5.9.3 Hole Preparation and Metallization

Hole preparation and metallization are also very important elements in the creation of a reliable plated through-hole PCB. A well plated through hole is highly reliable; a poorly plated though hole may not be. The hole preparation steps vary from process to process and can be quite extensive. In addition, metallization may be replaced with a coating of carbon or graphite film. The flowchart in Fig. 5.12 illustrates process flow for typical through hole preparation steps for electroplating.

5.9.4 Resist Coating

Resist coating of drilled and metallized boards can be accomplished by using any one of several methods. Screen printing has been a main-

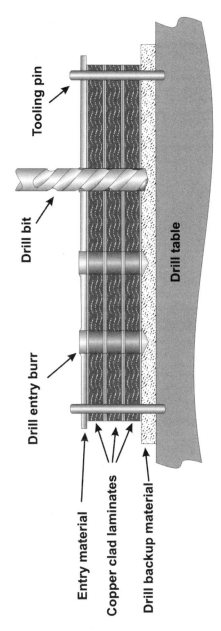

Figure 5.11 Stack drilling.

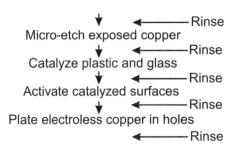

Figure 5.12 Basic electroless copper process flow.

stay process for many years; however, it is limited in terms of the features it can define. While it can be very cost effective, screen printing also has relatively long setup times compared with other resist coating methods.

The other methods for resist coating are normally associated with photoimagable resists. These methods include roll coating with dry film resist, flood printing with a liquid resist, and electrophoretic coating. The latter process is similar to electroplating except that a polymer film is plated rather than a metal.

Roll coating of dry film is one of the most popular methods, owing to its relative ease and cleanliness. It also is capable of producing fineline circuit images such as required by many of today's advanced products. In addition to variations in resist applications, there are two different forms of resist, based on their exposure mechanisms and chemistry. The variations are called *positive working resists* and *negative working resists*. The former type is less commonly used than the latter, although it is generally capable of resolving finer features. With positive working resists, the areas exposed to light are developed away whereas, with negative working resists, the opposite is the case (i.e., where the light hits, the resist will remain after development). See Fig. 5.13 for examples of sample imaging processes.

5.9.5 Imaging and Development

An imaging step is required for all photoimagable plating resists. The process is self descriptive; however, there are a number of potential variations on the exposure process. Perhaps the most common form of exposure is contact printing, wherein a film containing the desired circuit pattern is aligned with the drilled hole pattern, and the two items are brought into intimate contact by means of a vacuum frame. Exposure is done by means of a UV light source, and exposure length is determined by the type of resist and its thickness.

Other processing methods include off-contact printing and laser direct writing. In the former case, exposure is done by image projection

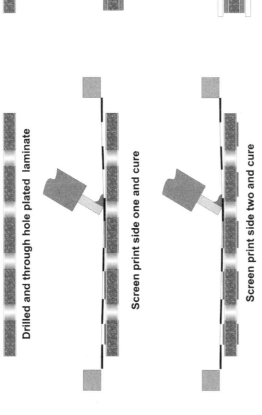

Figure 5.13 Imaging process comparison.

235

over either a short or long distance. In the latter case, a laser is used to scan the image onto the resist. These methods may see increased use in the future.

Development is normally carried out using a conveyorized process wherein the exposed panels are passed between spray banks of a solution designed to dissolve the unexposed resist (or exposed resist, in the case of a positive working resist). Presently, the solutions used are relatively benign and consist of heated dilute (~1 percent) solutions of sodium or potassium carbonate.

Care must be taken to neither over- nor underexpose and to similarly neither over- nor underdevelop the resist. Failure to do so can lead to problems in later processing steps.

5.9.6 Electroplating

Electroplating processes are used to bulk metallize the circuit patterns on the surface and through the holes. Because a negative image of the circuit is most commonly patterned onto the drilled and metallized panel, this process is called *pattern plating*.

In the process, the panel is clamped in a plating rack of some sort so that an electric potential can be applied. Because the metal ions in solution are almost invariably positively charged, a negative potential is placed on the rack. The panel is called the *cathode* in this case. The thickness of the copper and other plated metals is controlled by a combination of time and current. Normally, copper is plated to a total thickness of approximately 25 µm (0.001 in.), but thicker callouts are common in military boards. Other metals are often plated over the top of the copper. These metals include, tin, tin-lead solder, nickel, gold, and others. The thickness of these metals is normally substantially less, as they serve as finishing metals only.

5.9.7 Resist Stripping

After all the require metal plating layers are complete, the plating resist is no longer required, so it can be removed. This is accomplished by exposing the plating resist to a process that can safely remove the resist without damaging the metal finishes. Conveyor-based processes are common. Care must be taken to fully remove the resist, or problems can arise during the etching process. The greatest concerns are shorts and bridges between circuit traces due to left over resist.

5.9.8 Etching

An etching step follows the stripping process. The etching step removes all of the copper not coated with either an etch-resistant metal or polymer etch resist. When metals are used as the etch resist, they must obviously be resistant to the etching chemistry. For example, when tin-lead is used as the etch-resistant metal, ammonia-ammonium chloride and sulfuric-hydrogen etchants can be used, but cupric chloride etchants cannot. Etching must be controlled so as to neither overetch nor underetch the circuits. The effects of both are illustrated in Fig. 5.14.

5.9.9 Soldermask Coating

Solder mask coating operations vary with the type of mask being applied. For example, screen printing can be used to apply the solder mask everywhere except the areas where soldering is to take place. In contrast, dry film solder masks are roll laminated onto the circuit using heat and pressure. The coated circuit is exposed and developed in a fashion nearly identical to the process used for circuit patterning with dry film. Curtain coating is another method for applying soldermask. In this instance, the circuit panel is rapidly passed under a continuous and carefully controlled and monitored cascade of photoimagable resin. The uniformly coated panel is then dried, and the process is repeated if a second side coat is required for the panel. The panel is exposed and developed in a similar fashion to that used for dry film.

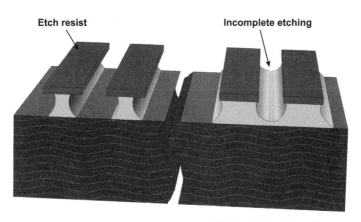

Figure 5.14 Control of the etching process is vitally important to the successful manufacture of a printed circuit. Over etched circuits may not meet designed performance requirements, while under etching can result in shorts. Close spacing of traces is often a factor in both cases.

5.9.10 Solderable Finishes

Following soldermask coating, the (normally) exposed copper surfaces are treated to ensure long-term solderability. The most common method is to coat the exposed copper areas of the panel with tin-lead solder. This is most often done by immersing the panel in molten solder and then removing the excess solder using a high-pressure blast of hot air, hence, the name *hot air solder leveling* or *HASL*.

Other solderability protectants can be and are used; these include organic coatings (often referred to as *organic solderability protectants* or *OSPs*), tin, and noble metals such as gold and palladium. These are applied by dipping in the former case, or by plating in the latter case.

5.9.11 Edge Card Contact Plating

Edge card contacts are a convenient way of interconnecting the circuit card to a next level or system board. The process of plating edge card circuits consists of a few modest steps. First, the area above the contacts is masked off using a suitable "plater's tape." If a metal etch resist was used, this is removed using an appropriate chemistry (i.e., one that does not aggressively attack copper metal). The exposed copper is overplated using first nickel and then, normally, gold. The nickel acts as a barrier layer to prevent diffusion of copper into gold. It also serves as an anvil of sorts, because both copper and gold are relatively soft metals and can be easily galled. The process steps are illustrated in Fig. 5.15.

5.9.12 Depanelization Processing

Once all major processing steps are completed, the panel can be separated into individual circuits. This process is called *depanelization*. There are two primary methods for separating out the circuits, punching and profiling. Punching requires a tool and die set and tooling holes to ensure that the parts are properly aligned for punching. Profiling is commonly done using a numerically controlled (NC) router. Again, tooling holes are required to ensure accuracy. Other numerically controlled cutting methods are potential candidates for depanelization of circuits, including the use of lasers and high-pressure water jet cutters.

When small circuits are being designed, it is normally advantageous to assemble the circuits in panel form. In such cases, the circuits are commonly cut only part way out of the panel. Following assembly, the circuits can be separated into individual parts by cutting the tabs that hold the circuits in the panel. For circuits that are rectangular, one- or two-side scoring of the panel can be used instead. This allows the as-

Printed Circuit Board Fabrication 239

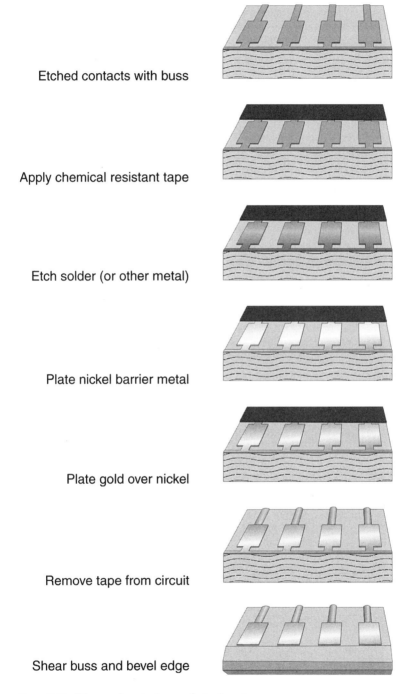

Etched contacts with buss

Apply chemical resistant tape

Etch solder (or other metal)

Plate nickel barrier metal

Plate gold over nickel

Remove tape from circuit

Shear buss and bevel edge

Figure 5.15 Edge card contact manufacturing steps.

sembly to be easily snapped apart in chocolate bar fashion. A word of caution is offered, however. One must exercise care, as it is possible to crack solder joints or components when snapping apart such assemblies. Figure 5.16 illustrates the general concepts just described.

5.9.13 Edge Beveling

Following depanelization, it is common practice to bevel the edges having contacts to facilitate insertion. Special tools have been developed that make this process very simple. The last step in Fig. 5.15 illustrates a beveled edge.

5.9.14 Inspection and Test

Inspection and test of printed circuits are important final steps in manufacturing PCBs. The cost of the components can be quite high, making assurance of assembly function and quality of great importance. Electrical testing is recommended for highly complex boards. The data for a test fixture or test program in the case of flying probe type testers can be relatively easily extracted from CAD data provided.

Cross-sectional analysis of the finished plated-through hole is normally desired to ensure that the applied metal platings are all within prescribed ranges. It is also possible to get a sense of the quality of the

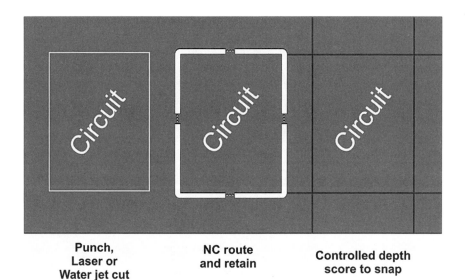

Figure 5.16 Methods for depanelizing finished circuits.

processing using this method. Figure 5.17 gives examples of the types of defects than might be encountered in examining a plated-through hole as received and post solder stress testing.

5.10 Additive and Subtractive Processing

There are two primary methods for creating metal circuit patterns on insulating base materials: additive and subtractive. There are also a number of potential variations on these two common themes; for example, there are processes that use combinations of these two, such as

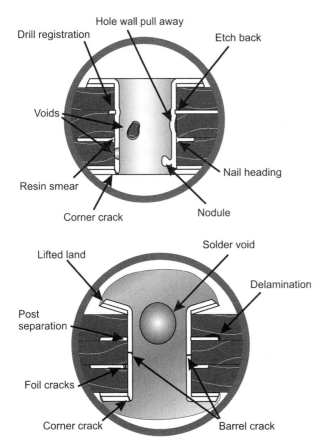

Figure 5.17 Examples of potential copper plating defects that might be encountered in the examination of plated-through hole cross sections. On the left is a hole before solder stress testing, while on the right is an example of a through hole after solder stress testing. Not all defects are cause for rejection; some are simply cause for concern as indicators of possible process control problems.

so-called semi-additive and semi-subtractive processes. A brief examination of some of these themes will help to clarify the differences.

5.10.1 Additive Processes

There are several approaches to making additive circuits, and there is thus potential for some confusion in the terms used for some of these manufacturing concepts that are similar in intent but very different in process. Some of the first circuits manufactured in volume used what must be considered, in practical terms, additive processing methods. These circuits were produced by screen printing circuits directly onto insulators. The insulators could be either organic (such as epoxy or polyester resin) or inorganic (such as ceramic or glass) in nature. Other methods for creating additive circuits include electroless copper plated circuits and transfer laminated circuits. Figure 5.18 illustrates the process steps for a simple two-sided additive board.

5.10.2 Subtractive Processes

Subtractive processes were also used very early in the development of circuit manufacturing methods. While the so-called "print-and-etch" process is the most common image of subtractive processing, there are a number of possible variations. These include punching of copper foil, embossing and milling, and various plate-and-etch processes. The "Subtractive Process (Panel Plate)" portion of Figure 5.18 illustrates the process steps for manufacture of a two-sided subtractive PCB.

5.10.3 Semi-Additive Processes

There is a crossover point at which the line blurs as to whether a process is additive or subtractive. Such processes are most often called *semi-additive*, but the term *semi-subtractive* is also used on occasion. The term is most often used in describing processes in which in a very thin layer of copper foil is used on the other layers. The chief advantages are that it allows very fine line circuits to be produced and that a lesser amount of materials, such as etching solutions, are consumed in processing. Figure 5.18 provides a simple example of this process for manufacturing a two-sided PCB.

5.11 Single-Sided Circuit Process Examples

Having earlier described a general process for a double-sided PCB, it is now possible to describe fairly easily some common single-sided cir-

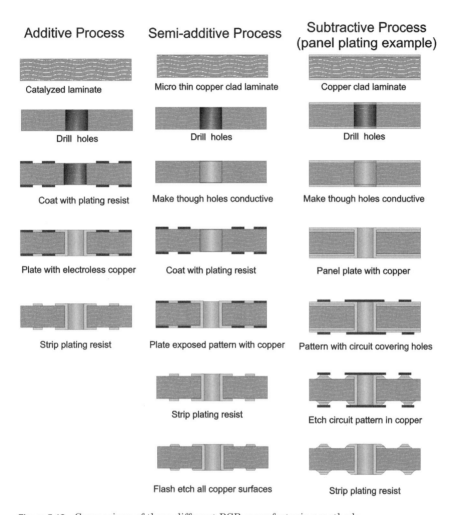

Figure 5.18 Comparison of three different PCB manufacturing methods.

cuit processes. There are a number of potential methods and variations on their themes. The following are a few examples.

5.11.1 Print-and-Etch Processing

Print-and-etch processing is perhaps the least complex method for making a simple printed circuit. In processing, the metal foil clad laminate is coated with a circuit pattern using any of the previously describe processes, and the circuit is etched. Holes, if needed, can be drilled before or after etching. Figure 5.18 shows a cross section of the simple construction.

5.11.2 Foil Routing

A novel method for producing single-sided (also double-sided with special processing) circuits is to route the metal clad laminate to a controlled depth, removing on the copper that is not needed for the circuit. Thus, each circuit trace is individually defined by routing completely around its periphery. The process is obviously slower than etching and allows only one circuit to be manufactured at a time, but it is nonpolluting and can be done in a small space. Figure 5.19 provides examples of product produced in this fashion.

5.11.3 Direct Printing with Conductive Ink

Direct printing of the circuit pattern with conductive ink is one of the oldest methods and was described briefly earlier in this chapter.

5.11.4 Flush Grinding

It is possible to emboss a plastic material with a circuit pattern and then metallize the result using one of several processes. The panel can then be ground flat to expose the raised portions of the polymer while the circuit pattern at the lower level remains.

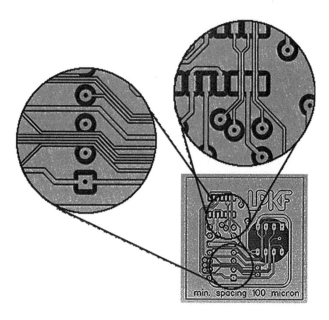

Figure 5.19 Examples of foil routed PCB (courtesy of LPKF Laser & Electronics AG).

5.12 Double-Sided Circuit Process Examples

Like single-sided processing, there are several different ways to create double-sided circuits. The following are brief looks at some of the more common processes.

5.12.1 Panel Plate, Tent-and-Etch

Panel plate, tent-and-etch processing can be very reliable and economical, depending on the approach and materials chosen. Ideally, a very thin copper foil should be used to start. This makes etching easier and more accurate. The method also obviates the need for an etch metal resist layer removal before solder mask coating. The process follows the steps outlined in Fig. 5.18 under the heading "Subtractive process."

5.12.2 Panel Plate, Pattern Plate

Panel plate, pattern plate processing is similar to the previous method except that, in place of tenting, a circuit pattern of etch-resistant metal is plated, and the circuit is etched. This method obviates the concern of holes in the etch resist or breaks in the resist over holes, which can result in all the metal being etched out of the through hole. Like the case above, a very thin copper foil ideally should be used to start to make etching easier and more accurate.

5.12.3 Pattern Plate

Pattern plating is a very common method used in circuit processing and most often follows the process steps described in Sec. 5.9. When a thin copper foil is used to start, a semi-additive process can be used. This allows the processor to avoid having to plate and later remove and etch resist metal layer. The process steps for pattern plating using the semi-additive method are shown in Fig. 5.18.

5.12.4 Printed-Through Hole

An unusual method for making connection from side to side on a PCB is to use conductive inks printed through the holes following a double-sided etch process. The ink is normally screened or stenciled on while the plane is placed on a vacuum table, which draws the ink down through the holes. The process is normally repeated from the other side to ensure full coating.

There are numerous ways to fabricate PCBs, and there are also many different possible constructions. Shown in Fig. 5.20 are a sam-

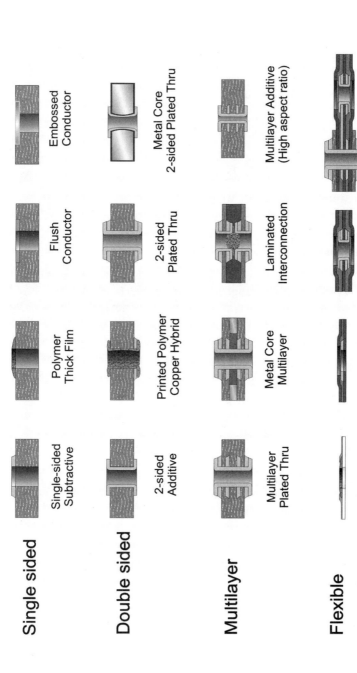

Figure 5.20 Examples of various printed wiring constructions.

pling of some of the constructions used. (Please note that the cross sections are not all drawn to a common scale.)

5.13 Standard Multilayer Circuit Process Example

There are many ways to produce a multilayer printed circuit, but only a few methods see any major use, and these are the products of many years of experiments and experience. The following are brief discussions of some of the key process steps in the creation of a multilayer printed circuit.

5.13.1 Inner Layer Image Requirements

In the construction of a multilayer circuit, it is necessary to carefully control every step of the manufacturing process. Thus, the images that are to be generated for the circuit patterns should adequately address the needs of production. This is manifest in several ways, depending on the complexity of the design, but it normally includes (at a minimum) the retention of copper around the circuit to improve dimensional stability and an adjustment of the circuit trace widths to compensate for the effects of the etching process. For contact exposure, a negative image is normally used (clear traces and opaque spaces), as negative acting resists are most common.

5.13.2 Inner Layer Material Preparation

Unless materials are purchased precut, the first step in the manufacture of a multilayer is to cut laminate sheets to the appropriate panel size. Normally, this is a size acceptable by all of the following processes. Depending on how the material is cut, the inner layer material (also commonly called *core* material) may at this time have the edges cleaned to minimize the potential for fragments or particles of glass or epoxy from causing imaging defects.

5.13.3 Tooling Hole Generation

Tooling holes are a vital part of multilayer manufacture. They are the key element of the registration system that will be required to successfully produce the multilayer circuits. Highly accurate tooling hole punching systems have been developed that allow all the tooling holes to be punched at one time. While the minimum number of tooling holes, in theory, is two, most tooling systems punch four holes. These holes are most often at the centers of the laminate sides, near the edge.

This approach has been developed with the intent of making the center of the panel the 0/0 point, minimizing runout in any on direction.

5.13.4 Resist coating

The resist coating process is similar to that described earlier, except that screen printing is rarely if ever used for such purposes, because of its accuracy limitations. The panels must be clean and free of any foreign materials that might be impervious to or alter the etching rate. Roll coating of dry film resist is very common at present; however, other methods based on electrophoretic deposition of resists are proving popular for very fine-line circuits.

5.13.5 Exposure methods

There are several methods that can be used to expose the resist-coated laminate. Contact printing remains popular at present, wherein the coated core material is mated to the film containing the correct circuit pattern on top and bottom, using the tooling pin system, and then exposed to UV light. While this method is still most widely used, off-contact and laser direct write methods are showing some promise.

5.13.6 Development

Development methods are identical to those described earlier. However, because the laminates are very thin, special care in handling is required.

5.13.7 Etching

The etching process is a critical step in the processing of a printed circuit inner layer. This is especially true in the case of circuits being designed for controlled-impedance applications. The processes used are the same as those described earlier in this chapter.

5.13.8 Resist Stripping

The resist stripping process step follows the etching step. The process is very simple, with the only major concern being that the stripping be complete and the panels be adequately rinsed.

5.13.9 Copper Circuit Surface Preparation

The surfaces of the copper circuits on the laminate are normally provided with an oxide treatment to improve the adhesion of the resin to the circuits. A number of different oxide treatments are available. Black oxide (cupric) was very common in the past, but newer methods have been developed that leave the exposed copper with a brown oxide (cuprous) finish that has some improved characteristics. The use of a so-called *double treatment* copper foil can obviate the need for such a process and is favored by some manufacturers.

5.13.10 Lay-Up for Lamination

Circuit lay-up for lamination is accomplished by laying down circuit layers in an ordered fashion on a carrier plate of steel that shares the same tooling hole layout as the punch. Plies of prepreg are placed between each of the inner layer cores until the stack is complete. The type of prepreg, in terms of cloth type and resin formulation, is prescribed by the circuit design needs. Normally, two plies or sheets of the chosen prepreg are used between each of the layer pairs. A number of multilayer circuit panels can be laminated at one time. To facilitate handling of copper foil when cap foil lamination is desired, it is possible to purchase the thin copper foil preattached to a thicker metal carrier. This method obviates the need for separator plates between multilayer circuits in the stack. Figure 5.21 provides an example of a possible circuit lay-up.

5.13.11 Lamination Methods

A number of different approaches to multilayer lamination have been developed over the years. Traditional lamination methods are still suitable for many products, but advanced high-density circuits can often be better served by other methods. One important method is to

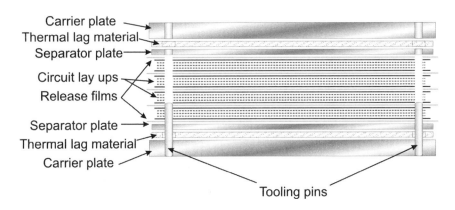

Figure 5.21 Example of a possible lay up for multilayer lamination.

augment the lamination process by the application of a vacuum during lamination. This approach help to expel air from the laminate and prevents its entrapment during lamination. Entrapped air can cause problems in processing and weaken the laminate. To assist in air removal, methods have been developed wherein lamination takes place in a vacuum. Other nonstandard methods have also been developed such as vacuum autoclave lamination. In vacuum autoclave lamination, the circuits are laminated by placing them in high-temperature bagging material, drawing a vacuum on the bag to expel the air, and then placing the bagged circuits into a pressure vessel. Pressure is applied by adding CO_2 gas and heating the gas to lamination temperature by means of a convention oven inside the pressure vessel. This method is very popular with flex circuit manufacturers.

Beyond the lamination processes described earlier, there is one other that is quite unusual. In lay-up, a double sheet of copper foil is rolled back and forth between the individual circuit panels, forming a continuous dual foil sheet. This assembly is then placed in a vacuum lamination chamber, and pressure is applied. Heat is generated by passing high electric DC current through the copper foil, which causes it to heat up to temperatures at which the prepreg resin can flow, fill and encapsulate circuit features, and cure.

Following lamination, the panels are normally trimmed to remove resin flash at the edges of the panel. Again, it is advisable that the edges be dressed so that they do not cause problems in later steps such as imaging of the outer layers.

5.13.12 Drilling

Accurate drilling of a multilayer circuit is integrally linked to the tooling system. The tooling holes serve as a collective datum for the drilling machine. As such, the tooling holes must accurately link to the circuit patterns that were imaged, etched, and laminated to become internal elements of the PCB and invisible to the unaided eye.

The process of drilling must be very well controlled to ensure its quality. It is common to reduce the drill stack height for multilayer circuits to ensure drilled hole quality. This is especially true when very small holes are being drilled or when design features are especially small.

5.13.13 Hole Cleaning and Etchback

Following drilling, it is fairly common to perform some sort of cleaning process to remove any resin that might have smeared over the surface of the copper inner layers. Failure to remove such resin smear could degrade the reliability of the plated-through hole interconnection to the inner layer, or it could completely block the interconnection, depending on the severity of the smearing. Some customers request etchback of the resin from the inner layer lands. Doing so creates a so-called *three-point interconnection*. This is commonly referred to as positive etchback and is normally deemed to be a more reliable interconnection. In contrast, there are some board users who suggest that negative etchback has an advantage. Here, the copper is etched back slightly and provides easily verifiable evidence of resin removal, because it would otherwise interfere with the copper micro etching process used as a part of the hole plating process. Figure 5.22 illustrates the difference between the two methods.

5.13.14 Subsequent processes for Multilayer PCB Manufacture

The process steps following hole cleaning are essentially the same as those used for double-sided processing and can be used for reference as benchmark processes.

5.14 Mass Lamination

Mass lamination is a term used to describe a technology wherein the laminate manufacturer fabricates multilayer panels in full sheet form, normally as a service to printed circuit manufacturers who either do not have lamination technology in house or whose lamination manufacturing capacity is limited.

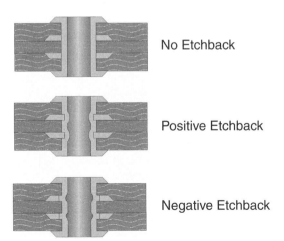

Figure 5.22 Etchback can be either positive or negative, as illustrated above.

The method is generally limited to use with relatively simple multilayers and is most often used for four-layer multilayer circuits; however, higher layer counts have been manufactured using modified approaches. Mass lamination differs from traditional lamination in that it is pinless. The inner layer cores are produced by accurately registering the top an bottom films of the inner core material (normally ground and power layers, which are not very complex) to the resist-coated panels and then exposing, developing, and etching them in the same fashion as used for normal multilayer cores. Targets are etched into the copper foil in the locations where the tooling holes normally would have been punched and after lamination. The targets are accessed by either carefully milling through the copper and laminate above to expose them for accurate drilling or by means of an x-ray assisted drill or punch. Once the tooling holes are drilled, the panels can be accurately registered of subsequent drilling of normal plated-through holes. The general process steps are illustrated in Fig. 5.23.

5.15 Metal-Core Printed Circuit Boards

Metal-core PCBs are a special subset of printed circuits. The name is adequately self-descriptive, and the purpose of these structures is normally to facilitate the rapid and efficient removal of heat from the board. These constructions are most desirable for use with products that will be generating large amounts of heat themselves or in conditions such as aerospace, where convection cooling is of limited potential or value. There are a number of potential constructions of these boards; however, if one is to keep true to the definition provided, there are only two general forms of metal-core PCBs: double-sided and multilayer. The following are descriptions of basic metal-core structures.

5.15.1 Double-Sided Metal-Core Structures

Double-sided metal core boards are the simplest form of the product. The fundamental structure consists of a piece of metal with holes, over which is a coat of insulation topped with metal circuit patterns. The insulation material can vary. Porcelain, baked, enamel and various organic coatings can be used. The choice is a function of the applications and the amount of heat being generated and dissipated. Figure 5.18 provides an example of a simple structure.

5.15.2 Multilayer Metal-Core Structures

Multilayer metal core structures are also found in a variety of formats. The most common again follows the precepts of the basic definition of

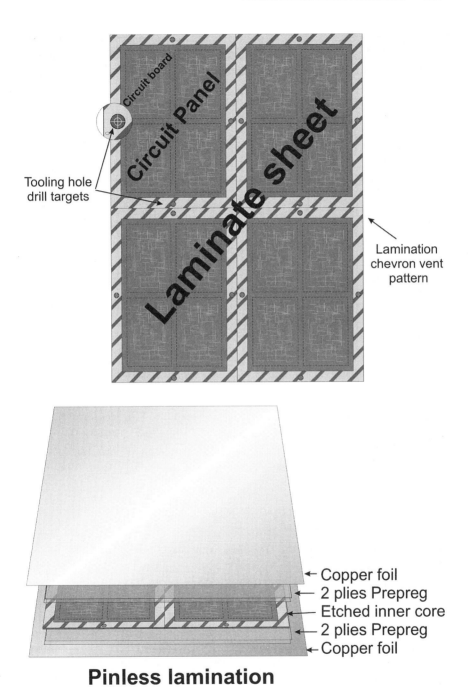

Figure 5.23 Mass lamination methods are used to laminate full sheets of circuit panels. After lamination the sheet is cut into panels and the tooling holes are drilled using the etched targets to assure proper placement.

a metal core, which consists of a multilayer circuit fabricated on either side of a central metal core. However, certain other constructions are included under the umbrella of metal core. These would include constructions where copper-clad invar is used to control the in-plane expansion of the board and where they might also serve as power and ground planes. An example of a basic metal core board is illustrated in Fig. 5.18.

5.16 Flexible Circuits

Flexible circuits are a unique and very important form of printed circuit. The IPC standards document IPC-T-50, "Terms and Definitions for Printed Boards," defines the as:

> A patterned arrangement of printed wiring utilizing flexible base material with or without flexible cover layers.

This is an accurate but somewhat limited definition, as it fails to consider fully how the technology can be applied. For example, flexible circuits can be used statically, in a "flex-to-fit" fashion, or they can be dynamically flexed either intermittently, such as when used in hinge applications, or nearly continuously, as when used in disk drive applications. Flexible circuits are most fundamentally a three-dimensional interconnection technology. Special design practices are required to create reliable flexible circuits, and the potential user is advised to become familiar with flex circuit design information before embarking on a flex circuit design to avoid potential problems. There are several different types of flexible circuits. These are described in the following sections.

5.16.1 Single-Sided Circuits

Single-sided flex circuits are the most commonly produced members of the family. They are the lowest-cost variant and the type best suited to dynamic flexing applications. They can be produced using either etched metal or printed conductive inks to create the circuit patterns. The metal can be accessed from one or both sides if desired. In the latter case, the polymer base film is removed by a suitable means such as a laser. A special case variation involves selective etching of the copper to create a circuit that has copper of different thickness along the length of the circuit patterns, thin in areas to be flexed and thicker in areas where interconnection is desired. A cover layer is commonly applied to protect the circuits and to improve their longevity in dynamic flexing applications. See Fig. 5.24 for an example.

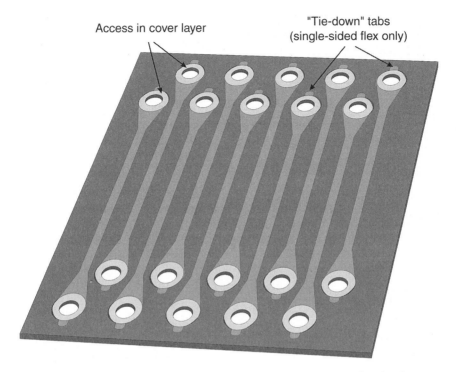

Figure 5.24 Example of the construction of a simple single metal layer flex circuit.

5.16.2 Double-Sided Circuits

Double-sided flex circuits are the second most common form of flex circuit and are used where higher-density interconnection is required. These flex circuits obviously have two metal layers, and they are normally interconnected by means of a plated-through hole, although this is not always necessary. For example, the so-called "Type 5" flex circuit is a non-plated-through hole flex in which both metal layers are accessed from the same side. See Fig. 5.25 for examples.

5.16.3 Multilayer Circuits

Multilayer flex circuits are used in applications where higher-density interconnections are required. They are similar to their rigid counterparts but are often more complex and very engineering intensive. Very fanciful looking but highly practical multilayer flex circuits have been fabricated to solve complex electrical and electronic interconnection problems. It is possible to build multilayer flex circuits with tentacles having varying numbers of layers broken out from a central section. See Fig. 5.26 for an example.

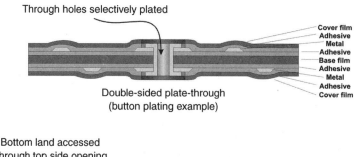

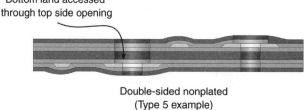

Figure 5.25 Two metal-layer flex circuits can be fabricated in one of the two basic ways illustrated above.

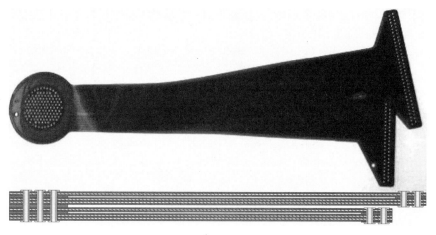

Figure 5.26 A deceptively simple-looking multilayer flex circuit with a complex construction, as can be seen in cross section. Single- and doubled-sided flex circuits are prefabricated, joined, and plated through, and those structures are joined and plated through again to create the finished circuit.

5.16.4 Rigid Flex

Rigid flex circuits are the final variation of flex circuits. As the name suggests, they are a hybrid construction that brings together construction elements of both rigid and flexible circuits. Most often, the rigid portions of the rigid-flex circuit are used to support connectors and/or components, and the flexible portions are used to interconnect the rigid sections. The technique won the favor of military product designers in the 1970s as a means of creating higher-reliability and lighter interconnection structures than the wire harness alternatives they often replaced. Like multilayer flex circuits, these circuits are frequently highly engineered and require a great deal of understanding of the manufacturing process for them to be designed properly. Figure 5.27 provides an example of a rigid flex construction.

5.17 High-Density Interconnection (HDI) Structures

Semiconductor manufacturing technology is the driving force behind the electronics industry. Semiconductor integrated circuit features continue to be reduced in size to meet the demand for more function in lesser amounts of space. This in turn has resulted in the creation of devices that operate at higher speeds and with greater efficiency with each new product generation. However, as more recent generations of IC devices pressed the limits of older traditional IC packaging technologies, new approaches had to be developed. The early 1990s saw a number of different attempts to create substrates to meet this challenge and provide solutions. These products were generally referred to

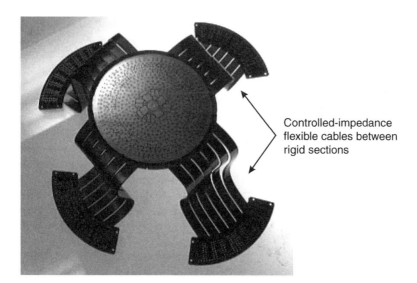

Figure 5.27 An example of a rigid flex circuit with integral controlled impedance sections between rigid elements of the design.

multichip modules, or MCMs. The substrates used for these structures were really the first of the HDI substrates, and they drew heavily from the technologies used to create ICs. They were, however, very expensive, and the final products were vexed by their inability to obtain reliable sources of known good die (KGD), without which their yields were low enough to be viewed as impractical (see Fig. 5.28).

About the same time, there was a surge of interest in a new IC packaging technology based on the used of printed circuit technology. The new approach, based on area array interconnections, was a response to the rising pin counts and the inability of industry to obtain good yields with the more traditional peripherally leaded packages. These devices are known now as ball grid arrays (BGAs). While the early structures were rather simple (beyond the need for relatively small holes) and presented little challenge to PCB manufacturers, as chip I/O counts increased, there was a need for additional layers of circuitry. The additional circuit layers were required to help redistribute the circuitry to a pitch more suitable to the needs of next-level PCB design, manufacture, and assembly. HDI concepts explored for MCMs were reexamined for use with these new packages but using advanced PCB technologies in place of the semiconductor technologies that were dominant in the earlier effort. This became a proving ground of sorts for a number of HDI manufacturing concepts.

5.17.1 HDI Substrate Construction Types

Presently, there are a significant number of different processes that have been proposed and/or developed for manufacturing high-density interconnection substrates. The IPC has attempted to bring some order to the matter by identifying and classifying, into general types, the various HDI PCBs that have been described in literature. Thus far, six general types of HDI structures have been identified. These are described in the following sections.

5.17.1.1 Type 1 Construction. The Type 1 HDI structure is typified by a circuit having a rigid core, which could have multiple circuit layers, on which microvia buildup layers and through holes are plated simultaneously. Normally, the microvia layers can be plated up on one or both sides of the circuit core (see Fig. 5.29).

5.17.1.2 Type 2 Construction. The Type 2 HDI structure is typified by a circuit having a rigid core which could have multiple circuit layers and which has holes plated through with copper. These holes are filled

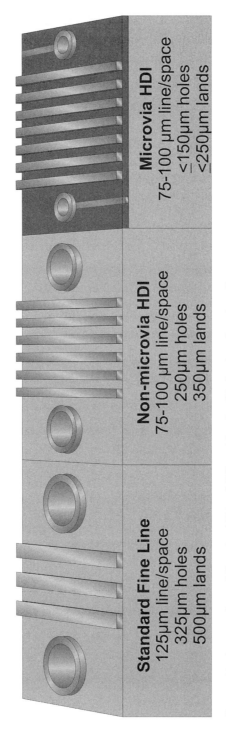

Figure 5.28 A comparison of the wireability of different levels of technology based on line width, space, hole diameter and land size.

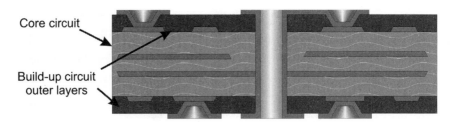

Figure 5.29 HDI Construction Type 1 (see text for full description).

with resin before further processing and thus become blind vias (or possibly semi-blind vias) upon completion of the fabrication process. Completion is effected by plating microvia build up layers on either one or both sides of the circuit core (see Fig. 5.30).

5.17.1.3 Type 3 Construction. The Type 3 HDI structure is typified by a circuit having a rigid core with buried vias (as in Type 2) and one or more microvia buildup layers on one side and two or more on the second side. These structures also have plated-through vias, which make direct connection from side to side (see Fig. 5.31).

5.17.1.4 Type 4 Construction. The Type 4 HDI structure is typified by a circuit having a rigid insulating or metal-core substrate with two or more buildup layers on each side. It also has plated-through vias connecting the two sides of the PCB (see Fig. 5.32).

5.17.1.5 Type 5 Construction. The Type 5 HDI structure is typified by co-laminated circuit structures with circuit layers interconnected during lamination to make vertical interconnection using conductive pastes or alloys. There are several variations on this basic theme (see Fig. 5.33).

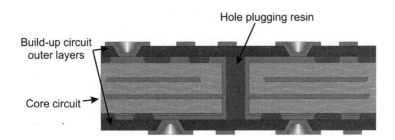

Figure 5.30 HDI Construction Type 2 (see text for full description).

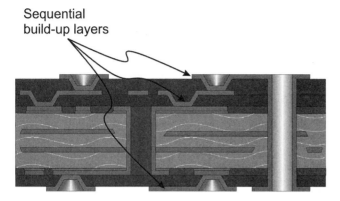

Figure 5.31 HDI Construction Type 3 (see text for full description).

Figure 5.32 HDI Construction Type 4 (see text for full description).

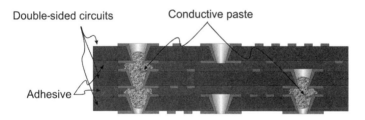

Figure 5.33 HDI Construction Type 5 (see text for full description).

5.17.1.6 Type 6 Construction. The Type 6 HDI structure is typified by circuits fabricated using the insulation-piercing features of integral metal or stenciled conductive polymer, which make interconnection during a lamination process (see Fig. 5.34).

5.17.2 Build-Up Board Example

The concept of the organic laminate build-up board was likely inspired by the technology used to fabricate hybrid circuits. Hybrid circuits are routinely fabricated by sequentially layering circuits and insulating

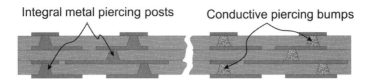

Figure 5.34 HDI Construction Type 6 (see text for full description).

materials to create the final product. However, hybrids generally employ inorganic materials in their construction and use a different set of manufacturing processes and equipment. The major attraction was hybrid circuit technology's small via structures that offered better routing space potential and improved electrical/electronic performance. As was illustrated in the foregoing discussion of HDI types, there are numerous variations on the basic HDI concept. Most of those are build-up type structures. The flow diagram in Fig. 5.35 provides a general illustration of the basic build-up board process.

5.17.3 Co-laminated Structure Example

Co-laminated HDI circuits also are likely to have taken inspiration from hybrid circuit manufacture for many of the same reasons; how-

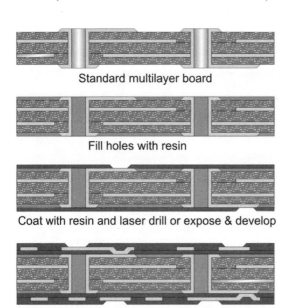

Figure 5.35 Simplified example of the basic process steps in manufacture of a build-up board.

ever, there are some significant differences as well. Co-laminated HDI substrates offer a two important advantages.

1. They do not require the plating of high-aspect-ratio holes.
2. The layers can be individually tested and yielded before lamination.

IBM, Tessera, and CTS Corporation have all described such methods. An example of one such process is provided in Fig. 5.36.

5.17.4 Sequential Laminated Structure Example

This approach to manufacture has been employed by at least two major Japanese companies, Matsushita and Toshiba. While there are significant differences between the two approaches, the final products are comparable. Toshiba uses an insulation piercing conductive bump to make connection between layers of copper foil at predetermined points during lamination. In contrast, Matsushita pre-punches or pre-drills the insulation layer and fills the holes with conductive paste before laminating on the copper foils. Packard-Hughes of Irvine, CA has described in a patent a concept similar to that used by Toshiba but they use integrally plated copper bumps to pierce the insulating layer during lamination. The flow diagrams for these processes can be seen in Fig. 5.37.

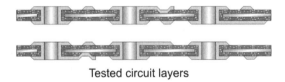
Tested circuit layers

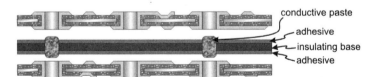
Align layers with programmed joining layer

Layers interconnected and joined in single step

Figure 5.36 Simplified example of a co-laminated HDI structure.

Insulation piercing process

Stencil conductive adhesive bumps on copper foil

Lay-up for lamination with bond ply material

Laminate, piercing bond ply with conductive bumps

Image and etch circuit pattern over conductive bumps

Repeat process steps as required to complete design

Conductive paste process

Drill or punch holes in partially cured laminate

Stencil conductive paste into holes

Laminate with copper foil curing resin & making connection

Image and etch circuit pattern over conductive posts

Repeat process steps as required to complete design

Figure 5.37 A number of different possible methods exist for manufacturing sequentially co-laminated structures.

5.18 Future Directions

The future of printed circuit manufacture is indefinite except for the fact that one can state, with some reasonable confidence, that there will always be a need for electronic interconnection substrates of some sort. New schemes of manufacturing printed circuits are in development at this time that could radically alter the traditional approaches to manufacture. For example, one company in England (TDAO, Ltd.) is known to be working on a technology that would allow for the direct imaging and plating of flexible circuits in a continuous fashion. This concept would bring the industry closer to realizing the dream of making circuits directly from a computer without the need for film or resist. Another example is a concept for creating metal core laminates with predrilled and metallized holes that can serve as the raw materials for structures such as represented by the circuits shown in Fig. 5.36.

While no one can predict the future with absolute certainty, the one thing that seems relatively certain in printed circuit manufacture is that the circuit traces and spaces and the holes that connect them will continue to get smaller with time. Of these trends, the one that is likely to be most important is the ability to make smaller holes. Small holes are the key to improved circuit routing.

5.19 Summary

Printed circuits have had a long and colorful history, and they remain one of the most important elements in electronic manufacturing. There are numerous types of printed circuits and numerous approaches to their manufacture. With the ever expanding electronics market and the myriad of products that are being produced, there will likely be a continuing increase in the number of different types of electronic interconnection substrates and printed circuits in the years to come.

5.20 References

1. IPC National Electronics Roadmap for the Year 2000.
2. Coombs, C. *The Handbook of Printed Circuits,* McGraw-Hill, New York, 1994.
3. Fjelstad, J. "Fundamentals of Printed Circuits," IPC Short Course Notes, 1993–2001.
4. Gilleo, K. "The History of the Printed Circuit," *PC FAB* vol. 22, no. 1, January 1999.
5. Gilleo, K., and Murray, J. "The Rise and Fall of the Father of the PCB," *PC FAB* vol. 22, no. 3, March 1999.
6. Fjelstad, J. "A Brief Overview of Microvia Interconnection Technologies," *On Board Technology,* September 2000.
7. Harper, C., *Electronic Packaging Handbook*, McGraw-Hill, New York, 2000.
8. Landis, J. "Shopping for High-Density PCBs," *Electronic Packaging and Production,* March 2000.

9. Singer, A "Microvia Technology: The Key to Using CSPs," *ChipScale Review,* January-February 1999.
10. Fjelstad, J., and DiStefano, T. "Where Are We Headed?" *Electronic Packaging and Production,* December 1999.
11. Holden, H. "Microvia Printed Circuit Boards: Next Generation of Substrates and Packages," *IEEE Micro,* vol. 18, July/August 1998.
12. Mahidhara, R. "Substrate Requirements for Chip-Scale Packaging" *ChipScale Review,* January-February 1999.
13. Fjelstad, J. *An Engineers Guide to Flexible Circuits,* Electrochemical Publications, Isle of Man, UK, 1997.

Chapter

6

Package and Component Attachment and Interconnection

Charles G. Woychik
IBM Corporation
Endicott, New York

6.1 Introduction

The challenges that are confronted in electronic packaging had their origin in the post-World War II era of the computer industry. In the late 1950s, with the development of the transistor, novel methods were required to effectively package the first semiconductor device. In the early days of packaging technology at IBM, a special grade of paper was used as the dielectric core in the card. In the first card-like technology, the IBM SMS program required the use of a pin-in-hole (PIH) type of device to house the semiconductor device. This device was then attached to the card using PIH assembly methods. As the number of circuits increase in a semiconductor, this usually corresponds to an increase in the number of I/Os on the device, resulting in an increased density of interconnects. As was the case then, and as still applies today, the challenge in electronic packaging is to increase circuit density. In addition, another challenge in electronic packaging is to effectively accommodate new chip designs that possess increases in the speed and circuit density of the semiconductor device, which, combined with increased circuit densities of the chip carrier and card, will result in an improvement in the overall performance of computer systems.

In electronic packaging, the primary focus is on the physical "packaging" of the semiconductor devices to allow for effective communications. By properly packaging a device, the maximum performance of the semiconductor can be achieved. A sequence of four different engineering areas needs to be addressed to achieve an optimized electronic package. These are as follows:

1. *Electrical performance.* The design needs to ensure that the electrical performance of the system is achieved. This is of primary interest in designing an electronic package.

2. *Thermal performance.* The heat generated in the package must be effectively removed so as to ensure proper performance of the semiconductor. A major problem that the heat generated causes degradation of the integrity of the package, usually due to differences in the coefficient of thermal expansion (CTE) of adjacent interfaces. Therefore, the thermal performance not only ensures that the device is properly cooled but also minimizes mechanical degradation.

3. *Mechanical integrity.* As a result of the mechanical degradation due to CTE mismatches in the package, suitable designs need to be used to ensure maximum reliability of the package. In addition, external forces applied to the package (e.g., ship-shock, handling stresses during assembly, and other types of externally applied forces) need to be modeled.

4. *Materials science/engineering.* This is the only variable that can be used to optimized all three of the above aspects. Here, the use of materials can used to optimize the electrical, thermal, and mechanical aspects of the package. This is really the independent variable available to accomplish this task.

In packaging technology, two main areas are of special interest. The first is related to the semiconductor and the surrounding package, commonly referred to as the *component*. The next area of interest is the card to which the component is attached. The interconnection technology used internally in the component, and that used to attach the component to the card, will affect the resulting component density and thus functionality of the system.

6.2 Levels of Interconnections

A schematic illustration of a surface mount component attached to a card is shown in Fig. 6.1. In this figure, one can identify the first-level interconnection, which is the interconnection between the semiconductor and the chip carrier. The second-level interconnection then corresponds to the interconnection between the component and the card.

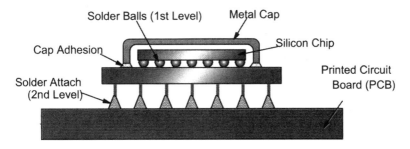

Figure 6.1 A schematic illustration of a surface mount type of component attach to a card.

There are many different methods to form both of these types of interconnections, and the details of all these will be presented in the following pages of this chapter.

The main function of a packaging interface is to provide an electrical interconnection between two levels of packaging. As the complexity of the semiconductor increases, so will the number of I/Os on the chip. In 1960, an IBM engineer E. Rent discovered a relationship between the number of I/O lines and the circuits in the cards of the IBM 1401 computer.[1] This relationship is commonly referred to as Rent's rule,

$$I = bC^p$$

where I = the number of I/O lines
 b = a constant
 C = the number of circuits on a package
 p = a positive exponent

In Rent's early work, p was found to be about 2/3, while b was about 2.[2] The number of I/Os on a chip has been increasing and continues to increase rapidly. The major challenge in packaging today is to be able to handle higher interconnect densities and continue to maintain package reliability requirements while meeting aggressive cost targets mandated by the market. This has not been as much of a challenge in the past as it is today, due to the more extensive consumer applications for electronics. As applications for electronics grow more numerous, so will the demand to provide low-cost packaging solutions.

6.3 Types of First-Level Interconnections

There are two basic types of first-level interconnections. A wire bond type of interconnection is still the most common method of interconnecting a chip to the carrier. Figure 6.2 shows an SEM image of a ball-

270 Chapter 6

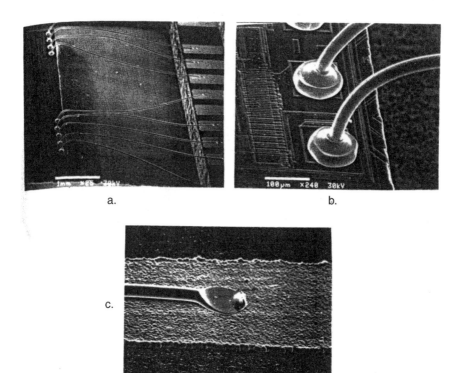

Figure 6.2 Scanning electron microscope of a commonly used ball-wedge type of wire bond.

wedge type of wire bond commonly used. For this type of bonding scheme, a gold (Au) wire is used, typically 1 mil in dia. A ball is formed at the tip of the wire, which is then bonded to an aluminum (Al) pad on the die. The wire is then fed through the collet on the wire bond tip, and a wedge type of bond is formed between the wire and the mating Au-plated pad on the chip carrier. An illustration of the process flow to produce this type of ball/wedge wire bond interconnection is shown in Fig. 6.3.

The next type of first-level interconnection is the flip chip. Figure 6.4 illustrates a flip chip type of solder interconnection. There are two types of flip chip attach. The first method uses a single solder alloy that completely melts and therefore allows the bumped die to self-align on the chip carrier when reflowed (see Fig. 6.4a). This type of single-metal system flip chip is most often referred to as a "controlled collapse chip connection," commonly called a C4 joint. C4 technology has been developed primarily for Pb-rich solders that have melting temperatures in excess of 300°C and are used on ceramic chip carriers.

Package and Component Attachment and Interconnection   271

Steps in performing a nail-head bond.

(a) Initial positioning of work;
(b) deformation of ball to form nail-head bond;
(c) positioning of capillary for second bond;
(d) formation of wedge bond with capillary edge;
(e) withdrawal of capillary from wedge bond;
(f) flame cutting of wire leaving ball.

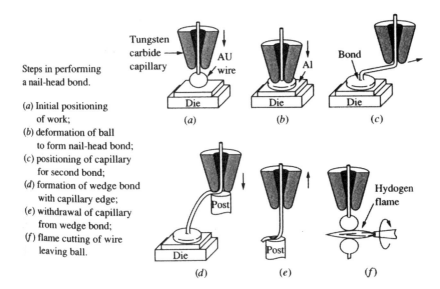

Figure 6.3 Process flow to produce a ball-wedge type of wire bond interconnection using Au wire.

a) Single-Metal FCA

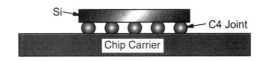

b) Dual-Metal FCA

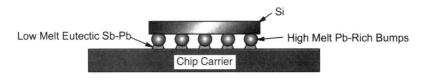

Figure 6.4 An illustration of a flip chip attach (FCA) type of first-level interconnection. (a) A completely collapsible type of low-melt FCA using a single alloy of eutectic Sn-Pb solder. (b) Dual-metal FCA. Here, a Pb-rich bump is attached to the chip carrier using low-melt eutectic Sn-Pb solder.

Low-melt C4 solder joints are formed using a eutectic Sn-Pb alloy that is commonly used for chip attach on organic chip carriers. The second type of flip chip configuration is a dual-metal system (see Fig. 6.4b). In this type of configuration, a high-melting-point bump on the die is attached to the chip carrier using a lower-melting-point solder. Typically, in the dual-metal system, a standard high-melt Pb-rich bump is deposited on the die and in turn attached to a chip carrier using eutectic Sn-Pb solder. This dual-metal system was the most common type of low-melt interconnection for organic chip carriers.

6.4 Types of Second-Level Interconnections

For second-level interconnections, there are two basic type of configurations. The first type is the standard pin-in-hole (PIH) type of solder joint, as shown in Fig. 6.5a. Here, a pin from the component is attached into a plated-through hole (PTH) in the circuit card using a eutectic Sn-Pb solder. The second type of interconnection is effected when the component is attached to a mounting pad on the surface of the card, as shown in Fig. 6.5b, which is commonly called surface mount technology (SMT).

6.4.1 Pin-In-Hole (PIH) Technology

The early semiconductor devices packaged a single-transistor device using a T018 type of transistor can. In this type of configuration, a semiconductor device is mounted on a ceramic substrate, and the device I/Os are attached to the substrate using a wire bond attach method. A can is then packaged over the ceramic device, and leads

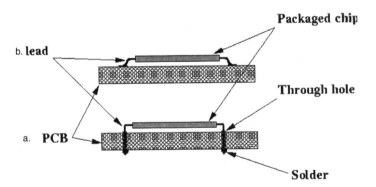

Figure 6.5 An illustration of the two types of second level interconnections. (a) A conventional pin-in-hole (PIH) type of second-level solder interconnection. (b) A surface mount technology (SMT) type of second-level solder interconnection.

then protrude from the bottom of the can device. These leads are then soldered into a PTH-type of printed circuit board (PCB). A PIH-type of solder joint is known to be very reliable, due to the large amount of contact area of the pin and the PTH with the solder to join both surfaces. As is apparent from this type of interconnection scheme, components can be attached only on a single side of the card. This severely limits the component density that can be achieved.

To achieve a reliable PIH-type of solder interconnection, both the surfaces of the pin and the PTH of the card must be highly solderable. This will allow for good wetting of the molten solder alloy with both surfaces to produce the necessary intermetallic phases at the interfaces to ensure good solder joint integrity. The region at the top of the solder fillet is where early cracking is known to occur.[3] Also, it should be mentioned that, for this type of interconnection, integrity of the signal and ground planes that intersect with the PTH in the card can also be a source of reliability problems. If the adhesion of the top and bottom planes with the PTH is degraded, the plane can separate from the barrel and result in "interplane separation." This is a common type of reliability problem when poor-quality circuit boards are used with a PIH-type of solder interconnection.

6.4.2 Surface Mount Technology (SMT)

For SMT, the component is attached to a mounting pad on the card. The major feature of using this interconnection approach is that components can be attached to both sides of the card. This capability can nearly double the component density and offers an dramatic improvement in performance. However, the major challenge when implementing a double-sided SMT was to develop an assembly process that can produce similar solder joint configurations for both the top- and bottom-side solder interconnections. This will ensure that the individual components on both sides have similar solder interconnection reliability.

To achieve a double-sided assembly, many different assembly methods were initially explored. As a result of this work, a double-sided double-pass (DSDP) assembly process was developed to populate SMT components on both sides of a card. The highlights of this process are as follows. First, we screen solder paste on the back side of the card, and the components are placed and then reflowed. The top side of the card is then screened, again followed by component placement and reflow. The only differences in the interconnection configuration is that, typically, the bottom-side solder joints can become slightly elongated as a result of the second reflow in the inverted configuration. This occurrence is more common when using larger-mass components. For small and light components, essentially no differences can be detected in the solder joint standoff on both sides. However, when the back-side solder joints are slightly extended, this is known to extend the reliability of the interconnection. However, for very heavy components (most often, large ceramic components), the surface tension of the molten solder may not be enough to hold the component in place during the second inverted reflow. This case will require a fixture to keep the component in place.

There are two basic types of surface mount components. The first type is a leaded component, as shown in Fig. 6.6. Here, both gull-wing and J-leaded configurations of the lead are commonly used. The gull-wing configuration is very susceptible to bending during transport and handling (see Fig. 6.6a). The J-leaded configuration is much more robust against handling damage and offers the flexible bending features of the gull wing (see Fig. 6.6b). A feature of the gull-wing lead configuration is that it can be easily inspected after soldering. The second type of surface mount configuration is the ball grid array (BGA) type of interconnection, shown in Fig. 6.7. Here, a solder ball is used to form the interconnection between the component and the card. Similar to the single- and dual-metal system for flip chip attach, a higher-melting-point, Pb-rich solder ball is typically used to maintain the standoff when using a heavier type of ceramic chip carriers (see Fig. 6.7a). When using lighter organic chip carriers, a single low-melt eutectic Sn-Pb solder ball that fully collapses during reflow is most often used (see Fig. 6.7b).

6.4.3 Ceramic vs. Organic Packages

Ceramic packages have been in the marketplace for many years and are well known for their high reliability. Many high-end flip chip ap-

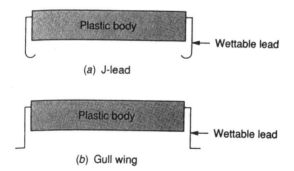

Figure 6.6 An illustration of two major types of leaded surface mount components. (a) A standard gull-wing type of surface mount component. (b) A J-leaded type of surface mount component.

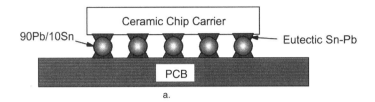

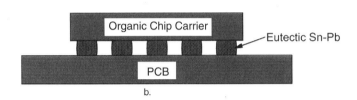

Figure 6.7 An illustration of two types of BGA components. (a) A dual-metal solder joint that uses a Pb-rich solder ball to maintain the appropriate standoff and eutectic solder to attach the ball to the component as well as to form the second-level interconnection. (b) A single-metal system that uses a low-melt eutectic Sn-Pb composition.

plications use ceramic chip carriers. However, there continues to develop a new market demand for organic chip carriers, especially for flip chip applications, which will be discussed in more detail later in this chapter. A simple schematic illustration of a flip chip configuration for both ceramic and organic packages is shown in Fig. 6.4. Most often, a fully collapsible high-melt C4 solder interconnection is used for a ceramic FCA interconnection (see Fig. 6.4a). Also, for an organic FCA interconnection, a dual-metal system consisting of a high-melt Pb-rich bump attached to the organic chip carrier using a low-melt eutectic Sn-Pb solder (see Fig. 6.4b).

In general, a ceramic flip chip package will have very high first-level reliability due to a relatively small change in CTE between the silicon and the chip carriers, but the second-level interconnection will have a lower reliability due to a large CTE mismatch between the ceramic carrier and the card. In the case of an organic flip chip package, the CTE between the die and the organic chip carrier is large. This creates tremendous challenges for the manufacturer of the component. Two major technical problems associated with the reliability confronting the high first-level CTE mismatch of an organic type of chip carrier are die cracking and underfill delamination. Today, most die are processed to eliminate any backside and dicing defects, which have dramatically reduced the occurrence for any type of die cracking after underfill cure and subsequent thermal exposure of the organic package. Finally, the second-level interconnection for the organic flip chip package is very high, which is due to the low CTE mismatch between the organic chip carrier and the card.

For a 32-mm and smaller ceramic chip carriers, one can use a noncollapsible Pb-rich solder ball to maintain the standoff. These packages are referred to as ceramic ball grid array (CBGA) packages and are shown in Fig. 6.8. Packages that are larger than 32 mm use a Pb-rich solder column; these packages are called ceramic column grid array (CCGA) packages and are shown in Fig. 6.9. The solder columns offer a higher standoff, which will reduce the shear strain in solder joints. A problem in using solder columns is that they are very difficult to handle due to the inherent softness of the Pb-rich solder alloy used in the column. Two major solderability problems arise when using solder columns: *stress fracture solder joints* and *solder grapes*. Stress fracture opens are a result of the excessive temperature gradients in the card and the weakness of the solder joint at solidification. These produce a clean 1-mil gap between the solder and the intermetallic phase on the card pad. This type of defect is very difficult to find, even with advanced tomography inspection methods. Figure 6.10 shows a micrograph with this type of failure.

Figure 6.8 A photograph of a 32-mm CBGA package that uses a Pb-rich solder ball attached to the ceramic chip carrier using eutectic Sn-Pb solder.

Figure 6.9 A photograph of a 42-mm CCGA package that uses a Pb-rich solder column to maintain the standoff between the ceramic chip carrier and the card.

Another type of solder defect is when the solder coalesces at the tip of the CCGA column but does not wet to mating card pad. This type of defect is referred to as a *solder grape*. This produces an open solder joint, which is easily detected by x-ray inspection methods. These two problems are eliminated when the organic package is used.

6.4.4 BGA Selected as the Strategic Packaging Technology

As is apparent from the previous discussion, a BGA type of package is more robust for second-level assembly. The organic package is also known to have superior electrical and mechanical reliability as compared with the ceramic counterpart. In addition, the organic package can be delivered at a lower cost than the ceramic package. All of these attributes for organic packages make this design the preferred offering for not only flip chip but also wire bond organic packages. Future demands are projected to continue to increase for organic packages, with a declining demand for ceramic. Ceramic applications will most

Figure 6.10 Photomicrograph of a stress fracture solder joint on the second-level interconnection of a CCGA type of module. Photo courtesy of Kevin Knadle, IBM Endicott.

likely continue for low-volume, high-reliability applications that are not restricted by cost objectives.

As with any change, there is never just one factor that contributes to making a transition from one technology to another. Some of the major factors that drove the rapid conversion to BGA technology are as follows:

1. High second-level assembly yields as compared with leaded surface mount components.

2. Lower cost of BGA solder ball vs. sophisticated Ni/Au plating on a kovar pin, such as the pins used on pinned ceramic grid array modules.

3. Less susceptibility to handling damage vs. leaded components, especially gull-wing lead configurations.

4. The ability to accommodate a full area array I/O with pitches as low as 0.8 mm.

5. Superior electrical performance as a result of the ability to bring in power beneath the center of the die.

These factors have contributed to the rapid conversion of leaded SMT and PIH components to BGA technology. At the same time, one needs to realize that there are some detractors to this technology, which are listed below:

1. BGA components are much more difficult to rework, especially large body sizes with high-density I/O configurations. This problem was offset by the fact that the assembly yields are very high.
2. The second-level interconnection reliability is not as high as that of either the leaded SMT component or the PIH type of configuration. However, for the applications specified, the technology is able to meet the customer requirements.

6.5 Ceramic Packages

BGA technology was initiated by converting the standard pin grid array (PGA) ceramic technology that was soldered into a PIH type of solder joint in the late 1980s. At this time, a ceramic PGA module was a state-of-the-art package for most high-end applications. To improve the overall performance of the system, it was necessary to increase the component density, which could be accomplished only by migrating to SMT. The major problem to overcome in this decision to migrate the ceramic PGA to SMT was to define a second-level interconnection configuration that would be able to meet the reliability requirements. When a eutectic solder ball was used to form the interconnection between the ceramic chip carrier and the card, the reliability was substantially lower than with the PIH type of solder joint. The fully collapsible nature of the eutectic ball did not produce a high enough standoff to reduce the amount of shear strain in the solder interconnection. Therefore, it became necessary to use a Pb-rich solder ball that would not collapse upon reflow to ensure that a higher standoff could be maintain so as to improve the reliability of the solder joint. Figure 6.11 illustrates the resulting configuration of the 90Pb-10Sn solder ball that is attached to the ceramic chip carrier, as well as to the card, using eutectic Sn-Pb solder. This type of package is commonly referred to as ceramic ball grid array (CBGA).

Today, most of the CBGA packages are used for high-end applications. For these applications, performance and reliability are the primary application requirements. The CBGA package offers excellent thermal performance, as a result of the flip chip configuration, and especially when a metal cap is used to more effectively conduct heat from the die to the heat sink. As a result of the full area array BGA matrix, power can be brought in underneath the die to provide for improved electrical performance. These two factors alone make the CBGA package very attractive. In addition, the package can be designed to be totally hermetic, which would eliminate any concerns about migration or corrosion in the critical flip chip interconnection region. Lastly, the relatively low coefficient of thermal expansion (CTE)

Figure 6.11 A photomicrograph of a CBGA type of solder interconnection. Note the 90/10 solder ball that does not collapse during reflow. Eutectic Sn-Pb solder is used on both the top and bottom sides of the solder joint.

between the die and the ceramic chip carrier makes for a very reliable first-level solder interconnection. As is apparent, the CBGA package has many of the attributes for high-reliability and high-performance chip carrier applications.

A major detractor of a CBGA package is the high cost. Multilayered ceramic chip carriers are very complicated to manufacture and thus are costly. In addition, the reliability of the second-level interconnection is not as high as that of a PIH type of solder joint or a leaded SMT interconnection. For 32-mm and smaller ceramic chip carrier sizes, the 35-mil-dia 90Pb/10Sn solder ball is used to maintain the standoff. However, for body sizes greater than 32 mm, a 90Pb/10Sn solder column is used to increase the standoff and therefore make the package reliability acceptable. A photograph of a ceramic column grid array (CCGA) module, showing the doped Sn-Pb eutectic joint used at the top of the column/substrate region and the eutectic Sn-Pb solder used at the bottom/card region of the column structure, is shown in Fig. 6.12. Even though there will continue to be applications for ceramic packages, there is a strong drive toward organic chip carriers. This new demand continues to erode the future growth for ceramic chip carriers.

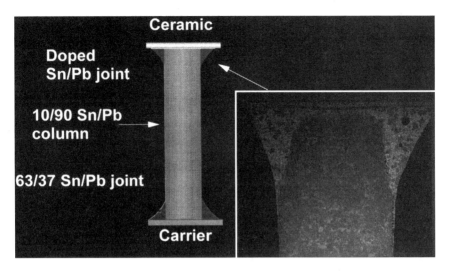

Figure 6.12 A photograph of a CCGA type of second-level solder interconnection.

6.6 Organic Packages

The trend toward organic chip carriers continues. This is mainly a result of the high electrical performance requirements of future high-end ASIC devices. Electrical modeling of these devices suggests an improvement in signal propagation as well as a reduction in noise effects. These electrical performance demands have established a new technology that requires a major effort to engineer a package that will remain intact through the various stages of module assembly, card assembly, and stress testing. Basically, the main technical issue is to maintain package integrity; i.e., no die cracking or delamination at any of the interfaces or, cohesively, in the internal structures of the underfill or the laminate chip carriers. Die cracking problems have become less of a problem in the past five years due to the improvement in dicing techniques, which have dramatically reduced the amount of defects on both the backside and side walls of the die. Delamination of the underfill to the die passivation, as well as delamination at other interfaces in the package, continue to be the greatest challenges today. Improvements in adhesion and maintaining adhesion when exposed to aggressive levels of preconditioning is the is the greatest issue involved in this technology.

The main obstacle in meeting this demand is that this is a very new technology, and only a few suppliers have been able to successfully deliver test results to show its success, particularly when using flip chip attach for the aggressive reliability requirements of high-end customers. The greatest challenge is to produce a reliable package that is manufacturable. Card assembly of this module is superior to conventional ceramic BGA packages. The advantages that the high-performance organic chip carriers have over CCGA are very significant, and this is a major driving factor for the conversion of high-performance organic chip carriers such as the IBM HyperBGA package, which is shown in Fig. 6.13.

The benefits of organic chip carriers are highlighted as follows:

1. *Improved electrical performance.* The signal speeds are improved with the finer metallizations and thinner cross sections. This design feature also greatly reduces the amount of noise.

Figure 6.13 A photograph of a HyperBGA package. This is a high-performance organic package.

2. *Outstanding second-level reliability.* There is essentially very little mismatch of the coefficient of thermal expansion (CTE) between the organic chip carrier and the card, which therefore translates to very low stresses generated in the second-level solder joint.

3. *Robust second-level assembly yields.* The fully collapsible BGA solder ball combined with the application of solder paste on the card contributes to a very high assembly yield. It is also thought that the organic chip carrier relaxes during reflow, due to the temperature of reflow exceeding the glass transition temperature T_g of the resin system.

4. *Lower cost.* The cost of manufacturing for an organic chip carrier is lower than that of the ceramic. This is because, for the organic packaging technology, one is using the existing manufacturing infrastructure to make these sophisticated chip carriers. However, it should be mentioned that the materials set to manufacture these new types of organic chip carriers are much more costly that conventional laminates used for cards and boards.

A major detractor of organic chip carriers is the process know-how required to manufacture the package to maintain package integrity. As cited earlier in this section, proprietary techniques are required to ensure that the interfacial adhesion is maintained throughout the package. As it stands today, this is the greatest challenge. Once the processes are defined to improve interfacial adhesion and have been confirmed by qualification (which shows that all of the interfaces can withstand the higher Pb-free soldering reflow temperatures along with the corresponding levels of JEDEC preconditioning moisture exposure), this technology will grow at even a faster rate than today.

6.7 Module-Level Assembly

In this section, there is a discussion of three different types of packaging configurations that represent the majority of packages offered today. They are (1) ceramic flip chip modules, (2) organic flip chip modules, and (3) organic wire bond modules. Both types of flip chip packages are most often used for high-performance applications, whereas the wire bond organic package is most often used for cost/performance types of applications. Of course, there are some high-performance applications that use fine-pitch wire bond configurations, in which wire bond pad pitches on the die are 60 microns, and the BGA pad pitch can be 0.8 mm. Therefore, one must realize that not all lower-end applications are wire bond but that, by utilizing a finer-diameter wire and tighter die pad pitches, a wire bond package can be designed to have both high performance and lower cost as compared with a flip chip design. One must realize that it is difficult to make general statements, since one must know the specific customer application conditions and how the package can best be designed to meet these requirements at the lowest possible cost. A general rule of thumb is that, if one can use wire bond and meet all performance requirements, the reliability and cost objectives will usually be met. There are still many applications today, and will continue in the future, that can use wire bond. A common mistake of an application engineer is to use flip chip attach (FCA) when it is not really required. FCA is still very costly and cannot compare with the proven low cost and high reliability of a wire bond package.

6.8 Ceramic Flip Chip Modules

In the late 1960s, IBM developed a flip chip package using a ceramic substrate. In this type of package configuration, a Pb-rich Sn-Pb solder bumped die is attached to the metallization on the ceramic chip carrier. Figure 6.14 illustrates a standard type of high-melt-solder ceramic FCA interconnection. Here, the 95Pb-5Sn (95/5) solder completely reflows to form a controlled collapse chip connection, commonly referred to as a C4. One of the major benefits of the C4 joint was that is was able to self align during reflow A major motivation for this type of design was that the ceramic chip carrier was able to withstand

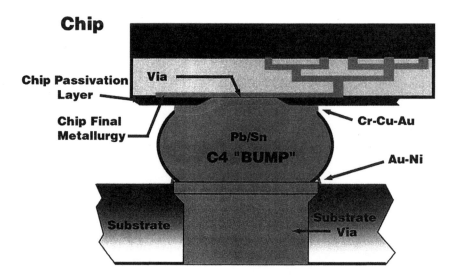

Figure 6.14 An illustration of a standard high-melt type of FCA interconnection used on a ceramic chip carrier.

higher reflow temperatures, which are about 320°C, when using a high-melt Sn-Pb solder. Therefore, the substrate would maintain its flatness during high-temperature exposure. As the die size continued to grow and the C4 pitch decreased, it was determined by engineers that a higher-Pb-content alloy with a composition of 97Pb-3Sn (97/3) would improve the mechanical properties of the C4 better than the earlier 95/5 alloy composition. Today, the 97/3 alloy continues to be the preferred alloy for a ceramic FCA interconnection. It should also be mentioned that another feature of an alumina type of ceramic chip carrier is that the CTE is about 6 ppm/°C. The CTE of Si is typically 2–3 ppm/°C, and thus the CTE mismatch between the die and the ceramic chip carrier is very low, and this is another contributor to the high reliability of the first-level FCA solder joint.

As the die continued to grow to greater than 15 mm, it became necessary to use an underfill to ensure that the C4 joint did not prematurely fail during thermal cycle stress testing. When an underfill is used, the configuration of the first-level interconnection changes. An illustration of a multilayered ceramic (MLC) package in which an underfill is used for a large die to ensure high reliability of the C4 joints is shown in Fig. 6.15. Here, the underfill can improve the package reliability substantially over an non-underfilled ceramic FCA design.[4] For

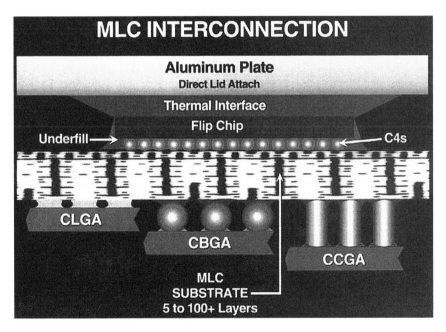

Figure 6.15 A schematic illustration of a ceramic FCA interconnection that uses an underfill to improve the reliability of the first-level solder interconnection.

high-end ASIC applications, most often the die exceeds 10 mm, and today most often an underfill is used to ensure high reliability of the C4 interconnection.[5]

The process flow to produce a ceramic flip chip package is outlined as follows:

1. The die is bumped with a 97/3 alloy.

2. Flux is applied to the surface of the C4 pads on the ceramic chip carrier.

3. The die is flipped and placed on the C4 sites.

4. The flipped die is then reflowed to form the first-level FCA interconnection.

5. The attached die is underfilled and cured.

6. A cap (either ceramic or metal) is placed over the die.

7. The BGA solder balls are attached to the back side of the module. Here, a 90Pb-10Sn (91/10) alloy is used along with a water soluble flux.

8. The water soluble flux is cleaned.

9. The package is electrically tested.

10. The fully assembled module is placed in JEDEC trays.

11. The modules in the JEDEC trays are baked.

12. The baked trays are sealed in metallized vacuum sealed bags and then stored.

An illustration of the final module having a metal cap is shown in Fig. 6.16. It should be mentioned that, when a ceramic substrate size exceeds 32 mm, a 90/10 solder column is used instead of a solder ball to meet second-level reliability requirements.

6.9 Organic Flip Chip Modules

A major difference when using an organic chip carrier, as compared with a ceramic chip carrier, is that a lower temperature is required during chip attach. The reflow temperatures used for a 97/3 solder will char and thus damage the organic chip carrier; therefore, the use of a lower melting point eutectic Sn-Pb solder alloy is required to attach the 97/3 bumped die to the chip carrier. The resulting type of dual-phase structure FCA interconnection is illustrated in Fig. 6.17. In this type of solder joint configuration, the 97/3 bumped die does not reflow;

Figure 6.16 A photograph of a ceramic package that employs a metal cap to provide enhanced thermal performance.

only the eutectic Sn-Pb solder melts to form the interconnection between the bumped die and the metallized pad on the organic chip carrier. In this package configuration, it is absolutely necessary to use an underfill to couple the die to the organic chip carrier and prevent any type of shear strain of the solder joint. As is apparent from this design, the strains generated in the FCA interconnection are very high due to the Si CTE of 2–3 ppm/°C as compared with the nominal CTE of 18 ppm/°C of the organic chip carrier. The underfilling of the FCA interconnection strongly couples the die to the chip carrier such that a bending mode is created in the chip carrier during thermal cycling. Moiré interferometry can be used to characterize the amount of bending during thermal cycling. Figure 6.18 shows a Moiré image of a flip chip organic package having a large die during a change in temperature from 90 to 20°C.

The process flow to produce dual-metal type of FCA on an organic flip chip package is outlined as follows:

1. The die is bumped with a 97/3 alloy.
2. Eutectic solder is applied to the chip carrier, reflowed, and coined.

Figure 6.17 A photomicrograph of a dual-metal type of low-melt FCA interconnection used on an organic chip carrier.

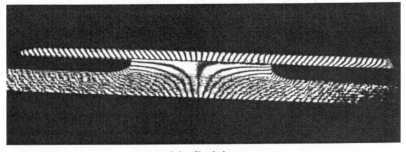

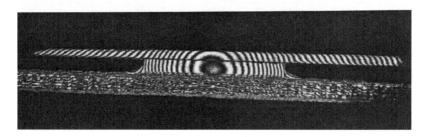

Figure 6.18 A Moiré image of a flip chip organic package during a single thermal ramp of 90 to 20°C. Image courtesy of Krishna Darbha, IBM Endicott.

3. Flux is applied to the surface of the C4 pads on the organic chip carrier.
4. The die is flipped and placed on the C4 sites.
5. The flipped die is then reflowed to form the first-level FCA interconnection.
6. The attached die is underfilled and cured.
7. A metal cap is placed over the underfilled die region of the package.
8. The BGA solder balls are attached to the backside of the module. Here, either a eutectic Sn-Pb solder alloy or a 2 percent Ag addition to a eutectic alloy is used for the BGA solder balls. The 2 percent Ag alloy is known to improve the BGA reliability.[6] A water-soluble flux is used.
9. The water-soluble flux is cleaned.
10. The package is electrically tested.
11. The fully assembled module is placed in JEDEC trays.
12. The modules in the JEDEC trays are baked.
13. The baked trays are sealed in metallized vacuum sealed bags and then stored.

Apparently, from this discussion, interfacial adhesion is crucial in this type of package configuration due to the excessively high stresses generated in the package during thermal stressing. The different interfaces must maintain integrity when the package is exposed to moisture and then subjected to multiple reflow cycles during card assembly and, potentially, module rework. As the die sizes continue to increase to 20 mm, organic FCA becomes even a greater challenge. Here, it becomes important to ensure that the underfill can cover the entire die region without any voiding. As the C4 pad/pitch changes from a 0.004-in. pad with a 0.009-in. pitch to 0.003-in. pad with a 0.006-in. pitch, the problems associated with proper underfill coverage greatly increase. Even with proper underfilling, as the C4 density increases for die approaching a size of 20 mm, the reliability of an organic FCA package can be dramatically reduced if the proper underfill materials and processes are not fully optimized.[7,8]

6.10 Organic Wire Bond Modules

An organic wire bond package is one of the most common methods to package a die. Here, a die is attached backside to a die pad on the laminate chip carrier using a thermally conductive adhesive, which is called the *die attach adhesive*. The adhesive is cured, and then the die can be wire bonded. The standard ground rules of wire bonding are at approximately 90–100 microns pad pitch on the die. As is well known, one of the major limitations of wire bond packages is that they can accommodate only a perimeter row of active pads on the die. The package is then overmolded and, lastly, the BGA solder balls are attached. A standard type of chip-up laminate wire bond package is illustrated in Fig. 6.19. Most applications can be met using this type of package configuration.

For higher electrical and thermal performance using wire bonding, a cavity package design is used. Here, the die is attached to a copper plate to achieve maximum thermal performance.

By using ultra-fine-pitch (UFP) ground rules for the circuitization of the laminate chip carrier, one may be able to achieve the same electrical performance as an FCA type of package design. The package design can have 1-mil lines on 1-mil spacing to achieve high circuit densities. Here, a die bond pad pitch of 60 microns is used, along with a 25-micron dia. gold wire. A schematic illustration of a cavity UFP wire bond package is shown in Fig. 6.20.

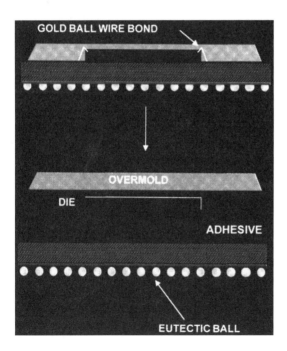

Figure 6.19 A schematic illustration of a standard chip-up wire bond package configuration.

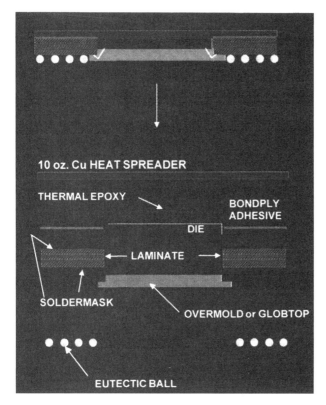

Figure 6.20 A schematic illustration of a cavity UFP wire bond package configuration.

The process flow to produce an organic chip-up wire bond package is outlined below:

1. Attach the back side of die to laminate chip carrier using thermally conductive adhesive.
2. Cure the adhesive.
3. Clean the surface of die to remove any contamination prior to wire bonding.
4. Wire bond die, most often a gold wire bond, is used. However, Al can also be used.
5. Inspect wire bonds to ensure the integrity of interconnection.
6. Overmold the package.
7. The BGA solder balls are attached to the backside of the module. Here, either a eutectic Sn-Pb solder alloy or a 2 percent Ag addition

to a eutectic alloy is used for the BGA solder balls. The 2 percent Ag alloy is known to improve the BGA reliability.[6] A water-soluble flux is used.

8. The water soluble flux is cleaned.

9. The package is electrically tested.

10. The fully assembled module is placed in JEDEC trays.

11. The modules in the JEDEC trays are baked.

12. The baked trays are sealed in metallized vacuum sealed bags and then stored.

Wire bond packages offer one of the lowest-cost methods to package a die. The reliability of this type of package is very high. These two attributes are very difficult to match using any other type of packaging scheme. There are still many customer applications that can be achieved using wire bonding.

6.11 Second-Level Assembly

In second-level card assembly, there are two basic processes. The first attach method is for PGA modules. Here, the pinned module is attached to the plated-through hole (PTH) using solder that is supplied by the back side of the through hole. The second assembly method does not use a PTH, but a surface mounting pad on the card. The module is then attached to this surface mount pad, and this is the basis for surface mount technology (SMT). There are three basic types of SMT attach methods: leaded components, nonleaded components, and BGA components.

6.11.1 PTH Attach Methods

For this attach method, either a wave soldering machine or a solder fountain is used to attach pinned leaded configuration module to a card through a PTH. For high-volume manufacturing, a wave soldering machine is used. Figure 6.21 shows a photograph of a state-of-the-art wave soldering machine sold by Electrovert. Either air or nitrogen environments are most often used. For low-end, high-volume PIH assembly, the processes will be most often defined using an air atmosphere to reduce the cost of the process. However, it is well known that the use of nitrogen can dramatically lower the number of solder joint defects and therefore offset the added cost.[9]

For low-volume specialty applications, a solder fountain is used for module attach. Here, the solder contacts only the localized PTHs that

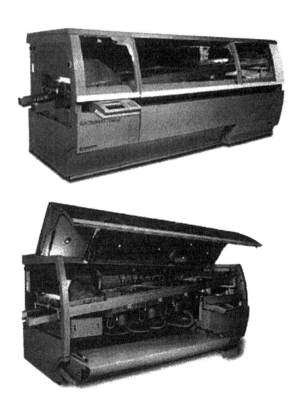

Figure 6.21 Photographs of a state-of-the-art wave soldering machine by Electrovert that is used to form a PIH type of solder interconnection.

will be soldered and thereby reduces any thermal stress on the other regions of the card. This method of solder attach for PIH components is also used to rework the modules.

Most often, a PGA module attached to a PTH is done on a thick board cross section. In certain applications, the board thickness can approach 0.250 in., with up to 40+ metal layers. When such a thick cross section is used, there is a concern that the soldering temperatures (which approach 240°C, most often using eutectic Sn-Pb solder alloy), can cause excessive stress on the PTH and make the barrel crack or the inner planes separate ("IP sep") from the barrel. Both barrel cracking and IP sep are well known failure modes on thick boards soldered with eutectic solder. Therefore, it is necessary to use a lower-melting-point solder, such as the eutectic Sn-Bi solder alloy, which melts at 138°C. Since molten Sn-Bi is widely known to produce a heavy oxide dross when exposed to air during soldering, it is necessary to do the soldering in a protective atmosphere such as nitrogen to ensure that good PTH solder fillets are formed. Another method to solder with eutectic Sn-Bi uses the immersion wave soldering processes. In this process, the entire card is soldered beneath a liquid glycerine/EDTA flux.[10] Lower-melting soldering is essential when soldering thick board cross sections.

6.11.2 SMT Attach Methods

The primary feature of a SMT component is to form the solder interconnection directly on the surface pad on the card. This solder joint configuration is easier to form than the earlier PIH type, which required a sophisticated wave soldering machine along with high wettability of the surface finishes used on both the pinned leads of the component as well the PTH of the card. The driving force for SMT was the ability to populate both sides of the card with components. PIH attach methods were limiting, since only a single side of the card could be populated. In SMT, there are three main types of solder joint configurations: leaded components, nonleaded components, and BGA components.

6.11.2.1 *Leaded components.*

As mentioned earlier, both gull-wing and J-leaded configurations are used for leaded components. These component lead configuration types were very commonly used in the 1980s but started to becomes less pervasive by the early 1990s when the BGA technology started to gain more popularity. Both of these lead types are known to produce very reliable interconnections. However, the gull-wing lead configuration is susceptible to lead damage during handling and can therefore affect the integrity of the resulting solder joint. Leaded components are very limited in that they can populate only the perimeter of the package and therefore limit the I/O density. As lead pitches migrated down to 0.4 mm and body sizes increased to 40 mm, it became very difficult to achieve acceptable yields at card assembly.

6.11.2.2 *Nonleaded components.*

In this type of solder joint configuration, only a pad on the chip carrier contacts the solder on the card. This nonleaded type of solder joint configuration is acceptable for most organic chip carrier designs in which very little CTE mismatch occurs between the chip carrier and the mating card during thermal cycling.

6.11.2.3 *BGA Components.*

The use of BGA modules has increased substantially over the past decade. In general, BGA assembly encompasses many different types of packages. The ceramic ball grid array

(CBGA) modules were the first type of BGA package used in the electronics industry, followed by ceramic column grid array (CCGA) packages, which were able to increase the reliability of the interconnect by increasing the height of the interconnection. The organic BGA packages also became available in a number of different package configurations; however, all of these packages are referred to as plastic ball grid arrays (PBGAs). Most PBGA packages are wire bond chip-up and cavity designs that are protected with either overmold or glob top organic materials to encase the crucial wire bond interconnections.

A unique type of PBGA package is the tape ball grid array (TBGA) design, which can accommodate either a perimeter die that is thermocompression bonded to the package, or a flip chip version of the package. Both types of packages are known to be highly reliable and offer superior electrical performance. In the case of a wire bond first-level interconnection, a gold-plated lead on the TBGA substrate is thermocompression bonded to a gold bumped die. For a flip chip interconnection, a conventional 97/3 bumped die is transient liquid phase (TLP) bonded to the Au plated pad on the substrate. In the TLP bonding process, a Pb-rich binary eutectic alloy composition is formed by the TLP reaction. A subsequent anneal is done to diffuse the Au throughout the Pb-rich bump to produce a high-melting solder joint that does not melt upon exposure to conventional Sn-Pb second-level soldering processes. This type of flip chip attach process is commonly referred to as solder attach tap technology (SATT) joining and is outlined in detail in Refs. 11 and 12. The BGA ball used for this package is a 90Pb-10Sn alloy that is spot welded to the pad on the TBGA chip carrier. This is one of the few organic packages that use a Pb-rich solder ball for the BGA.

One of the distinct differences between the ceramic BGA technology and the organic BGA is the type of solder ball. All of the ceramics use a Pb-rich solder that is attached to the module and the motherboard using low-melt eutectic solder. Most of the organic BGA packages, except for TBGA, use a low-melt eutectic Sn-Pb solder ball. From an assembly and rework concern, the CBGA technology can be considered to be more challenging, especially during rework.

BGA modules cover a performance range from the low end, suing plastic types of packages, to CBGA/CCGA modules that are high-density, high-performance surface mount packages. Since the array of balls cannot be inspected once the module is attached to the card, it is very important to optimize the package design and the assembly process so as to achieve both high yields and reliability of the package. A major reason for the rapid acceptance of BGA technology is that these types of packages achieve very high yields, typically in the range of 1–3 ppm/lead, and forming the required interconnection configuration to ensure that the reliability is met.[13]

Ceramic ball grid array assembly. A CBGA package is a high-reliability, high-performance package. Lidless CBGA packages are not moisture sensitive, and lidded packages are of either JEDEC level 2 or 3 moisture classification. A JEDEC level 1 classification is capable, when using a solder seal seam, to attach the cap to the ceramic chip carrier. They have very good shelf life and maintain BGA coplanarity very well during shipment, due to the ball type of BGA configuration. Figure 6.11 shows a photomicrograph of a CBGA type of solder interconnection using a 0.89-mm (0.035-in.) dia. 10/90 Sn-Pb solder ball that is joined to the ceramic substrate and the printed wiring card with eutectic Sn-Pb solder. The high-melt solder ball does not reflow during card assembly and therefore creates a predetermined standoff of 0.89 mm. The CBGA module is mounted on a card using a "dog bone" pad design, which has been shown to be necessary to produce a reliable solder joint.[14] A solder dam between the landing pad and the via is required to prevent loss of solder paste to the via. A non-solder mask defined pad is used, which has been shown to produce a high reliability in the BGA interconnection. The nominal pad diameter is 0.72 mm (0.0285 in.) at the top of the pad, which is required to ensure joint reliability. Warping of the card is also considered to be a major contributor to the z-axis tolerances for all SMT devices, and especially the BGA modules. Generally, by using the optimized paste printing processes, the card warpage is overcome during the reflow operation. It is recommended to consider the following design, which is aimed at preventing local warpage that can occur during assembly and rework processes:

- Symmetrical card cross sections are required to minimize warping of the card, especially during the reflow operations.
- Minimize local card warpage created by adjacent components that "anchor" the PCB such as large PIH connectors.
- Maximize uniformity of assembly thermal mass across the PCB.
- Consider form factor of the card: large thin cards are more likely to warp during processing and may require fixtures.
- Establish a reliable PCB supplier that delivers quality product and not "potato chip" cards. It is very difficult to specify and enforce a card flatness specification other than the IPC standard.

A flatness requirement of 0.025 to 0.076 mm (0.001 to 0.003 in.) average is common for a 32-mm CBGA site on the card. There are other considerations that need to be considered in the design of the card. When fine-pitch components are integrated on the card and adjacent to a CBGA module, enough clearance between the two modules sites is required for a step-down stencil. When tented or plugged vias are used on the card, they can become entrapped with contaminants, such as flux residue, which can be very difficult if not impossible to remove.

CBGA packages are assembled to printed circuit boards using standard SMT tool and processes. A single-sided process flow is outlined below:

1. Apply solder paste on the card.
2. Perform solder paste verification.
3. Place components.
4. Reflow solder.
5. Clean, if required.
6. Test.
7. Rework component, if required.
8. Attach heat sinks.

CBGA packages are also capable of being attached in a double-sided component attach configuration. An important consideration here is the ability of the inverted reflow position to maintain the module on the card. For the molten solder to be able to hold the inverted module during reflow, a minimum amount of molten solder is necessary to ensure that the surface tension per pad is greater than the weight of the module per pad.[15] Although solder joint structures of the back-side and the front-side module are not that much different, the reliability is reduced when modules are back to back with a shared via in card configuration.[16] A double-sided assembly processes is as described below:

1. Perform solder paste application on backside.
2. Perform solder paste verification.
3. Place components.
4. Reflow solder.
5. Clean, if required.
6. Perform solder paste application on front side.
7. Perform solder paste verification.
8. Place components.
9. Clean, if required.

10. Test.
11. Rework, if required.
12. Attach heat sinks.

In both of these assembly processes, the solder paste screening operation is the critical process step. Most often, either a type 3 or 4 solder paste is recommended, depending on whether additional fine pitch components are used on the card. The solder alloy is a eutectic Sn-Pb, which is 50 percent by volume of the paste mixture, or 90 percent by weight. CBGA assembly is known to work very well for both no-clean and water-soluble types of solder pastes. For CBGA assembly, the process needs to be optimized such that a reliable solder interconnection is formed. Process development is therefore required for the specific module design to ensure that the solder fillet will guarantee that the solder joints reliability requirements are achieved.

After screen printing, the paste deposit needs to be inspected, especially for the high-reliability requirements for ceramic types of BGA packages. Here, the paste volume is measured on a fully automated tool such as a Synthetic Vision Systems (SVS) tool. After screening, the module is placed on the solder paste deposit. The CBGA type of module is very forgiving during placement. As long as the solder balls contact the paste, they will self-align during reflow. Both the top eutectic on the substrate side of the BGA ball and the solder paste reflow, which allows the high-melt BGA ball to float, and to equilibrate between both the substrate and card pads. This ability to float better enables the module to accommodate warpage in the card.

During the solder reflow operation, it is important to maintain the solder profile recommended by the paste supplier. A recommended reflow profile for no-clean type of solder paste is shown in Fig. 6.22. The peak temperature specification of 220°C is driven by the requirement to limit the amount of Pb from the BGA ball into the molten eutectic alloy. The cool-down rate is well known to affect the yield and reliability of the interconnection.[17] Two conditions of concern are:

1. *Card pickup during oven exit.* Here, the solder joints are still molten and, if an operator manually picks up the card to remove it from the conveyor, he can distort the resulting solder joint. Cards should not be removed from the conveyor until the solder is solidified.

2. *Uneven cooling of the top and bottom card surfaces.* This case can produce warping of the card, which can potentially cause balls to lift off the pads during solidification. This is also known to produce column cracking on CCGA modules.

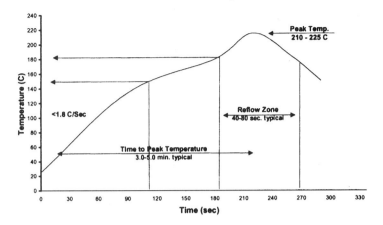

Figure 6.22 A reflow profile for a commercially available no-clean type of solder paste.

8.3.1 *Ceramic ball grid array rework.* As was mentioned earlier, CBGA modules have a very high assembly yield. However, the rework operation is not as robust as the initial assembly operation. It is important to consider the same critical requirements in initial assembly and apply these to the rework operation. Again, most important steps are the solder paste volume, card warpage, and control of the reflow profile. The CBGA rework process is outlined as follows:

1. Site thermal profiling
2. Module removal
3. Site dress and touch-up
4. Cleaning, if required
5. Solder application
6. Module placement
7. Module reflow
8. Cleaning, if required
9. Inspection and electrical test

Specialized rework tools are required that have the following capabilities:

- hot gas heating
- computer-controlled temperature profiles

- calibrated vision system
- automatic vacuum pickup and component placement
- menu-driven software control
- complete PC data logging

The key process variables are preheat temperature, peak joint temperature, and solder reflow time. To remove the module, a hot gas reflow tool with a bottom card heater (bias bay) is required. The entire carrier should be preheated to between 75 to 125°C prior to the application of hot gas from the top heater. Preheating is a critical step that minimizes the card warping during removal and limits the thermal shock to the card. The maximum preheat temperature is about 10°C below the glass transition temperature of the card. Typically, the adjacent module is limited to less than 150°C. Moisture-sensitive components require a bake-out of 24 hr at 125°C prior to rework to prevent any type of moisture-driven damage to the component during the rework cycle, such as popcorning.

Thermal profiling is required for module removal, site flattening, and module reflow. A requirement in thermal profiling is to have a thermocouple read the joint temperature. The thermocouple can be attached with eutectic solder and coated with a thermally conductive epoxy to ensure contact. Sometimes, one will drill a hole through the back side of the component, into the solder joint, and fill the drilled hole with thermally conductive adhesive. The latter method is highly recommended. Recent experiments have shown that the amount of supercooling can be as great as 20° below the melting temperature of eutectic solder.[18]

A hot gas tool with vacuum pickup is used to remove the module, and a full carrier preheater is required. In this step, the component is removed while minimizing any type of thermal damage to the card, such as pad lifting or warping. After module removal, the site on the card must be dressed. On most pad sites, the 10/90 ball is still on the card pad, held in place with the eutectic solder. All of these Pb-rich balls need to be removed. The remaining eutectic solder on the card pads is nonuniform in coverage. The desired method to dress the card pads is to use a solder vacuum. In this process, a pad-by-pad removal of excess solder is done using a vacuum. The resulting surface is similar to a HASL card surface. The manual solder vacuum technique is known to impart very little damage to the card. Cleaning is then required only if water-soluble flux is used during the dressing operation.

Eutectic solder must be applied to either the card pads or to the balls on the module. Solder paste is preferred, since it provides tack to hold the module in place during reflow and compensate for any z-axis variation. As mentioned earlier, the eutectic solder volume needs to be precisely controlled to achieve an optimized solder joint configuration. There are several ways to apply eutectic solder: screen paste on the CBGA module balls, screen paste on the card pads, use solid preforms, or syringe dispense paste on the card pads. Module screening is a preferred process to apply eutectic solder on the new module that will be reattached to the card. A unique clamshell type of screening fixture is required to accomplish this operation. The weight of the solder paste applied is a critical process control parameter.

When the solder paste has been applied to the module, a split-optics prism method is used to place the component precisely on the card pads. Typically, the module is placed with enough force to ensure that the solder paste contacts at a minimum 50 percent of the pad surface, and it must not contact any of the vias. The applied eutectic solder must be reflowed so as to allow to achieve the proper metallurgical reactions. Preheating the entire card is critical to minimize any card warping. The thermal profile needs to use the same requirements as specified for initial attach. Finally, the adjacent modules must be protected from exposure of overheating. Thus, the need for proper thermal profiling prior to reworking actual product. The same inspection techniques that were implemented for initial attach are also used for rework.

6.12 Migration to Pb-Free Solders and Soldering Processes

In the early 1990s, legislation was proposed in the U.S. Congress to abolish Pb used in solders for electronics assembly. The looming prospect of government regulations initiated research on Pb-free solders in the U.S.A. and other countries. A notable study was that headed by Dr. Duane Napp, from The National Center for Manufacturing Sciences (NCMS), who established a consortium consisting primarily of U.S.-based corporations. From this report, three solder alloy families were recommended, all of a eutectic compositions: Sn-Bi, Sn-Ag, and Sn-Ag-Bi.[19] In the mid-1990s, representatives of the electronics industry testified before Congress and convinced the government not to pursue legislation banning the use of Pb in solders used for assembly of electronics packages.

During the past two years, there has been a strong drive to convert to Pb-free solders in the global electronics industry. This trend was initiated by legislation in the European Union (EU), with the Waste Electrical and Electronic Equipment (WEEE) Directive, which has caught the attention of the industry worldwide. In this document, emphasis is directed at dangerous substances, which is stated as follows in the Explanatory Statement of the European Parliament Draft Report, dated 5 February 2001:

> Lead, mercury, cadmium, hexavalent chromium and the flame retardants polybrominated biphenyl (PBB) and polybrominated diphenyl ether (PBDE) are dangerous substances which are a major contributory factor to the environmental problems occurring during the disposal of electrical and electronic equipment; they have given rise to a number of measures at community level in the past. An effective date of 1 January 2006 will give the sector sufficient time to adjust to the need for substitution.

As a result of this statement, there has been a concentrated effort in three main geographies of the world: domestic U.S.A., Asia/Japan, and Europe. Both private industry and as government-sponsored consortia have begun to address the issue of Pb-free solders used in all levels of soldering in the electronics industry. The National Electronics Manufacturing Initiative, Inc. (NEMI) Pb-free Project was initiated 1Q99, having 22 North American companies participating. The NEMI chose Sn-8.0Ag-0.6Cu ±0.2 percent (SAC) for reflow soldering and Sn-0.7Cu for wave soldering. The SAC alloy represents a composition in close proximity to the ternary eutectic point in the Sn-Ag-Cu alloy system. Figure 6.23 shows the section of the ternary Sn-Ag-Cu phase diagram with the identified ternary eutectic composition and the shaded region has alloys with a freezing range < 10°C.[20] The global direction for Pb-free seems to be converting to the NEMI recommended SAC alloy for BGA solder balls as well as for solder paste. Both the SAC and Sn-Cu wave alloys require a maximum solder reflow temperature of 260°C, as compared with the conventional maximum soldering temperatures of 220°C imposed when using eutectic Sn-Pb solder. As is becoming more apparent from the Pb-free development activity, one of the major challenges facing the industry is how to make the moisture sensitivity level (MSL) of the Pb-free packages meet the same level as qualified using conventional Sn-Pb solders. This may prove to be the greatest challenge and is the subject of the remainder of this section.

When using conventional Sn-Pb solder alloys, 220°C is the maximum temperature cards and boards encounter during the soldering operation. When using the SAC alloy, the current JEDEC specifications set the maximum temperature at 260°C. The 40°C increase in the soldering temperature has major implications on the integrity of the electronic package assembly. Notably, delamination of internal layers is well known to occur in many laminated structures, especially those with high moisture update. The combination of internally trapped moisture and a 40°C increase in temperature contribute to extensive delamination using most conventional laminates. To make organic packages compatible with the higher reflow temperatures imposed by Pb-free sol-

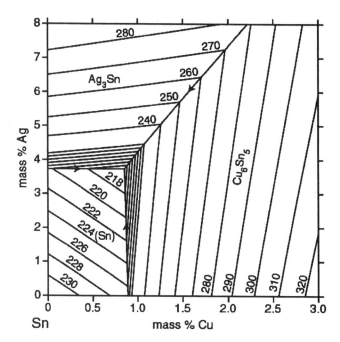

Figure 6.23 A section of the Sn-Ag-Cu phase diagram that identifies the ternary eutectic point as determined by scientists at NIST.[20]

dering, lower-moisture-uptake materials and improved interfacial adhesion are essential. Recent improvements in a few select materials have shown promise for Pb-free application; however, most materials are far from achieving success with the higher reflow temperatures.

As just cited, card assembly poses many challenges with Pb-free solders. Reworking packages with Pb-free solders, most likely using the SAC alloy, will pose even greater challenges. The 40°C increase in reflow temperature imposed on Pb-free card assembly is a challenge today. Rework of large-body surface mount components, especially organic packages, will pose even greater challenges. It will most likely be mandatory to develop very highly controlled soldering profiles for rework to minimize high-temperature exposure. Such challenges are just starting to be investigated.

6.13 References

1. C. Radke, "A Justification of and an Improvement on a Useful Ruse for Predicting Circuits to Pin Ratios," *Proc. 1969 Design Automation Conf., IEEE, Computer Society*, pp. 257–267.

2. D. P. Seraphim and Alan L. Jones, "Interaction of Materials and Processes in Electronic Packaging Design," Chapter 2 of "Principles of Electronic Packaging," edited by D. P. Seraphim, R. Lasky and C. Y. Li, McGraw-Hill, 1989, pp. 39–42.
3. C. G. Woychik and A. K. Trivedi, "A Comparison of the Solderability of Wrought and Plated Nickel Surfaces: The Selection of a Reliable Pin for Second Level Packaging," *Jnl. Electronic Packaging*, vol. 112, Sept. 1990, pp. 223–233.
4. G. Shook, "Flip Chip Assembly for BGA semiconductor packages," *Proceedings of Nepcon West 96*, vol. 3, Anaheim, CA, USA, Feb. 1996, pp. 27–29.
5. C. P. Yeh, W. X. Zhou, and K. Wyatt, "Parametric finite element analysis of flip chip reliability," *Int. J. Microcircuits Electron. Packag. (USA)*, vol. 19, no. 2, 1996, pp. 120–127.
6. C. G. Woychik, D. L. Hawken, J. R. Wilcox, and P. J. Brofman, "Extending Flip Chip Ball Grid Array Field Life," *SEMICON West 99, Semiconductor Packaging Symposium*, July 12–14, 1999, pp. C1–C8.
7. M. Ranjan, L. Gopalakrishnan, K. Srihari, and C. Woychik, "Die Cracking in Flip Chip Assemblies," *Electronic Components and Technology Conf., 1998, 48th IEEE, 1998*, pp. 729–733.
8. L. Gopalakrishnan, M. Ranjan, Y. Sha, K. Srihari, and C. Woychik, "Encapsulant Materials for Flip Chip Attach," *Electronic Components and Technology Conf., 1998. 48th IEEE, 1998*, pp. 1291–1297.
9. M. Theriault, P. Blostein, and A. Rahn, "Nitrogen and soldering: Reviewing the issue of inerting," *Surf. Mount Technol. (USA)*, vol. 14, no. 6, June 2000, pp., 52–52,58.
10. C. G. Woychik and R. C. Senger, "Joining Materials and Processes in Electronic Packaging," Chapter 19 of "Principles of Electronic Packaging," edited by D. P. Seraphim, R. Lasky and C. Y. Li, McGraw-Hill, 1989, pp. 577–619.
11. U.S. Patent no. 6, 130,476. Semiconductor Chip Packaging Having Chip To Carrier Mechanical/Electrical Connection Formed Via Solid-State Diffusion, Oct. 10, 2000.
12. U. S. Patent Number 6, 162,660. Method for Joining a semiconductor chip to a chip carrier substrate and resulting chip package, Dec. 19, 2000.
13. S. Dunford, P. Viswanadham, C. Clark, "Enhancing board level BGA assembly and reliability," *SMTA International Proceedings of Technical Program. Conference Proceedings 1999*, pp. 76–96.
14. S. Yee, and H. Ladhar, "The influence of pad geometry on ceramic ball grid array solder joint reliability," *Nineteenth IEEE/CPMT International Electronics Manufacturing Technology Symposium, Proceedings 1996 IEMT Symposium (Cat. no. 96H35997), 1996*, pp. 267–273.
15. P. Viswanadham, K. Ewer, R. Aguirre, and T. Carper, "Package to board interconnection and reliability of BGA packages over extended temperature range," *Pan Pacific Microelectronics Symposium, Proceedings of the Technical Program 1998*, pp. 115–128.
16. C. G. Woychik, D. L. Hawken, A. Migliore, and J. Potenza, "Reliability Enhancements of Organic Flip Chip Ball Grid Array Packages," *8th Annual SEMICON Singapore 2000 Test, Assembly and Packaging*, May 10, 2000, Singapore, pp. 202–209.
17. J. S. Hwang, and Z. Guo, "Reflow cooling rate vs. solder joint integrity. I," *Proceedings of the Technical Program, National Electronic Packaging and Production Conference, Nepcon West '94*, 1993, vol. 2, pp. 1095–1120.
18. D. Henderson, C. Woychik, private communications, IBM Endicott, Endicott NY 13760.
19. D. Napp, "NCMS lead free electronics interconnect program," *Surface Mount International Conference and Exposition, Proceedings of the Technical Program*, 1994, pp. 425–432.
20. Sn-Ag-Cu Phase Diagram, Metallurgy Division of the National Institute of Standards and Technology, Database at NIST: www.metallurgy.nist.gov/phase/solder/agcusn.html.

Chapter 7

Solder Materials and Processes for Electronic Assembly Fabrication

Jennie S. Hwang, Ph.D.

H–Technologies Group, Inc.
Cleveland, Ohio

7.1 General Trends

What drives the end-use market are the continued convergence of computing, communication, and entertainment as well as the relentless growth of the wireless, portable, hand-held digital electronics.

In terms of growth, take the cellular phone as an example. Its annual growth rate spans the range of 40 to 110 percent, varying by country. Worldwide sales of cellular phones rose 51 percent, to 163 million units, in 1998 and are projected to reach 1 billion units by 2005.

The blitz-like speed of advancements in computers is vividly reflected by the fact that desktop PCs have exceeded the specifications of U.S. government export rules, making PCs fall into the supercomputer category. According to the U.S. government export rules, a special license is required for the export of computers rated at more than 2000 millions of theoretical operations per second (MTOPS) to institutions with connections to the militaries of the restricted countries and at 7000 MTOPS for commercial sales to the restricted countries. Recently, the government had to change the export rule for high-end computers. The U.S. Commerce Department has just published regulations in the *Federal Register* to implement the change in U. S. export control, easing controls on high-performance computers (August, 1999). The new rules raise the individual licensing levels from 2000 to 6500 MTOPS for military end-users and from 7000 MTOPS to 12,300 MTOPS for civilian end-users.

7.1.1 Semiconductor Chip Level

It is generally recognized that microprocessor and memory chips are two primary technology drivers.[1,2] Circuit lines continue to decrease (0.25 to 0.18 to 0.13 µm and finer). Interconnect RC delay, which increases as the square of the minimum feature size, thus determines the IC chip performance. For instance, interconnects account for over 70 percent of the signal delay in 0.25 micron chips. The clock speed of microprocessors is moving toward multiple gigahertz levels. The industry consensus is that IC technology is advancing beyond Moore's Law, which has served the industry well for over two decades by predicting that the density and function doubles every 1.5 years.

Wafer size is going through the transition from 200 to 300 mm. The anticipated yield boost and cost reduction propel the implementation of 300-mm wafer production. For example, it is expected that, typically, a 200-mm wafer can produce 198 good die, and 300-mm wafer 471 good die. However, the capital expenditure, infrastructure, and other factors remain as the residual momentum for 200 mm. The year 2002 may well be a transition year for 200- to 300-mm wafers.

7.1.2 IC Packaging

For the most part, ICs have been traditionally packaged. The primary reasons to package ICs are to

- Protect silicon die
- Supply power to the die
- Facilitate heat dissipation
- Provide interconnections
- Test the die

However, each interconnection adds parasitics (capacitance, resistance, and inductance) to the circuitry. Intrinsically, silicon is capable of transmission speeds in the range of a picosecond. For most end-use electronics, such as computers, a typical transmission speed is in the range of a nanosecond. This is a gap of three orders of magnitude, so the packaging or interconnection remains a bottleneck for achieving the fastest transmission speed. Thus, when designing a system to fully utilize the capability of silicon transistors, the aim is to minimize the parasitics, with the ultimate goal of approaching silicon's intrinsic speed. Achieving the shortest path and the highest conductivity are still the fundamentals used to accomplish the ultimate speed.

Has the IC packaging technology been keeping pace with silicon technology? As shown in Fig. 7.1, across the decades of the 1980s and 1990s, the industry has evolved from the dual in-line (DIP) package, pin grid array (PGA) to 0.050-in. (1.00-mm) pitch surface mount leadless ceramic chip carriers (LCCCs), plastic lead chip carriers (PLCCs), to fine-pitch, small outline ICs (SOICs) and quad flat packs (QFPs). More recently, ball grid array (BGA), chip-scale packages (CSPs) and direct chip attach (DCA) have been the focal points.

Each technology differs in its design of circuitry and interconnections. To implement the advanced packages into applications, different types of packages demand a different set of requirements in processes and equipment. Yet they all fall into the realm of surface mount manufacturing in terms of connections to the main circuit board.

7.1.3 Bare PC Boards

High-density and low-cost PCBs continue to demand the major development effort. Surface build-up and microvia technology are keys to the materialization of a high-density PCB. Microvia development on all three techniques (plasma etching, laser drilling, and photo via processing) and their improved versions are among the major tasks for establishing future PCB fabrication processes.

Along with the ability to achieve smaller vias and finer lines, the associated processes in PCB fabrication for the betterment of quality, economy, and environment are equally important. The interdependency between microvias and the use of high I/O flip chips and CSPs has been increasingly recognized. Without the sound, high-density board technology, high-I/O flip chip and CSP cannot be fully utilized.

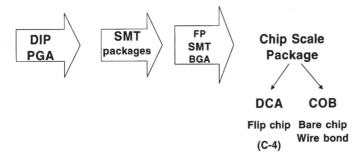

Figure 7.1 Evolution of IC packages.

Fundamentally, substrate materials that provide high frequency operation, a low dielectric constant approaching 2.00, a high glass transition temperature, and low moisture absorption are always in demand. Other desirable performance properties include designs and materials that provide efficient thermal management and dimensional stability plus low thermal expansion.

All of these physical properties and design characteristics of the PCB affect the integrity of solder interconnections, imposing higher demands on the performance level of solder interconnections during the product's service life.

7.1.4 SMT Assembly

An ongoing effort will be made to maximize yield, minimize defect rates, and improve performance properties by utilizing the knowledge and the state-of-art equipment and materials that have evolved. Defect-free manufacturing becomes increasingly important as the interconnection density continues to increase. Production yield sensitively depends on the defect rate. Figure 7.2 illustrates the dependency of the first-time yield on the defect rate and the number of solder joints per assembly.

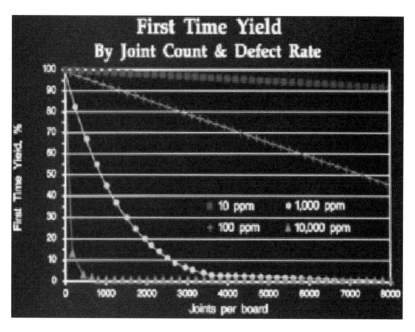

Figure 7.2 Dependency of the first-time-yield on the defect rate and the number of solder joints per assembly.

Constant assessment of new IC packages in conjunction with board design will become a part of the board-assembly business. Solder paste will remain the primary interconnection material, characterized by its fitness to automated manufacturing, established infrastructures, and its metallic nature. Solder paste will not only work for SMT interconnections but also for certain through-hole components (paste-in-hole). Other solder deposition techniques including solder jetting will be assessed for specific packaging and assembly operations. Automation, SMT fitness (e.g., pick-and-place), and cost will be the determining factors for the viability and vitality of a new technology.

With the introduction of new packages and the increased number of package types for PCB assembly, the manufacturing process, reflow profile in particular, warrants further attention. Reflow profile not only affects the production defects and therefore the yield but also the overall reliability of the assembly. To accommodate the inherent heating disparity across the board, a slower heating rate (<2°C/s, and ideally <1°C/s in the preheating zone) makes a more compatible reflow profile. The same principle should also apply to rework and repair; using preheating and a top/bottom heat source will facilitate the process and minimize any potential damage that may occur during rework. The use of inert atmosphere (N_2) soldering with low-N_2-consumption reflow ovens will also find niches.

The accuracy and speed of placement equipment will continue to improve. In addition, the "gentle" placement capability to work with small and fragile CSPs will also be on demand, including reliable feed mechanisms and vision capability. To handle CSPs in 0.50-mm (0.020-in) pitch, positioning accuracy of ±0.002 in. is required. Printing and dispensing equipment for the application of solder paste, underfill, adhesives, and coatings will be characterized by increased automation and precision. New functional features continue to emerge, added to the equipment to facilitate production operation and to enhance the end results.

7.2 Solder Materials

Solder has served as the interconnecting material for all three levels of connections: die, package, and board assembly. In addition, tin/lead solder is commonly used as a surface coating for component leads and PCB surface finishes. Soldering is the technique that uses solder alloys to accomplish the vital function of providing electrical, thermal, and mechanical linkages between two metallic surfaces. Considering the established role of lead (Pb) technically and otherwise, solder materials can be classified as either lead-containing or lead-free.

7.2.1 Solder Paste[3]

Solder in paste form is of particular importance to the industry because of its unique virtues. Solder paste, with its deformable viscoelastic form, can be applied in a selected shape and size and can be readily adapted to automation. Its tacky characteristic provides the capability of holding parts in position without additional adhesives before the permanent metallurgical bond is formed. The metallic nature of solder paste offers relatively high electrical and thermal conductivity. The combined features in adoption-to-automation, tackiness, and high conductivity make solder paste the most viable material for surface mount assembly manufacturing. It provides electrical, thermal, and mechanical interconnections for electronic packages and assemblies.

Paste technology is an interplay of several scientific disciplines as depicted in Fig. 7.3. From the concept of paste technology, many commercial product lines are derived. These include thick film materials, polymer thick film products, conductive adhesives, EMI shielding materials, brazing pastes, and other products that are composed of metallic or oxide particles uniformly distributed and embedded in the organic/polymeric matrix. Each of these product lines has its unique performance requirements and processing parameters in its production process and its end-use function. However, they have one thing in common: paste technology. A dedicated discussion can be found in the literature.[3]

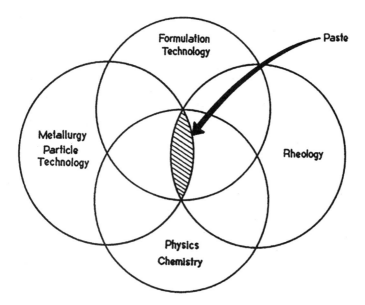

Figure 7.3 Paste technology intersection of several scientific disciplines.

7.2.2 Solder Powder[4,5]

Alloy powders can be produced by one of several common techniques: chemical reduction, electrolytic deposition, mechanical processing of solid particulates, and atomization of liquid alloys.

Alloy powders made from chemical reduction under high temperature are generally spongy and porous. The fine particles of noble metal powders are frequently precipitated by reduction of the salts in aqueous solution with proper pH. The precipitate slurry is then filtered, washed, and dried under highly controlled conditions. A mechanical method is generally used to produce flake-like particles. The metals possessing high malleability—such as gold (Au), silver (Ag), copper (Cu), and aluminum (Al)—are most suitable for making flakes.

The electrolytic deposition process is characterized by dendrite particles, and it produces high-purity powders. The resulting particle sizes are affected by the type, strength, and addition rate of the reducing agent and by other reaction conditions. The characteristics of the particles are also affected by current density, electrolytes, additives, and temperature. The principle of atomization is to disintegrate the molten metal under high pressure through an orifice into water or into a gaseous or vacuum chamber. The powders produced by this method have relatively high apparent density, good flow rate, smooth surface, and a spherical shape, as shown in Figs. 7.4 and 7.5. Powders to be used in solder paste are mostly produced by atomization because of its desirable inherent morphology and the shape of the resulting particles. Hence, the discussion that follows is concerned with the atomization technique only.

An inert gas atomization system, with options of a bottom-pouring system and a tilting crucible system, normally consists of a control cabinet, vacuum induction furnace, tundish, argon (or other inert gas) supply line, ring nozzle, atomization tower, cyclone, and powder collection container. The alloy is melted under inert gas at atmospheric pressure to avoid the evaporation of component ingredients. A high melt rate can be achieved. The molten material is then charged into the atomization tower. The melt is disintegrated into powder at atmospheric pressure by an energy-rich stream of inert gas. The process, conducted in a closed system, is able to produce high-quality powder.

In addition to inert gas and nitrogen atomization, centrifugal and rotating electrode processes have been studied extensively. The atomization mechanisms and the mean particle diameter are related to the operating parameters (diameter of electrode D, melting rate Q, and angular velocity ω of the rotating electrode) and to the material parameters (surface tension at melting point g, dynamic viscosity h, and density at melting point r of the atomized liquid). The relationships among these parameters are interdependent.

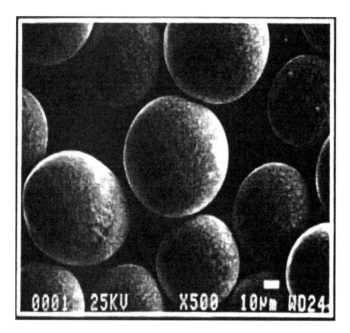

Figure 7.4 SEM micrograph of 63Sn/37Pb solder powder (500×).

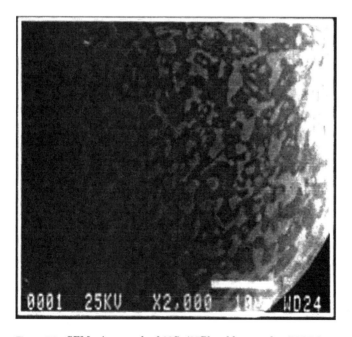

Figure 7.5 SEM micrograph of 63Sn/37Pb solder powder (2000×).

It has been found that the mean volume-surface diameter d is proportional to the surface tension of the atomized liquid and the melting rate, but inversely proportional to the angular velocity of the rotating electrode, the diameter of the electrode, and the density of the atomized liquid.

7.3 Physical Properties[1]

Although other properties can be contributing factors to the overall performance of solder, five physical properties are discussed in this section, due to their special importance to today's and future electronic packaging and assembly.[4] These are

- phase transition temperature
- electrical conductivity
- thermal conductivity
- coefficient of thermal expansion
- surface tension

Metallurgical phase transition temperatures have practical implications. In application, the liquidus temperature can be considered to be equivalent to the melting temperature, and the solidus to the softening temperature. At a given composition, the temperature range between liquidus and solidus is referred to as the *plastic range* or *pasty range*.

It is apparent that the solder alloy selected as the interconnecting material must be compatible with the service temperature to the worst-case condition. It is advisable that the alloy have a liquidus temperature at least two times higher than the upper limit of the expected service temperature. As the service temperature gets closer to the liquidus temperature, solder generally becomes mechanically and metallurgically "weaker."

The electrical conductivity of solder interconnections contributes to the performance of transmission of electrical signals. Solder, a metal, can be viewed as an assembly of positive ions immersed in a cloud of electrons and metallic crystals held together by the electrostatic attraction between the negatively charged electron cloud and the positively charged ions. By definition, electrical conductivity is the result of movement of electrically charged electrons or ions from one location to another under an electric field. Electron conductivity is predominant in metals, whereas ionic conductivity is responsible for oxides and nonmetallic metals. For solders where electrical conductivity is primarily contributed by electrons, the resistivity (the reciprocal of electrical conductivity) increases with increasing temperature due to the reduction of electron mobility, which is directly proportional to the mean free path of electron motion as temperature increases. The electrical resistivity of solders also can be affected by the amount of plastic deformation; resistivity increases with increasing amount of plastic deformation.

The thermal conductivity of metals normally correlates well with electrical conductivity due to the fact that the electrons that carry both thermal and electrical energy are primarily responsible for thermal conductivity as well as electrical conductivity. For insulators, however, the phonon activity predominates. The thermal conductivity of solders decreases with increasing temperature.

CTE issues have been under observation and an area of effort by the industry since the inception of surface mount technology in the printed circuit industry. This is due to the large difference in CTEs of materials that are interconnected. A typical assembly consists of FR-4 board, solder, BGA, CSP, and other leadless and leaded components. Their respective CTEs are FR-4, $16.0 \times 10^{-6}/°C$; 63Sn/37Pb, $23.0 \times 10^{-6}/°C$; Cu leads, $16.5 \times 10^{-6}/°C$; and Al_2O_3 leadless components, $6.4 \times 10^{-6}/°C$. Under temperature fluctuation and power on-off, this mismatch in CTE increases the stress and strain imposed on the solder joint, which may consequently shorten its service life and lead to premature failure. Two major material properties dictate the magnitude of CTE, namely crystal structure and melting point. When materials have similar lattice structure, the thermal expansion of the materials varies inversely with their melting points.

The surface tension of molten solder is a key parameter related to the wetting phenomenon, thus also to solderability. The relative strength of attraction forces acting between molecules of the surface is weaker than that of molecular forces in the interior due to the broken bonds at the surface. Thus, the free surface of a material has higher energy than the interior. Surface tension is a direct measure of the intermolecular forces acting at the surface. A simple but important concept is that the wetting/spreading occurs when the free energy of the newly formed system after wetting is lower than before wetting. In other words, for molten solder to wet the substrate, the substrate surface must have higher surface energy than the molten solder. In view of this requirement, the lower the surface energy of the molten metal or the higher the surface energy of the metal substrate, the more favorable is the wetting process. It should be noted that fluxing is intended to maximize the surface energy of the metal substrate, not to lower its surface energy as occasionally misrepresented in the literature. In conjunction with the proper metallurgical reaction, this is how the flux/fluxing plays a role in wetting.

7.4 Metallurgical Properties[4]

Under the environment and conditions to which solder interconnections are exposed during their service life on the circuitry, the metallurgical phenomena commonly occurring in solder materials include plastic deformation, strain hardening, recovery, recrystallization, solution hardening, precipitation hardening or softening, and superplastic deformation.

When solder is exposed to an applied force, be it the result of mechanical or thermal stress, solder deforms irreversibly. This irreversible deformation is called *plastic deformation*. Plastic deformation is usually initiated through shearing on a number of parallel planes of its crystal structure. The plastic deformation may proceed globally or locally within the solder joint, depending on the stress level, strain rate, temperature, and material characteristics. Continued or cyclic plastic deformation eventually leads to solder joint fracture.

As a result of plastic deformation, solder may be hardened (strain-hardening), as often observed in the stress vs. strain relationship.

A counteracting phenomenon to strain hardening is the recovery process, which is a softening process. The solder tends to release the stored strain energy. The recovery process is driven by thermodynamics. This energy-releasing process, which starts at a rapid rate and proceeds at a slow rate, is called *recovery*. During the recovery state, the physical properties that are sensitive to joint defects tend to be restored to their original value; however, this does not impart any detectable change in microstructure.

The recrystallization process is another phenomenon often observed in a Sn/Pb solder joint during its service life. It usually occurs at relatively high temperatures and involves a larger amount of energy release from the strained materials than the recovery process. During recrystallization, in addition to the energy release, a new set of essentially strain-free crystal structures is formed, which obviously involves both a nucleation and a growth process.

The effect of solid solution alloying results is an increase in yield stress. A typical example of solution hardening is that Sn/Pb compositions are strengthened by an Sb addition. Solution hardening can occur at an even larger extent in the well designed Pb-free solder alloys. Another strengthening effect can come from a structure with a well distributed fine precipitates (precipitation hardening).

In general, for a system with liquid to wet the solid substrate, the spreading occurs only if the surface energy of the substrate to be wetted is higher than that of the liquid to be spread.

As the molten solder solidifies during cooling to form solder joints, aspects of the cooling process, such as the cooling rate, have a direct bearing on the resulting solder joint as to its microstructure and void development.

7.5 Mechanical Properties

Three fundamental mechanical properties of solders include stress vs. stress behavior, creep resistance, and fatigue resistance. Although stress can be applied by tension, compression, or shear force, most alloys are weaker in shear than in tension or compression. Shear strength is important, because most solder joints are subjected to shear stress during service.

Creep is a global plastic deformation that results when both temperature and stress (load) are kept constant. This time-dependent deformation can occur at any temperature above absolute zero. However, creep phenomena only then become significant at "active" temperatures.

Fatigue is the failure of alloys under alternating stresses. The stress that an alloy can tolerate under cyclic loading is much less than that under static loading. Therefore, the yield strength, a measure of the static stress that solders will resist without permanent deformation, often does not correlate with fatigue resistance. The fatigue crack usually starts as several small cracks, which grow under repeated applications of stress, resulting in a reduction of the load-carrying cross-section of the solder joint.

Solder in electronic packaging and assembly applications normally undergoes low cycle fatigue (a fatigue life less than 10,000 cycles) and is subjected to high stresses. Thermomechanical fatigue is another test mode used to characterize the behavior of solder. It subjects the material to cyclic temperature extremes, i.e., a thermal fatigue test mode. Either method has its features and merits, yet both impose strain cycling on solders.[5]

In addition to the intrinsic bulk material strength, the strength of solder joints is often affected by joint configuration, metallurgical reactions, interfacial wettability, interfacial effect, and the characteristics of other materials incorporated in the assembly. For example, alloys of Sn/Ag, Sn/Sb, and 5Sn/85Pb/10Sb are found to impart high creep resistance. This is primarily attributed to solution hardening as substantiated by their high strength and low elongation. When load is applied, the deformation is hindered by means of either interaction of solute atoms with dislocations or interaction with the formation and movement of vacancies, resulting in the impediment of the dislocation movement.

Alloys 10Sn/90Pb and 5Sn/95Pb, however, benefit from the high melting point of their microstructural continuous phase, resulting in the more sluggish steady-state creep. This is attributed to lower self-diffusion, although the alloys are ductile and have moderate strengths. 62Sn/36Pb/2Ag has the highest creep resistance. Its mechanism, whether through the impediment of grain-boundary sliding due to silver segregation or the result of high activation energy for the dislocation movement, is not substantiated.

Bismuth alloys, 42Sn/58Bi and 43Sn/43Pb/14Bi, although having high tensile strength, are found prone to creep. This may be primarily due to their low melting temperature and the predominance of the diffusion-controlled process. The low melting point of their microstructural continuous phase is considered a main factor.

7.6 Solder Alloy Selection—General Criteria

Generally, the alloy selection is based on the following criteria:

- Alloy melting range in relation to service temperature
- Mechanical properties of the alloy in relation to service conditions
- Metallurgical compatibility, consideration of leaching phenomenon, the potential formation of intermetallic compounds
- Rate of intermetallic formation in relation to service temperature
- Other service compatibility (considerations, such as silver migration)
- Wettability on specified substrate
- Eutectic versus noneutectic compositions
- Ambient environment stability

7.7 Lead-Free[6]

The top three criteria for a viable lead-free alloy composition are

1. Its mechanical properties are to be equal to or better than the established reference (63Sn/37Pb).
2. Its physical properties are comparable with the reference.
3. Its application characteristics are compatible with the practical SMT manufacturing infrastructure.

From the simplest binary system alloy to incrementally complex systems containing more than two elements, lead-free materials have been thoroughly explored, studied, and designed. With more than a decade of effort and progressive research and development work, seven systems with the specified compositions stand out for their performance merits. These seven systems are

- Sn/Ag/Bi
- Sn/Ag/Cu
- Sn/Ag/Cu/Bi
- Sn/Ag/Cu/In
- Sn/Cu/In/Ga
- Sn/Ag/Bi/In
- Sn/Ag/Bi/Cu/In

Each of these systems will have its optimal compositions. The selected compositions from each of the systems are also compared with the pertinent known lead-free alloys as well as with 63Sn/37Pb. An overall comparison will then be provided among these seven systems, leading to the final ranking.

7.7.1 Viable Alloys and Rankings for Surface Mount[6]

Among a large number of compositions, the first criterion used to select the viable alloys is for their mechanical properties to be equal to or better than the reference (63Sn/37Pb)—with one exception, which is Sn/Cu eutectic. Then these viable lead-free alloys are ranked by the melting temperature as summarized in Table 7.1 along with mechanical properties—specifically represented by a low cycle fatigue resistance (Table 7.2). It should be noted that the results on fatigue life vary with the specific testing parameters; nonetheless, the data obtained under a set of consistent testing conditions are valid for comparison purposes. Among the candidate elements, there are some concerns about lead-free compositions that contain one or more of the three elements Ag, Bi, and Sb, on account of various beliefs and/or rationale. For the convenience of reference, Table 7.3 separates the selected alloys into various groups based on melting temperature as well as the elements contained in the alloy compositions.

Figures 7.6 and 7.7 compare Sn/Ag/Bi system with 63Sn/37Pb in stress vs. strain behavior and low cycle fatigue, respectively; Figs. 7.8 and 7.9 compare Sn/Ag/Bi/In; Figs.7.10 and 7.11 compare Sn/Ag/Bi/Cu., Figs. 7.12 and 7.13 compare Sn/Ag/Cu/In, and Figs. 7.14 and 7.15 compare Sn/Cu/In/Ga.

It is indicative that 99.3Sn/0.7Cu is inferior to 63Sn/37Pb in mechanical properties. With its binary simplicity, some applications have been successful in using Sn/Cu eutectic in wave soldering. One exam-

Table 7.1 Ranking of Viable Alloy Compositions by the Melting Temperature

Alloy	Melting temperature	N_f
85.2Sn/4.1Ag/2.2Bi/0.5Cu/8.0In	193–199	10,000–12,000
88.5Sn/3.0Ag/0.5Cu/8.0In	195–201	>19,000
93.3Sn/3.1Ag/3.1Bi/0.5Cu	209–212	6,000–9,000
91.5Sn/3.5Ag/1.0Bi/4.0In	208–213	14,000–16,000
92.8Sn/0.7Cu/0,5Ga/6.0In	210–215	6,000–8,000
92Sn/3.3Ag/4.7Bi	210–215	3,850
95.4Sn/3.1Ag/1.5Cu	216–217	6,000–9,000
96.2Sn/2.5Ag/0.8Cu/0,5Sb	216–219	6,000–9,000
96.5Sn/3.5Ag	221	4,186
99.3Sn/0.7Cu	227	1,125

Table 7.2 Ranking of Viable Alloy Compositions by Fatigue Resistance

Alloy	Melting temperature	N_f
88.5Sn/3.0Ag/0,5Cu/8In	195–201	>19,000
91.5Sn/3.5Ag/1.0Bi/4.0In	208–213	
92.8Sn/0.7Cu/0.5Ga/6.0In	210–215	10,000–12,000
85.2Sn/4.1Ag/2.2Bi/0.5Cu/8.0In	193–199	
93.3Sn/3.1/Bi/0.5Cu	209–212	6,000–9,000
96.2sn/2.5Ag/0.8Cu/0.5Sb	216–217	
95.4Sn/3.1Ag/1.5Cu	216–217	
96.5Sn/3.5Ag	221	4,186
92Sn/3.3Ag/4.7Bi	210–215	3,850
63Sn/37Pb	183	3,650
99.3Sn/0.7Cu	227	1,125

Table 7.3 List of Viable Alloy Compositions by Elements

	Melting temperature	N_f
Melting T < 205°C, no Bi		
87.4Sn/3.0/0,5Cu/8In	195–201	>19,000
Melting T < 215°C, no Ag, Bi, Sb		
93Sn/0.5Cu/0,5Ga/6In	209–214	6,337
Melting T < 215°C, no In, Cu		
92Sn/3.3Ag/4.7Bi	210–215	3,850
Melting T < 215°C, no In		
93Sn/3.1Ag/3.1Bi/0.5Cu	209–212	6,522
Melting T < 215°C, no Cu		
91.5Sn/3.5Ag/1.0Bi/4.0In	208–213	14,000–16,000
Melting T > 215°C, no Bi, In		
96.2sn/2.5Ag/0.8Cu/0.5Sb	216–218	8,850
95.4Sn/3.1Ag/1.5Cu	216–217	6,000–9,000
96.5Sn/3.5Ag	221	4,186
99.3Sn/0.7Cu	227	1,125

ple is the Panasonic VTR controller card shown in Fig. 7.16 using Sn/Cu for waving soldering assembly.

In addition to the above systems and corresponding compositions, the eutectic for the binary Sn/Ag and Sn/Cu also are viable compositions. It should be noted that within an alloy system, only the specified compositions deliver the desired performance. This is not different from the Sn/Pb system—the specific composition (63Sn/37Pb) is the optimum. In summary, viable lead-free compositions are listed below:

- Sn/3.0-3.5Ag/1.0-4.8Bi
- Sn/3.0-3.5Ag/0.5-1.5Cu
- Sn/3.0-3.5Ag/3.0-3.5Bi/0.5-0.7Cu
- Sn/3.0-3.5Ag/0.5-1.5Cu/6.0-8.0In
- Sn/0.5-0.7Cu/5.0-6.0In/0.4-0.6Ga
- Sn/3.3-3.5Ag/1.0-3.0Bi/1.7-4.0In
- Sn/3.0-4.1Ag/2.2Bi/0.5Cu/8.0In
- 96.5Sn/3.5Ag
- 99.3Sn/0.7Cu

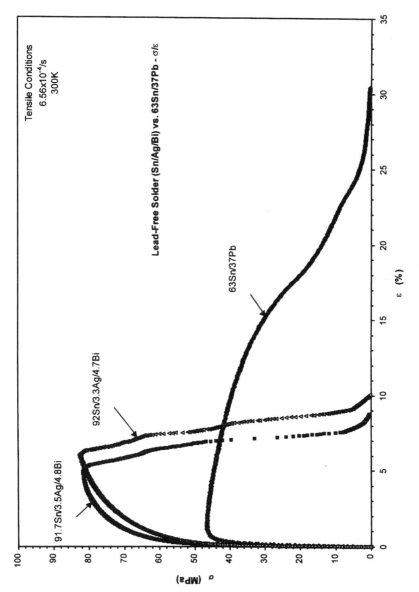

Figure 7.6 Tensile stress (σ) vs. strain (ε) at 300 K and 6.56×10^{-4}/s for Sn/Ag/Bi alloys and 63Sn/37Pb.

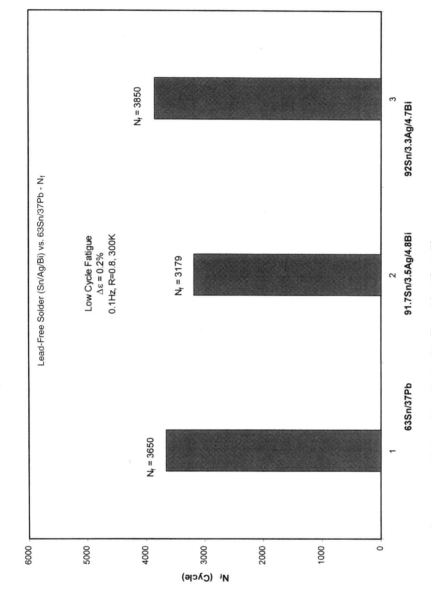

Figure 7.7 Comparison of fatigue life of Sn/Ag/Bi alloy with 63Sn/37Pb.

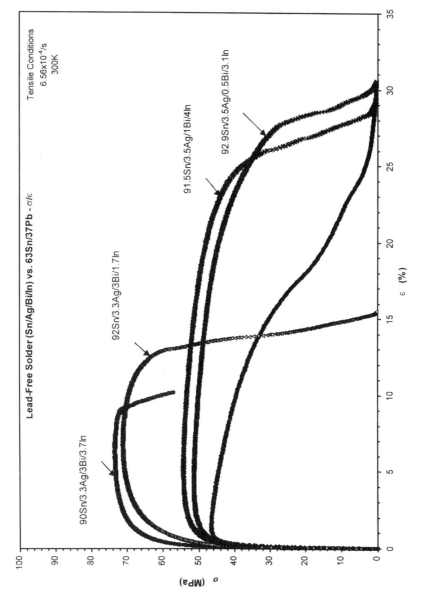

Figure 7.8 Tensile stress (σ) vs. strain (ε) at 300 K and 6.56×10^{-4}/s for Sn/Ag/Bi/In alloys and 63Sn/37Pb.

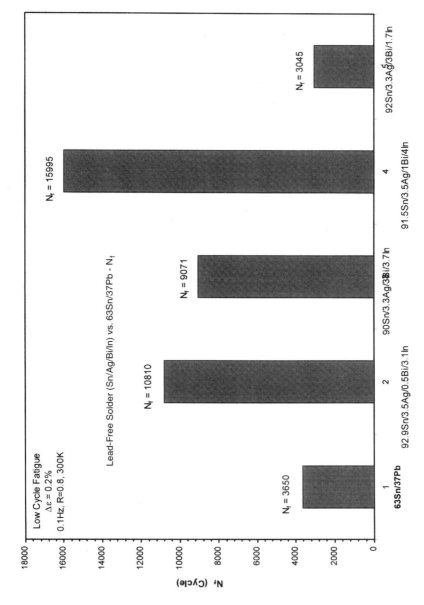

Figure 7.9 Comparison of fatigue life of Sn/Ag/Bi/In alloys with 63Sn/37Pb.

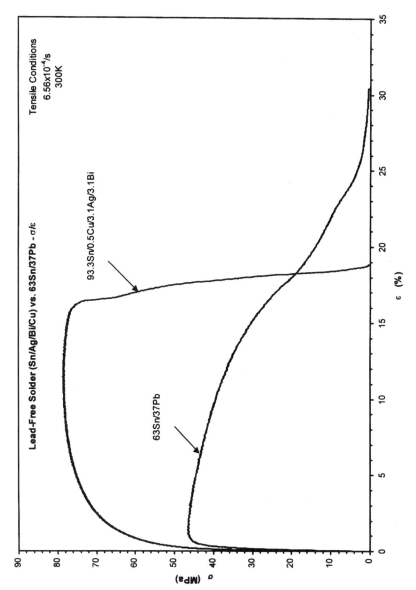

Figure 7.10 Comparison of fatigue life of Sn/Ag/Cu/Bi alloys with 63Sn/37Pb.

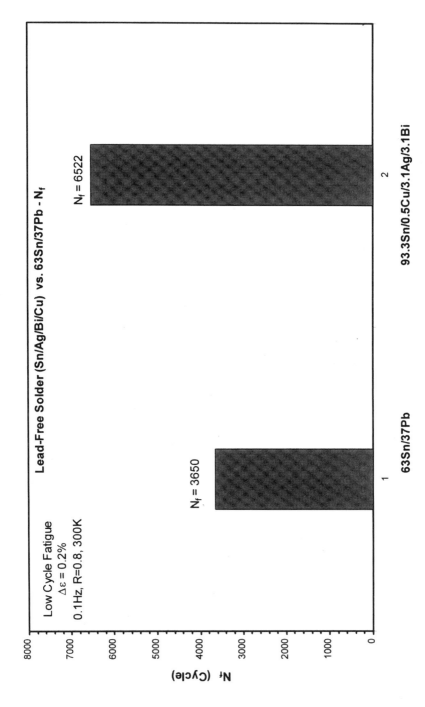

Figure 7.11 Comparison of fatigue life of Sn/Ag/Cu/Bi alloys with 63Sn/37Pb.

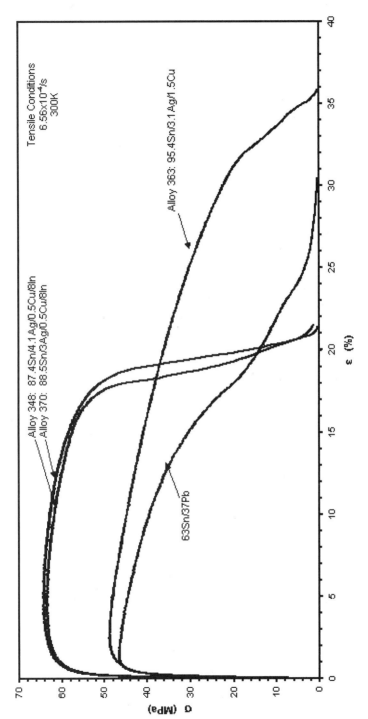

Figure 7.12 Comparison of fatigue life of Sn/Ag/Cu/In alloys with 63Sn/37Pb.

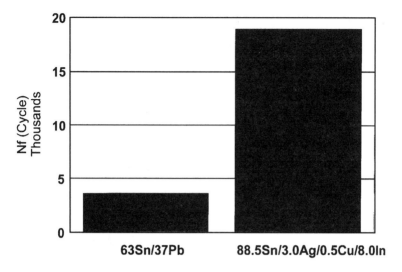

Figure 7.13 Comparison of fatigue life of Sn/Ag/Cu/In alloys with 63Sn/37Pb.

Figure 7.17 is a Panasonic image acquisition card used in the Panasert line of SMT assembly equipment using Sn/Ag/Bi/In for surface mount reflow and Sn/Cu eutectic for wave soldering.

Figures 7.18, 7.19, and 7.20 exhibit the Panasonic product evolution from player (MJ30, MJ70) to player/recorder (MR100) products, and Figures 7.21, 7.22, and 7.23 are the corresponding main circuit cards. All circuit boards were produced in lead-free with Sn/Ag/Bi/Cu alloy compositions.

7.8 Reflow Soldering

During soldering, a series of reactions and interactions occur in sequence or in parallel. These can be chemical or physical in nature in conjunction with heat transfer. The mechanism behind fluxing is often viewed as the reduction of metal oxides. Yet, in many situations, chemical erosion and dissolution of oxides and other foreign elements acts as the primary fluxing mechanisms. Using a more complex fluxing process in solder paste as an example, the primary steps are represented by the flowchart in Fig. 7.24.

Several events occur during this stage, as shown in Steps (II) and (IV) of the flow chart. These include temperature set to fit the specific flux activation temperature of the chemical system of the paste, and the time at heat to fit the constitutional make-up of the paste. Improper preheating often causes various problems, e.g., the spattering problem, which manifests itself as discrete solder balls and component

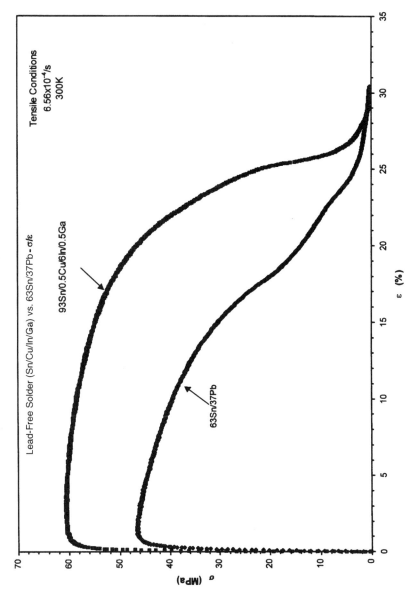

Figure 7.14 Tensile stress (σ) vs. strain (ε) at 300 K and 6.56×10^{-4}/s for Sn/Cu/In/Ga alloys and 63Sn/37Pb.

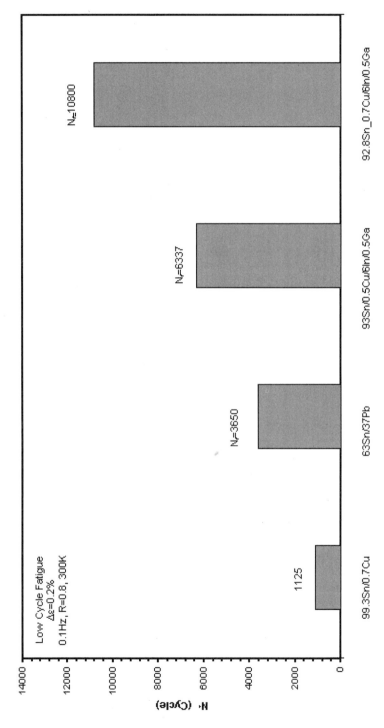

Figure 7.15 Comparison of fatigue life of Sn/Cu/In/Ga alloys with 63Sn/37Pb.

Figure 7.16 An example of manufacturing using wave soldering of Sn/Cu to successfully produce the Panasonic VTR controller cards.

Panasonic Panasert P861

Figure 7.17 An example of manufacturing using a reflow solder of Sn/Ag/Bi/In and a wave solder of SnCu to successfully produce the Panasonic Panasert image acquisition card.

damage. Too high a temperature or too long a time at the elevated temperature results in insufficient fluxing and/or over-decomposition of organic, causing solder balling or hard-to-clean residue (if the no-clean route is adopted). The third stage is to spike quickly to the peak reflow temperature at a commonly practiced rate of 1.0 to 4.0°C/s. The purpose of temperature spiking is to minimize the exposure time of the organic system to high temperatures, thus avoiding charring or overheating. Another important characteristic is the dwell time at the peak temperature. The rule of thumb in setting the peak temperature is 20 to 50°C above the liquidus or melting temperature; e.g., for the

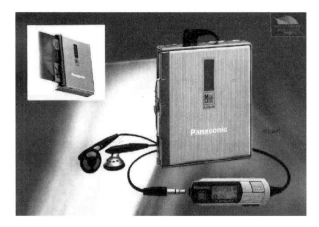

Figure 7.18 Panasonic minidisk player MJ30.

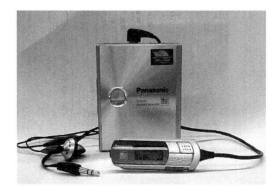

Figure 7.19 Panasonic minidisk player MJ70.

Figure 7.20 Panasonic minidisk player MR100.

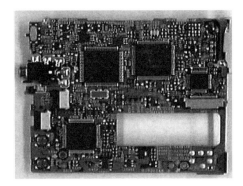

Figure 7.21 Main circuit card of Panasonic minidisk player MJ30.

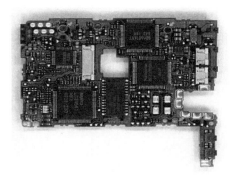

Figure 7.22 Main circuit card of Panasonic minidisk player MJ70.

Figure 7.23 Main circuit card of Panasonic minidisk player MR100.

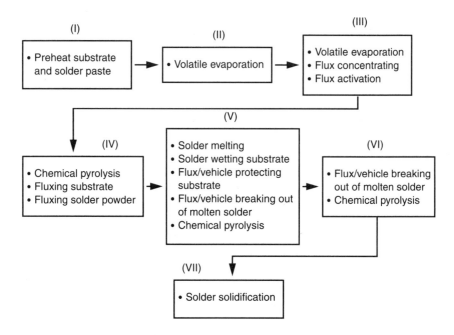

Figure 7.24 Key steps in the reflow process.

eutectic Sn/Pb composition, the range of peak temperatures is 203 to 233°C.

The wetting ability is directly related to the dwell time at the specific temperature in the proper temperature range and to the specific temperature being set. Other conditions being equal, the longer the dwell time, the more wetting is expected, but only to a certain extent; the same trend applies at higher temperatures. However, as the peak temperature increases or the dwell time is prolonged, the extent of the formation of intermetallic compounds also increases. An excessive amount of intermetallics can be detrimental to long-term solder joint integrity. Peak temperature and dwell time should be set to reach a balance between good wetting and to expel any non-solder (organics) ingredients from the molten solder before it solidifies, thus minimizing void formation.

With the prevalence of oven reflow, a few more words about oven heating profile and operating parameters are pertinent. It should be stressed that the reflow is a dynamic heating process in that the condition of the workpiece is constantly changing as it travels through the furnace in a relatively short reflow time. The momentary temperature that the workpiece experiences determines the reflow condition; therefore, the reflow results.

It is ultimately important to establish a correlation between setting the temperature of a given oven, the measured temperature of the workpiece at each specified belt speed, and the soldering performance. The resulting correlation between soldering performance and temperature setting or profile provides a "workable range" for the assembly.

Under mass reflow operation, both heating and cooling steps are important to the end results. It is generally understood that the heating and cooling rate of reflow or soldering process essentially contributes to the compositional fluctuation of the solder joint. This is particularly true when there are significant levels of metallurgical reactions occurring between the Sn/Pb solder and substrate metals. In the meantime, the cooling rate is expected to be responsible for the evolution of the microstructure.

7.8.1 Process Parameters

The key process parameters that affect the production yield as well as the integrity of solder joints include the following:

- preheating temperature
- preheating time
- peak temperature
- dwell time at peak temperature
- cooling rate

For a given system, cooling rate is directly associated with the resulting microstructure, which in turn affects the mechanical behavior of solder joints.[7]

It was found that the microstructural variation and corresponding failure mechanisms of solder joints that were made under various reflow temperature profiles are complex. Nonetheless, some correlation between the cooling rate and the basic properties can be obtained.

7.8.2 Reflow Temperature Profile

Reflow temperature profile representing the relationship of temperature and time during reflow process depends not only on the parameter settings but also on the capability and flexibility of equipment. Specifically, the instantaneous temperature conditions that a workpiece experiences is determined by:

- temperature settings to all zone controllers
- ambient temperature
- mass per board

- total mass in the heating chamber (load)
- efficiency of heat supply and heat transfer

7.8.3 Effects of Reflow Profile

The reflow profile used for surface mount manufacturing has a direct bearing on manufacturing yield, solder joint integrity, and the reliability of the assembly. Specific areas that are affected by reflow profile follow. Each area may be affected, to a different degree, by one or more of the three heating stages in the following ways:

- temperature distribution across the assembly
- plastic IC package cracking
- solder balling
- solder beading
- wetting ability
- residue cleanability
- residue appearance and characteristics
- solder joint voids
- metallurgical reaction between solder and substrate surface
- microstructure of solder joints
- board warpage
- residual stress level of the assembly

7.8.3.1 Uniformity of temperature distribution. In a normal reflow environment, temperature differential across the assembly is inevitable. This is due to the large disparity in mass and in the characteristics of the components coupled with the relatively short total reflow time (the entire cycle lasts only several minutes). A large temperature differential causes uneven soldering, resulting in localized cold joints or overheated joints. These problematic joints may contribute to manufacturing defects or jeopardize the long-term integrity of the solder joints under service conditions if they are not detected as manufacturing defects and corrected.

For a given oven, the rate of natural warm-up (°C/s) and the intended preheating temperature and time are the main factors that control temperature uniformity across the assembly. A slower heating rate in the warm-up state is desired to reach a more uniform board temperature distribution.

7.8.3.2 Plastic IC package cracking.[8]   Along with factors such as die size, the moisture sensitivity of the molding compound and its thickness, reflow profile plays an important role in causing or preventing plastic IC package cracks. When the IC package (e.g., BGA, QFP, SOIC) absorb a certain level of moisture during storage, handling, or in transit (without proper dry pack), the absorbed moisture may cause package cracking during reflow. Setting a proper reflow profile can mitigate the cracking problem; the heating rate from ambient temperature to 140 to 150°C is most critical.

7.8.3.3 Solder balling.   Elevated temperatures or excessive time at those temperatures during the warm-up and preheating stages can result in inadequate fluxing activity or insufficient protection of solder spheres in the paste, causing solder balling. In addition to the quality of solder paste, the presence of solder balls may be essentially related to the compatibility between the paste and the reflow profile. On the other hand, inadequate preheating or heating too fast may cause spattering, evidenced by random solder balls. The two heating stages preceding the spike/reflow zone are primarily responsible for this phenomenon.

7.8.3.4 Solder beading.   Solder beading refers to the occurrence of large size solder balls (usually larger than 0.005 in. in diameter) that are always associated with small and low-clearance passive components (capacitors and resistors). This problem will occur even when the paste may otherwise perform perfectly, i.e., free of solder balls at all other locations (components) on the board and with good wetting. The trouble with solder beading is that it may occur in most or all board assemblies, rendering the first-time yield to nearly zero. The current remedy on the production floor is to manually remove the beads.

  The formation of solder beads near or under capacitors and resistors is largely attributed to the paste flow into the underside of the component body between two terminations aided by capillary effect. As this portion of paste melts during reflow, it becomes isolated away from the main solder on the wettable solder pads, forming large discrete solder beads. In addition to other factors, reflow profile is one of the areas that contribute to this phenomena. The practice of adopting a slower preheating rate and a lower reflow peak temperature can reduce solder beading. However, if the reflow profile is at its optimum, and the problem still persists, a new paste with a strengthened chemistry is the solution. In addition, the pad design and the volume of the paste deposit that is in turn determined by the stencil design are other factors contributing to this beading phenomenon.

7.8.3.5 Wettability.
The temperature setting and time spent in both preheating and spike/reflow affect wettability. However, each stage works by a separate mechanism. In the preheating stage, the range of temperature and the time spent in this range directly affect the activity of flux. Wettability, in turn, is affected through the fluxing action. However, in the spike/reflow zone, wetting on the "cleaned" surface is influenced by the peak temperature because of the intrinsic wetting ability of molten solder alloy. This ability increases on a wettable substrate with higher temperature. With all other conditions being equal, a longer dwell time can, to a limited degree, further enhance wetting. Modification of the spike/reflow zone may sometimes solve a minor wetting problem.

7.8.3.6 Cleanability.
For solder paste designed to be cleaned, particularly water cleaned, excessive heat may make it difficult for the residue to be removed, rendering a normal cleaning process ineffective. In this case, all stages of the reflow profile can be contributors.

7.8.3.7 Residue appearance and characteristics.
The importance of the compatibility of the solder paste's chemical composition with the reflow profile can be readily demonstrated when using a no-clean soldering process. For instance, if the paste was reflowed with a temperature profile below the heat requirement, a higher amount of residue than expected will remain. In addition, the characteristics of that residue may range form being tacky to ionically active.

7.8.3.8 Solder joint voids.
Incomplete outgassing (gases entrapped in the solder joint) is the main cause of voiding. In addition to design factors, the compatibility between the reflow profile and the chemical makeup of solder paste is important. There should also be sufficient dwell time in the molten state (above 183°C for 63Sn/37Pb) to ensure that the gases have enough time to separate and escape from the molten solder. With respect to other factors contributing to voiding in solder joints, readers are referred to the literature.[9]

7.8.4 Optimal Profile[10,11]

The heat transfer from the surrounding hot air to the various components on the board, such as leaded packages, array packages, and discretes, differs during the process where a thermal equilibrium hardly exists. This disparity can be compensated for by either setting a reflow profile with a higher heat supply rate and higher temperature or one with a slower heating rate and lower temperature. On most manufacturing lines, unfortunately, a reflow profile with a higher heating rate and higher temperature has been used, as shown in Fig. 7.25.

This disparity in the heat transfer may be heightened as the large or heavy array packages are incorporated. Although increasing the temperature has accommodated most reflow results, the approach will not work well with heat-sensitive components or with PCBs that contain increasingly versatile components, particularly BGA, CSP, and flip chips.

The initial warm-up state plays a far more influential role in the quality and reliability of assembled boards than was first thought. An initial heating rate at less than 1°C/s in conjunction with the heating rate for the rest of profile at not more than 3°C/s is considered most beneficial and is recommended. Under SMT environments, the small degree of reduction in heating rate would not be a bottleneck for production throughput. By using the slower rate in the warm-up and preheating stages prior to reaching 183°C, the peak temperature can be maintained in the range of 210 to 215°C, in contrast to 215 to 230°C. The total dwell time above the liquidus temperature (183°C) falls in the range of 30 to 90 s.

Reflow profiles based on slower heating rates and cooler temperature (as for example, shown in Fig. 7.26) will be more in line with to-

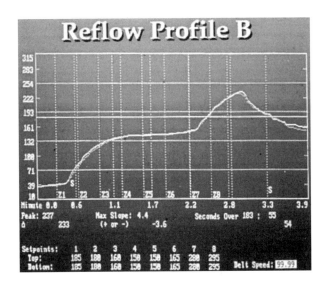

Figure 7.25 Reflow profile—with faster initial heating rate.

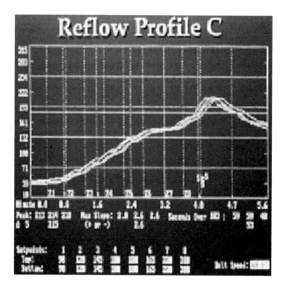

Figure 7.26 Reflow profile—with slower initial heating rate.

day's complex assemblies, minimizing in-process heat exposure as well as residual stress.

7.9 Inert and Reducing Atmosphere Soldering[12–18]

At the soldering temperature, the atmosphere surrounding the workpiece protects or interacts with the surface of substrates, the solder alloys, and the chemical ingredients in the flux-vehicle system. These interactions determine the chemical and physical phenomena in terms of volatilization, thermal decomposition, and surface-interfacial tension. A controlled atmosphere is expected to deliver a more consistent soldering process.

In addition to consistency, the inert or reactive atmospheres may potentially offer further merits. These include:

- improved temperature uniformity
- solderability enhancement
- solderability uniformity
- minimal solder balling
- irregular residue charring prevention
- reduced residue

- polymer-based board discoloration prevention
- wider process window
- overall quality and yield improvement

The inert and reactive atmospheres are expected to facilitate conventional fluxing efficiency during soldering. It should be noted, however, that performance results rely greatly on the specific atmospheric composition and its compatibility with the solder material, substrate, and chemicals incorporated in the system, which must also be compatible with the soldering temperature profile. Figure 7.27 shows that the solderability under N_2 atmosphere is significantly improved, as solder balls that are formed under ambient air are eliminated.

7.9.1 Process Parameters

The additional process parameters for inert and reducing atmosphere soldering include

- gas flow rate
- humidity and water vapor pressure
- oxygen level
- belt speed
- oven temperature settings

7.9.1.1 Gas flow rate. The gas flow rate required to achieve a specific level of oxygen in the dynamic state of the reflow oven is largely con-

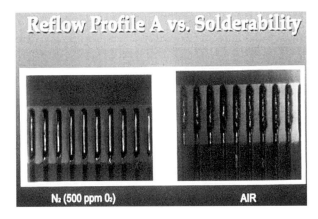

Figure 7.27 Solderability under N2 atmosphere is significantly improved.

trolled by the type of oven, categorically closed-system, semi-closed system, and open system. For a given oven, the required flow rate increases when the allowable oxygen level is lowered. At a given flow rate, when the air tightness in oven construction is reduced, the achievable oxygen level will be higher.

As expected, for a given reflow system, the oxygen level is inversely related to gas flow rate. The gas flow rate also affects the temperature distribution and temperature uniformity of assembly. The temperature gradient is reduced when the gas flow rate is increased. However, the downside of using high flow rate goes to the higher gas consumption, therefore increasing cost. The cost impact may be mitigated when the design of oven is capable of internal gas recirculation in an efficient fashion.

7.9.1.2 Humidity and water vapor pressure. Water vapor pressure inside the soldering govern can be contributed by

- the composition and purity of atmosphere
- the reaction product of flux/vehicle chemical system with metal substrates
- the moisture release from the assembly including components and board
- the ambient humidity

Because water vapor is essentially oxidizing to metal substrates that are to be joined by soldering, its partial pressure in the oven affects the overall function of the atmosphere.

The partial pressure of water vapor in an atmosphere gas is conveniently expressed as dew point, that is, the temperature at which condensation of water vapor in air takes place. The dew point can be measured by a hygrometer or dewpointer by means of fog chamber, chilled mirror aluminum oxide technique.

The purity of incoming gas in terms of moisture is normally monitored by measuring the dew point.

7.9.1.3 Belt speed. For an evenly spaced loading on the belt, the belt speed not only determines the throughput but also affects other operating parameters that can alter the soldering results.

As examples, the parameters that are affected by the change of belt speed include:

Peak temperature. At fixed temperature settings, increasing belt speed results in the decrease of peak temperature.

Atmosphere composition. While other conditions are equal, the change of belt speed may alter the oxygen level (including moisture content).

7.9.1.4 Temperature settings.
The operating temperature or temperature profile is an integral part of soldering process. It affects the physical activity and chemical reaction of the organic system in solder paste or flux. The operating temperature, particularly peak temperature, changes the wetting ability of molten solder on the metal substrate; wetting ability generally increases with increasing temperature. Chemical reactions and thermal decompositions respond to the rising temperature and the temperature profile.

7.9.1.5 Oxygen level.[17]
Various studies have focused on the application of no-clean process and on the determination of the maximum oxygen level allowed for using nitrogen-based no-clean soldering process in solder paste reflow and in wave soldering.

Each study was performed with a specific solder paste and flux or with a selected series of paste and flux. Tests were conducted with specific equipment and under a designated process. In view of the continued introduction of new equipment and the diversity of processes coupled with the versatility of solder paste and flux compositions, the test results are expected to represent the specific system (paste, oven, process, assembly) and, at best, to provide a guideline reference point. For example, a solder paste from the Vendor I, to be used with Process A, may require a maximum of 50 ppm oxygen level to obtain good solderability, grossly solder-ball free results, and acceptable after-soldering residue. To achieve the similar outcomes, the same paste to be used with Process B, may only need a maximum of 300 ppm oxygen. The same could be true for a different paste used in the same process.

The precise oxygen level requirement for a no-clean soldering depends on the characteristics of a system. The general principle and trends in the relationship between the performance feature and the allowable maximum oxygen level can be derived. Section 7.9.2 illustrates the range of the optimal oxygen level.

7.9.2 Optimal O_2 Level[17]

In general, when higher than 2,000 ppm O_2, the effect of nitrogen may hardly be detected. Below 20 ppm O_2, the process will become difficult to control and, needless to say, too costly. For a given oven and process, the O_2 level required is essentially controlled by the chemistry and makeup of the solder paste. For example, a solder paste from Supplier A may require a maximum level of 800 ppm O_2 to obtain the desirable results (good wetting, no solder balls, etc.). To achieve similar results, solder paste from Supplier B may need a maximum of 200 ppm O_2. In practice, O_2 levels in the range of 20 to 2,000 ppm should accommodate most applications.

Soldering under nitrogen poses two additional demands—more stringent process control and higher operating cost. However, its potential effects on solderability, heat transfer, PCB materials, and the process window may bring benefits in mounting large area and heavy BGAs, as well as in connecting small and delicate CSPs onto complex PCBs.

7.9.3 Temperature Measurement[18]

A low mass, direct, and firm contact without the need for extraneous attaching material is one way to meet the criteria for achieving accurate temperature measurement.

7.9.4 Lead-Free Reflow Profile

Considering the fact that the melting temperature of the viable lead free compositions is higher than that of Sn/Pb eutectic (or 183°C), a proper reflow profile is particularly important to minimizing the reflow temperature requirement.

Based on the existing constraints under the SMT production establishment and infrastructure including the temperature tolerance level of components and the PCBs, the peak reflow temperature should be kept under 235°C. Consequently, this poses a limit on the melting temperature of lead-free alloys to be below around 215°C. However, the peak temperature required to reflow a lead-free solder paste with melting temperature of 215°C can be reduced, to some extent, by the optimized temperature-time relation prior to the melting zone, that is by the preheating stage of the profile.

7.10 Printing

In addition to the selection of solder paste, the printer, and the settings of printing parameters, major factors contributing to the printing results of solder paste include stencil thickness versus aperture design, stencil aperture versus land pattern, and stencil selection.

7.10.1 Stencil Thickness vs. Aperture Design

When printing solder paste, the design of the relative dimensions of stencil thickness and stencil aperture is to achieve the balance between the printing resolution and the proper amount of solder deposit so as to avoid starved solder joints or pad bridging. For a selected stencil thickness, too small a stencil aperture width leads to open joints or starved joints. Too large an aperture width causes pad bridging. Table 7.4 provides guidelines for designing stencil thickness in relation to aperture.

Table 7.4 Guidelines in Designing Stencil vs. Aperture Size

Component lead pitch		Aperture width		Maximum stencil thickness	
(inch)	(mm)	(inch)	(mm)	(inch)	(mm)
0.005	1.26	0.023	0.58	0.0140	0.35
0.025	0.63	0.012	0.30	0.0075	0.19
0.020	0.50	0.010	0.25	0.0063	0.16
0.015	0.38	0.007	0.18	0.0043	0.11
0.008	0.20	0.004	0.10	0.0025	0.06

7.10.2 Stencil Aperture Design vs. Land Pattern

To make solder joints by one-pass printing process, the stencil thickness must be selected for transferring a sufficient amount of paste onto the non-fine-pitch solder pads while avoiding the excessive amount of paste deposited onto the fine-pitch pads. There are several options to achieve the deposition of a proper amount of solder paste on the land pattern to accommodate a mix of sizes of solder pads. These are as outlined below.

1. Step-down stencil

    This is commonly achieved by chemically etching the non-fine-pitch pattern area from one side of the stencil while etching the step down area for fine pitch pattern to the other side during double-sided etch process. Alternatively, step-down area is etched in one foil and non-fine pitch pattern etched in the other foil, then we register and glue the two foils together.

    The practical step gradient is 0.002 in. (0.05 mm); some common combinations are

    0.008 in. (0.20 mm) for non-fine pitch

    0.006 in. (0.10 mm) for fine-pitch

or

0.006 in. (0.15 mm) for non-fine pitch

0.004 in. (0.15 mm) for fine-pitch

2. Uniform reduction on four sides of apertures

   The dimensions of the fine-pitch aperture on stencil are reduced by 10–30 percent in relation to those of the land pattern. This reduces the amount of paste deposition on the fine-pitch land pattern, and it also provides some room for printing misregistration and paste slump, if any.

3. Staggered print

   The opening of the stencil is only one-half length of the solder pad and arranged in an alternate manner, as shown in Fig. 7.28.

   For tin-lead coated solder pads, when the paste starts to melt during reflow, the molten solder is expected to flow to the other half of the pad, making the complete coverage. With a bare copper or nickel surface, the molten solder may not flow out to cover the area upon which the paste is not printed.

4. Length or width reduction

   The dimensions of the stencil opening are reduced along the length or along the width by 10–30 percent in relation to that of solder pads, achieving the reduction of the amount of paste deposited.

5. Other shapes

   The stencil openings are made with selected shapes, such as a triangle or a teardrop, to achieve the reduced solder paste deposition on fine pitch pattern.

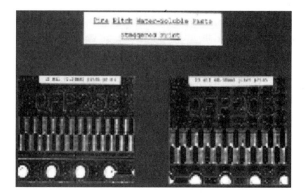

Figure 7.28 Staggered print.

6. Compromised stencil thickness

    Instead of using the specific thickness that is considered as the most suitable for a specific land pattern, select a thickness that is practical to both fine-pitch and non-fine-pitch patterns.

7.10.3 Stencil Selection

The performance of stencils is primarily driven by the foil metal and the process being used to create the printing pattern. Currently, five types of stencil materials are commercially available—brass, stainless steel, molybdenum, alloy-42, and electroformed nickel. The processes for making the stencils may involve chemical etching, laser-cutting, electropolishing, electroplating, and electroforming. Each type of foil or fabricating process possesses inherent merits and limitations. The key performance of a stencil is assessed by the straight vertical wall, wall smoothness, and dimensional precision. In addition, durability, chemical resistance, fine opening capability, and cost are also important factors.

Table 7.5 compares various stencil materials, and Table 7.6 summarizes the relative performance characteristics of stencil-making techniques.[19]

7.11 PCB Surface Finish

For making sound interconnections, the characteristics and properties of circuit board surface finish are as important as the component leads and termination.

Hot air solder leveled (HASL) SnPb has been successfully used as the surface finish for surface mount and mixed PCBs. As the achievement of a flat surface with uniform thickness becomes increasingly important to forming consistent and reliable fine-pitch solder joints, the HASL process often falls short. Alternatives to HASL include immersion Sn, electroplated SnPb (reflowed or nonreflowed), electroplated Au/Ni, electroless Au/electroless Ni, immersion Au/electroless Ni, immersion Pd, immersion Pd/electroless Ni, electroplated Sn-Ni alloy, and organic coating. When selecting an alternative surface finish for PCB assembly, the key parameters to be considered are solderability, ambient stability, high-temperature stability, suitability for use as contact/switch surface, solder joint integrity, wire bondability of those assemblies that involve wire bonding, and cost.

Ideally, a PCB surface finish fulfills four functions:

- Solderability protection
- Contact/switching

Table 7.5 Comparison of Various Stencil Materials

Performance	Material				
	Brass	Stainless steel	Molybdenum	Alloy 42	Ni (Electroforming)
Mechanical strength	Unfavorable	Favorable	Favorable	Favorable	Favorable
Chemical resistance	Unfavorable	Favorable	Unfavorable	Favorable	Favorable
Etchability	Favorable	Less Favorable	Favorable	Favorable	N/A
Sheet stock availability	Favorable	Favorable	Unfavorable	Favorable	N/A
Cost	Favorable	Less Favorable	Unfavorable	Less favorable	N/A
Fine pitch	Favorable	May need electropolishing	Favorable	May need electropolishing	Most favorable
Unique feature	Lowest cost	Durable	Self-lubricating smooth wall	Durable	Finest opening

348

Table 7.6 Comparison of Stencil-Making Techniques

Techniques	Characteristics	Superior capabilities or features
Chemical etching	Most established process; sensitivity of fine pitch capability to process and control; sensitivity of aperture size and vertical wall control	Versatile, economical
Laser cut	Grainy wall surface; sequential cut, not concurrent formation of openings; higher cost; difficulty in making step stencil	Fine pitch capability; no photoresist needed
Electroforming	Additive process via electrodeposition; concern about fine foil strength; difficulty in making step stencil; suitable for stencil of less than 0.004 in.	Gasket effect, minimizing bleeding, no need for electropolishing
Electropolishing	Complementary step to produce smooth wall surface	Smooth wall surface
Ni plating on aperture wall (polished or unpolished)	Reducing aperture opening; smooth surface	Finer opening

- Wire bonding
- Solder joint interface

However, in practice, some surface finish systems are designed primarily for solderability protection.

For solderability retention and protection, hot air solder leveled (HASL) SnPb has been successfully used as the PCB surface finish for surface mount and mixed PCB assemblies. As the industry continues to evolve, the following driving forces are primarily behind the development of HASL alternatives:

- Increased demands for flat and uniform solder pads
- Increased demands for consistent thickness of surface finish
- Obtaining the same metal system and process for contact/switch
- Less thermal stress process for temperature-vulnerable PCBs, such as PCMCIA
- Elimination of Pb

7.11.1 Solderability Factors

Several factors affect the solderability of PCB solder pads:

- Pad surface composition
  - Copper
  - Tin-lead-coated copper
  - Antioxidant-coated copper
  - Gold/nickel/copper
  - Palladium-nickel-copper
- Surface conditions
  - Oxides, sulfides content
  - Organic contaminants
  - Intermetallics
  - Other contaminants
- Thickness of coating
  - Determination of a proper thickness in relation to storage conditions
- Storage conditions
  - Time
  - Temperature
  - Humidity

Although the required coating thickness may vary, a proper thickness is the one that is compatible with the time and condition of storage to avoid the excessive formation of intermetallics and the exposure of intermetallics to the ambient environment. Generally, the lower the temperature and the humidity, the less degradation of solderability with time.

7.11.2 Basic Processes

Three basic techniques used to deposit metallic surface finish are electroplating, electroless plating, and immersion.

Inherently, electroplating utilizing an electric current is able to economically deposit thick coating up to 0.000400 in. The exact thickness depends on metal and process parameters. Electroless plating, requiring the presence of a proper reducing agent in the plating bath, converts metal salts into metal and deposits them on the substrate. The immersion plating process, in the absence of both electric current and the reducing agent in the bath, deposits a new metal surface by replacing the base metal. In this process, plating stops when the surface of base metal is completely covered; thus only a limited coating thickness can be obtained through an immersion process. For both electroless and immersion processes, the intricate chemistry and the control of kinetics are vital to the plating results. Furthermore, the designed process parameters and chemistry, including pH and chemical ingredients, must be compatible with the solder mask and PCB materials.

To summarize, the respective characteristics of these three processes are as follows:

- Electrolytic
  - Requires electric current
  - Economical
  - Wide operating window
  - Can serve as etch resist

- Electroless
  - Requires abundant reducing agent in plating bath
  - Reducing agent is metal-specific, e.g., Ni: hypophosphite, dimethylamine borane; Cu: formaldehyde
  - Chemistry is critical
  - Higher cost
  - Uniform coating

- Immersion
  - No electric current used
  - Not requiring reducing agent
  - Limited thickness
  - Chemistry is critical
  - Temperature important

7.11.3 Metallic System

Available metallic surface finishes on copper traces include Sn, SnPb alloy, SnNi alloy, Sn/Bi, Sn/Sb, Au/Ni, Au/Pd, Pd/Ni, and Pd. The systems containing noble or semi-noble metals, such as Au/Ni, Au/Pd, Pd/Ni, Pd/Cu, are capable of delivering the coating surface with uniform thickness. Those systems imparting a pure and clean surface also provide a wire-bondable substrate. In addition, wire bonding generally requires thicker coating, namely, more than 0.000020 in. A unique feature of an Au/Ni system is its stability toward elevated temperature exposure during the assembly process as well as during its subsequent service life. When in contact with molten solder of SnPb, SnAg, or SnBi, surfaces coated with Sn and SnPb are normally associated with better spreading and lower wetting angle than others. Of the metallic systems, those containing a Ni interlayer are expected to possess the more stable solder joint interface; in these systems, solder is expected to wet on Ni during reflow. For a phosphorus-containing plating bath, a balanced concentration of phosphors in electroless Ni plating is needed. When the P content is too high, wettability suffers; when it is too low, thermal-stress resistance and adhesion strength are sacrificed. In addition, noble metals in solder need to be noted so as to prevent any adverse effects in solder joining integrity.

Another characteristic that is important to solderability is the porosity on the surface. Thinner coating is more prone to porosity and therefore the porosity-related problems, although the surface density and texture can be controlled, in part, by the chemistry and kinetics.

The following list a number of specific systems, their characteristics, and the functions fulfilled:

- Electroless Pd/Cu
  - 4–20 μin Pd
  - solderability protection
  - contact/switch

- Electroless Pd/Ni/Cu
  - Pd thickness a function of dwell time, deposition rate dependent on temperature
  - 4–20 μin Pd, 150–200 μin Ni
  - processing temperature at 60–70°C
  - solderability protection
  - contact/switch
  - wire bonding

- Immersion Pd/Cu
  - dissolve Cu to deposit Pd
  - 2.5-3.0 μin Pd
  - controlled base metal plating bath
  - dense and low-porosity coating achievable
  - can reach maximum deposit thickness in two minutes at 50°C
  - porosity related to the plating time and the strength of the solution
  - surface may discolor after multiple passes in reflow over, but solderability retained
  - solderability protection

- Immersion Au/electroless Pd/Cu
  - <0.1 μin Au, 10–25 μin Pd

– solderability protection
  – contact/switch
  – wire bonding

- Immersion Au/electroless Pd/electroless Ni/Cu
  – <0.1 µin Au, 10–25 µin Pd
  – 125–150 µin Ni
  – solderability protection
  – contact/switch
  – wire bonding
  – stable solder joint interface

- Immersion Au/electroless Ni/Cu
  – 3–5 µin Au, 50–200 µin Ni
  – vulnerable to Au porosity
  – solderability protection
  – contact/switch
  – wire bonding
  – stable solder joint interface

- Electrolytic Au/electrolytic Ni
  – harsh plating condition to solder mask
  – difficult to get adequate Ni plating in small vias in thick board
  – avoids "black pad" problem for some BGA assemblies
  – solderability protection
  – contact/switch
  – wire bonding
  – stable solder joint interface

- Immersion Au/electrolytic Ni
  – good solderability with electrolytic Ni (vs. electroless)
  – solderability protection
  – contact/switch
  – wire bonding
  – stable solder joint interface

- Electrolytic Au/electroless Ni
  – characteristics of electrolytic Au (10–100 µin)

- Immersion Bi
  – 4–6 µin Bi
  – flat surface

- Immersion Sn
  - Sn deposited on Cu while Cu is transferred into solution
  - 2.5–35 μin Sn
  - incorporating an organo-metallic complex
  - plating bath temperature around 65°C
  - plating solution with stannous sulfate or chloride
  - Sn/Cu intermetallics
  - flat surface
  - not wire bondable
  - not a good switch material
  - cost competitive

- Immersion Ag
  - Ag deposited on Cu while Cu is transferred into solution
  - About 5 μin Ag
  - incorporating an oxidation inhibitor layer
  - plating solution with silver nitrate
  - flat surface
  - may be wire bondable
  - not a good switch material at thin deposition (3–9 μin)

- HASL Sn/Cu eutectic
  - operating temperature higher than Sn/Pb
  - potential effects of higher temperature on PCB to be considered

Table 7.7 Operating Parameters of HASL Sn/Cu vs. HASL Sn/Pb[20]

	Sn/Pb	Sn/Cu
Bath temperature	250	280
Air knife temperature	250	280
Oil temperature	230	255
Air heat exchanger	250	300
PCB preheat	150	200

7.11.4 Organic System

Benzotriazole has been well recognized as an effective Cu antitarnish and antioxidation agent for decades. Its effectiveness, attributable to the formation of benzotriazole complex, is largely limited to ambient temperature.

As temperature rises, the protective function disintegrates. Azole derivatives, such as Imidazole (m.p. 90°C, b.p. 257°C) and Benzimidazole (m.p. 170°C, b.p. 360°C) have been used to increase the stability under elevated temperature. SMT assembly of mixed boards involves three stages of temperature excursion-reflow, adhesive curing, and wave soldering. The reflow step, however, is considered to be potentially most harmful to the intactness of organic coating, because it is the step with the highest temperature and longest exposure time. In addition to its vulnerability to high temperature, this family of chemicals has an appreciable extent of solubility in alcohol and water.

Although the performance of organic coating varies with the formula and process, the general behavior of organic coating falls in the following regimen:

- It needs a compatible flux (generally more active flux).
- It may need more active flux in wave soldering for mixed boards.
- Thicker coating is more resistant to oxidation and temperature, but may also demand more active flux.
- Organic coating needs to be preformed as the last step of PCB fabrication.
- At temperatures higher than 70°C, coating may degrade. However, the degradation may or may not reflect on solderability.
- It may be sensitive to PCB presoaking process (e.g., 1250, 1 hr to 24 hr).
- For no-clean chemistry, it may require N_2 atmosphere or higher solids content in no-clean paste.
- The steam aging test is not applicable.
- It is not suitable for chip on board where wire bonding is required.

Nonetheless, when fluxing activity and the process parameters are compatible, an organic coating can be a viable surface finish for PCBs. An additional bonus effect is that the bare copper appearance of the organic coated surface enhances the ease of visual inspection of peripheral solder fillets.

7.11.5 Comparison of PCB Surface Finish Systems

When selecting an alternative surface finish for PCB assembly, the key parameters in terms of solderability, ambient stability, high-temperature stability, suitable for the use as contact/switch surface, solder joint integrity, and wire bondability for those assemblies that involve wire bonding, as well as cost, are to be considered.

Table 7.8 summarizes the relative performance of PCB surface finish systems. Regardless of other deficiencies, however, HASL provides the most solderable surface. When comparing metallic systems with HASL, the HASL process subjects PCBs to high temperatures (above 200°C), producing inevitable thermal stress in the PCB. HASL is also not suitable for wire bonding. To make a choice in replacing HASL, many variables are to be assessed. Understanding the fundamentals behind each variable in conjunction with setting proper priority of importance among the variables for a specific application is the way to reach the best, balanced solution.

Table 7.8 Relative Performance of PCB Surface Finishes

HASL	Au/Ni	Pd/Ni	Pd/Cu	Organic
Pros				
Most solderable	Uniform thickness Most stable to T.	Uniform thickness Wire bondable	Uniform thickness Wire bondable	Uniform thickness Low cost Easy inspection
Cons				
Nonuniform thickness	Higher cost	Higher cost	Higher cost (thicker coating)	Unsuitable for COB
Potential IMC problem				Flux and reflow process sensitive
Unsuitable for COB				Hi T degradation
PCB exposed to high T.				Cu reaches upper
				Limit in solder bath Required as a last Board fabrication step

Wetting spread is compared among various PCB surface finishes for three solder alloys, under convection air reflow. As shown in Table 7.9, less spread is associated with the lead-free alloys in comparison with 62Sn/36Pb/2Ag, and the extent of wetting spread was essentially similar except that Sn/Ag/Cu on immersion Pd is the lowest among the systems tested. Equivalent tests were conducted under vapor phase soldering, which provided a somewhat protective atmosphere during soldering. Table 7.10 shows that the wetting spread improved over air

Table 7.9 Wetting Spread of Three Alloys on Various PCB Surface Finishes Under Convection Air Flow[21]

PCB surface finish	62Sn/36Pb/2Ag	Sn/3.8Ag/0.7Cu	Sn/3.3Ag/3.0Bi/1.1Cu
OSP	4.5	4.2	4.0
Immersion Ag	4.7	4.55	4.6
Immersion Pd	4.4	3.9	4.4
Ni/Au	5.0	4.4	4.7

** on relative scale of 1 to 5 with 5 the complete spread.

Table 7.10 Wetting Spread of Three Alloys on Various PCB Surface Finishes Under Vapor Phase Reflow[21]

PCB surface finish	62Sn/36Pb/2Ag	Sn/3.8Ag/0.7Cu	Sn/3.3Ag/3.0Bi/1.1Cu
OSP	5.0	4.3	4.5
Immersion Ag	4.7	4.8	5.0
Immersion Pd	4.7	3.9	4.7
Ni/Au	5.0	5.0	5.0

**on relative scale of 1 to 5 with 5 the complete spread.

reflow, and still the Sn/Ag/Cu alloy on immersion Pd showed the lowest spread. Furthermore, Ni/Au showed the most consistent wetting spread for all three alloys.

7.12 "Green" Manufacturing

In addition to technological advances, the awareness of environmental and safety/health issue is intensifying. Environment-friendly manufacturing and the delivery of environmentally benign end-use products that are ultimately safe at the end of the product life cycle will become essential to technology business competitiveness. To achieve this goal, electronics manufacturers need to eliminate the use of highly toxic materials from products and encourage recycling. This is a continuing challenge to the industry.

At this time, most noble elements that are considered to be unfriendly to environment include lead in solders, bromide, chloride, and phosphorus in PCB fabrication. Other issues include VOC reduction,

Wastewater recycling; these have been managed and implemented with qualified success. The elimination of lead and lead-containing solders remains a pending issue and is expected to be actively pursued.

In the global market, the record shows that eliminating the use of CFCs once was deemed "impossible" by various industries, yet the Montreal Protocol turned out to be a "glorious success." Within the electronics industry, the no-clean process, once considered disruptive, eventually evolved to deliver the desired performance. This level of performance did not happen until a period of development and refinement has passed. If implementing no-clean process is a precedent and an example, with effort, equal success for lead-free substances will also materialize in the foreseeable future. Overall, environmentally friendly production, ranging from design to material to process, will be critical to the future manufacturing. Environmental stewardship is becoming an integral part of corporate policies.

7.13 References

1. Singer, P., DRAMs Microprocessors to Drive Technology in the 90's, *Semiconductor International,* February 1993, 17.
2. In-Stat, Incorporated, Scottsdale, Arizona.
3. Hwang, J. S., Solder Paste in Electronics Packaging—Technology and Applications for Surface Mount, Hybrid Circuits and Component Assembly, Von Nostrand Reinhold, New York, 1989, Chapters 2 and 3.
4. Hwang, J. S., *Modern Solder Technology for Competitive Electronics Manufacturing,* McGraw-Hill, New York 1996, Chapter 3.
5. Hwang, J. S., Low-Cycle Fatigue vs. Thermo-mechanical Fatigue, *Surface Mount Technology,* 1995, January.
6. Hwang, J. S., Environment-Friendly Electronics—Lead Free Technology, *Electrochemical Publications,* Great Britain, 2001
7. Hwang, J. S., *Modern Solder Technology for Competitive Electronics Manufacturing,* McGraw-Hill, New York 1996, Chapters 6 and 12.
8. Hwang, J. S., Practical consideration to minimize BGA cracks, *Surface Mount Technology,* August 1996.
9. Hwang, J. S., Voids in Solder Joints, *Surface Mount Technology,* September 1996.
10. Hwang, J. S., Effects of Reflow Temperature profile, *Surface Mount Technology,* June 1996.
11. Hwang, J. S., Optimal mass reflow profile, *Surface Mount Technology,* July 1996.
12. Hwang, J. S., Nitrogen Atmosphere soldering—passe or future, *Surface Mount Technology,* August 1998
13. Hwang, J. S., Controlled Atmosphere Soldering—Principles and Practice, *Proc. NEPCON West,* 1539–1546, 1990.
14. Cox, N. R., The influence of Varying Input Gas Flow on the Performance of a Nitrogen/Convection Oven, *Proc. NEPCON East,* 1994, 323.
15. Hwang, J. S., Soldering and Solder Paste Prospects, *Surface Mount Technology,* October 1989.
16. Ford and Lensch, P. J., Cover Gas Soldering Leaves Nothing to Clean off PCB Assembly, *Electronic. Packaging and Products,* April 1990.
17. Hwang, J. S., Optimum oxygen level for nitrogen atmosphere soldering, *Surface Mount Technology,* March 1995.

18. Hwang, J. S., Reflow Profiling—Temperature Measurement, *Surface Mount Technology,* May 1997.
19. Hwang, J. S., *Modern Solder Technology for Competitive Electronics Manufacturing,* McGraw-Hill, New York 1996, Chapter 10.
20. Snowdon, K., Lead free—the Nortel Experience, *Conference Proc. IPC Works'99,* October 1999, S-05-1-4
21. Feldmann, K., and Reichenberger, M., Assessment of lead-free solders for SMT, *Conference Proc. APEX, 2000,* P-MT2/2-3.

7.14 Suggested Readings

United States Patent #5,229,070
United States Patent #4,778,733
United States Patent #5,102,748
United States Patent #5,538,686
United States Patent #4,879,096
United States Patent 5,527,628
United States Patent #4,670,217
United States Patent # 5,520,752
United States Patent # 5,985,212
United States Patent # 6,176,947
International Patent: WO 94/25634
Hwang, J. S., Design and use of solder paste for system reliability, *Surface Mount Technology,* March 1997.
Hwang, J. S., *Ball Grid Array and Fine Pitch Peripheral Interconnections,* Electrochemical Publications, Great Britain, 1995, 157–162.
Hwang, J. S., *Ball Grid Array and Fine Pitch Peripheral Interconnections,* Electrochemical Publications, Great Britain, 1995, Chapter 7.
Hwang, J. S., *Environment-Friendly Electronics—Lead Free Technology,* Electrochemical Publications, Great Britain, 2001, Chapter 4.
Hwang, J. S., Z. Guo, Lead-Free Solders for Electronic Packaging and Assembly, *Proc. SMI Conference, 1993,* 732.
Hwang, J. S., Overview of Lead-Free Solders for Electronic Microelectronics, *Proc. Surface Mount International, 1994,* 405.
Hwang, J. S., Koenigsmann, H., New Developments of Lead-Free Solders, *Proc. Surface Mount International,* 1997.
H-Technologies Group—Internal Reports, 1996, 1997, 1998.
The list of 54 references in *Modern Solder Technology for Competitive Electronics Manufacturing,* McGraw-Hill, New York, 1996, Chapter 15.
Lucey, G. K., Wasynczuk, J. A., Clough, R. B., and Hwang, J. S., Composite Solders, United States Patent: 5,520,752, 1996.
Vianco, P. T., Rejent, J. A., Tin-Silver-Bismuth Solders for Electronics Assembly, United States Patent: 5439639, August 8, 1995.
Seelig, K. F., Lead-Free and Bismuth-Free Tin Alloy Solder Composition, International Patent: WO 94/25634, November 10, 1994.
Anderson, I. E., Yost, F. G., Smith, J. F., Miller, C. M., Terpstra, R. L., Iowa State University Research Foundation, Inc. and Sandia Corporation, Pb-Free Sn-Ag Ternary Eutectic, United States Patent 5,527,628, June 18, 1996.
Hwang, J. S., Z. Guo, Lead-Free Solders for Electronic Packaging and Assembly, *Proc. SMI Conference, 1993,* 732.
Allenby, Ciccarelli, J. P., Artaki, I., Fisher, J. R., et al, An Assessment of the Use of Lead in Electronic Assembly, *Proc. SMI Conference,* 1992, 1.
Hwang, J. S., Lucey, G., Clough, R. B., and Marshall, J., Futuristic Solders—Utopia or Ultimate Performance, *Surface Mount Technology,* September 1991, 40.
Hampshire, W. B., The Search for Lead-Free Solders, *Proc. SMI Conference, 1992,* 729.
Greaves, J. B., Evaluation of Solder Alternatives for Surface Mount Technology, *Proc. NEPCON West,* 1993, 1479.

Gilleo, K., The Polymer Electronics Revolution, *Proc. NEPCON West*, 1992, 1390.
Benton, D. H., Lead Removal From a New Aqueous Cleaning Agent Before Discharge, *Proc. NEPCON West*, 1994, 1346.
Dawson, C., Schultz, C. J., An Assessment of Lead Exposure to Wave Solder Machine and Solder Pot Preventive Maintenance Workers in an Electronics Assembly Operation, *Proc. NEPCON West*, 1994, 619.
Warwick, J. H., Vincent, P. G. Harris, S. R. Billington, et al., Screening Studies on Lead-Free Solder Alloys, *Proc. NEPCON West*, 1994, 874.
Napp, D., NCMS Lead-Free Electronic Interconnect Program, *Proc. SMI, 1994.*
Hwang, J. S., Innovation, Leadership and Competitiveness, *Surface Mount Technology*, May 1993.
Grivas, K. L. Murty, and Morris, J. W., Deformation of Sb-Sn Eutectic Alloys at Relatively High Strain Rates, *Acta Metall.* 27, 1979, 731.
Tribula, D. G., and Frear, R. D., Microstructure Observation of Thermomechanically Deformed Solder Joints, *Welding Research Supplement*, October, 1978, 404s.
Hwang, J. S., Vargas, R. M., Solder Joint Reliability—Can Solder Creep?, International Symposium on Microelectronics, *ISHM, 1989.*
Summece, J. W. M., Isothermal Fatigue Behavior of Sn-Pb Solder Joints, *Transactions of ASME*, vol. 112, June 1990, 94.
Guo, A. F. S., and Conrad, H, Plastic Deformation Kinetics of Eutectic Pb-Sb Solder Joints in Monotonic Loading and Low-Cycle Fatigue, *Journal of Electronic Packaging, Transactions of ASME*, vol. 114, June 1992.
Vaynman, M. F. F., and Jeannott, D. A., Low-Cycle Isothermal Fatigue Life of Solder Materials, *Solder Mechanics*, The Minerals, Metals and Materials Society, Chapter 4, 155.
Frear, D., Thermomechanical Fatigue in Solder Material, *Solder Mechanics*, The Minerals, Metals and Materials Society, Chapter 5, 192.
Marshall, J. L., Walter, R. and S., Fatigue of Solders, *The International Journal for Hybrid Microelectronics, 1987,* 261.
Materials for Electronic Packaging, *Proc. SMI*, 1993, 662.
Hwang, J. S., *Solder Paste In Electronic Packaging-Technology and Applications in Surface Mount, Hybrid Circuits, and Component Assembly*, Von Nostrand Reinhold, New York, 1989, Chapter 9, 282.
Nordyne, Lead Products, *Ceramic Bulletin*, vol. 70, no. 5, 1991, 872.
Silverstein, K., Lead Issues Gain Steam at State and Federal Levels, *Modern Paint and Coatings*, April, 1993, 30.
Hwang, J. S., A Strong Lead-free candidate—SnAgCuBi, *Surface Mount technology*, August 2000.
Hwang, J.S., Another Strong Lead-free candidate—SnAgBiIn, *Surface Mount technology*, September 2000.
Artaki, A. M. J., and Viance, P. T., Fine Pitch Surface Mount Assembly with Lead-Free, Low Residue Solder Paste, *Proc. SMI*, 1994.
Viano, P. T., Artaki, I., and Jones, A. M., Assembly Feasibility and Reliability Studies of Surface Mount Boards Manufactured with Lead-Free Solders, *Proc. SMI Conference, 1994.*
Melton, C., Skipor A., and Thome, J., Material and Assembly Issues of Non-Lead Bearing Solder Alloys, *Proc. NEPCON West*, 1993, 1489.
Hosking, M. et al, Wetting Behavior of Alternative Solder Alloys, *Proc. SMI Conference, 1993,* 476.
Vincent, et al., Alternative Solder for Electronics Assemblies, *Circuit World*, vol. 19., no. 3m, 1993, 32.
Harada, R. S., Mechanical Characteristics of 96.5 Sn/3.5 Ag Solder in Microbonding, *IEEE Tr. Comp. Hybrids Mfg. Tech.*, vol. CHMT-13, no. 4, 1990, 736.
Getty, The Effect of Power and Temperature Cycling on Tin/Bismuth and Indium Solders, *IPC Technical Review*, December 1991, 14
Kwoka, M.A., Foster, D. M., Lead Finish Comparison of Lead-Free Solders versus Eutectic Solder, *Proc. SMI Conference, 1994.*

Abbott, D. C., Brook, R. M., McLelland, M., and Wiley, J. S., Palladium as a Lead Finish for Surface Mount Integrated Circuit Packages, *IEEE Transaction on Components, Hybrids and Manufacturing Technology,* vol. 14, no. 3, September, 1991, 567.

Grudul, H.E., and Carlson, R. R., Low Temperature Lead-Free Wave Soldering for Complex PCBs, *Electronic Packaging & Production,* September, 1992, 31.

Hwang, J. S., *Environment-Friendly Electronics—Lead Free Technology,* Electrochemical Publications, Great Britain, 2001, Chapter 19.

Nikkei Sangyo, Tokyo, Japan, November, 1998.

Fan, C., Abys, J. A., and Blair, A., Wirebonding to Palladium Surface Finishes, *Conference Proc. NEPCON West, 1999,* 1505.

Guy, J., Solder Joint Reliability Impact of Using Immersion Metallic Coatings, *Conference Proc. NEPCON West, 1997,* 1540.

Baliga, J., Tin for no-lead Solder, *Semiconductor International,* July, 1999, 74.

Ormerod, D. H., The development and use of a modified immersion tin as a high performance solderable finish, *Conference Proc. NEPCON West, 1999,* 1515.

Ormerod, D. H., Production Application to Flat Tin Finishes, *Conference Proc. NEPCON West, 2000,* 860.

Cullen, D. P., New generation metallic solderability preservatives: Immersion Silver Performance Results, *Journal of SMT,* October, 1999, 17.

Parker, L. L., The Performance and Attributes of the Immersion Silver Solderability Finish, *Conference Proc. NEPCON West,* 1999, 444.

Stafsrom, E., Wengenroth, K., OSPs: The Next Generation, *Conference Proc. NEPCON West, 2000,* 875

Feldmann K., and Reichenberger, M., Assessment of lead-free solders for SMT, *Conference Proc. APEX, 2000,* P-MT2/2-3.

Chapter 8

Printed Wiring Board Cleaning

William G. Kenyon
Global Center for Process Change, Inc.
Moatchanin, Delaware

8.1 Introduction

This chapter is intended to cover preassembly [i.e., *bare board* or printed wiring board (PWB)] substrate cleaning. It is applicable to designs ranging from yesterday's plated-through hole (PTH) boards up to modern packages such as ball grid arrays (BGAs) and chip scale packages (CSPs).

Previously, when only tin-lead or bare copper metallization was used, substrate fabrication was a simpler process with fewer options. The advent of lead-free processes and many more specialized substrate finishes has made substrate cleaning more complex, while the use of "no-clean" assembly processes has imposed stricter preassembly cleanliness requirements.

Designs with relatively wide line traces on PWBs and larger component pitches made the removal of troublesome residues less demanding and more forgiving from a performance standpoint. Importantly, less than adequate cleaning of substrates prior to use by the assembler did not pose a significant problem, because the completed assemblies were also cleaned.

Aqueous cleaning processes were used for fabrication cleaning, as well as aqueous etching and plating processes. The only processes using halogenated (chlorinated) solvents were the developing (1,1,1-trichloroethane) and stripping (methylene chloride) processes for primary imaging and for solder mask application. These were replaced with aqueous processes as soon as suppliers made aqueous processed products available.

Semi-aqueous, two-step cleaning systems (e.g., solvent wash followed by water rinse or solvent in water emulsion followed by water rinse) and aqueous inorganic based saponifiers have been introduced relatively recently for certain cleaning steps.

Environmental issues and worker safety were not nearly as fully understood or regulated as they are today. Volatile organic component (VOC) regulations, for example, were just beginning to affect operations and, overall, were not as demanding as they are today. An analogous situation existed for waste disposal and closed-looping, with the additional complication that treatment processes were in the process of being developed for this field. With the phase-out of ozone-depleting solvents and avoidance of solvents with significant global warming potentials, fabricators have minimized some regulatory concerns.

While the world of fabrication cleaning needs has become significantly more challenging and complex, it can safely be said that modern cleaning technologies are up to the task. Recent years have also witnessed important improvements in the efficacy of cleaning equipment. Wastewater treatment and closed-looping are now fairly well understood technologies. Much has been learned and published in recent years regarding bare board cleaning. Work on generating a relevant cleanliness test method for PWBs has resulted in an industry consensus test method.

8.1.1 Purpose

The purpose of this chapter is to provide a survey of PWB fabrication cleaning methods and processes. The cleaning process prior to any printed board fabrication step is the single most important function over which there must be complete control. Also, it is paramount to remember that, with any *build-up* type of manufacturing, each step taken, no matter how trivial it is perceived, may involve multiples in cost added. This is especially true with printed wiring boards where the myriad of steps require complete cleaning procedures. Cleaning before solder mask application is critical, because there will not be any further opportunity to remove extraneous matter after the application step. In general, the use of a class 10,000 or better clean room is desirable to reduce particulate contamination. Because of the differences in preparing and cleaning boards with copper and tin-lead, they are described separately.

8.1.2 Scope

This chapter is restricted to the fundamentals of current and developing PWB fabrication cleaning issues and is not intended to function as a comprehensive treatise on all areas of electronics cleaning. It covers the sources and nature of fabrication residues as well as various cleaning steps that occur during fabrication of PWBs (or, more correctly, *substrates*). Fabrication cleaning takes into consideration the specific needs associated with a number of new surface finishes. It is important to note that, for bare boards destined for no-clean assembly, the cleanliness requirements for finished bare boards have increased dramatically in importance. This is particularly the situation for hot air solder leveling (HASL) processed boards and solder masked boards.

Therefore, this chapter is intended to cover all aspects, both method and degree, of cleaning associated with the preparation of PWBs prior to solder mask application. It continues with the prudent control of the cleanliness level during the solder mask application and cure processes. It includes maintenance of the cleanliness level of solder masked boards during preassembly processes and/or storage time prior to assembly. Finally, based on the procedures used in the previous steps, it deals with maintaining the soldered assembly at a degree of cleanliness consistent with the end use.

While the primary focus of this chapter is PWB fabrication cleaning, key related subjects are included. Modern cleaning cannot be properly discussed without also addressing the issues of worker safety and the environment. In areas of fabrication cleaning where recent detailed industry consensus (IPC) documents/manuals already exist, the reader is urged to seek out the relevant sections. A listing of such documents is provided in Section 8.7.

8.1.3 Current and Emerging Cleaning Issues

The overall need for cleaning in the manufacture of modern electronics has not diminished. Cleaning processes either remained where they were or may have shifted upstream. In some cases, new needs, such as cleaning during the solid solder deposit (SSD) process, now recognized as a high value alternative surface finish, have emerged. Those who are knowledgeable about low-residue assembly teach that "no-clean" process does not mean "zero cleaning." Cleaning plays a key role in the successful implementation of a no-clean process. Low-residue assembly has simply shifted cleaning from the assembly stage back to the bare board fabrication, and even the component manufacturing stages.

Continuing developments in the area of surface mount technology (SMT) have brought about additional cleaning needs. Spaces on electronic assemblies have become increasingly confined. Fabrication cleaning processes have managed to meet this challenge. Currently, cleaning between BGA attachment pads on PWBs does not appear to be a problem, because of their relatively coarse pitch. If difficulties do arise in the future, they will occur because BGA I/O spacing moves toward a finer and finer pitch. The same trends will be observed with flip chip and CSP.

It has been estimated that approximately 70 percent of SMT solder defects at the component attachment step are due to solder paste printing problems. SSD processes have been implemented to eliminate these defects by applying all the solder needed to make the joint during the PWB fabrication process. The PWBs are then reflowed and washed to eliminate any residues and solder balls. Similarly, cleaning is still an important operation for the optimum adherence and performance of conformal coatings.

Environmental and worker safety issues are very much a part of the today's cleaning picture. Fabricators must consider factors such as volatile organic compounds (VOCs), biological oxygen demand (BOD), chemical oxygen demand (COD), wastewater treatment, heavy metals, close-looping, and pH control. Because of demanding official regulations (federal and/or local), one or more of these factors may determine cleaning process selection.

8.1.4 Substrates to Be Cleaned

The substrates fall into two general classes: standard and specialty. In addition, there are many metallizations that need to be considered. Standard substrates would be FR4, flexible, ceramic, G10, and paper phenolic. Specialty subtrates could include polyimide, tetrafunctional outer layer, Aramid reinforced, and Teflon,. In addition, there are combinations of laminates such as rigid-flex construction. Non-bromine-containing flame retardants are now under development to minimize the environmental impact of lower-than-optimal temperature incineration of discarded electronics. Another development will be substrates capable of withstanding the significantly higher temperatures needed for soldering with lead-free alloys. The new alloys will have melting points in the range of 217–221°C instead of the 183°C melting point of eutectic tin-lead. While these are the melting points, the actual working temperatures will be higher, thus causing concern over the ability of the substrate to survive the soldering process. Substrate metallization includes copper, tin, tin-lead, gold, nickel, palladium, silver, and their alloys. The elimination of lead from electronics manufacturing will remove the "standard" eutectic and near eutectic tin-lead alloys from this lead. These alloys will be replaced with tin-silver-copper alloys for reflow and wire soldering, while wave soldering will be replaced with a tin-copper alloy. Readers should realize that, where tin-lead is cited, a tin-lead alternative may well replace it in the near future.

8.1.5 Soils to Be Removed

Soils fall into three general classes within two groups. The categories are

1. ionic
2. nonionic
3. articulate

Examples of these classes are given below. Some are found in all processes, but some are very process specific.

Table 8.1 PWB Fabrication Contaminants (Residues)

Category 1	Category 2	Category 3
Reflow fluid and HASL Activators	Reflow fluid carrier	Resin and reinforcement
Reflow fluid and HASL Activator residues	HASL fluid carrier	Debris from drilling, laser ablation or punching operations
HASL process salts	Oils	(Reinforcement may be
Fingerprints (sodium and potassium chlorides)	Grease	fiberglass, aramid, or
Residual plating salts	Waxes	other materials)
Residual etching salts	Synthetic polymers	Metal and plastic chips
Neutralizers	HASL oils	from machining and/or
Ethanolamines	Metal oxides	trimming operations
Surfactants (ionic)	Fingerprints (skin oils)	Dust
	Polyglycol degradation by-products	Fingerprints (particulate)
	Hand creams	Lint
	Lubricants	Insulation
	Silicones	Hair/skin
	Surfactants (nonionic)	

8.1.6 Terms and Definitions

Terms and definitions used herein are in accordance with IPC-T-50, except as otherwise specified. Any definition denoted with an asterisk (*) is a reprint of the IPC-T-50 definition.

- *Solvent cleaning.** The removal of organic and inorganic soils using a blend of polar and nonpolar organic solvents.
- *Wash or washing.* The primary cleaning operation that removes undesirable impurities (contaminants) from surfaces by chemical and physical effect.
- *Rinse or rinsing.* A cleaning operation (usually following the wash step) in which fresh cleaning medium replaces any residual contamination, leaving surfaces wet with pure cleaning medium.

- *Dry or drying.* The process of removing any residual cleaning medium on the surface of the washed and rinsed parts.
- *Assembly.* A number of parts of subassemblies or any combination thereof joined together.

8.2 Board Fabrication

8.2.1 General

The interconnection of electronics into assemblies brings together various types of preassembly materials, e.g., printed wiring boards, flexible circuits, ceramic substrates, components, and attachments. Printed wiring boards and flexible and ceramic substrates are platforms that support the components and attachments.

While the application of flexible circuitry as an alternative to rigid printed wiring substrates is developing relatively slowly, cleanliness concerns can be expected to be virtually identical for rigid, flexible, and rigid-flex platforms.

To address the cleanliness concerns, familiarity with packaging materials, fabrication of electronic devices, and processing of printed wiring laminates and circuits is desirable. Contaminant residues can be introduced to the final electronic assemblies by their initial presence on preassembly devices and materials independent of the assembly process. (Refer to Sec. 8.2.4 for cleanliness test methods and recommended limits.) If contaminants are not removed from the components and printed wiring before assembly, there is no assurance they will be removed in assembly cleaning. In particular, if "no-clean" techniques are used for assembly, it is essential that the boards and components be uncontaminated. This is even more important with high-density interconnecting structures (HDISs), as any residual contamination would be very highly stressed because of the small spacing (1998 baseline: 25 μm).

8.2.1.1 Flow paths. No two board fabricators manage their processes in exactly the same manner. It is therefore impossible to give here a detailed review of all the steps where contamination may be introduced. However, there are many common stages in most manufacturing plants, and these are summarized below.

Table 8.2 shows most of the operations that are more or less common to all types of manufacture, although the order may vary according to specific requirements. Table 8.3 shows additional steps for conventional multilayer circuits, and Table 8.4 indicates the additional steps for four different methods of manufacturing HDIS (microvia) circuits. The majority of these steps require cleaning prior to the next operation in the manufacturing sequence.

Table 8.2 Common PWB Manufacturing Operations

Step number	Manufacturing step description
1.	Laminate
2.	Shear/panelization
3.	Drill
4.	Deburr
5.	Electroless copper
6.	Clean
7.	Apply resist
8.	Develop (photoresist)
9.	Etch
10.	Strip resist
11.	Oxide
12.	Pattern plate
13.	Reflow tin-lead
14.	Solder mask/legend
15.	Leveling
16.	Cut

Table 8.3 Additional Steps for Conventional Multilayer Circuit Manufacturing

Step number	Manufacturing step description
A.	Surface preparation
B.	Oxidize
C.	Stack
D.	Press
E.	Desmear/cutback

Table 8.4 Additional Steps for HDIS (Microvia) Circuit Manufacturing

Method A	Method B	Method C	Method D
Add copper/dielectric	Add dielectric	Add photo-dielectric	Add copper/dielectric
Laser drill copper	Laser ablate	Expose/develop	Etch resist
Laser ablate	Surface preparation	Surface preparation	Develop
Electroless copper	Add copper	Add copper	Etch
			Plasma/laser ablate
			Electroless copper

Table 8.5 Additional Steps for Solid Solder Deposit (SSD) Circuit Manufacturing

Step number	Manufacturing step description
AA.	Clean
BB.	Apply dry film solder mask
CC.	Image with SSD Optimized Aperture Phototool
DD.	Develop
EE.	Wash
FF.	Dry
GG.	Apply water-soluble solder paste
HH.	Reflow
II.	Inspect
JJ.	Wash (remove residues and solder balls)
KK.	Hot roll flatten solder into solder bricks
LL.	Apply "tacky" flux
MM.	Apply Mylar® cover sheet

Board fabrication requires sequential performance of a number of separate operations which include drilling, desmearing, resist application and removal, electroless copper, plating, and etching. These operations expose the laminate to many different chemicals and conditions. The residues require removal to prevent interaction with subsequent process steps. Ideally, the residues will be completely removed by cleaning before the laminate enters the next stage. In practice, no cleaning method is 100 percent efficient, and all that can be hoped for is to reduce contaminants to an acceptable level. In addition, laminate surfaces will undergo progressive deterioration as unreacted or partially reacted resin constituents are selectively removed. Therefore, it is of great importance that cleaning methods and chemicals be compatible with the soils to be removed while minimizing solvent exposure in terms of time and temperature. This applies both to process chemicals and cleaning agents.

Note: certain modern circuits have had problems with trapped electrostatic charges. Such charges appear to be trapped on or within the substrate and the solder mask. Fabricators should be aware of this phenomenon and take precautions as warranted.

8.2.2 Typical Fabrication Operations

8.2.2.1 Laminate. This refers to the incoming laminate in sheet format.

Laminate type. Paper-based laminates are difficult to clean because of their high absorbency. Therefore, they are typically used in end-use applications that usually do not demand stringent cleanliness requirements. Glass reinforced laminates with polyester, epoxy, improved epoxy, polyamide, and PTFE resin systems are easier to clean but will probably have to meet stricter requirements in terms of performance.

Incomplete polymerization. A completely polymerized resin is remarkably resistant to attack by most cleaning agents. However, complete polymerization is rarely achieved.

The causes of poor polymerization are

1. Undercure

2. Poor mixing of resin components

3. Non-stoichiometric formulation of resin and hardener constituents in the polymer system

Incomplete polymerization results in some areas of the polymer being soluble in one or another of the agents to which it is exposed during processing. By definition, the outer layer of a cross-linked resin system is inevitably incompletely polymerized where it interfaces with the copper or release film, by boundary effect at a molecular level. Subsequent processing after etching will remove most monomeric materials, but the surface may remain slightly porous. This incomplete cure may lead to the release of sodium chloride that is a by-product of the epoxy reaction, which is normally safely embedded in the polymer matrix. Subsequent exposure to moisture could result in mobile chloride ions being present on the laminate surface. In addition, the resin surface may become absorbent and susceptible to trapping or holding contaminants. Incomplete polymerization, which is also characterized by a lower glass transition temperature (T_g), can also allow further entrapment of organic contaminants during fabrication and assembly. When due to causes (1) and (2), it cannot be remedied during the fabrication process.

Absorbed layers. Laminate resins are hydrophobic and repel water. In the presence of laminate, materials such as long-chain surfactants will form monolayers of surfactant molecules on the laminate surface. Since the hydrophilic ends of the chain face outward, wetting will be excellent, giving the appearance of a clean surface—but surface insulation resistance will be low under humid conditions. These surfactants should be avoided for high-reliability products. If a surfactant must be used, its bond strength to the resin should be weak enough to allow removal with a water rinse.

Attack on coupling agents. To ensure good glass-to-resin bonding, glass fiber surfaces are normally treated with proprietary silane-based coupling agents. When glass fibers coated with silane are exposed at resin surfaces or in plated-through holes, they are susceptible to hydrolysis accelerated by hydrochloric acid (HCl) based reagents. This reaction allows water penetration along glass fiber bundles and can lead to defects such as measling and conductive metal filament growth. These defects affect the reliability of the circuit board.

Contaminants on raw material. Contaminants such as antitarnish agents, smearing or exudation from release films or packing, oils from caul plates, and other diverse handling soils may be present on the raw material surfaces.

Contaminant prevention. As a general rule, most fabricators do not take any measures at this stage, as the sheet sizes are too large for convenient handling and for most cleaning machinery.

8.2.2.2 Shearing/panelization.
This consists of dividing sheets into conveniently sized working panels, often by, for example, guillotining them into perhaps four or six equal panels and punching or drilling registration holes.

Contaminants from raw material processing. Shearing and punching may introduce metal and laminate debris.

Surface preparation. This is really the first opportunity to clean the board surface of contaminants. Mold release agents and other oils used in the lamination process may be present on the copper-clad panels when received from vendors. Some suppliers will use antitarnish agents to protect the copper surfaces; these should be removed before continuing the process.

Prevention. At this stage, some fabricators do not perform a specific cleaning operation, except for simple "print-and-etch" circuits that are drilled after etching.

8.2.2.3 Drilling metal. Drilling metal and laminate debris can be formed during the drilling of holes or the punching of copper clad panels. In addition, packaging residues and particles from the environment are possible sources of contamination. In particular, laser ablation of microvias may leave residues around each hole.

Drilling contaminants. All of the contaminants introduced by drilling are particulate.

Prevention. Incomplete removal of debris will affect the quality of subsequent stages. It is therefore important to prevent debris from being transmitted onward in the process. Vacuum or compressed air cleaning is possible, but removal in a solvent or aqueous solution is also possible, conceivably with ultrasonics. If deburring is carried out, it may be considered preferable to wait until after deburring.

8.2.3 Cleaning Processes Specific to Board Fabrication

8.2.3.1 Mechanical cleaning. This refers to the physical removal of the outer surface to be cleaned by some mechanical process, thereby exposing the fresh, clean surface underneath. The methods used to clean in this manner are abrasive in nature, with the difference being how the abrasion is accomplished. With all abrasive cleaning processes, there is a real risk of implanting abrasive particles into the metal surface and the insulating substrate. In the metal surface, the main problem is that the solderability may be reduced, with consequent dewetting. This may be mitigated by giving the metal a light surface etch (e.g., 3 μm removal with a persulfate solution, if the metal is copper) followed by a very energetic wash and rinse to displace the released abrasive particles. There are four main methods used on nonsolderable surfaces.

1. Abrasive brushes
2. Pads
3. Impact slurries
4. Slurries with soft brushes

Note: do not use abrasive cleaning on solderable surfaces. A surface that is tinned, reflowed, leveled, or plated with a metal, such as gold, is considered to be a solderable surface, and abrasive residues may affect solderability or surface insulation resistance.

These cleaning techniques are applicable to laminate before primary imaging or to circuit patterns before solder mask application. They cannot be used after any components are attached, owing to the abrasive nature of the process.

Bristle brushes. Bristle brush cleaners consist of long, horizontal brushes rotating in contact with the board surface. The brushes are made of stiff nylon bristles with bits of silicon carbide embedded in the bristles. As the board moves under the brush, the bristles rotate against it, abrading away surface contaminants and some of the surface itself. Water flushes serve to lubricate and cool the brushes as well as to wash away the removed contaminants. The biggest advantage of mechanical brushing is its ease of use and relatively low cost as compared with other cleaning methods. It is the most efficient and cost-effective way to clean large volumes of panels in the least amount of time. The equipment is compact, self-contained, uses very little floor space, and does not involve monitoring and disposal of chemicals.

Despite the convenience and low cost, there are disadvantages to this cleaning method. It is very easy to remove too much metal, especially from the edges of holes, where the brush pressure is the highest. It is possible to stretch inner layer cores enough to throw the tooling marks out of register if extreme care is not taken. Flexible material such as polyimide is frequently damaged or destroyed by wrapping around the brushes.

Bristle brushes should not be used on any soft metal (tin, tin-lead, etc.) or on boards with a solder mask, because of severe effects on the surface. Also, the component side of populated boards cannot be cleaned with bristle brushes, due to potential damaging of components (i.e., ink markings removed, etc.). An additional disadvantage to using bristle brushes is that, when scrubbing boards prior to solder mask, slivers may break away from the edges of the circuits. These are extremely small and cannot be seen without the aid of a 100× microscope. This is typically not a problem until you are producing a board with less than 0.2 mm (7.9 mil) spacing. Then, they cause low resistance and intermittent shorts.

Another problem is that brush-abrasive cleaning may smear dielectric onto the conductors, causing a number of potential problems in subsequent processes. Brushes used for etched circuits should not be used for cleaning unetched copper-clad laminate. Finally, if too much brush pressure is used, or if cooling water flow is inadequate, the nylon brush material can be "smeared" on the board surface, leaving an invisible residue that is almost impossible to see or remove.

Pad brushes (compressed pad brushes). Pad brushes are porous, felt-like materials with abrasive grits bonded to the surface. They operate in equipment much like used for bristle brushes, but, because the pad contains water in the porous structure, cooling at the scrubbing surface is less of a problem.

Pad brushes have advantages over other methods; because the fibers are closer together, more abrasives are presented to the work surface. Also, the compressed disks that form the brush allow more pressure to be applied to the work surface. Pad brushes tend to give a more uniform finish than bristle brushes. For this reason, they are sometimes used for inner layers.

As for disadvantages, frequent checks for wear on the pad brushes are required for uniformity and timely replacement. As the pad wears, fibers may be deposited in holes that might be present on the board surface. Consequently, a fairly high spray pressure (up to 1.4 MPa or 200 psig) is necessary to remove these fibers from the holes. Like bristle brushes, pads may cause distortion of the surface being cleaned, and they cannot be used to clean any soft metal such as tin, tin-lead, solder mask, or surfaces that are coated with an organic coating.

Impact slurries. This cleaning process uses a slurry of pumice or metal oxides in water that is pumped through nozzles to strike a surface to provide abrasive action. The grit size of pumice used is usually 3F or 4F. This fine grit is hard to rinse, and a spray pressure of about 1.4 MPa (200 psig) is required to adequately rinse a board. Compared with either bristle or pad brushes, the major advantage of impact slurry process is the ease of control. This is because

- No separate water coolant is required.
- There is no chance of nylon smear from brushes.
- Brush pressure is not critical.
- There are no pad or brush particles to get caught in a drilled hole.

In addition, the process provides a more uniform surface than that obtained from brushes. This is most important where finer resolution is required (0.10 mm [0.004 in.]). Furthermore, the impact slurry tends to reduce distortion of the board, which again is an advantage with finer conductor lines and spaces.

However, there are disadvantages to this process. Rinsing of the boards can be a problem, and high-pressure water sprays (@1.4 MPa or 200 psig) are necessary to remove the fine particles of pumice and possible embedded particles from metal surface. Another disadvantage is that impact slurries cannot be used for very thin material. With pumice, significant amounts of solid waste are generated.

Slurries with brushes. This process uses the same kind of slurry as the impact spray but has soft nylon brushes that oscillate on the board surface to impart abrasive action from the slurry.

This process has the same general advantages as the impact slurry, plus equipment maintenance is less of a problem, because no high-pressure pumping is necessary.

A rinsing problem is the disadvantage of this process, such that water sprays alone are not satisfactory. A water rinse with similar soft brushes for agitation is necessary to remove all particles.

8.2.3.2 Chemical cleaning. In this section, cleaning processes will be covered in two groupings.

- Cleaning and etching boards using harsh chemicals
- Cleaning and etching boards using plasma

Cleaning bare boards for oxide removal using harsh chemicals. This is frequently referred to as *surface preparation* and is an alternative to the mechanical cleaning previously described. It is used very commonly where the circuit pattern has already been formed and mechanical scrubbing could damage the pattern. Solutions currently used are usually hydrogen peroxide/sulfuric acid, cupric chloride, or ammonium/potassium persulfate. All of these solutions work by dissolving copper oxides and/or other copper compounds on the surface, along with a small amount of the underlying copper surface. The resulting surface is extremely reactive and may be passivated with a solution of 3–5 percent citric acid/5 percent sulfuric acid, or something equivalent, to prevent surface oxidation. In other situations (as for inner layers or before applying a permanent solder mask), forming a light coating of some copper oxide (brown, red, or black) is desirable to provide better adhesion. In either case, a final, thorough rinsing with water, finishing with deionized water, and complete drying is necessary to avoid water spotting and stains.

Plasma processing for cleaning and etching surfaces. Plasma processing (i.e., the use of an ionized gas to interact mechanically or chemically with a surface) is a dry processing technique that is applicable for the treatment of surfaces to modify their surface energy and/or reactivity. Plasma processing, when executed properly, is the one cleaning process that leaves a surface microscopically free of organic contaminants (but will not be effective on inorganic contamination or particulates). By using proper plasma treatment, most polymer surfaces can be modified to be either hydrophobic or hydrophilic, acidific or basic. Adhesion and/or wettability can also be tailored to meet subsequent requirements. An example of a commonly used plasma treatment in the printed circuit industry is an oxygen plasma surface treatment of Kapton® to enhance the application of a liquid adhesive. This plasma treatment improves the wettability of Kapton and eliminates surface voids caused by spotty adhesive coverage. Plasma processing can also etch (chemically react with) polymer surfaces to remove residues left by other process steps. One example is the removal of polymer smeared onto the sidewalls of a multilayer board by the through-hole drilling operation. An oxygen plasma reacts with carbon polymer chain, removing carbon as carbon monoxide or carbon dioxide. Also, after photoprinting and developing, a thin residue may remain over areas that should be free of photoresist. Oxygen plasma can be used to remove the residues and clean the underlying surface. This is a very successful technique for cleaning surface mount pads. When a low-pressure gas is subjected to a high-energy input, typically radio frequency or microwave energy, the gas dissociates and ionizes through collisions with high-energy electrons. This mixture of electrons, unreacted gas, and neutral and ionic atomic and molecular fragments is termed a *plasma*. Even though the starting gas may be chemically inert, the fragments can be highly reactive, allowing a variety of chemical processes to occur. For example, although the reaction of molecular oxygen with polymers at room temperature is generally slow, oxygen plasma at room temperature can "burn," or remove polymer with a reaction that is possible only at much higher temperatures for molecular oxygen. [See Eq. (8.1).]

$$O_2 \rightarrow 2O^*$$

$$C_xH_y(\text{polymer}) + (2x+y)O^* \rightarrow xCO_2 + (y/2)H_2O \qquad (8.1)$$

where O^* = oxygen ion

Thus, plasma processing is often described as being "low temperature," since the reactive fragments themselves generally have temperatures of, at most, a few hundred degrees Celsius (and often quite close to ambient).

8.2.3.3 Aqueous-based cleaning. These operations usually follow hot air solder leveling or solder reflow operations. Contamination to be removed is typically a water-soluble fusing fluid or flux, so a water or water/detergent solution is used for cleaning. The equipment is an automated horizontal belt conveyor that goes through a cycle such as wash, rinse, blow-off, and dry. Agitation is by liquid sprays on the board, but occasionally brushes are used. Additional rinses and blow-offs can be added where better cleanliness is required.

8.2.4 Cleanliness Test Methods and Recommended Limits

Final bare board cleanliness (i.e., the cleanliness of the bare board as it leaves the premises of the fabricator) is a more important issue today than it has ever been in the past. The reason for this significant increase in interest has been the expanded use of low-residue or no-clean soldering process. The success of no-clean process depends directly on the degree of cleanliness of incoming bare boards. Currently, an IPC specification for final bare board cleanliness does not exist. A new ionic test method for bare PWBs has been prepared and published, but it does not contain a "pass/fail criterion." The PWB cleanliness acceptance level is determined by PWB users as a function of their reliability requirements.

Document IPC-6012 gives a total ionics specification for non-coated rigid boards prior to solder resist application. That specification does not, and was not intended to, address the cleanliness requirements of bare boards to be utilized in a low-residue soldering process. This is not a good situation, given the importance of this subject in the practice of low-residue assembly. It is our understanding that individual corporations have indeed set company standards for the cleanliness of incoming parts, and these standards vary depending on the assembly being built. However, the assembly community is generally at a loss for not knowing just how clean the incoming parts have to be for low-residue assembly processes. To provide guidance on the desired cleanliness for bare boards, an IPC task group prepared a test method for measuring bare board contamination (see above paragraph of Sec. 8.2.4).

8.2.5 Displacement Drying

The normal sequence of drilling, followed by electroless copper, and pumice scrubbing prior to application of the primary resist film in a fabrication "yellow room" suffers from one drawback. Extra copper must be plated onto the PWB surface so that enough remains after the pumice scrubbing to provide an oxide-free surface. To optimize the process, many fabricators began using a CFC-113-based displacement drying step after electroless copper. After electroless plating of the PWBs in baskets, the baskets are moved to the two-tank displacement dryer. The first tank contains a trace of special surfactant, which enables the heavier-than-water solution to "lift off" or displace any water from the surface and out of the holes in the PWBs. Immersion in a second tank of pure solvent removes any surfactant. The clean, water-free PWBs are dried during their exit through the solvent vapor blanket. The process was rapid and, since it was performed in an oxygen-free environment, yielded PWBs with oxide-free copper surfaces. Thus, the pumice scrubbing step and associated waste were eliminated; the PWBs were sent directly to the "yellow room" for application of the primary imaging resist. Furthermore, since overplating was not needed, the time in the electroless copper bath was minimized. While the process using CFC-113 was phased out as a result of the Montreal Protocol, suppliers of partially fluorinated replacements have commercialized displacement drying fluids based on the alternative solvents. Overall, this process shortened cycle times, eliminated pumice scrubbing and its waste stream while saving costs associated with overplating.

8.3 Solder Mask over Bare Copper

In this process, an etch resist, which is usually plated tin-lead, is removed so that the plated copper traces are left bare. Solder mask is applied over bare copper to permit solder leveling, reduce solder pot contamination, protect electrical integrity, and eliminate mechanical damage. The removal of etch resists, precleaning and post-cleaning, are explained in this section. See Fig. 8.1 for the process flow chart.

Note: materials of equipment construction should be checked for compatibility with all solutions to be used in them.

8.3.1 Metallic Resist Stripping

Two techniques are used for stripping tin-lead and tin, the key metallic resists, from circuit boards. They are distinguished by the use of either a peroxide or a nonperoxide (acid-based) solution and are described in the following paragraphs.

8.3.1.1 Peroxide. Peroxide is an aggressive oxidizer; thus, all equipment exposed to it should be of plastic construction. In addition, many

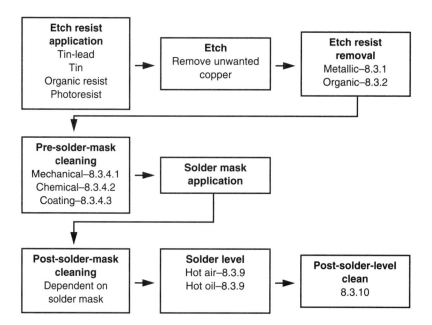

Figure 8.1 Solder mask over bare copper process.

products also contain ammonium bifluoride, which will etch glass. Since the peroxide reactions are exothermic, cooling coils are recommended. Laminate attack can result from the heat and the ammonium bifluoride. Compatibility of laminates with the solutions should be checked before setting process guidelines.

8.3.1.2 Nonperoxide. Nonperoxide strippers are nitric or fluoroboric acid-based solutions. They are used in either single- or two-step processes. Single-step processes usually involve spray techniques, whereas two-step processes can involve either soak or spray techniques. Nonperoxide strippers are generally not exothermic and will not attack laminate. However, a specific laminate should be checked for compatibility with the particular stripper being used. In general, nonperoxide strippers are slower than peroxide. With fluoroboric acid-based solutions, spray equipment used must not contain titanium, as it is readily attacked by fluoroboric acid.

8.3.1.3 Detection of residuals. Incomplete tin-lead stripping is evidenced by a dull, gray-colored copper or white powdery residue. This indicates incomplete removal of the intermetallic layer or lead fluoride.

Note: at times it may be possible to have a board that visually appears to be free of residuals but actually still has an intermetallic layer that can lead to nonwetting. This condition can be identified using surface analytical techniques such as x-ray fluorescence or Auger electron spectroscopy.

8.3.1.4 Removal of residuals. Removal of residuals varies with the type of stripping technique used. In peroxide stripping, lead fluoride residues are removed by an extended dwell time in the stripper or by immersing the parts in an acid cleaner. Residues may also indicate that a need exists for replenishment/replacement of the stripper solution. In nonperoxide stripping, the incomplete removal of a copper-tin intermetallic layer may require extended dwell time in the second-step stripper solution or the replenishment/replacement of the stripper solution.

8.3.2 Stripping Organic Resists

Panels made with the tent-and-etch process, or any other process in which a photoresist or other organic resist is used as the final etch resist, must be stripped to bare copper to ensure proper solder mask adhesion and to prevent blisters during subsequent high-temperature processes.

8.3.2.1 Solvent-processed resists. Solvent-processible resists are generally stripped in methylene chloride containing an azeotropic amount of methanol. Properly exposed, these films strip very quickly and leave almost no residue. The biggest potential problem is an insufficient supply of fresh solvent, which leads to resist residues building up in the final solvent chambers. Evaporation of the dirty solvent then leaves a film of resist on the surface, which must be removed through a chemical cleaning process as described below.

8.3.2.2 Aqueous-processed resists. Aqueous and semi-aqueous resists are stripped in a caustic solution that sometimes contains an organic solvent. The caustic can be either organic (mono-ethanolamine) or inorganic (sodium hydroxide). In general, these films do not strip as cleanly as the solvent products and frequently leave a thin film of organic residue on the surface of the copper that must be removed with a chemical cleaning process.

8.3.2.3 Removing resist residues. The chemical cleaning of copper stripped of resist can be done in either conveyorized spray equipment or process tanks. The choice of equipment will probably depend on the method of stripping. There should be little or no hold time between the stripping and the cleaning process, since the resist residues can dry and become firmly attached to the copper surface, making them more difficult to remove. The chemical cleaning process generally consists of three parts, each separated by a thorough water rinse.

1. *Detergent cleaner.* A detergent cleaner is used to loosen the resist residues and soften them so that the microetchant chemicals can penetrate and attack the copper to which they are attached. If the equipment is a spray system, it must be a nonfoaming, nonionic cleaner such as those used in automatic dishwashers. The temperature is usually in the 60–71°C (140–160°F) range, depending on the limitations of the processing equipment. Residence time in a spray system is typically 1 min; in a tank system, 4 min. Cleaners can be either alkaline or acidic, but alkaline detergents are the most common. Some acidic detergents contain a microetch. These should be used with caution, as the etch characteristics can be erratic, and etchant residues can be left on the surface.

2. *Microetch.* The microetch is the step that actually removes the resist residues and exposes the fresh copper surface. In cases of severe surface contamination, thin films of resist can be seen floating in the etchant after a short period of use. For this reason, a filter system should be considered for this solution. Sufficient copper should be etched to ensure that the entire surface is fresh uncontaminated copper, typically 1.8 µm (70 µin). The etchant should be a sulfate-based system (sodium persulfate, hydrogen peroxide/sulfuric acid, etc.) to ensure that the etchant rinses cleanly.

3. *Sulfuric acid rinse/anti-tarnish.* The microetch should be followed by a 5–10 percent sulfuric acid rinse to remove any residual etchant salts. The anti-tarnish is optional, and the solder mask vendor should be contacted for recommendations. As an alternative to the anti-tarnish, a light scrubbing will serve to passivate the surface by mechanically removing any trace salts, thus keeping it from becoming excessively oxidized.

Removing resists in this manner leaves a very active copper surface, so care should be taken in maintaining proper storage conditions.

8.3.2.4 Effectiveness of organic resist stripping. To determine the effectiveness of the cleaning process in removing photoresist residues, a test panel can be run in a manner similar to that used for determining the effectiveness of a preplate cleaning process. Prepare a panel with resist, leaving a section of bare copper exposed. Expose or cure the resist, and send it through the stripping and cleaning processes. When finished, there should be no difference between the portion that had resist on it and the portion that never contacted the resist. As a final test, the panel should be plated in a copper sulfate bath for 10–15 min without additional cleaning. There should be no evidence of peeling or hazy, dull, or pitted plating, which would indicate an organic contamination; nor should there be any line of demarcation between the treated and untreated areas.

8.3.3 Board Handling after Resist Stripping

Boards should always be handled by the edges only. The proper choice of gloves can provide some handling benefit. Improper glove selection can provide a false sense of security, since some gloves contribute fibers to the board. Some will also absorb hand oils that can be transferred to the board. Racking is the preferred storage technique for boards in process. Boards should be stored in an atmosphere without high humidity or the presence of process chemicals. The longer boards are stored, the more the copper can oxidize. Excessive oxidation could require changes in the cleaning cycle to ensure a proper copper surface.

8.3.4 Pre-solder Mask Surface Preparation

To ensure maximum adhesion of the solder mask to the board surface, conditioning of the surface is required before the solder mask is applied. In most cases, the term *surface preparation* refers to the roughening of the copper in the circuit traces and ground planes to provide the greatest amount of surface area for solder mask adherence. The major concern is the exposed metal, which tends to be fairly smooth at this stage of the process. The laminate is usually rough enough to give good adhesion without any further treatment and needs only to have surface contamination removed. The most commonly used methods for surface conditioning prior to solder mask application over bare copper are mechanical brushing, pumice scrubbing, chemical microetching, and the application of a chemical coating, or a combination of these methods.

8.3.4.1 Mechanical methods. Mechanical cleaning has been used in most fabrication processes, but caution should be exercised to be sure that this method is suitable for a particular application. Most mechanical cleaning is done on conveyorized machines that generally fall into two categories. The first is a brushing machine or scrubber that uses a rapidly rotating, abrasive brush to remove the surface contamination (along with some of the surface copper) and roughen the surface of the remaining copper for better adhesion. The second type of machine does the same thing but uses pumice slurry, applied either as a spray or with a brush.

Mechanical scrubbing. The most common mechanical method is the use of a scrubbing machine with a rotating abrasive brush. The available equipment is usually a free-standing machine with a brush section to do the surface preparation. This is followed by a water rinse to remove debris from the panel surface and a dryer to remove water droplets and excess moisture. The three available types of brushes are the compressed wheel, the nylon bristle, and the composite wheel. The compressed wheel is most aggressive and is the fastest, but fibers can be torn away to leave debris in holes and shorten brush life. Nylon brushes have the longest life and do not leave unwanted debris, but they are less aggressive and can lead to nylon smear on circuitry. Composite wheels use a finer grit and are aggressive, but debris is still a problem. While a wide variety of brushes is available, the choice can be narrowed down somewhat for this application. Nylon bristle brushes should be avoided over copper, since the nylon can leave an invisible smear that is detrimental to solder mask adhesion if too much pressure is used. A fine-line or high-resolution compressed or composite wheel is much better suited for this surface conditioning, since these have a much finer abrasive grit than other wheels and are less likely to leave a deep gouge in the copper that the solder mask cannot penetrate. The actual grit size for these brushes is around 600 or smaller (higher number means smaller particle size). The proper brush pressure is generally less than would be used to scrub a board prior to the application of the etch or plating resist. Excessive pressure can cause epoxy to smear on to copper surfaces when laminate is not completely cross-linked. The proper pressure can easily be determined by using the stripe test. A copper clad board of approximately the same width as the boards to be masked is run over (or under) the brush to be adjusted and the conveyor stopped. The brush is brought into contact with the board and allowed to run against it for several seconds, and the brush pressure meter reading is noted. The brush is raised, and the board removed from the scrubber. A narrow stripe of polished copper is left where the brush made contact. The proper brush pressure is when the width of this stripe is between 6.3 and 9.5 mm (1/4 and 3/8 in.). The test board is then run through the scrubber again, and the brushes adjusted until the same reading is obtained on the pressure meters. The machine is now ready for the boards to be solder masked.

In machines where the conveyor cannot be stopped while the brush is rotating, a correlation of brush pressure and conveyor speed with end use properties should be run on test panels. A water break test may be used to indicate the adequacy of the machine setting. A properly scrubbed copper clad board, equivalent in thickness to the board to be processed, will hold an unbroken film of water for 30 s while held vertically after proper scrubbing. This procedure, although widely used, can be misinterpreted, especially with copper surfaces that are overscrubbed. Consistent results and maximum life can be ensured by keeping the brushes round. Brushes wear unevenly if panels are all loaded in the same spot on the conveyor with parts of the panels being undercleaned while other areas seeing too much pressure. Dressing is fairly easy with compressed and composite wheels and can be done on the machine with dressing plates provided by the machine or brush manufacturer. Nylon bristle brushes are much more difficult to dress on the machine without ruining the brush. It is suggested that the machine manufacturer or brush supplier recommendations be followed when these brushes need dressing.

Care must be taken to avoid contaminating equipment to be used for cleaning copper with tin or tin-lead residues. A single scrubber must not be used for cleaning both copper and tin or tin-lead panels. In the same way, all traces of tin, tin-lead, and intermetallics must be removed before scrubbing the copper surface. Because the brushes can leave debris on the board surface, good rinsing is necessary. Most brushing machines have an integrated rinse section that is usually adequate, as long as the water pressure is at least 1.3 bar (20 psi). Following rinsing, the board should be dried using warm, oil-free, forced air. Most brushing machines have an integrated dryer that will do the job. The use of infrared dryers that dry by evaporating the moisture on the board surface should be avoided, as any contaminants that are carried by the water remain on the surface after the water evaporates and will be trapped under the solder mask. These could lead to oxidation and reduced adhesion resulting in blistering.

Pumice scrubbing. This method is also a traditional and effective mechanical cleaning method for copper prior to solder mask application. Pumice is either acid activated or plain. The acid-activated type is used with hand scrubbing, while the plain type is primarily used as a slurry in spray machines. Spraying should only be done in equipment especially made for that purpose. Care should be taken to maintain the slurry per the manufacturer's recommendation. Slurries are good in removing oxides and previous process residues. This process leaves the surface clean and roughened for good solder mask adhesion. However, it is very important to have adequate rinsing, since pumice residues under the solder mask can cause peeling and blistering during subsequent processing. Since pumice is not soluble in water, high-volume, high-pressure spray rinses are required. Boards should be dried with oil-free, hot air and processed quickly to avoid oxide buildup. Chemical cleaning can precede the pumice scrubbing. Pumice scrubbing is not effective in post-solder-mask cleaning. Not only would it abrade the surface of the solder mask, but it would not clean inside the holes where cleaning is needed.

8.3.4.2 Chemical cleaning. Chemical conditioning methods can be used to clean the surface of the copper, and this is most efficiently accomplished in conveyorized spray equipment. A typical sequence of process stations would consist of a cleaner to remove organics (e.g., fingerprints) from the board surface, a water rinse, a microetch to roughen the surface of the copper, a water rinse, and finally a warm air dryer. The use of an anti-tarnish agent after the microetch may be needed, since the microetched copper surface will oxidize very quickly, especially if there is any delay between the cleaning and the application of the solder mask. Compatibility of the anti-tarnish agent with the solder mask must be ensured (see Sec. 8.3.4.3). There are many proprietary, sprayable cleaners available, both acidic and alkaline, for the first cleaning step. For best results, it is recommended that a cleaner formulated for spray applications be used, since foaming can be a major problem with these types of cleaners. It is not necessary to remove much copper in the microetching step, since only a surface roughening is needed. In most cases, a total removal of 1–2 µm (40–80 µin) is more than sufficient to achieve the desired roughening. The most common microetch baths are either based on persulfate, hydrogen peroxide/sulfuric acid, or ferric chloride.

- *Persulfate microetchants* are available as a proprietary mixture from most chemistry suppliers. Persulfate baths generally give a slightly rougher surface and remain more stable over a period of time. It is essentially a batch process, with the bath needing to be disposed of when the copper loading reaches specified limits.

- *Peroxide/sulfuric microetchants* are also available as proprietary formulations for both tank and spray operations. The peroxide solution needs to be strongly stabilized so as to prevent the consumption of peroxide from becoming excessive. The major advantage of a peroxide/sulfuric bath is that it can be regenerated. Chilling and precipitating it as copper sulfate can remove copper buildup in the bath. Additions of hydrogen peroxide and sulfuric acid are then made to replace losses, and the bath is put back into service. This process can be regenerated through the use of a chiller/crystallizer or done in a batch operation.

4. *Ferric chloride microetchants* are also available as proprietary formulations for both tank and spray operations. When ferric chloride is used as a microetchant, it is important to analyze the board surface for iron chloride and other chloride residues. Rinsing is extremely important, especially when using an alkaline cleaner, and/or persulfate or ferric chloride microetchants. When using a persulfate microetch, it would be useful to follow the microetch with an acid rinse of 2–3 percent sulfuric acid to be sure that all the persulfate residues are removed. After the final rinsing step, the boards should be dried using a warm, forced-air dryer.

8.3.4.3 Chemical coating. Although the mechanical and chemical cleaning techniques are the predominant methods of preparing copper circuitry for solder mask application, there are several chemical coating methods available that leave either a permanent or a temporary chemical treatment. Their use is more restricted, and the process condition "window" is narrower, so care must be taken for proper use. Compatibility of these coatings with the solder mask must be confirmed.

Copper oxides. The use of an oxide bath, either brown or black, to put a controlled layer of 0.5–1.0 µm (19–39 µin) copper oxide on the copper surfaces can further enhance the adhesion of the solder mask. The crystalline structure of the oxide layer tremendously increases the surface area for improved mask adhesion. A brown oxide is generally more desirable, since it has a smaller crystal structure and can be applied at a lower temperature. It is not necessary (as is typically done on multilayer inner layers) to put a coating on the copper so thick that a deep brown or black color results. For solder mask application, a bronze or light brown color is all that is needed. Most oxide applications are done in dip tanks, although conveyorized equipment is available. The process steps to prepare the copper for the oxide layer are the same as the chemical conditioning described in Sec. 3.4.2, with the exception that no anti-tarnish is applied. The suppliers of oxide chemistry will have complete step-by-step instructions for their chemistries.

Antioxidants/anti-tarnish agents. If there is delay between cleaning and the application of solder mask, antioxidant/anti-tarnish agents can be used. There are many different types, and all are used in low concentrations. They are applied as a spray or an immersion dip, followed by a water rinse. Exact concentration, time, and temperature are according to vendor's recommendations. Some solder masks can be applied directly to the treated surface, and the antioxidant/anti-tarnish is later removed from the other areas for further processing. Removal is usually done with mild mineral acids, following vendor's recommendations for time, temperature, concentration, and rinse cycle. Most agents must be removed prior to solder leveling, since solderability is greatly impaired on treated surfaces. Compatibility testing with adhesion of solder mask must be done prior to putting antioxidant/anti-tarnish agents in production, since blistering and peeling of the solder mask can occur. Test panels can be run to determine if the antioxidant/anti-tarnish agent should be removed just prior to solder mask application, or the vendor can be contacted for recommendations.

8.3.4.4 Final rinse. Final rinse requirements may be one of the most overlooked aspects of presolder mask cleaning. Many factors are involved, including reliability requirements for the board and the spacing. Depending on the water source, tap water (routinely monitored) may be adequate as a final rinse. However, due to variations in water hardness and temperature (which can happen very quickly), deionized water with a resistivity of 3–5 MΩ-cm is preferred for the final rinse.

8.3.4.5 Drying. Adsorbed water and water residues are frequent causes of blisters and loss of solder mask adhesion. It is, therefore, essential that all standing water be removed from the board surface and holes and that the boards be thoroughly dried before solder mask application. Water should first be physically removed (blown off) from the surface and from via holes before using a heated evaporative drying cycle with a high-volume, hot air knife and/or turbine dryer with filtered, oil-free air. If the first step is omitted, contaminants carried by the rinse water could be deposited on the board during evaporation. Also, drying by evaporation may not remove all water from holes and may leave residues and water spots on the surface. On typical FR-4 panels, absorbed water can be removed by a 20-min minimum bake in an oven set at temperatures of 110 ± 5°C (230 ± 9°F) or by a conveyorized infrared machine with user-established operating parameters. Boards with substrates other than FR-4 glass/epoxy or boards with different constructions (i.e., thickness, thermal mass, etc.) may require different drying conditions. If boards are cleaned in solvent only (no aqueous cleaning or water rinse), the turbine drying may be omitted, but drying is still important. Cleaning and drying in nonflammable solvent is straightforward, but accomplishing the same steps using flammable solvents requires attention to a number of safety considerations. Drying ovens should be dedicated pieces of equipment to prevent cross-contamination of panels. The oven design should guarantee an even temperature profile at all locations within the oven. Forced-air ovens with good external exhaust to remove vapors should be used to maximize drying efficiency.

8.3.5 Pre-solder Mask Cleanliness

The cleanliness of the boards following cleaning can be evaluated by several techniques. Both ionic and organic residues are possible from previous processing. Their presence would require additional cleaning if higher than acceptable customer limits. It is important that a representative board sample be used for testing to ensure that process changes give adequately cleaned and tested panels. A residue may contain mixtures of ionic and nonionic materials, which are usually deposited on the board surfaces together. The ionic portion of the residue can then provide an indicator of overall surface cleanliness. However, to be certain that cleanliness requirements are met, both ionic and organic testing should be conducted.

8.3.5.1 Water break test. This is not a cleanliness test and is best used on copper clad panels to indicate the adequacy of mechanical scrubbing. See Sec. 8.3.4.1.

8.3.5.2 Solderability test. A solderability determination is made to verify that the board is free of surface contamination that would affect solder mask adhesion. Following test procedures in J-STD-003, a printed board that is contaminant free will exhibit a characteristic known as *wetting*. Wetting is the formation of a relatively uniform, smooth, unbroken, and adherent film of solder to a base metal. A board that has surface contamination will exhibit dewetting or nonwetting characteristics. Dewetting is a condition that results when the molten solder has coated the tested surface and then receded, leaving irregularly shaped mounds of solder separated by areas covered with a thin solder film (base metal not exposed). Nonwetting is a condition whereby molten solder has contacted a surface but has not adhered to all the surface (base metal remains exposed).

8.3.5.3 Ionics. Since the presence of ionic residues on bare printed boards prior to application of solder mask can significantly affect the board quality, residue measurements should be run periodically. Ionic materials left on the board can contribute to corrosion of metallic materials, even when the ionic contamination is overcoated with solder mask. Since a solder mask layer can, in time, be permeated by moisture and oxygen, it is insufficient to protect surfaces beneath from atmospheric corrosive effects, particularly if hygroscopic ionic materials are present. Automated equipment is sometimes used to determine ionic cleanliness. Care should be taken to obtain appropriate equivalence factors between different equipment brands and models.

Ionic extraction. To measure the ionic residues quantitatively, it is first necessary to extract the ions into aqueous solution. Since residues often contain components that are not readily water soluble, a mixture of water and an alcohol, e.g., isopropanol, is used for extracting ions from board surfaces. In this way, both the water-soluble ionic component and the alcohol-soluble nonpolar component can be solubilized to allow measurement of the ionic components.

Conductivity meter. Most ionic testing procedures follow the methods (developed at the Naval Avionics Center, Indianapolis, IN) specified in MIL-P-28809. A 75 percent isopropanol/25 percent water mixture is flushed over the board surface to dissolve the ionic materials. The drippings or runoff from the board surface are collected and measured separately using a conductivity or resistivity meter. Reference 2.3.25 of IPC-TM-650 for test methods.

Static volume extraction. An automated system using a similar static volume extraction method immerses the sample into a fixed volume of an alcohol-water mixture. After a suitable time is allowed for the extraction, typically 15 min, the resistivity of the solution is read. This value can be converted to equivalent amounts of an ionic material, usually sodium chloride, and expressed as the total amount of ionic material extracted from the sample. Reference Sec. 2.3.26.1 of IPC-TM-650 for test methods; see Sec. 8.7.1.

Dynamic ionic extraction. Another instrument utilizes a dynamic system of ionic extraction in which the water/ alcohol solution is pumped continuously in a closed loop from the sample tank through an ion-exchange column and then back to the sample tank. When the sample is immersed into the tank, the conductivity rises to a peak and then falls back to the original pre-immersion level. Measurement and integration of the conductivity gives values that are linear functions of the amounts of ions taken into solution. Reference 2.3.26 of IPC-TM-650 for test methods; see Sec. 8.7.1.

8.3.5.4 Organics. The presence of organic, nonionic contaminants on the bare board surface due to board fabrication processes can be detrimental to solder mask adhesion. The nondestructive detection and quantification of such contaminants is covered in IPC-TM-650, Test Methods 2.3.38/2.3.39.

Surface organic contaminant detection test (in-house method). This is a qualitative method of determining the presence or absence of a wide spectrum of possible organic contaminants using a solvent, acetonitrile, as the extracting medium. This method does not identify the specific contaminants present or separate contaminant mixtures into the individual constituents. The visual limit of detection via this method is approximately 10 $\mu g/cm^2$. Reference 2.3.38 of IPC-TM-650 for test methods; see Sec. 8.7.1.

Surface organic contamination identification test (infrared analytical method). This is an analytical method of quantitatively analyzing a wide spectrum of nonionic organic contaminants and identifying the individual constituents in the contaminants. This technique, which also uses the solvent acetonitrile, is more involved than Test Method 2.3.38 and requires an experienced spectroscopist to effectively utilize the multiple internal reflectance (MIR) infrared spectrophotometric equipment required for the test. Another analytical test for identification of total organic contamination is *electron spectroscopy for chemical analysis (ESCA)*. Since acetonitrile is a solvent for numerous organic compounds, including many constituents of the printed wiring board manufacture, initial characterization of the extract obtained from the less technical Test Method 2.3.38 may be required. The more technical, more sensitive Test Method 2.3.39 may be utilized to identify those organic constituents in the extract that truly represent undesirable contaminants from those constituents that are an inherent part of the printed wiring board. After characterization, the utilization of Test Method 2.3.38 may be used as a sampling or in-process check. Reference 2.3.39 of IPC-TM-650 for test methods; see Sec. 8.7.1.

Surface organic contamination test (HPLC method). This method was designed to provide a quantitative monitoring method for assessing the post-fabrication cleanliness level of incoming telecommunication boards. It has been shown to be extremely effective at measuring fusing fluid, HASL fluid and rosin flux residue levels. Reference 2.3.27.1 of IPC-TM-650 for test methods; see Sec. 8.7.1.

8.3.5.5 Moisture and insulation resistance (M&IR). A sensitive method for the detection of the presence of some surface contaminants on a laminate board is the measurement of surface insulation resistance (SIR). This test provides a measurement of the susceptibility of a circuit board to develop electrical leakages between adjacent conductors on the surface. If traces of hygroscopic materials are contained in contaminant residues, the absorption of moisture from the atmosphere, particularly when exposed to elevated humidity/temperature conditions, will result in surface electrical leakages when voltages are applied. This can adversely affect the electrical performance of many circuits or even be catastrophic to the functioning of high-impedance circuits. The procedure is conducted on a test board that has been put through the identical processing conditions as the batch of boards being manufactured. This is done to provide, from test to test, patterns with identical geometries (such as a standard comb pattern) and to obtain results that can be meaningfully compared to each other. This is not feasible if one attempts to measure SIR between adjacent lines of assorted functional circuit patterns. To overcome this problem, a board is designed with a small test comb pattern added to the side of the functional circuitry for the specific purpose of this test. A similar approach is to design a small break-off portion of the board that can be removed and independently tested after the board processing. Still another method is to run one or more separate test boards along with a production batch being processed. These test boards, such as the IPC-B-25 pattern, can then be considered as representative of the effects of the processing and materials on the production group. The measurement of M&IR is a very complex and time-consuming process that requires much attention to get consistent and reproducible results. Typically, a standard comb pattern that had been processed is mounted in a temperature/humidity chamber with electrical connections made to contact pads. A bias voltage is applied between adjacent lines of the comb pattern, and the system is allowed to condition for extended periods of time, usually up to a week or more. At periodic intervals, measurements are made of the resistance between sections of the interdigitated comb pattern. The original bias voltage, if used, is removed, and a specified measurement voltage is applied. The minute currents can be measured with a very high-sensitivity ammeter and the resistance value calculated using Ohm's law. The resistance can also be read directly using a high-resistance meter. It must, however, have the capability of measuring resistance at a predetermined, specified value of measuring voltage. The resulting measured value is then designated as the M&IR of the test that is very specific to the conditions of the test, and the results can vary widely from one set of test conditions to another. Some of the parameters affecting the M&IR results are as follows:

1. Geometry of board used
2. Board material

3. Measurement temperature
4. Measurement humidity
5. Measurement voltage
6. Original state of cleanliness of the sample
7. Board processing conditions
8. Types, amounts, and distribution of contaminants left on board
9. Test technique
10. Test equipment

If the effects of variables 7 or 8 on M&IR are to be evaluated, it is imperative that all of the others be rigorously controlled and reproduced from test to test. The conditions used in the IPC testing of M&IR are specified in Method 2.6.3.1 of IPC-TM-650.

8.3.6 Storage Prior to Solder Mask Application

With cleaned panels, the allowable maximum storage time will vary with the method of cleaning, type of copper, and the atmosphere of the storage area, but generally should be minimized to reduce oxidation of the bare copper and contamination of the board. In general, less than two hours are preferred between cleaning of the board and application of the solder mask. Extended hold times may require recleaning. The laminate can adsorb water and other volatiles when not adequately protected by moisture-proof packaging. To overcome this, the bare boards should be baked for a minimum of one hour at 150°C (302°F), and then recleaned. Finger contact with the printed surface must be avoided. Due to the corrosive nature of hand oils, operators should always handle boards by their edges only and wear clean gloves. See Sec. 8.3.3 for a discussion of gloves.

8.3.7 Application of Solder Mask

The methods, procedures, and techniques for applying and curing solder masks are beyond the scope of this chapter. The application of solder mask onto clean copper circuitry should be in agreement with the individual solder mask vendor's recommendations.

8.3.8 Pre-solder Leveling Problems

Table 8.6 is a list of cleaning-related problems and solutions related to cured solder mask on bare copper.

Table 8.6 Cured Solder Mask on Copper

Problem	Cause	Solution
Poor solder mask adhesion (general)	Contamination on board surface prior to solder mask application	Follow cleaning procedures outlined in Sec. 8.3.4. Assure proper cleanliness level (Sec. 8.3.5) Recheck for residuals in resist stripping processes and rerun, if needed.
	Moisture trapped under solder mask that leads to circular blisters	Briefly oven dry boards at 71°C (160°F) just prior to mask application. Check mask application room for high humidity or presence of aqueous processing equipment. Review storage time prior to solder mask application and determine maximum hold time.

8.3.9 Solder Leveling Processes

Solder leveling is a process that leads to copper printed circuit boards with the improved solderability and longer storage life usually provided by tin-lead boards. The process eliminates some less desirable traits, e.g., variable metal adhesion, swollen or bridged circuit traces, flux entrapment, and a tendency toward poorer insulation resistance and electromigration performance, which are often associated with the tin-lead printed circuit boards.

8.3.9.1 Solder leveling processes, background. This process is faster than tin-lead plating and uses far less tin-lead to provide equivalent hole and pad solderability. Its use is being further accelerated by the push to finer lines and more densely packed printed circuit designs, which are not possible with tin-lead plated systems. Various equipment and materials have been developed in an effort to improve and optimize this process. This includes the use of both hot oil (hot oil solder leveling, or HOSL), and hot pressurized air (hot air solder leveling, or HASL) methods for removal of excess solder from hole barrels and pad surfaces. Other variations included dipping and extracting boards in both the vertical and horizontal modes as well as the use of different angles, configurations, speeds, and dwell times to improve leveling characteristics. The HASL operation has evolved as the more widely accepted of the two. It is necessary that control of processes and materials be carefully maintained to reduce contamination. All forms of solder leveling can only be characterized as dirty, corrosive, and contamination-prone operations, with the following steps being common to all processes.

8.3.9.2 Inspection and preclean. These two critical steps make precleaning the single most important function of the operation. The first step is to predict success by inspection and by employing whatever cleaning is necessary to remove contaminants that might interfere with removing oxide layers or prevent the solderability of the copper (Ref. IPC-SM-839 April 1990). The second step is the successful removal of those oxide layers and is usually done with a mildly active microetchant type of acid preclean. Generally, a cleaning agent capable of removing surface oils, and a microetch, are chosen. If required, this can be done in two stages, although good cleaners do exist for single-stage cleaning. The worst panels should be run first to evaluate the process. The after-rinse is important. It will determine the amount of precleaner and free copper that may be dragged into the flux and then into the solder pot. Proper precleaning will allow the choice of fluxes to be based solely on soldering criteria rather than on cleaning capabilities. It is recommended that a room-temperature, water-soluble, low-smoking, non-carbonizing, and low-foaming flux be used. A good preclean allows complete solderability with the least aggressive fluxing systems available, thus producing a more easily cleanable board after the solder leveling operation. This, in turn, provides a solderable and more easily cleaned board after assembly.

8.3.9.3 Fluxing. The first requirement of this process is to produce a dry panel just prior to fluxing. This is a benefit with the new breed of fluxes offering better wetting characteristics. This also reduces the solder leveling dwell time, since water does not have to be boiled off in the preheat (if used) or in the solder-dipping application. A stage to strip, squeegee, or remove the excess flux is also helpful in reducing the dwell time and residues which could be left inside the solder pot. Trade-offs exist between amount of flux applied, solderability, and residues. There must be sufficient flux on the board to prevent solder webbing and balling. Moreover, a certain amount of excess flux aids in leveling the panel and minimizes oxidation on the surface of the solder pot. Proper maintenance of the pot will prevent carbonized or degraded flux residues. Room-temperature fluxes are used due to their lower surface tension, which is maintained more consistently through multiple operations. However, if preheat is used, it is important to control heater temperatures while being certain that the proper activation temperature of the flux is achieved at the board surface (often depending on board design). It is important to avoid high temperatures that could promote residues that might contaminate the solder pot, might be extremely difficult to remove (clean) from the board, or most importantly, could cause a fire from autoignition of the fluxes.

8.3.9.4 Solder coating. The application of solder is straightforward. The board is submerged into the molten solder at a temperature that may vary from 232°C (450°F) to 260°C (500°F), depending on the tin-lead ratio and purity and equipment efficiency. An optimum range is 238–250°C (460–480°F). The dwell time will vary from 2 to 8 s, depending on the heat sink capability of the board (determined by panel thickness, number of metallic layers, and percent of ground plane area) and the purity of the solder. The copper content of the solder should be kept below 0.3 percent by weight, with solder analysis being performed on a regular basis.

Note: these temperatures will change with the adoption of lead-free solder alloys.

8.3.9.5 Solder leveling. The stripping off of the excess solder is accomplished by using either specialized air jets, hot air, or a liquid—the latter a high-temperature-resistant oil with stabilization against decomposition (hot oil). With hot air, the air pressure used for the removal of excess solder can be adjusted relative to the results desired. The thickness, brightness, and evenness in the holes and feature flatness of the solder coating are effected by air pressure, air temperature, air configuration, and angle of immersion. In HOSL, adjustments are dependent on the liquid used. Some controversy remains about which leveling choice is best for each feature, but satisfactory results are accomplished in most cases.

8.3.10 Post-solder Leveling Cleaning

The type and degree of cleaning is dependent on the previous steps of the solder leveling process. The solder must be in a solid state (i.e., completely cooled) prior to washing. This cleaning step removes the flux residues from the panels for most water-soluble fluxes. A two-stage warm water wash with a mild detergent is followed by multiple-stage cascade counterflow spray rinsing and complete drying. A soft, nylon-bristle brush can be used in the first stage if needed. Water-soluble fluxes may not require the use of detergent. Deionized water heated to 60°C (140°F) can sometimes be used. It is important to spot-test and attempt to control these processes. "Appearance" is not the prime criteria, since improper processing at this stage can lead to failures due to high surface conductivity, while active flux residues will quickly deteriorate a stored panel. (Sections 8.3.5.3, 8.3.5.4, and 8.3.5.5 address cleanliness testing.)

8.3.11 Post-Solder-Leveling Problems

Table 8.7 is a list of problems and solutions occurring with cured solder mask over bare copper after solder leveling.

8.4 Solder Mask over Tin-Lead

In this process, an etch resist, usually tin-lead, is reflowed (fused) so that the copper circuitry is covered with solder. Since various types of fusing fluids can be used to aid this process, it is necessary to completely remove them to have a clean, adherent surface for maximum solder mask adhesion. Section 8.4 is a detailed explanation of these methods. (See Fig. 8.2 for a process flow chart.)

8.4.1 Fusing Fluid Cleaning

The cleaning should be directly in line with the fusing unit to effectively remove the water-soluble fusing fluids that are typically used for fusing of bare boards. Multiple-stage cascade counterflow spray rinsing is best, with the final rinse being deionized water. Additional

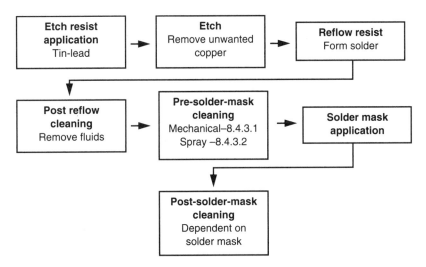

Figure 8.2 Solder mask over tin-lead.

Table 8.7 Cured Solder Mask after Solder Leveling

Problem	Cause	Solution
Plated-through hole not soldered	Dry film solder mask developer residues	Determine efficiency of post-developer rinsing.
Pads not completely soldered	Solder mask residue on pads	Use acid precleaner that produces low microetch and will not cause attack on previously soldered boards.
Dewetting of solder	Copper surface may be contaminated with organic residues	Run test of Sec. 8.3.5.4 to detect the presence of organics. Rerun cleaning procedure in Sec. 8.3.4 or change to one that doesn't leave organic residues.
Solder mask blisters/ poor adhesion	Copper surface improperly cleaned	Determine presence of intermetallic, and/or organic residues. Change cleaning method such as brush type, brush pressure, rinsing, and drying. If not used, add microetch before scrubbing.
Poor electrical properties following solvent post-clean	Incomplete removal of fluxes that contain nonionic and ionic contaminants	1. Use a chlorinated hydrocarbon-alcohol azeotrope to remove polar and nonpolar contaminants. 2. Avoid partially fluorinated fluorocarbon solvents, since they can be less efficient. 3. Check cleaning equipment for plugged spray nozzles or other mechanical failures. 4. Completely remove fluxes within 10 min after soldering.
Poor electrical properties following aqueous post-clean	Incomplete removal of fluxes that contain nonionic and ionic contaminants	1. Maintain pH of saponifier solution between 10.5–11.8, since low values cause inefficient cleaning and high values will degrade laminate, components, and coatings. 2. Maintain efficient high-pressure sprays to flush away non-saponified rosin acids. 3. Replace spent cleaning fluid. 4. Add rinse aids to rinse water to help remove residues.

cleanliness can be obtained using rinsing agents in a recirculating spray following the recommendations of the vendor.

8.4.1.1 Residues and problems. The two types of problems experienced with solder masked tin-lead boards are solderability and cleanliness under the solder mask. *Solderability* refers to the ability of the solder to wick up through the holes and onto the pads and is determined by IPC-J-STD-003. Flux residues left in the holes causing solderability problems could be polar (ionic) or nonpolar (organic).

Heavy solder oxides could also cause solderability problems. Cleanliness must be determined prior to solder mask application, because boards not thoroughly cleaned will have a short life expectancy and functional difficulties. Flux residues that are ionic (acid) will also cause corrosion. Nonpolar residues can be hygroscopic and thus become polar upon water absorption, which subsequently causes failures.

8.4.1.2 Test methods for residual detection. The two types of test methods are ion extraction (see Sec. 8.3.5.3) and surface insulation resistance (See Sec. 8.3.5.5).

8.4.1.3 Method of removal. All boards should be thoroughly rinsed in cascading counterflow spray rinses. If additional cleanliness is indicated by the test methods, then alkaline-based rinsing agents can be added to the post-fusing cleaning system. If scrubbing is indicated, it should be done only with a soft, non-abrasive brush so as not to smear the solder.

8.4.2 Board Handling

Boards should always be handled by the edges only. The proper choice of gloves can provide some handling benefit. Improper glove selection can provide a false sense of security, since some gloves contribute fibers to the board. Some gloves will also absorb hand oils, which can be transferred to the board. Racking is the preferred storage technique for boards in process. Boards should be stored in an atmosphere without high humidity or the presence of process chemicals.

8.4.3 Pre-solder Mask Cleaning

To ensure adhesion of the solder mask to the fused board, conditioning of the surface is required before the solder mask is applied. This can be achieved by mechanical or chemical methods. Pumice should not be used on fused solder. Scrubbing is also not recommended. Boards with heavy oxide buildup could be cleaned with a solder conditioner. If this is done, it is important to follow vendor's guidelines and to rinse thoroughly. Final rinsing after these procedures should follow recommendations under Sec. 8.3.4.4.

8.4.3.1 Mechanical methods. Tin-lead is usually too soft for abrasive cleaning, but it can be done when extreme care is taken. Cleaning equipment with soft, non-abrasive brushes can be used if contact pressure is minimized to avoid damage to the circuitry and smearing of the metal. Never use a scrubber that is used for copper on tin-lead, since subsequent cross-contamination can occur.

8.4.3.2 Spray cleaning. The appropriate spray equipment is recommended that allows a contact time with the cleaning agent of at least 45 s followed by at least 15 s of a clean spray rinsing. Spray cleaning can be accomplished by either solvent or aqueous processes.

Solvent. The best results are obtained when solvents are used in both liquid and vapor phases. Typical degreasing solvents such as hydrochlorofluorocarbons (HCFCs), hydrofluorocarbons (HFCs), hydrofluoroethers (HFEs), or formulations based on them, as well as trichloroethylene (TCE) may be used. The use of methylene chloride (MC) for cleaning is not recommended, because this solvent will attack the base laminate. Cleaning cycles of longer than 60 s can damage the metal surface or the base laminate.

Aqueous. Aqueous cleaning agents combined with a water rinse should provide a clean surface. Rinsing should incorporate high pressure, fan-type spray nozzles operating at 2–3 bar (30–45 psi) with warm 15–35°C (60–95°F) water. Rinse duration should be at least 30 s. (Refer to Secs. 8.3.4.4 and 8.3.4.5).

8.4.4 Pre-solder Mask Cleanliness Testing

See Sec. 8.3.5 for a complete description of the appropriate test methods.

8.4.5 Application of Solder Mask

The methods, procedures, and techniques for applying and curing of solder mask are beyond the scope of this guideline. The screening, coating, or lamination of solder mask onto clean tin-lead circuitry should be considered with the individual solder mask vendor's recommendations.

8.4.6 Troubleshooting Guide—Cured Solder Mask

Table 8.8 is a list of cleaning-related problems and solutions related to cured solder mask on fused tin-lead.

Table 8.8 Cured Solder Mask on fused Tin-Lead

Problem	Cause	Solution
Poor solder mask adhesion (general)	Contamination on board surface prior to solder mask application Moisture trapped under solder mask that can lead to circular blisters	Follow cleaning procedures outlined in Sec. 8.4.3 Assure proper cleanliness level (Sec. 8.3.5). Check mask application room for high humidity or presence of aqueous processing equipment.
Poor solvent or cleaner resistance	Incompatible cleaning chemistry	Change to a compatible mask/cleaner combination.
Poor electrical properties	Contamination on board surface prior to solder mask application	Follow defined cleaning procedures. Perform cleanliness testing prior to applying solder mask. Recheck resistivity of process rinse waters.
Poor cosmetic appearance pinholes, craters or fisheyes	Oily contamination on surface prior to solder mask application.	Thoroughly remove all grease and oils (see Sec. 8.3.3).
Discoloration	Incomplete removal of residues *(solvent post-clean)*	1. Use a chlorinated hydrocarbon-alcohol azeotrope to remove both polar and nonpolar contaminants. 2. Avoid partially fluorinated fluorocarbon solvents, since they can be less efficient. 3. Check cleaning equipment for plugged spray nozzles or other mechanical failures. 4. Completely remove fluxes within 10 min after soldering.
	Incomplete removal of residues *(aqueous post-clean)*	1. Maintain pH of saponifier solution between 10.5–11.8, since low values cause inefficient cleaning, and high values will degrade laminate, components, and coatings. 2. Maintain efficient high-pressure sprays to flush away non-saponifiable rosin acids. 3. Replace spent cleaning fluid. 4. Thoroughly post-clean boards within 10 min after soldering to minimize cleaning difficulties. 5. Add rinse aids to rinse water to help remove residues.

8.5 Environmental Controls and Considerations

8.5.1 Introduction

The emissions from fabrication cleaning machines are or can be

- Waste aqueous cleaning agents that are contaminated with plating salts, etching salts, HASL fluxes, fusing fluids, and other soils
- Waste solvent cleaning agents that are contaminated with plating salts, etching salts, HASL fluxes, fusing fluids, and other soils
- Rinse water that contains small amounts of plating, etching, and cleaning agents
- Vapors of plating chemicals, etching chemicals, aqueous cleaning agents, water, and solvents
- Solid waste from stripping of primary resist film ("skins") upon completion of primary imaging and stripping
- Contaminated ion exchange resins, filters, carbon canisters, and membranes used in the various processes
- Air emissions of volatile organic compounds (and hazardous air pollutants, if any)

Each waste stream must be handled according to applicable national and local regulations via the appropriate environmental controls. Semi-aqueous cleaning processes are used as an example in this section to illustrate how environmental controls are implemented in PWB fabrication cleaning. If fabricators are planning on using an aqueous only process, the paragraphs discussing the organic semi-aqueous cleaning agent can be disregarded. An overview of environmental regulations, imposed by regulatory bodies upon fabricators, is provided in Sec. 8.6.

Various governmental bodies have established rules and regulations for transporting wastes, discharging wastes into the air, and discharging wastewater into sewer or septic systems and ultimately into rivers, lakes, and oceans.[1]

Users should carefully consider these protective regulations when they implement any new fabrication cleaning process. Users should contact all appropriate air, wastewater, and solid waste authorities when implementing any new fabrication processes.

---

[1] In the U.S.A., the Clean Air Act, the Clean Water Act, Resource Recovery and Conservation Act (RCRA), etc., detail the responsibilities both user and manufacturer have for fabrication cleaning agents. Similar regulations may be applicable in other nations.

Spills, as defined by local and national standards, should be reported to the appropriate authorities. *Material safety data sheets* (MSDSs) contain information about proper methods and equipment to contain and clean up spills.

8.5.1.1 Discussion of Examples. While the use of ordinary water is fairly straightforward, compliance with applicable regulations, when aqueous cleaning agents or cleaning agent emulsions are used, can be more complex for users upgrading their fabrication processes. For example, organic-based cleaning agent emulsions have been shown to be equivalent to solvents for the removal of non-aqueous fusing fluids from boards without the customary degradation experienced with water-soluble fluids and water removal. Subsequent paragraphs will discuss the environmental issues associated with such processes.

Organic-based cleaning agents or their emulsions used in combination with water (e.g., semi-aqueous) are usually classified as biodegradable and do not inhibit bacteria in water found in publicly owned treatment works (POTWs). The MSDSs or similar documents will contain information specific to their products. Since local water regulations differ significantly, users should consult their local POTW authorities.

8.5.2 Waste Semi-aqueous Cleaning Agent

Depending on the type of cleaning machine and semi-aqueous cleaning agent used, there are two sources of waste organic semi-aqueous cleaning agent. The first is the semi-aqueous cleaning agent tank itself. Ideally, the concentration of the flux residue in the semi-aqueous cleaning agent tank will remain at an equilibrium value due to the drag-out of semi-aqueous cleaning agent from the semi-aqueous cleaning agent tank and replenishment with fresh cleaning agent. If air knife efficiency at the exit of the semi-aqueous cleaning agent tank is too high, the concentration of soil in the semi-aqueous cleaning agent could become too high and force the entire semi-aqueous cleaning agent chamber to be replaced occasionally. The second, and by far the most likely, source of spent semi-aqueous cleaning agent is either the organic layer in the decanter or the concentrate from water recycling processes. The exact disposal route will depend on the flash point of the waste, the amount of hazardous material dissolved in the semi-aqueous cleaning agents, and the water content.[1]

---

[1] In the U.S.A., these limits are dictated by RCRA rules; in other countries, they are generally governed by legislation on the disposal of hazardous wastes.

Flux residues can contain limited quantities of lead and other heavy metals. This is particularly true if aggressive fluxes are used. Since Type I semi-aqueous cleaning agents are pH neutral, they do not remove heavy metals by chemical reaction. Some semi-aqueous cleaning agent vendors have established programs to pick up spent semi-aqueous cleaning agents for correct disposal or recycling. Such services may be geographically limited in remote areas or in countries other than the nation of origin.

For semi-aqueous cleaning agents that are soluble in water, the option for separating the bulk of the spent solvent from the rinse water in a decanter or coalescer is not available. The material in the wash tank is handled in the same manner as water-insoluble materials.

All cleaning processes are capable of removing solder balls, solder splash, and other solder bits, plus other heavy metal sources, from the printed wiring boards during fabrication. The cleaning equipment should be fitted with suitable filters to remove such solid metal particulate from the cleaning agent.

8.5.3 Rinse Water

The calculated concentrations are low enough that the dilute rinse waters from the process may be discharged to POTWs, depending on local regulations. (Local regulations often include pretreatment of the waste stream prior to discharge.) All semi-aqueous cleaning agents are biodegradable when suspended or dissolved in water and are noninhibitory toward bacteria commonly found in the POTW facilities. Users should be aware that some nations have a national standard for wastewater to drain that has different limits on metals and other parameters for each province or state. Also, each POTW may set specific limits on either or both the volume and individual contaminant loading. Users should inform themselves on such issues prior to plant construction or process changes.

8.5.3.1 Rinse water from water-insoluble (Type I) semi-aqueous cleaning systems. In the U.S.A., for example, water from subsequent rinses in the rinse systems with decanter can often be discharged directly to POTW facilities via a connected facility sewer system. This should be verified with the local POTW authority or similar body in the nation of operation. More information on this matter is covered in the section on environmental controls (see Sec. 8.6).

8.5.3.2 Rinse water from water-soluble (Type II) semi-aqueous cleaning systems. Rinse water from a water-soluble semi-aqueous cleaning systems will contain a higher level of organics. There are several ways to dispose these rinse water streams, as discussed below. If open loop disposal (discharging to drain or to POTW) is restricted or prohibited, evaporators could be use to concentrate the waste streams. When the waste streams are concentrated, the water is removed, and most or all of the water-soluble chemistry and contaminant are usually left for alternative disposal. Also, open-loop processes may generate VOCs and HAPs, which are regulated emissions.

8.5.3.3 Rinse water discharge paths. The discharge from an aqueous cleaning system can follow several pathways. The first is discharge directly to drain, which may or may not flow into a POTW or other waste acceptor. The second is known discharge to a POTW. The third is discharge to drain that is piped to a underground injection control (UIC) well. The fourth is direct discharge to surface waters. The fifth is one or a combination of the above, followed by removal by drum or tanker car. All or any of these methods would constitute the methods of discharge.

Rinse water discharge to drain. Direct discharge of rinse water to drain is the easiest to set up. However, environmental and regulatory considerations need to be taken in account. Facilities should first clarify where the drain is going. Some drains go to POTWs.[1]

The waste streams from these systems will be mostly water, with a little water-soluble chemistry and an even smaller amount of contaminant from the parts cleaned. Water-soluble chemistry is biodegradable and has a neutral or slightly basic pH. Usually, a water permit will allow the small amount of water-soluble chemistry found in the waste stream to go to drain. The amount of chemistry in the waste stream can be estimated by the amount of dragout and tested to verify the levels. Some locations do not allow discharging to drain or limit the amount of liquid allowed to go to drain. For these places, other options must be considered.

Rinse water discharge to a POTW. Rinse water from an aqueous PWB fabrication cleaning system may or may not be able to be discharged directly (or indirectly) to a POTW. In many locations, the water-solu-

---

[1] Facility drains not connected to POTWs generally go to septic systems or other Underground Injection Control (UIC) wells. The U.S. Environmental Protection Agency (US-EPA) regulates the discharge of fluids or wastewater to subsurface UICs that may endanger underground sources of drinking water. Facilities with UICs should contact their state UIC Program authorities to determine compliance requirements. Users in other nations will have similar regulations to follow.

ble materials will be regulated but are allowed to be discharged to a POTW with the payment of a small surcharge fee.

8.5.4 Evaporator to Waste Drums

When an evaporator is used to concentrate wastewater streams, the users may periodically empty the evaporator to waste drums, or they may operate the evaporator continuously. The advantage of periodically emptying the evaporator to waste drums is that more chemistry is contained in the evaporator and less is discharged to the environment as a VOC. The disadvantage is more waste drums for disposal. This option slows down the evaporation rate and uses a demister to help concentrate the chemistry in the evaporator. The disadvantage of slowing down the evaporation rate is that the concentration of chemistry in the chemical isolation section is increased, causing more chemistry to reach the subsequent rinse stage through drag-out. If carbon/ion exchange beds are used in a subsequent water recycling process, they will be depleted slightly faster because of the additional chemistry in the rinse stage that must be removed (see Fig. 8.3).

8.5.4.1 Steady-state evaporator.
When an evaporator is operated at a steady state, it can be run a long time without emptying the evaporator to a waste drum. Waste stream from the chemical isolation or the first rinse section can be concentrated at a continuous basis.

The advantage of a steady-state evaporator is that it minimizes the amount of waste going to drums, and it allows the evaporator to run longer between shutdown and cleanout. The disadvantage is that the amount of VOC discharged to the environment is higher. This option increases the evaporation rate of the evaporator and does not use a demister. The disadvantage of increasing the evaporation rate and not using the demister is that the amount of water used in the chemical isolation section is increased. The advantage of this is that the concentration of chemistry in the chemical isolation section is lowered, which lowers the amount of chemistry in the subsequent rinse stage. This means that, if carbon/ion exchange beds are used in the subsequent water recycling, they will last longer, because they will not have to remove as much chemistry.

8.5.5 Water Recycling

Several methods have been developed to remove the last traces of semi-aqueous cleaning agent from the rinse water so that it can be reused. The advantages of recycling water are reduced water consumption, elimination of water disposal, and control of incoming water

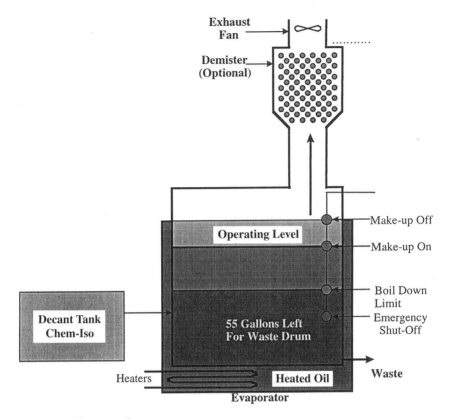

Figure 8.3 Schematic of evaporator system.

quality. Users should examine such systems as are already in place for their proposed application to ensure that the system will work for them.

8.5.5.1 Carbon/ion-exchange bed. This option uses carbon/ion-exchange beds to remove the last traces of cleaning chemistry. This option redirects the suction of the final rinse pump to take a suction on the chemical isolation tank, pump it through the carbon and ion-exchange beds, and discharge it to the final rinse stage via the spray nozzles. If this system is used to recycle rinse water from a water insoluble semi-aqueous system, a portion of the semi-aqueous agent that is dispersed in the rinse water can be removed from the water by using an oil separator prior to the carbon beds.

The carbon bed removes the organics in the liquid, and the ion-exchange bed removes any ions. As carbon adsorption can be almost molecule-size-specific, it may be necessary to have a series of carbon beds, each with a different quality of carbon. Incorrect adsorption may result in an accumulation of hygroscopic nonionics that may have a deleterious effect on the electrical qualities of the assemblies.

This process greatly reduces the contaminant loading in the waste stream being sent to drum-up or to drain. The contaminants are removed or substantially reduced, leaving only relatively uncontaminated water to be forwarded for disposal. Users should be aware that some POTWs may refuse to accept the relatively pure waste stream, just because the water volume would overcome the capacity of the POTW facility. (Such a volume problem can be encountered in areas where the construction of the POTW has been followed by unanticipated industrial expansion.) The cost of the carbon/ion-exchange beds is high, and they will be exhausted frequently. This will probably be the most expensive long term option and therefore should be considered as an intermediate or last step solution to the problem. Figure 8.4 shows this option in combination with a spray-under-immersion, in-line cleaning system. While this is a spray-under-immersion system designed for semi-aqueous cleaning, the consecutive wash-rinse-dry steps would be the same for saponifier or plain water processes.

8.5.5.2 Reverse osmosis. A reverse osmosis system works similarly to the carbon/ion-exchange option, except it has a higher initial cost but lower operating cost. It takes the waste stream and separates the water-soluble (Type II) chemistry and contaminants using membranes. Membranes are essentially extremely fine filters that are capable of separating materials on a molecular level. Figure 8.5 is a diagram of how membranes work. Contaminated solutions are pumped under pressure through membranes that are designed to allow small water molecules to pass and to restrict the flow of ions and large molecules, such as those found in some semi-aqueous cleaning agents. Contaminants are concentrated in the reject stream and reduced in the permeate stream.

Well chosen membranes can achieve better than 99 percent separation. Membranes are selected based on compatibility and separation with a specific cleaning agent. A membrane that is suitable with one semi-aqueous cleaning agent may not be compatible with other cleaning materials. Damaged membranes can be expensive to replace. Semi-aqueous material suppliers can provide advice about the utility of this process for their semi-aqueous cleaning agents. While this technology showed great promise in the early days of semi-aqueous (Type I) technology implementation, it turned out to be more complex than many process personnel were capable of managing successfully. Thus, it is rare to find this technology in use today.

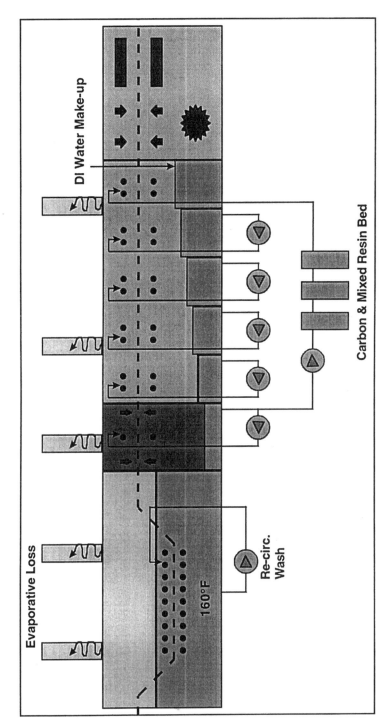

Figure 8.4 Spray-under-immersion, in-line cleaning machine with carbon/ion-exchange.

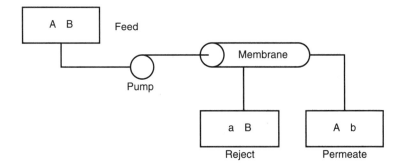

Figure 8.5 Reverse osmosis concept schematic.

Figures 8.6a and 8.6b are diagrams of reverse osmosis systems that have been incorporated into semi-aqueous cleaning machines. The rinse water from rinse 2 flows to a feed tank on the membrane system and from there is pumped through the membrane. Water with low mineral and organic content may be polished using carbon/ion-exchange beds and sent back to the final rinse stage as make-up water. The carbon/ion-exchange beds used in conjunction with a reverse osmosis system will last significantly longer. The concentrate from the membrane may be sent back to the feed tank, where the concentration of the contaminant will build up slowly. At some point, part of

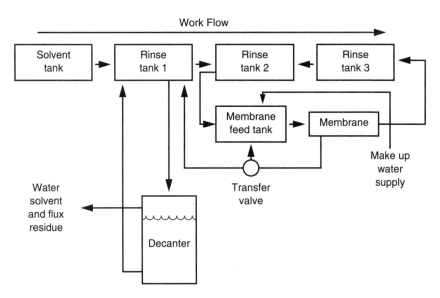

Figure 8.6a Membrane rinse water recycling system concepts for semi-aqueous cleaning processes with cleaning agents that are insoluble in water (Type I).

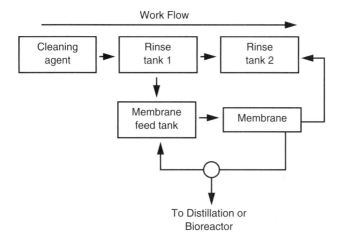

Figure 8.6b Membrane rinse water recycling system concepts for semi-aqueous cleaning processes with cleaning agents that are soluble in water (Type II).

the reject stream is removed to prevent excessive contaminant buildup.

With water-insoluble (Type I) materials, the reject stream is usually returned to the decanter on the cleaning machine. Once in the decanter, the organic solvent will separate from the water. Working in tandem, the decanter and the hold tank for the reverse osmosis system will naturally reach the same low equilibrium concentration. Typically, the optimal concentration in both systems is two percent or lower. Reverse osmosis units are rarely used in Type II semi-aqueous processes due to compatibility problems with the membranes. Figure 8.7 depicts a spray-under-immersion, in-line cleaner fitted with the reverse osmosis system option.

In water-soluble processes, these materials are disposed of in one of three ways. They may go to drain as the first step, followed by several other options as described in Sec. 8.5.3.3. See Sec. 8.6 for environmental regulations and considerations for these disposal practices.

8.5.6 Volatile Organic Compounds (VOCs)

Since both water-soluble and water insoluble types of semi-aqueous cleaning agents are hydrocarbon-based materials, they all are classified as volatile organic compounds (VOCs). Many countries are extremely concerned about VOCs or their equivalents and have enacted legislation to limit VOC emissions. (See Sec. 8.6.4 for VOC regulatory considerations.)

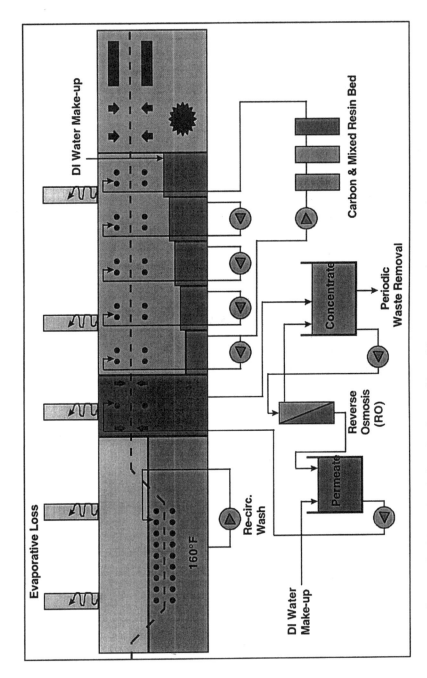

Figure 8.7 Spray-under-immersion, in-line cleaning machine with reverse osmosis system option.

The amount of VOC that is emitted is directly related to the vapor pressure of the solvent, the ventilation flow rate from the equipment, the temperature of the cleaning solvent, and the temperature and concentration of cleaning agent in the rinse sections. VOC emissions will be lowest when

- Low vapor pressure solvents are used.
- The solvent temperature is minimized.
- The velocity and volume of ventilation are maintained as low as possible to provide safe and comfortable working conditions around the cleaning machines.
- The temperature and concentration of material in the water in the rinse sections are as low as possible.

These VOC concerns must be balanced with cleaning requirements. The vapor pressures for most semi-aqueous cleaning agents are much less than 1 mm of mercury, and significantly lower than vapor pressures of common alcohols and other lower-boiling materials. The amount of semi-aqueous cleaning agent that escapes as VOC material is relatively small and variable. The exact amount depends on the type and size of equipment and the operating conditions. The best method to determine the loss due to evaporation is to consider the mass balance of the cleaning process. The simplest way to calculate the daily loss of VOC material is to first calculate the daily weight of cleaning agent lost, then multiply that times the percent of VOC classified material in the cleaning agent formulation, which can be obtained from the MSDS. The daily cleaning agent loss is equal to the amount of semi-aqueous cleaning agent added to the cleaning machine minus the amount of cleaning agent that leaves the machine through the decanter and in the rinse water.

Actual VOC emitted (kg/day) = %VOC in cleaning agent × cleaning agent lost (kg/day)

where

$$\text{cleaning agent lost as kg/day} = A - B - C$$

where A = cleaning agent added, kg/day
B = cleaning agent removed from decanter, kg/day
C = cleaning agent discharged in rinse water, kg/day

where

$$\text{cleaning agent discharged in rinse water, kg/day (C)} = D \times E$$

where D = cleaning agent concentration in rinse water, kg/liter
E = rinse water volume discharged, liters/day

The concentration of the semi-aqueous cleaning agent in the rinse water can be easily measured by determining the chemical oxygen demand (COD) of the rinse water. The COD is proportional to the concentration. For most semi-aqueous cleaning agents, the concentration of semi-aqueous cleaning agent, in parts per million, is determined by dividing the measured COD value by roughly 2.5.

VOC emissions vary widely depending on the cleaning machine and the operating parameters. A typical in-line machine may emit about half a kilogram of VOC per hour. VOC emissions can be reduced with standard VOC control technology including demisters, dead zones in the exhaust lines, scrubbers, and condensers. Users must carefully weigh the cost of VOC compliance when considering cleaning process choices. Inorganic cleaning chemicals based on materials such as carbonates and sulfates do not have a vapor pressure, so they would not be a factor in the emission profile of a cleaner that uses them.

8.5.7 Greenhouse Effect

Semi-aqueous cleaning agents do not emit significant quantities of greenhouse gases. Thus, they are not regulated in the U.S.A. but may either be or become regulated in other nations. Minimization of VOC emissions should also minimize any potential concern.

8.5.8 Other Issues

Users should be aware of certain other issues for safe operation of a fabrication cleaning process. These are

1. The flash point can be an issue; pick materials with higher flash points if possible.

2. Containment basins around cleaning equipment can be used to keep cleaning agents and process residues out of drains.

3. Choose explosion-proof wiring if flammable materials are used as cleaning agents.

Use care in choosing control/compliance technologies. For example, the rule of thumb for economical use of carbon adsorption systems to remove cleaning agent vapors from the air is a minimum of 500 ppm in the air. Thus, if the allowed level in the air is 350 ppm, using this technology to control vapors is very unwise.

8.6 Regulatory (OSHA and EPA) Considerations

8.6.1 Introduction

Although this section is specifically concerned with U.S. legislation and codes of practice, most other nations have similar regulations to protect both the environment and the workplace. These may be considerably more or less severe and may be applied differently. The general theme of this section is universally applicable, but it should be read with local regulations and an attention to detail in mind. This section provides a general overview of the regulations that are typically imposed on individual fabricators.

8.6.1.1 Chemicals. Organic and inorganic cleaning agents are chemicals. As such, they are subject to the rules and regulations governing the use of chemicals in the work place. Over the past several years, Material Safety Data Sheets (MSDSs) have been expanded. Conscientious suppliers provide an abundance of information to users on toxicity testing, health hazards, waste disposal, safe work practices, protective equipment, material reactivity and flammability, etc. These documents should be read and used. The law requires that any personnel who may work with cleaning agents and any other chemicals have access to MSDSs. Most suppliers now have these available on their web sites for "around the clock" access.

8.6.1.2 Chemicals in the workplace. The work area should be well ventilated, and the workers' exposure to chemicals should be minimized. Proper equipment should be available and used when dispensing chemicals or cleaning equipment. Recommended protective equipment is listed in the MSDS that suppliers ship with the cleaning agents.

8.6.1.3 Cleaning process waste streams. The cleaning process generates air emissions, solid waste, and wastewater. A discharge permit is required, in many instances, for very small or infrequent industrial wastewater discharges from cleaners such as batch or dip tanks. The three main groups of regulatory agencies that a user must review are federal, state, and local (or their equivalents in other countries). Users should be aware there are often multiple state or local regulatory agencies with environmental jurisdiction over a waste stream. Each company must meet the minimum federal standards. The state regulations may be the same or even more restrictive than the federal. Finally, the local community's regulations will be at least as restrictive as state regulations.

8.6.2 Environmental Considerations

In the United States, there is a full complement of regulations, including the Clean Air Act (CAA), Clean Water Act (CWA), and Resource Conservation and Recovery Act (RCRA). In summary, these laws make the user and the manufacturer responsible for air, water, and solid waste emissions and scrap materials used to manufacture printed wiring from cradle to grave. This has prompted suppliers to provide better documentation about the environmental and safety issues associated with their cleaning agents. Users are better informed and are demanding safer chemicals, more disposal routes, and closed-loop technologies.

In the selection and use of a PWB fabrication cleaning agent and process, manufacturers must consider the effect of the cleaning agent on air emissions, wastewater discharge, and solid waste generation. Permits for both air and wastewater are usually required. Permits and special handling of solid waste are very often required as well. Waste streams can be grouped into the following three classes with subordinate concerns:

1. Air emissions
   - VOCs (volatile organic compounds)
2. Wastewater
   - lead (and other heavy metals)
   - pH (acid or basic condition)
   - COD/BOD
   - other (temperature, suspended solids, etc.)
3. Solid waste
   - lead (and other heavy metals)
   - used solvents from cleaning or stripping[1]
   - primary imaging film waste ("skins") after stripping[2]

1Even though spent solvent is not a solid, the EPA classifies liquids, semi-solids and solids as a solid waste (see Code of Federal Regulations, Protection of the Environment, Parts 260-299, July 1, 1991, p.23)

8.6.3 Air Emissions

The Clean Air Act (CAA) requires states to meet six National Ambient Air Quality Standards (NAAQSs) for specific pollutants. States and local governments are empowered under the Clean Air Act to enforce local regulations that are more stringent than national standards so as to attain the NAAQS. When any new chemicals, including cleaning agents, are introduced into manufacturing processes, the user must seek guidance and rule interpretation from local authorities before receiving the required permits. The approval process is lengthy, so users should attend to the permit application process early in the selection and implementation stages of installing any new cleaning process.

8.6.4 Volatile Organic Compounds (VOCs)

The Clean Air Act (CAA) requires states to meet six National Ambient Air Quality Standards (NAAQSs) for specific pollutants. NAAQS establish the maximum concentration allowed for each pollutant in the background air in all areas of the country. Those pollutants include nitrogen oxides (NOx), carbon monoxide (CO), and ozone. The CAA's NAAQS provisions apply to areas of the country that do not currently attain an NAAQS for a particular pollutant (these areas are called *non-attainment areas*). If an enterprise is located in a non-attainment area, and the facility emits a non-attainment pollutant, the enterprise may have to meet stricter requirements for air pollution control for that pollutant than if the facility were located in an attainment area.

Electronic manufacturers are most likely to be affected by requirements in the non-attainment areas for ground-level ozone (also called *smog*). Ozone is a pollutant that is not discharged to the environment but is created by the reaction of ozone precursors (volatile organic compounds, or VOCs) and sunlight. VOC emissions are directly related to ozone production. Cleaning processes are likely to emit VOC, which combine with daylight to form smog.

8.6.5 Hazardous Air Pollutants (HAPs)

Section 112 of the Clean Air Act (CAA) lists 189 hazardous air pollutants (HAPs) for which the EPA has set National Emission Standards. Some solvents, such as methylene chloride, perchloroethylene, trichloroethylene, carbon tetrachloride, and chloroform are included in the CAA Section 112 HAP list.

---

[2]There has been continuing concern over the photoresist "skins" stripped off the PWBs and often found in the aqueous waste stream. Fabricators should check with IPC and local agencies to ensure they are in compliance with applicable regulations.

EPA issues Maximum Achievable Control Technology (MACT) standards to reduce emissions of hazardous air pollutants. Each MACT standard controls emissions of one or more air toxics from a specific air pollutant source. The MACT is the best control technology already in use for a given type of industry.

Under the CAA's National Emission Standard for Hazardous Air Pollutant (NESHAP) program, "major" sources of air toxics must use the MACT-specified technology. Facilities are considered "major sources" if their potential emissions are 4.5 metric tons per year (10 tons per year) or more of a HAP or 11.4 metric tons per year (25 tons per year) of a combination of HAPs.

NESHAP standards require affected users to employ certain control technologies to reduce emissions. Typically, record keeping, reporting, and emission control technologies may need to be upgraded to meet the requirements of these standards.

In 1994, the EPA finalized the Halogenated Solvent Cleaning NESHAP. This standard is published in 40 CFR Part 63, Subpart T, and affects new and existing halogenated solvent cleaning operations. For guidance, consult EPA publication EPA-453/R-94-081, EPA Guidance Document for the Halogenated Solvent Cleaner NESHAP. It is anticipated that the use of halogenated organic solvents used as solder mask strippers might fall within the boundaries of this regulation.

8.6.6 Ozone-Depleting Substances

The CAA amendments also contain provisions for controlling the use of chemicals believed to destroy stratospheric ozone, a layer of atmosphere that protects the Earth by filtering harmful sun rays. EPA has established two lists of such substances: a list of Class I substances and a list of Class II substances. Class I refers to substances containing chlorofluorocarbons (CFCs) or 1,1,1-trichloroethane (TCA). Class II substances include hydrochlorofluorocarbons (HCFCs). Many of the primary and secondary imaging films were developed and or stripped with such solvents. Such formulations have been replaced with new materials that are usually developed with highly diluted inorganic carbonates and stripped with highly diluted inorganic hydroxides.

In 1992, the EPA issued its final rule implementing Section 604 of the Clean Air Act. That section implemented the United States' obligations under the Montreal Protocol by limiting the production and consumption of substances with a potential to deplete stratospheric ozone. Carbon tetrachloride, trichlorotrifluoroethane (CFC-113), and 1,1,1-trichloroethane (methyl chloroform or TCA) were identified as Ozone Destroying Substances (ODSs). This rule, published as 40 CFR Part 82, Subpart A, required the phase-out of production of these products for emissive uses by December 31, 1995, in Article 2 parties to the Protocol or "developed" nations. In general, Article 5.1 parties to the Protocol or "developing" nations, have extended timelines for compliance. Article 2 fabricators with Article 5.1 partners may find that their partners may have been required to comply with an accelerated phase-out, if the vast majority of ODS is not used by locally owned enterprises.

Existing supplies of these solvents (produced prior to the phase-out), and recycled product can continue to be sold into emissive markets. Exceptions to the ban include product manufactured for transformation (chemical intermediate) processes, and "essential" uses as approved by the EPA (such as asthma inhalers).

Additionally, 40 CFR Part 82, Subpart E requires that products containing or manufactured with ODSs must be specially labeled identifying the presence of an ODS. Electronics manufacturers building goods in Article 5.1 countries may find that U.S. regulations require that all products must be labeled if either containing or made with ODSs.

The Clean Air Act also required the EPA to establish a program to identify alternatives for ODSs and to publish lists of acceptable and unacceptable substitutes. This program, the Significant New Alternatives Policy (SNAP) program, is described in 40 CFR Part 82, Subpart G. Once the EPA's acceptability list is established, it is illegal to replace a regulated ODS with any substance determined as harmful to human health or environment, if other substitutes with reduced overall risk are currently or potentially available. Acceptable replacements for CFC-113 and 1,1,1-trichloroethane include aqueous cleaners, semi-aqueous cleaners, oxygenated solvents, terpenes, trichloroethylene, perchloroethylene, and methylene chloride.

The above paragraphs of Section 6.6 apply to Article 2 ("developed") nations of the Parties to the Montreal Protocol or Article 5.1 ("developing") nations on the same phase-out schedule. The rest of the Article 5.1 nations usually have a "grace period" that allows several more years prior to completion of their Class I ODS phase-outs.

8.6.7 Waste Water

The Clean Water Act is a comprehensive law that sets forth the requirements for reporting spills, lists priority pollutants, dictates water quality criteria, and requires permits for discharges to POTWs or water bodies. In addition, discharges to underground wells that could endanger groundwater quality are regulated. Because local regulations are often more stringent than federal regulations, it is imperative to consult with local regulatory agencies when introducing new chemicals or chemical processes to a PWB fabrication facility.

In the past, solvent cleaning in boiling solvent machines did not always result in an aqueous discharge unless they were fitted with a water separator. Water separators were commonly found on chlorinated solvent machines and CFC-113 machines that did not contain an alcohol azeotrope. Today's solvent machines are often fitted with dehumidification coils to provide a layer of dense dry air over the boiling solvent–air interface. The water removed by this process may contain an extremely low level of solvent when the water is discharged to drain. The goal of these extra coils is to maximize solvent retention in the cleaning equipment, thus minimizing any that might escape to the environment.

In view of the relative simplicity of waste water compliance for such solvent processes, existing solvent process users may not be informed as to the importance of water quality requirements faced by aqueous process users.

Local POTWs regulate such parameters as biological oxygen demand (BOD), chemical oxygen demand (COD), pH, total suspended solids (TSS), and metal concentrations. Such POTWs can require special pollution control equipment when issuing new permits. Thus, local authorities should be consulted early in process of upgrading an aqueous process or conversion from a solvent to an aqueous process to ensure full compliance with all applicable regulations.

8.6.8 Solid Waste

RCRA legislation has established cradle to grave control for all chemical wastes. The rules cover all sorts of "solid" wastes (including liquids, solids, and gases), both hazardous and nonhazardous. There are two ways to determine that waste is classified as hazardous. It is hazardous by definition if it is listed as a hazardous material on the commercial-chemical, specific-source, or nonspecific source lists that are found in the U.S. *Federal Register.* A waste is also hazardous if it meets any of four characteristics: ignitability, toxicity, corrosivity, and reactivity. Materials with flash points below 60°C are by definition hazardous. Some cleaning agents are hazardous by this definition, but others are not. Cleaning agents as formulated may not be corrosive but can become corrosive when loaded with corrosive residues. None of the semi-aqueous cleaning agents are classified as toxic in their own right, and none of the semi-aqueous cleaning agents are reactive enough to be considered hazardous on their own. It is the responsibility of the enterprise to determine if any solid waste is considered hazardous. Often, the supplier can assist with this determination.

8.6.8.1 Heavy metals. One special concern is heavy metals, particularly lead. In practice, heavy metals will remain with water-insoluble emulsified cleaning agents or with certain constituents of water-soluble cleaning agents because they have complexed with soils dissolved in the cleaning agent. Any particulate lead will be filtered out of the cleaning agent during normal machine operation. The filters should be tested to determine if they are hazardous waste. All waste streams from the process should be monitored to ensure that they are in compliance with applicable regulations. As lead is phased out of PWB fabrication, users will have to determine the status of other materials within their facility.

8.6.8.2 Disposal of chemical and auxiliary goods. The disposal of spent or waste cleaning chemicals is regulated in most countries as hazardous waste. Spent cleaning chemicals, filter cartridges, activated carbon, and others must be analyzed and characterized by someone knowledgeable in hazardous waste generation, storage, and shipment regulations. In the U.S.A., the federal Toxic Characteristic Leachate Procedure (TCLP) test is the minimum that must be performed. Applicability of the hazardous waste rules subject a facility to extensive regulations on personnel training, storage, and handling requirements and record keeping. Even if hazardous waste rules are not applicable, local or state rules on solid waste may apply. Some states offer technical assistance to aid in this determination. Documentation of test results and records must be kept to assure the regulatory agency of compliance with today's "cradle-to-grave" requirements. Complete records of the amount of regulated wastes generated, transported, and disposed are required.

8.6.8.3 Sludge. The solid (or sludge) waste from any of the above processes may be classified as hazardous or regulated waste, depending on the federal (TCLP), state, and local regulations. The saponifiers will produce sludge or solid waste that probably will be considered as hazardous because of the amount of lead accumulated in the tank and other factors. Whether it is hazardous or must be regulated can vary depending on the actual chemical cleaner being used.

8.6.8.4 European considerations. In some European countries, all industrial sludge is considered hazardous and must be treated accordingly, with deposit in a chemically approved landfill. If the sludge is of organic compounds, it may be incinerated in an approved cement kiln.

Some kilns are also capable of accepting sludge containing small percentages of heavy metals, especially cement kilns designed for chemical incineration.

In addition to analog regulations, there are very severe rules in Europe simply for the transport of hazardous waste and, for that matter, any toxic material. Furthermore, most regions operate licensed liquid and solid chemical disposal sites that are specially constructed to protect the air, surface, and ground waters (especially phreatic sites, which are sites susceptible to volcanic activity) and the surrounding soil. Every load of waste must be clearly labeled as to the nature of the contents. Generally, it is a wise precaution to make sure that a site can accept a given waste before dispatching it. Combustible waste is usually incinerated in special kilns. In particular, special licenses are required to transport waste across a national frontier, according to the Basel Convention.

8.6.8.5 Resource Conservation and Recovery Act Considerations. The Resource Conservation and Recovery Act of 1976 (RCRA) directed the EPA to establish regulations that would manage the generation, transport, treatment, storage, and disposal of hazardous wastes while simultaneously ensuring the protection of human health and the environment. The statute addresses the potential for contamination from the point of waste generation to the point of final disposal or destruction.

RCRA has been amended several times, most importantly by the Hazardous and Solid Waste Amendments of 1984 (HSWA). Under HSWA, RCRA became focused on waste minimization and a national land disposal ban program. To accomplish these goals, the following objectives were set forth:

1. Proper hazardous waste management

2. Waste minimization

3. Reduction in land disposal practices

4. Prohibition of open dumping

5. Encouragement of state authorized RCRA programs

6. Encouragement of research and development

7. Encouragement of recovery, recycling, and treatment alternatives.

RCRA regulations first targeted large companies, which generate the greatest portion of hazardous waste. Business establishments producing less than 2,200 pounds of hazardous waste in a calendar month (known as *small-quantity generators*) were exempted from most of the hazardous waste management regulations published by EPA in May 1980. Under HSWA, however, the EPA was directed by Congress to establish new requirements that would bring small-quantity generators (those who generate between 220 and 2,200 pounds of hazardous waste per calendar month) into the hazardous waste regulatory system. EPA issued final regulations for small quantity generators on March 24, 1986.

RCRA's "cradle-to-grave" rules require small businesses that generate hazardous waste to follow stringent requirements for storage, record keeping, pre-transportation, and emergency response and preparedness.

Subtitle I of the HSWA amendments addressed the problem of leaking UST systems. Subtitle I includes requirements for tank notification interim prohibition, new tank standards, reporting and record keeping requirements for existing tanks, corrective action, financial responsibility, compliance monitoring and enforcement, and approval of state programs. In 1986, Congress passed the Superfund Amendments Reauthorization Act that amended Subtitle I to provide federal funds for corrective actions on petroleum releases from UST systems. Detailed documentation on "EPA Office of Solid Waste and Emergency Response," "Understanding the Hazardous Waste Rules: A Small Handbook for Small Businesses," and the EPA "Office of Underground Storage Tanks" can be found on the Internet (see applicable web sites below).

Small Business Considerations. Small businesses often have special issues to consider so as to maintain compliance with ever-changing complex regulations. Such users are advised to consult the following sites frequently for in-depth information on RCRA regulations and compliance.

- http://www.smallbiz-enviroweb.org/lawepalinks.html#RCRA
- http://www.epa.gov/epaosver/hotline/training/gen.pdf
- http://www.epa.gov/epaosver/hazwaste/spghand.htm

8.6.9 Regulatory Disclaimer

While the regulatory and related information were current at the time this chapter was drafted, changes may take place at any time. Readers are responsible for obtaining the most current version of applicable regulations and complying with them.

8.7 Applicable Documents

This section contains references to industry standards, federal regulations, test methods, and vehicles that are applicable to cleaning of printed boards and assemblies. Not all of these are cross-referenced in the text of this chapter. They are listed below for the convenience of readers.

8.7.1 IPC[1]

IPC-A-600 Acceptability of Printed Boards
IPC-A-610 Acceptability of Printed Board Assemblies
IPC-AC-62A Post Solder Aqueous Cleaning Handbook
IPC-B-24 Surface Insulation Resistance Test Board
IPC-B-25 Multipurpose 1 & 2 Sided Test Board
IPC-B-36 Cleaning Alternatives Benchmark Test Printed Wiring Assembly
IPC-CH-65A Guidelines for Cleaning of Printed Boards and Assemblies
IPC-CS-70 Guidelines for Chemical Handling Safety in Printed Board Manufacture
IPC-D-249 Design Standard for Flexible Single and Double-Sided Printed Boards
IPC-FC-231 Flexible Bare Dielectrics for Use in Flexible Printed Wiring
IPC-FC-232 Specification for Adhesive Coated Dielectric Films for Use as Cover Sheets for Flexible Printed Wiring
IPC-FC-241 Flexible Metal-Clad Dielectrics for Use in Fabrication of Flexible Printed Wiring
IPC-PC-90 General Requirements for Implementation of Statistical Process Control
IPC-SA-61A Post Solder Semi-Aqueous Cleaning Handbook
IPC-SC-60A Post Solder Solvent Cleaning Handbook
IPC-SM-840 Qualification and Performance of Permanent Polymer Coating (Solder Mask) for Printed Boards
IPC-T-50 Terms and Definitions
IPC-TR-580 Cleaning and Cleanliness Test Program Phase 1 Test Results
IPC-6012 Qualification and Performance Specification for Rigid Printed Boards
IPC-TM-650 Test Methods Manual:
    2.3.1.1 Chemical Cleaning of Metal Clad Laminate
    2.3.25 Detection and Measurement of Ionizable Surface Contaminations by Resistivity of Solvent Extract (ROSE)

2.3.25.1 Ionic Cleanliness Testing of Bare PWBs (Modified ROSE Test)
2.3.26 Ionizable Detection of Surface Contaminants (Dynamic Method)
2.3.26.1 Ionizable Detection of Surface Contaminants (Static Method)
2.3.27 Cleanliness Test—Residual Rosin
2.3.27.1 Rosin Flux Residue Analysis—HPLC Method
2.3.28 Ionic Analysis of Circuit Boards, Ion Chromatography Method
2.3.38 Surface Organic Contaminant Detection Test
2.3.39 Surface Organic Contaminant Identification Test (Infrared Analytical Method)
2.5.27 Surface Insulation Resistance of Raw Printed Wiring Board Material
2.6.3.1 Moisture and Insulation Resistance Polymeric Solder Masks and Conformal Coating
2.6.3.3 Surface Insulation Resistance, Fluxes
2.6.9.1 Test to Determine Sensitivity of Electronic Assemblies to Ultrasonic Energy

8.7.2 Joint Standards[1]

J-STD-001 Requirements for Soldered Electrical and Electronic Assemblies
J-STD-002 Solderability Tests for Component Leads, Terminations, Lugs, Terminals and Wires
J-STD-003 Solderability Tests for Printed Boards (formerly IPC-S-804)
J-STD-004 Requirements for Soldering Fluxes
J-STD-005 Requirements for Soldering Pastes
J-STD-006 Requirements for Electronic Grade Solder Alloys and Fluxed and Non-Fluxed Solid Solders for Electronic Soldering Applications

8.7.3 Telcordia Technologies[2] (Formerly Bellcore)

GR-78-CORE Issue 1, September 1997

8.7.4 Occupational Safety and Health Administration (OSHA)[3]

OSHA 29 CFR 1910.1000 Air Contaminants
OSHA 29 CFR 1910.134 Respiratory Protection
OSHA 29 CFR 1910.106 Flammable and Combustible Liquids

8.7.5 Environmental Protection Agency (EPA)[4]

EPA 40 CFR 355.30 (b) Emergency Release Notification
EPA 40 CFR 370.20 Hazardous Chemical Reporting: Community Right To Know—Applicability
EPA 40 CFR 370.21 Hazardous Chemical Reporting: Community Right To Know—MSDS Reporting
EPA 40 CFR 370.25 Hazardous Chemical Reporting: Community Right To Know—Inventory Reporting
EPA 40 CFR 423 Steam Electric Power Generating Point Source Category
EPA 40 CFR 433 Metal Finishing Point Source Category

8.7.6 National Fire Protection Association (NFPA)[5]

NFPA 35 Standard for the Manufacture of Organic Coatings

8.7.7 American National Standards Institute (ANSI), American Society for Quality Control (ASQC)[6]

ANSI/ASQC Z-1.15 Generic Guidelines for Quality Systems

8.7.8 Addresses

1. IPC, 2215 Sanders Road, Northbrook, IL 60062

2. Telcordia Technologies, 445 South Street, Morristown, NJ 07960-6438

3. OSHA, Public Affairs Office, Room 3647, 200 Constitution Avenue, Washington, DC 20210

4. EPA, 401 Main Street SW, Washington, DC 20460-0003

5. NFPA, 1 Batterymarch Park, Quincy, MA 02269-9101

6. ANSI, 11 West 42nd Street, New York, NY 10036

8.8 Source and Reference Materials

Most of the source material for this chapter has been taken from a number of IPC documents; especially documents listed in Section 8.7.1

I am grateful to the IPC executive staff for their kind permission to use this material, as well as for the opportunity to participate in many of the task groups that prepared the documents. Dr. Kenneth T. Dishart, developer of the hydrocarbon-based semi-aqueous process, has contributed greatly by generating the methodology for VOC calculations in Section 8.5.6, in addition to proofreading the entire manuscript.

In addition, I would also like to commend the following landmark book on the cleaning of PWBs and PWAs to the attention of readers: *The Contamination of Printed Wiring Boards and Assemblies,* by Dr. Carl J. Tautscher, 1976, Omega Scientific Services, Bothell, WA 98011. Library of Congress Catalog Card Number 75-38181.

8.9 Acknowledgements and Thanks

I would like to commend all the members of the IPC Cleaning & Coating Committee that I have had the pleasure of working with over the years. It has been a pleasure to watch them grow professionally and contribute to the electronics industry.

Chapter 9

Board Coating Materials and Processes

John Waryold
Edward B. Mines

Humiseal Division
Chase Corporation
Woodside, New York

9.1 Introduction

Conformal coating technology has become a vital part of many electronic manufacturing operations to assure reliable, long-life performance of the finished electronic assembly under adverse conditions, particularly high atmospheric humidity. Closer conductor spacings, tighter packaging densities, and increasing performance requirements demand the protection provided by conformal coatings.

This chapter was developed as a primer of conformal coating essentials for experienced industry personnel as well as for new users of these products. All of the coatings covered are applied as wet films, which become dry coatings by curing (chemical cross-linking) and/or drying (evaporation of solvent).

The data and guidelines presented are general. Hence, the user is advised to verify, by appropriate testing, that the coating formulation and application method selected yield acceptable performance for the specific requirements for the electronic assembly. It is the user's responsibility to determine the suitability of a particular conformal coating formulation for a particular application.

9.2 Purpose and Function of Conformal Coatings

When uncoated printed wiring assemblies (PWAs) are exposed to humid air, thick films of water molecules form on their surfaces. The thicker the water films, the lower the surface resistance (commonly measured as surface insulation resistance or SIR). The lower the SIR, the greater the effects on electrical signal transmission. Typical results are cross talk, electrical leakage, and intermittent transmission, which may lead to permanent termination of the signal, i.e., a shorted circuit.

Moisture films on uncoated circuits also provide favorable conditions for metallic growth and corrosion. Other common adverse impacts are decrease in dielectric strength and effect on high-frequency signals. Dust, dirt, and other environmental pollutants that settle on assembly surfaces trap moisture and magnify these effects. Conductive particles like metal chips can cause electrical bridging.

Properly applied conformal coatings prevent the occurrence of all of these detrimental effects.

Conformal coatings act as semipermeable membranes and allow some moisture to penetrate the coating film. Hence, the insulation resistance will drop over time with exposure to moisture, normal behavior of all coating films.

9.3 Specifications that Address Conformal Coatings

- **IPC-CC-830** Qualification and Performance of Electrical Insulating Compounds for Printed Board Assemblies (Material Specification for the Industry).
- **MIL-I-46058C** Insulating Compound, Electrical (Material Specification U.S. DOD).
- **MIL-P-28809 Printed** Wiring Assemblies (Inspection Criteria for Conformally Coated Assemblies, U.S. DOD).
- **UL746E** Polymeric Materials—Use in Electrical Equipment Evaluations (Underwriters Laboratories recognition).

9.4 Generic Types of Coating Materials

Both the IPC and MIL specifications classify the common types of coatings generically as follows:

- AR (acrylic resin)

- ER (epoxy resin)
- SR (silicone resin)
- UR (polyurethane resin)

Both specifications also list an XY type (parylene, paraxylylene), which is a vapor-deposited coating outside the scope of this chapter. Some of the new, so-called "high-tech" coating formulations fall into more than one generic category.

9.4.1 Acrylic (Type AR)

Acrylic conformal coatings are easy to apply. They dry to the touch at room temperature in minutes, have desirable electrical and physical properties, and are fungus resistant. They have a long pot life and low or no exotherm during cure, which prevents damage to heat-sensitive components. They do not shrink. Their main disadvantage is solvent sensitivity, but this also makes them easiest to repair. Traditional acrylic formulations are now available in volatile organic compounds (VOC) exempt solvents. Conformal coatings made with waterborne acrylics and polyurethane-acrylic hybrids are not as solvent-soluble as thermoplastic acrylic resins; hence, they are not as easy to repair.

9.4.2 Epoxy (Type ER)

Epoxy systems are usually available as "two-component" compounds. These rugged conformal coatings provide good humidity resistance and high abrasion and chemical resistance. They are, however, virtually impossible to remove chemically for rework, because any stripper that will attack the coating also dissolves epoxy-coated or epoxy-potted components and the epoxy-glass printed circuit board itself. The only effective way to repair a board or replace a component is to burn through the epoxy coating with a knife or soldering iron.

One-part, ultraviolet-curing (UV-curing) epoxy conformal coatings are also available and have become popular for high-volume coating production for such products as electronic ballast and surface mount assemblies.

9.4.3 Polyurethane (Type UR)

Polyurethane conformal coatings are available as single-component, two-component, UV-curable, and water-borne systems. As a group, all provide excellent humidity and chemical resistance plus outstanding dielectric properties for extended periods.

Traditional single-component polyurethane formulations require careful application procedures and close control of the coating and curing environments. Moisture-curing polyurethanes may form many minute "champagne-like" bubbles if the coating is too thick or the coating environment is too humid. Air-curing polyurethane takes about a month to cure in the presence of fresh air (the coating reacts with oxygen) or an overnight bake in a recirculating oven (tack-free dry time for these coatings is less than an hour). Air-cure polyurethane conformal coatings may crack or form "alligator-skin" surfaces if the coating is too thick.

Two-component formulations require tight control of humidity during application.

Polyurethane conformal coatings can be burned through with a soldering iron, making component replacement fairly easy. Chemical stripping of these coatings varies from easy to difficult, depending on the formulation.

UV-curable polyurethanes require a secondary cure mechanism to harden the coating beneath components and in other areas that are shielded from the UV light. Water-borne formulations are slower drying than ordinary solvent-borne coatings and have lower chemical resistance and less desirable wetting characteristics.

9.4.4 Silicone (Type SR)

Silicone conformal coatings are particularly useful for high-temperature service, up to about 200°C. They provide high humidity and corrosion resistance along with good thermal endurance, making them desirable for PWAs that contains high-heat-dissipating components such as power resistors. Silicone coatings are susceptible to abrasion (low cohesive strength) and have high coefficients of thermal expansion.

These coatings are available as 100 percent solids heat-curing liquids, 100 percent solids moisture-curing liquids, 100 percent solids UV-curable liquids, and solvent-borne reactive coatings. The 100 percent solids heat-curing versions are popular for high-volume production such as automotive electronics.

Repair of assemblies coated with silicones has been difficult in the past, because these formulations are not solvent-soluble and do not vaporize with the heat of a soldering iron. Chemical strippers are now available, however, that effectively dissolve silicone coatings.

9.4.4.1 New developments.
Industry efforts to shorten production cycle times and to reduce or totally eliminate solvent emissions have led to

the development of new coating formulations. These include many UV-curable coatings and those that use VOC-exempt solvents.

The user should carefully weigh the promised advantages of these new coatings against such known disadvantages as reduced wetting, adhesion properties, and chemical resistance. Perhaps most important to the user is their limited "track record," which requires testing to establish acceptable performance.

9.5 Criteria for the Selection of a Generic Coating Type

Coating selection is influenced by engineering or performance characteristics of the cured coating as well as by application or processing characteristics, which affect the physical and chemical properties of the coating as a liquid.

Selection of the proper conformal coating is a major task because of the great number of materials on the market. When in-depth experience is lacking, the user should turn to the manufacturers who specialize in formulating such coatings. Their printed information is responsive to the needs of circuit design and packaging engineers and contains the necessary data for selection.

9.6 Engineering or Performance Aspects

The environment the coating must endure and the characteristics of the assembly to be protected by the conformal coating are the two principal variables that must be considered. Typical properties are

- *electrical* (volume and surface resistivity, dielectric constant, dissipation factor, arc resistance, dielectric strength)
- *thermal* (temperature endurance, thermal expansion, thermal conductivity, flame resistance)
- *humidity* (absorption and vapor transmission)
- *mechanical* (resistance to cracking from thermal changes, abrasion resistance)
- *chemical* (chemical and fungus resistance, hydrolytic stability)

9.6.1 Electrical Properties

One of the main purposes of a coating applied to a printed wiring assembly (PWA) is to provide electrical insulation. Therefore, the cured coating must have sufficient dielectric strength and insulation resistance to satisfy design requirements over the operating temperature range of the assembly and its anticipated operating environment. In certain applications, the dielectric constant and the loss factor (Q) may also become important selection parameters. Typical electrical properties are shown in Table 9.1.

Table 9.1 Typical Performance Characteristics

	AR	ER	SR	UR
Electrical properties				
Dielectric strength short time, 23°C volts/mil at 1 mil	3500	2200	2000	3500
Surface resistivity at 23°C, 50%RH, Ω-cm	10^{14}	10^{13}	10^{13}	10^{14}
Dielectric constant at 23°C, 1 MHz	2.2–3.2	3.3–4.0	2.0–2.7	4.2–5.2
Thermal properties				
Resistance to heat continuous, °C	125	125	200	125
Linear coefficient of thermal expansion, μin/in/°C	50–90	40–80	220–290	100–200
Thermal conductivity, 10^{-4} cal/s/cm^2/cm/°C	4–5	4–5	3.5–8	4–5
Chemical resistance				
Resistance to weak acids	B	A	B	A
Resistance to weak alkalis	B	A	B	A
Resistance to organic solvents	D	A	B	A
Abrasion resistance	B	A	C	B
Humidity resistance	A	B	A	A
Humidity resistance (ext. per.)	B	C	A	B

Ratings: A through D are in descending order with A as optimal.

9.6.2 Thermal Properties

The coating material selected must have an operating temperature range that matches the assembly on which it is used. Within this range, the coating, in addition to meeting the minimum electrical performance requirements, must remain free from physical degradation such as embrittlement, cracking, or excessive shrinking. Coatings characterized by good flexibility usually resist such physical deterioration. Flame resistance to meet UL requirements is also important. Typical thermal properties are shown in Table 9.1.

9.6.3 Humidity Resistance

Coatings should have low moisture permeability and low water absorption. While most PWA conformal coatings have more than adequate resistance to humidity, where unusual or extended conditions of high humidity are anticipated, minimum insulation resistance becomes the governing factor. Humidity resistance ratings are given in Table 9.1.

9.6.4 Mechanical Properties

Hardness is related to abrasion resistance, an important property for protection against physical abuse. Most conformal coatings are flexibilized to meet the thermal shock requirements of the military specification. General ratings for abrasion resistance are given in Table 9.1.

9.6.5 Chemical Resistance

When operating conditions expose the coating to harsh chemicals, salt spray (marine environments) or warm-moist environments that promote growth of microorganisms such as fungus, the coating's resistance becomes an overriding factor in selection. The user should determine the specific chemicals that the coating must withstand and test for resistance. General ratings of chemical resistance are given in Table 9.1.

9.6.6 Thermal Humidity Aging

All military qualified conformal coatings are required to pass a specified test with extended time, humidity, and temperature.

9.7 Processing or Application Aspects

Manufacturing concerns are overall cost, processing time, safety, health, and environmental impact issues. Process volume will determine if the coating application should be manual or automated. In turn, cost models can be used to establish return on investment (amortization of the cost of automatic equipment). Automated systems lend themselves to fast drying/fast curing formulations, while turnaround time may be a factor for manual operations.

Ease of in-process rework (stripping and recoating) and repairability are significant attributes, but solvent and chemical resistance may become important trade-offs. These conflicting needs require negotiation between design engineering and production. Similar trade-offs apply to safety, health, and environmental factors.

Shelf life (the length of time a coating can be stored in an unopened container) and pot life (the length of time a coating can be used after being opened or mixed) are also considerations to be weighed. Short pot life can lead to substantial material waste and, more importantly, contributes to inconsistent coating thickness due to rapid buildup of viscosity (resistance to flow) in the pot. Two-component systems typically exhibit short pot life. Single-component systems, on the other hand, usually have an extended pot life that approaches shelf life.

Viscosity must be adjusted to permit adequate flow of the liquid coating over and around components to ensure complete coverage. Low-viscosity formulations are particularly important when coating high-density board assemblies. High-viscosity coatings, however, are desirable when the assembly's components have sharp corners and protruding leads from which the material tends to roll off. Coating viscosity must be established by trial and error to achieve optimal coverage for a specific board configuration. The viscosity of traditional solvent-borne formulations can readily be reduced by addition of an appropriate thinner. One hundred percent solids formulations can be adjusted only with heat. Viscosity also determines the ultimate thickness of the coating, an important economic consideration.

Solids content represents the portion of a coating that cures into a film; the balance is solvent, which evaporates. Solids content thus has a bearing on the coating thickness per application and an impact on the cost of the overall coating operation. Solvent-type coatings are usually supplied with solids content ranging from 10 to 50 percent. Solventless systems are rated as "100 percent solids" coating materials.

Ease of application is another important economic consideration. It is obviously advantageous for limited production runs to use a coating that does not require specialized equipment, skilled labor, or complex production controls. Single-component formulations are simpler to use than two-component coatings, because they do not need metering and mixing equipment that can be a source of operator error.

Handling and cure times are major production factors. Long cure times increase overall cycle times, tying up facilities and equipment. Boards coated with formulations requiring lengthy air cures must be racked when the boards are dry to the touch.

Another factor is the ability to verify complete coverage. Most MIL-I-46058C-qualified coatings applied as liquids contain an optical brightener that is visible with UV light, thus assuring rapid and effective visual inspection.

Typical processing characteristics of four generic types of coating materials are listed in Table 9.2.

The important parameter that must be considered is the supplier. For some applications, the selection of the right coating material is

Table 9.2 Typical Processing Characteristics

	AR	ER	SR	UR
Application	A	C	A	A
Removal (chemical)	A	D	B	B
Removal (burn-through)	A	C	C	A
Pot life	A	D	A	B
Optimal cure, room temp.	A	B	A	B
Optimal cure, elevated temp.	A	B	A	B

Ratings are in descending order, with A = best and D = worst.

simple; for others, the choice is not obvious. Often, complex application and curing processes must be factored in. The supplier selected should be able to provide technical support for the initial selection as well as follow-up assistance during production start-up and process troubleshooting. Because of the ever broadening line of application-specific conformal coating formulations, the user should choose a supplier that offers a complete line to assure objective recommendations that lead to optimal performance and cost effectiveness.

9.8 Basic Prerequisites for Optimal Coating Performance

9.8.1 Parts to Be Coated Must Be Clean and Dry

Process residues from board fabrication (platings, etch and resist residues, fluxes, and fingerprints), from component manufacturing (mold release compounds, fingerprints), and from assembly soldering operations (drosses, fingerprints), if left on assemblies, will affect wetting by the liquid coating and result in reduced or total loss of adhesion or dewetting of the coating. Water-soluble residues will be activated by absorbed moisture, resulting in vesication ("mealing" of the cured surface) that degrades the coating (see Fig. 9.1).

Selection of the chemical cleaners should consider the solubility of the residues to be removed. In many cases, both polar (ionic) and nonpolar (oily) residues need to be removed. Wash and rinse media that remove both types of residues should be used. Mechanical action, such as spraying and ultrasonic agitation and increasing wash and rinse temperatures (when permissible) will increase cleaning efficiency. Cleaned boards are frequently baked to remove all traces of wash and rinse materials. Most assemblies may be dried at 60°C for 2 hr without damage to components. However, some assembly configurations may trap rinse media and thus require longer drying times.

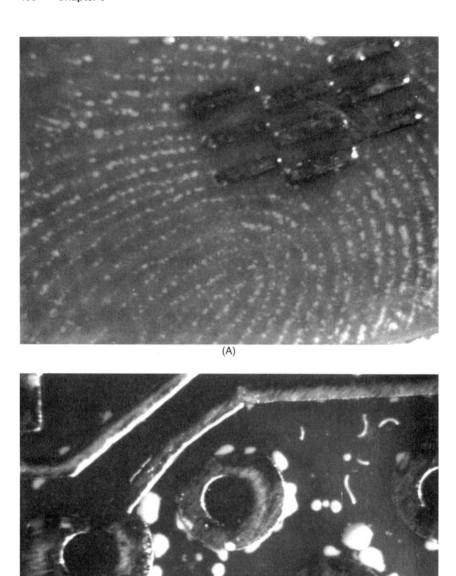

Figure 9.1 Moisture may become "sealed in" when an assembly is not dry at the time of coating and may also react with some coating materials (polyurethanes), causing bubbles or inhibiting UV cures (cationic epoxies).

When tested, the cleaned assembly should meet the requirements of MIL-P-28809 (ionic residues less than 1 µg of sodium chloride equivalent per square centimeter). After coating and humidity exposure, the completed assembly should not exhibit vesication or any other loss of adhesion.

Guidelines for cleaning, drying, and general decontamination control of electronic assemblies prior to coating are available from the HumiSeal Division of Chase Corporation (HumiSeal Publication No. 1297). These guidelines allow the choice among many options that have been developed as alternatives to cleaning with ozone-depleting solvents, and these are successfully used in industry.

9.8.2 Processing Areas Must Be Clean

After cleaning and drying, parts to be coated must be kept clean and dry until coated. Drying ovens and coating, masking, and curing areas must be also spotless; a light, positive air pressure in the critical process areas and "sticky mats" at their entrances will minimize airborne fallout. Clean shop coats, hair covers, and gloves for processing personnel are mandatory. Also prohibited in coating areas are hand lotions and cosmetic products with silicones, as well as silicone-containing oils, greases, and mold release agents. If cleaned assemblies are not to be coated immediately, they should be stored in a desiccator cabinet or sealed antistatic bags.

9.8.3 Assembly Preparation

If masking is required, special conformal coating masking tape should be used rather than conventional "pulp-type" tapes. These special tapes are less porous and more resistant to solvent attack during the coating procedure. Special precut shapes of conformal masking tape such as dots and rectangles are available. Thixotropic latex formulations can also be used as liquid masks. Snap-on rubber masking boots in various configurations are also available to speed production.

9.8.4 Coatings and Substrates Must Be Compatible

The degree of adhesion obtained with a specific coating varies with the substrate to which it is applied. Conformal coatings exhibit varying degrees of adhesion to various solder masks.

Occasionally, materials on the board prevent reactive coatings from curing. For example, additives in some PVC plastics used to shield capacitors prevent silicones from curing. The user should verify the compatibility of coating/board combinations prior to the final selection of coating.

9.8.5 Coating Coverage Must Be Complete and of Proper Thickness

The applied coating must cover the entire assembly with electrical contact areas and test points masked by a removable material. Masking materials are removed after the coating becomes tack free.

Coating thickness is an important parameter. Recommended thickness is as follows:

- Types AR, ER and UR   0.001 to 0.003 in.
- Type SR   0.002 to 0.008 in.

These dry film thicknesses, as specified in MIL-I-46058C, apply to flat and unobstructed surfaces only. These requirements are waived for fillets adjacent to component bodies. Thin coating films reduce protection against humidity effects. Thick coatings may contribute to mechanical stresses on components and solder joints, causing conformal coating to crack. Thick coatings may not cure completely.

9.8.6 Coating Materials Must Be of Consistent Quality and Stored Properly

High quality and uniformity of coating formulations, including diluents, are key prerequisites for consistent production and reliable performance of coated printed wiring assemblies (PWAs).

Coating materials must be stored properly to maintain optimal properties. Liquid coatings should always be stored in their original sealed containers, away from fire, open flames, and sparks. Open containers of coatings that are sensitive to moisture (various polyurethanes, silicones, and UV-curable formulations) should be purged with dry gas before closing. "Age control" (use of material within its specified shelf life as indicated on accompanying documents from the supplier) is important from both economic and production consistency standpoints.

9.8.7 System Compatibility

System compatibility can be achieved by proper design and good production practices in cleaning and contamination control. Anomalies in processing can be encountered and, if left unaddressed, can affect overall system performance. In particular, anomalies sometimes occur with photoresists, solder masks, UV-cured coatings, finish coats on metals, and "no clean" processes. Variations in assembly geometry can also contribute to incompatibilities.

9.8.8 General Process Flow

Figure 9.2 shows typical process flow for precoating, coating, and postcoating operations.

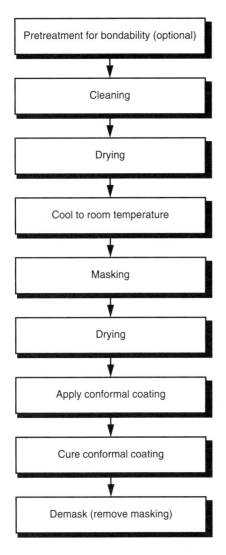

Figure 9.2 Typical process flow for pre-coating, coating, and post-coating operations.

9.9 Conformal Coating Application Methods

The liquid material may be applied by manual, semi-robotic, or fully robotic techniques. Four common methods are used.

1. Dip (manual or semi-robotic)
2. Spray (manual or semi-robotic)
3. Select coat (robotic application to selected areas)
4. Brush (manual)

9.9.1 Dip Method

This method, in which the masked assembly is immersed in a tank of liquid coating material and withdrawn, ensures uniform coverage and predictable, repeatable film thickness. Variables include rates of immersion and withdrawal and viscosity of the liquid material. The assembly should be immersed slowly to allow the coating material to displace the air surrounding components, remain immersed until bubbling has ceased, and then be withdrawn. Typical immersion and withdrawal speeds are 2 to 12 in/min for the first dip, with subsequent dips made at higher speeds. Immersion and withdrawal speeds depend on the size and complexity of the assembly. Timing may be manual or automated. Figure 9.3 shows dip coating equipment.

The thickness of conformal coatings applied by dipping depends on the coating solution viscosity, immersion and withdrawal rates, dry time, and component population density. A slight wedge occurs with dip coating; i.e., the coating is slightly thicker on the bottom. It is necessary to reconstitute dip-coating solutions with thinner as solvent evaporates. The time to do this can be determined by measuring the viscosity of the coating solution at regular time intervals. It is advisable to use a dry nitrogen blanket over the bath surface of moisture-sensitive coatings to avoid reaction with atmospheric moisture. The dip tank should be made of a material that does not react with the liquid coating, such as stainless steel.

9.9.2 Spray Method

Spraying, manual or robotic, is the most popular and the fastest method for applying conformal coatings. With the proper combination of solvent dilution, nozzle pressure, and pattern, reliable and consistent results are obtainable. For high-volume production, spray coating is readily automated. Figure 9.4 shows robotic spray coating equipment.

The viscosity of solvent-borne coatings is adjusted by addition of appropriate thinners. One hundred percent solids coatings must be sprayed at the consistency supplied.

If there are sharp edges on one side of the board (typically on the solder side), the board should be placed horizontally with component leads facing down after it is sprayed. This allows "stalactites" to form on the tips of such projections. It may be necessary to repeat the spray coating several times to cover such projections and also to preclude pinhole "line-up."

Coatings should be sprayed onto assemblies using clean, dry gas at the minimum pressure necessary to provide good atomization. Compressed air and nitrogen from a cylinder are common propellants. If

Figure 9.3 Dip coating equipment.

Figure 9.4 Spray coating equipment.

compressed air is used, caution should be exercised to avoid oil and water contamination in the spray equipment.

The assembly should be sprayed while holding the spray gun at a 45° angle. After each back-and-forth pass, the assembly should be rotated 90°, and the spraying should be repeated until all four sides are sprayed. If the assembly has closely spaced components, a spray from directly above should be applied. Spraying should be continued until the target coating thickness is obtained. The build (amount of coating thickness) per pass is operator dependent and depends on the distance of the nozzle from the object being sprayed and speed of hand movement.

Spraying should be performed in a well ventilated area with good illumination and explosion-proof fixtures.

9.9.3 Selective Coating Methods

New robotic equipment is available that uses a digital control system for computer-controlled application of coating material to selective areas of assemblies, eliminating labor-intensive masking operations. Figures 9.5 and 9.6 show selective coating workstations.

9.9.4 Brush Method

Brushing is the least efficient application method because of difficulty in achieving uniform coverage (hence, uniform coating thickness) and controlling bubbling. While this operator-dependent method is not practical for high-volume production, it is suited to short runs, prototypes, and touch-up after repair/rework. Particular attention should be paid to the underside of components and lead wires. Care must be exercised to avoid deposition of brush fibers in the coating.

9.9.5 Other Methods

Other less frequently used methods include various forms of curtain coating, spin coating, meniscus-flow or wave coating, as well as the use of ultrasonic energy for coating selected areas on a PWA.

9.10 Multiple Coats

Defects are most likely to be covered if several applications are used (see Fig. 9.7). Each coating method has its strengths and weaknesses. A combination of dipping and spraying is superior to either method alone.

Figure 9.5 Selective coating workstation.

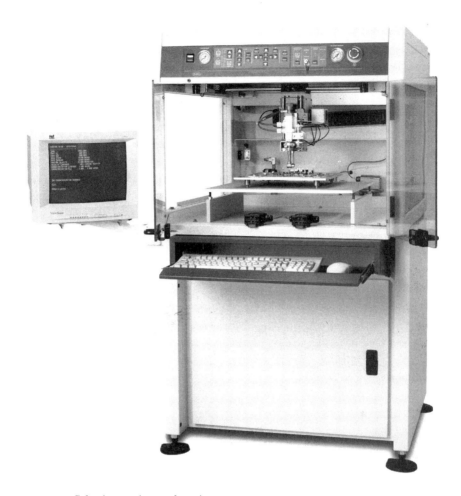

Figure 9.6 Selective coating workstation.

9.10.1 Drying and Curing of Coatings

Boards coated with traditional solvent-borne coatings are usually allowed to dry at room temperature until they are tack free to the touch and then baked in an oven. However, it is not necessary to bake coated boards to remove the remaining solvent, which typically will egress from unbaked coatings (air dry only) in about 1 week, with most escaping in 24 hr. Coaters with high-volume production sometimes force dry "wet" conformal coatings in a tunnel oven using progressively hotter zones. Care must be exercised; too much heat applied too fast causes wet conformal coatings to run and bubble.

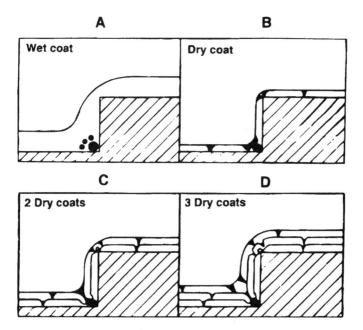

Figure 9.7 A, B: illustration, on an exaggerated scale, of the formation of pinholes and the stretching of the coating over sharp edges and protrusions after a single application. C: a substantial reduction in the number of pinholes after a second application. D: the same component with three coats, showing practically pinhole-free insulation.

When elevated temperatures are specified for curing, the thermal capacity of the oven is a factor to be considered. A high mass placed suddenly into a small oven can cause overload, requiring a prolonged time for the assembly to be brought up to the specified temperature. UV-curing ovens must be specified to meet the specific requirements for irradiation curing of the particular material used, e.g., wavelength and intensity.

For moisture-curing systems, a minimum of 30 percent relative humidity is required for the curing areas. For this reason, plant locations in hot and dry climates should avoid selecting conformal coating formulations, which require moisture-cure and the associated equipment, and operating expenses of humidification.

Oxygen-curing coatings react slowly with the oxygen in the air and take about a month to "cure" (as opposed to dry) at room temperature. (These coatings usually dry to the touch in less than an hour, however.) This cure can be accelerated by baking the boards in an oven overnight in a stream of fresh air, which can be supplied in an oven with a vent.

9.10.2 Repair of Coatings

Repair of coated circuitry usually consists of coating removal, desoldering with vacuum, resoldering, cleaning, and recoating. The repairability of a PWA is an important consideration for product maintenance and economics. Repair and rework is often performed "in the field," and so simple procedures are favored. When soldering through a coating, coating film thickness and general criteria for aesthetics are influencing factors.

9.10.2.1 Acrylics. Type AR coatings can be soldered through or easily removed by most organic solvents including alcohols, common terpenes, conformal coating thinners, and semi-aqueous cleaning solutions.

9.10.2.2 Polyurethanes. Repair of UR coated circuitry is generally by the solder-through method. Chemical strippers are available that dissolve most UR coatings without attacking assembly components. Type AR coatings can be used to repair type UR coatings in the field.

9.10.2.3 Epoxies. Chemical removal of epoxy conformal coatings is not currently practical, as epoxy strippers also attack board substrates. Decomposition with the heat of a soldering iron is possible but difficult; furthermore, the charred residue must be removed.

9.10.2.4 Silicones. Silicone conformal coatings cannot be soldered through, generally requiring removal by abrasion or other mechanical action. Strippers for silicone conformal coatings are now available that do not attack components. Thorough rinsing is necessary to prevent any residue from stripped coating from being trapped beneath the recoated area. Silicone entrapment in solder joints needs to be watched carefully, as it causes poor electrical connections.

9.11 Health and Safety Considerations Government Regulations

9.11.1 VOC (Solvent) Emissions

Solvent emissions are regulated in areas with strict air pollution controls. Volatile organic compounds (VOC) and hazardous air pollutant (HAP) emissions can be limited. VOC-compliant counterparts to several traditional solvent-borne conformal coatings are available. Spray booths and related equipment may require licensing by local authorities. Solvent emission regulations are enforced on state and local levels.

9.11.2 Waste Management

For optimal process economics, practices that minimize waste should be implemented. Expired and contaminated materials must be discarded in compliance with state and local regulations.

9.11.3 Health and Safety Issues

Material safety data sheets (MSDSs) for coatings and related materials must be available to operators and are required by law. All appropriate personnel must be trained in chemical safety. This is a legal requirement in the U.S.A.'s Occupational Safety and Health Administration's (OSHA's) HAZCOM Standard (HCS) 29 CFR 1910.1200. Special attention must be paid to the labeling, storage, and handling requirements of flammable solvents, material exposure limits, and personal protection equipment. Chemical exposure limits are enforced by the federal government.

Drying and curing areas should be properly ventilated to remove solvent vapors and other volatile emissions. Spraying operations, in particular, should be performed in particularly well ventilated areas and be equipped with explosion-proof lighting fixtures. Drying and curing ovens should also be specified with explosion-proof features.

9.12 References

1. Guidelines for Cleaning and Contamination Control Prior to Application of HumiSeal Coatings, HumiSeal Div. Of Chase Corp., Woodside, NY, 1997.
2. DeBiase, J., LaCroce, S., and Landolt, R., Compatibility of PWB Coatings with Assembly Processes and Materials, *Electronic Packaging and Production*, February 1996.
3. Fredrickson, G. and Krawlec, S., A Test Procedure for Coatings Flux Compatibility, *SMT*, March 1995.
4. Tautscher, C. J., *Contamination Effects on Electronic Products*, Marcel Dekker, New York, NY 1992.
5. Tautscher, C. J., SIR Profile Use of the Prediction of Electrical Performance of PW Dielectric Systems at Elevated Temperature, *Nepcon West, Anaheim, CA,* February 1991.
6. Licari, J.J. and Hughes, L.A., *Handbook of Polymer Coatings for Electronics*, Noyes Publications, Park Ridge, NJ, 1990.
7. Tautscher, C.J., Assembly Cleanliness—The Critical Prerequisite for High Quality Conformal Coating, *Parylene 87 Up-Dated Conference*, Indianapolis, IN, April 1987.
8. *Requirements for Conformal Coating and Staking of Printed Wiring Boards and Electronic Assemblies,* NHB5300.4(3J), National Space and Aeronautics Administration, April 1985.

9. Sohn, J. How to Avoid Metallic Growth Problems on Electric Hardware, IPC-TR-476, Blue Ribbon Committee Report, The Institute for Interconnecting and Packaging Electronic Circuits, Northbrook, IL, 1984.
10. Coombs, C.F., Jr., *Printed Circuit Handbook*, Chapter 12, "Conformal Coatings," by Waryold, J., McGraw-Hill Book Co., New York NY, 2nd edition, 1979.
11. Waryold, J., How to Select a Conformal Coating for Printed Circuit Boards, *Insulation/Circuits,* July 1974.
12. Tautscher, C. J., Causes and Prevention of Blisters in Conformal Printed Circuit Coatings, *Insulation/Circuits*, June 1972.
13. Brown, E. R., Effects of Coating and Cleaning on Corona and High Voltage Breakdown, Bendix Report No. BDX-613-1181, Bendix, Kansas City, MO, February 1972.

Chapter 10

Flexible and Rigid Flexible Fabrication Process

Martin W. Jawitz
Jaw-Mac Enterprise
Las Vegas, Nevada

Michael J. Jawitz
Boeing Satellite Systems
El Segundo, California

10.1 Introduction

This chapter was prepared as a guide for designers and fabricators, detailing the fabrication processes that are required to produce flexible and rigid flexible printed wiring boards for use in the electronic industry. Some of the information included in this chapter is also available from other sources but is included here to provide a single convenient source for the necessary fabrication information. The increasing demand for flexible and rigid flexible printed wiring boards can be attributed to

1. Unique designs that capture both the interconnection and wiring aspects of flexible printed wiring packages

2. The ability of board manufacturers to produce these systems with a high production yield

3. Weight and space saving

4. Simple, error-free installation

10.2 Classification

Flexible and rigid flexible printed wiring boards are subject to classification by intended end item use. Classification of producibility is related to the complexity of the design and the precision required to build that particular flexible circuit.

10.2.1 Board Types

Type 1 is a single-sided flexible printed wiring board containing only one conductive layer with or without a stiffener (Fig. 10.1).

Type 2 is a double-sided flexible printed wiring board containing two conductive layers with plated-through holes with or without a stiffener (Fig. 10.2).

Type 3 is a multilayer flexible printed wiring board containing three or more conductive layers with plated-through holes with or without stiffeners (Fig. 10.3).

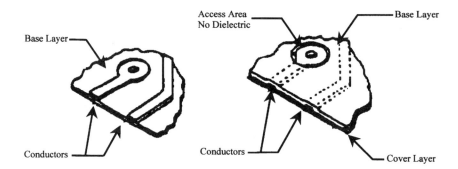

Figure 10.1 Single-sided PWB.

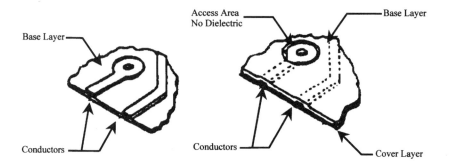

Figure 10.2 Double-sided PWB.

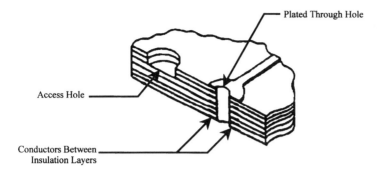

Figure 10.3 Multilayer PWB.

Type 4 is a multilayer with rigid and flexible material combination containing three or more conductive layers with plated-through holes (Fig. 10.4).

Type 5 is a flexible or rigid flexible printed wiring containing two or more conductive layers without plated-through holes (Fig. 10.5).

10.2.2 Performance Classes

Three general end-product classes have been established by the Association Connecting Electronics Industries (IPC) to reflect progressive increases in sophistication, functional performance requirements, and testing/inspection frequency. It should be recognized that there may be an overlap of equipment between classes. The written contract shall specify the performance required and note any exceptions to specific parameters, where required. These classes are as follows:

- *Class 1—General Electronic Products.* This includes consumer products, some computer and computer peripherals, as well as general military hardware suitable for applications in which cosmetic imperfections are not important and the major requirement is function of the completed board or assembly.

- *Class 2—Dedicated Service Electronic Products.* Includes communication equipment, sophisticated business equipment machines, instruments and military equipment in which high performance and extended life are required and for which uninterrupted service is desired but not critical. Certain cosmetic imperfections are allowed.

- *Class 3—High Reliability Electronic Products.* Includes commercial equipment and military products in which continued performance or performance on demand is critical. Downtime cannot be tolerated and equipment must function when required such as for

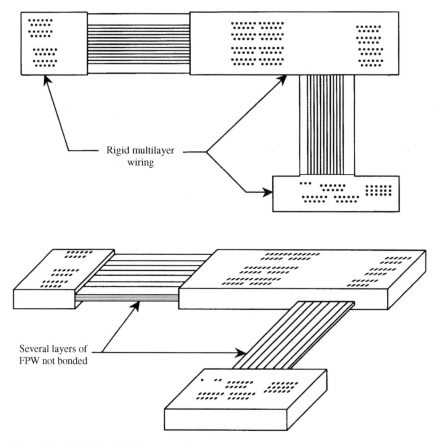

Figure 10.4 Rigid flex PWB.

life support items or critical weapons systems. Board assemblies in this class are suitable for application where high levels of assurance is required and service is essential.

10.3 Flexible Materials

Materials used in the fabrication of either flexible or rigid flexible printed wiring boards are divided into three parts: the dielectric (either rigid and/or flexible), adhesives, and the conductor.

10.3.1 Dielectrics

The majority of single- and double-sided flexible circuits fabricated today are produced using either copper clad polyester or polyimide films

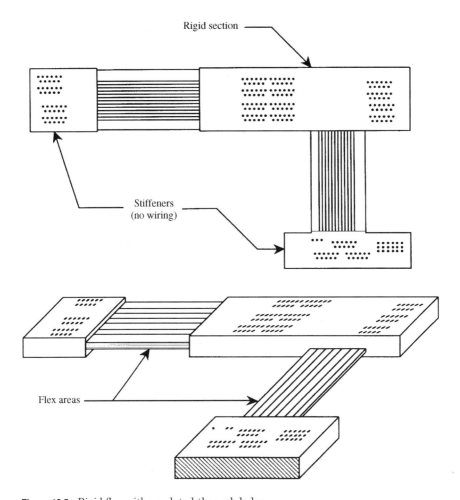

Figure 10.5 Rigid flex with no plated-through holes.

as the dielectric. However, other materials that can be used as dielectrics are Teflon®, polyethylene naphthlate (PEN), and polyethylene terephthalate (PET).

10.3.2 Covercoat

Covercoat (also called *coverlay*) materials are generally the same generic material (dielectric) as the copper clad dielectric. These films are applied over the etched copper patterns and act as a protective or insulating barrier. They are used to insulate the conductors against unintended contact with other conductors. Covercoat films are almost identical to the adhesive plus dielectric films that are used as a base for laminated metal clads, with the difference being in the adhesive thickness. More adhesive is required in the coverlay material, because the adhesive is required to flow into and fill the voided spaces in the etched pattern. Coverlay materials used for single- and double-sided flexible circuits traditionally have adhesive coating on only one side. The average adhesive thickness is generally in the range of 0.0015 to 0.003 in. A overcoat with adhesive coatings on both sides of the dielectric film is generally used when the covercoat material is used as both a covercoat and an adhesive, such as in multilayer and rigid flex construction. Covercoat materials also perform in the same manner as a solder mask coating does for rigid boards; that is, it provides a well defined termination point for the flow of solder during assembly. The bond plies are used to form multilayer circuits, rigid flexes, or for bonding stiffeners, heatsinks, and shields onto flexible printed wiring circuits. This material is simply a layer of adhesive on a release sheet that can be cut into any size or shape, lifted from the release paper, and inserted into the stack-up.

10.3.3 Adhesive

In single- and double-sided flexible circuitry, the adhesive is mandatory and is unavoidable in the use of covercoats and laminated copper clads. In multilayer and rigid flexible fabrication, the use of adhesiveless clad laminates is preferred in higher layer-count designs.

Although its presence is largely overlooked, adhesive (and the adhesive surrounding flexible printed wiring conductors), not the dielectric, is the primary insulator in the flexible printed wiring system. The adhesive and the dielectric film form the dielectric system for the flexible circuit, and neither performs well without the other. Flexible printed wiring adhesives are plasersized to provide good peel strength with low flow properties. However, the intentionally reduced cross-linked density in traditional flex circuits adhesives results in a degraded thermal performance, such as lowered use temperature and increased coefficient of thermal expansion along with reduced resistance to chemical attack and increased moisture absorption.

10.3.4 Conductors

Copper foils that are used as conductors are produced by either the electrodeposition process (called *ED* copper) or by rolled reduction (called *RA* copper). ED copper foils provide high strength but tend to crack if severely flexed, because the grain structure is coarse and vertically oriented. ED foils are used mostly with polyester-based circuitry. However, rolled (RA) foils have better properties than ED. Rolled reduction creates a horizontally aligned grain structure that resists cracking when flexed but has a smooth surface that is not easy for adhesives to adhere to. This necessitates the use of special treatments to create a stable microstructured surface for lamination.

10.4 Fabrication Processes

This section covers four fabrication processes:

- Section 10.4.1 covers the fabrication process for a single-sided flexible circuit.
- Section 10.4.2 covers the fabrication process for a double-sided flexible circuit with plated-through holes.
- Section 10.4.3. covers the fabrication process for a multilayer flexible circuit with plated-through holes.
- Section 10.4.4 covers the fabrication process for a rigid flex printed wiring board with plated-through holes.

Each section is a stand-alone fabrication process. Details used to fabricate one type of flex circuit will be repeated in another section so as to make each section a complete process.

10.4.1 Single-Sided Flexible Circuit

A flow diagram for the fabrication of a single-sided flex is shown in Fig. 10.6. Details of this operation are given below.

10.4.1.1 Select the material. Select the materials (coverlays and copper clad laminates) that are required for fabrication and, using a sharp cutter, cut them into the required panel size. Store all cut material in a temperature (40 to 80°F) and humidity-controlled (below 70 percent) chamber (e.g., nitrogen cabinet, dry box, etc.) until ready for use.

10.4.1.2 Drilling the laminate. Stack the copper clad laminate (copper side up) between sheets of entry material (this is the top sheet and is usually 0.005 to 0.015 in. thick aluminum) and back up board (this is the bottom sheet and is usually 0.062 to 0.093 in. thick aluminum clad particle board). The maximum number of sheets in the stack should not exceed ten. Attach the stack to the drill table, making sure it is flat and firmly secured to the drill table. Drilling parameters should be es-

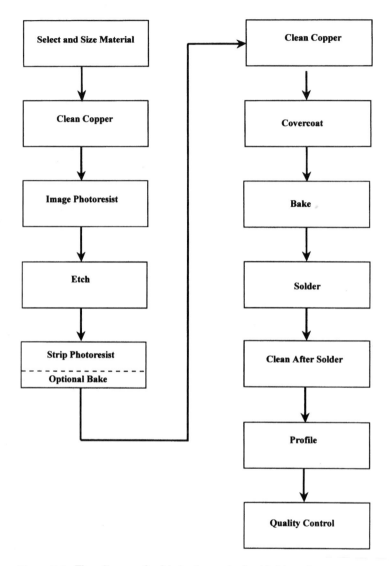

Figure 10.6 Flow diagram for fabricating a single-sided board.

tablished (chip load, drill speed, feed rate, etc.) prior to drilling the laminate. Carbide drills should be used to drill the copper clad laminate.

10.4.1.3 Deburring and hole cleaning. The drilled holes must be deburred prior to further processing. The most commonly used method is to hand deburr using a hand sander with fine grit wet and dry sandpaper. Apply light pressure so as not to distort the laminate. After deburring, rinse the laminate in running water and then dry thoroughly.

10.4.1.4 Cleaning before applying photoresist.
All copper clad laminates have an anti-tarnish coating applied to the exposed copper surfaces. This coating must be removed prior to applying the photoresist. Cleaning the anti-tarnishing agent from the copper surface can be accomplished using either a mechanical or chemical cleaning process.

Mechanical cleaning. Hand scrubbing, using a rotary brush with a 3F grit pumice slurry, is generally used to clean the copper surfaces. After scrubbing, rinse the copper thoroughly in running water to ensure complete removal of the pumice. The copper surfaces should be water-break free. After scrubbing and rinsing, completely dry the copper surfaces so as to prevent them from reoxidizing.

Chemical cleaning. Chemical cleaning using a combination of cleaners and microetchants is probably the better of the two cleaning methods, causing the least distortion to the copper laminate. The following sequence is recommended:

- Soak clean in either an acid (preferred) or alkaline cleaner.
- Rinse thoroughly in running water.
- Soak in a heated 10 percent H_2SO_4 (sulfuric acid) solution for 1 min at 120°F.
- Rinse thoroughly in running water and thoroughly dry the laminate.

10.4.1.5 Applying and developing the photoresist.
Using the proper equipment (air pressure, roll pressure, roll speed, air and roll temperature), apply the dry film photoresist to the cleaned copper surfaces as soon as possible (but within 4 hr) after the surfaces have been cleaned and dried. Place the artwork on top of the laminate and expose it in a vacuum frame with the proper vacuum, pressure, and light source as recommended by the material or equipment supplier. After the laminate has been exposed, it is then developed in a solution suitable for that photoresist.

10.4.1.6 Copper etching.
Using any of the three basic etching solutions (ammoniacal, cupric, or ferric chloride), etch away all exposed copper

surfaces not covered by the photoresist. After etching, rinse the panel thoroughly in warm running water (65 to 80°F) and dry. *Note:* to minimize shrinkage of the panel as a result of etching, it is recommended that copper borders be placed around the perimeter of the panel prior to etching. Typical borders should be 0.5 to –2 in. wide. Some border designs are shown in Fig. 10.7.

10.4.1.7 Stripping photoresist. Strip the photoresist by immersing the laminate in a solution that is recommended for the specific type of photoresist being used. After the photoresist is removed, rinse the laminate thoroughly in warm running water and then bake at 250°F for approximately 30 min. Do not stack the laminates on top of each other. Hang each laminate separately in the oven.

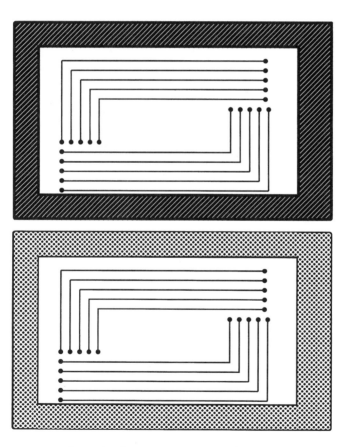

Figure 10.7 Copper borders.

10.4.1.8 Covercoating. A flow diagram for applying the covercoat to the etched single-sided board is shown in Fig. 10.8. Prepare the covercoat as described below.

Drilling access holes. Stack the coverlay material with the adhesive side up, between sheets of entry material (this is the top sheet and is usually 0.005 to 0.015 in. thick aluminum) and back-up (this is usually the bottom sheet and is 0.062 to 0.093 in. thick aluminum clad particle board) material. The maximum number of sheets in the stack should not exceed 15. Attach the stack to the drill table, making sure it is flat and firmly secured to the drill table. Drill parameters should be established (chip load, drill speed, feed rate, etc.) prior to drilling the covercoat material. Access holes should be 0.005 to 0.010 in. larger than the hole in the clad laminate.

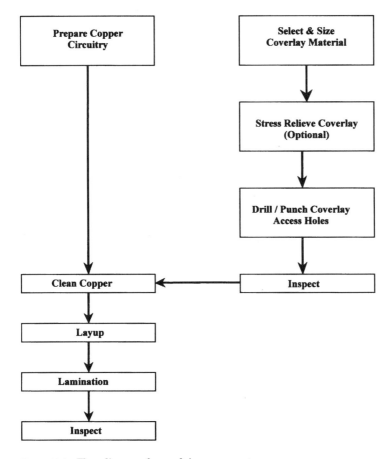

Figure 10.8 Flow diagram for applying covercoat.

Cleaning the copper clad laminate. The preferred method for cleaning the clad laminate is to use the chemical procedure rather than the mechanical process. The chemical method will cause the least distortion to the circuitry. Chemically clean the copper clad laminate using a combination of cleaners and microetchants. The following sequence should be used:

- Soak in an alkaline or acid (preferred) cleaner.
- Rinse thoroughly in running water.
- Soak in a 10 percent H_2SO_4 solution (sulfuric acid) for 1 min at 125°F.
- Rinse thoroughly in running water and then dry.

Laying up the covercoat. Remove the protective polyester film from the covercoat material and place the covercoat (adhesive side down) on top of the copper clad laminate. Usually, tooling pins will assist in registering the covercoat film to the etched laminate. If registration is not a problem then, if you can heat tack the four corners of the covercoat film to the etched laminate (using a soldering iron), the covercoat film will stay in place.

Laminating the covercoat to the laminate. Any one of several press types can be used. The original laminating presses were hydraulic presses, which are quickly being replaced by the vacuum type presses.

Hydraulic press. Typically, a covercoat stack can be laminated using either a cold or hot press. When a cold press is used, place the stack in the press (at less than 150°F) and then heat it up to laminating temperature (360 to 390°F). Full pressure (300 to 350 psi) is applied immediately and held for 60 min. The stack is left in the press to cool under pressure. When a hot cycle is used, the stack is loaded into the hot press, and full pressure (300 to 350 psi) is applied. The stack should not reach full temperature (375°F) in less than 20 to 30 min. After curing, the stack is removed from the hot press and placed in another press to cool down under pressure.

Vacuum bags or vacuum frames. Vacuum frames and vacuum bags are used with hydraulic presses. The stack is placed in a vacuum bag or frame which is then placed in the hydraulic press. Full pressure (250 to 300 psi) and temperature (360 to 390°F for 60 min minimum) is applied as soon as full vacuum is achieved (hold vacuum for 10 min minimum before applying full pressure and temperature). Cool the parts under pressure.

10.4.1.9 Solder coating. After the covercoat has been applied, clean the exposed copper surfaces using an alkaline or acid soak cleaner

(Sec. 10.4.1.4), follow this with a water rinse, and then dry thoroughly in an oven at 220°F for 10 to 30 min. Remove the part and then coat the exposed copper with either an organic preservative (OSP) or solder coat (HASL).

10.4.1.10 Routing the flex. Rout the flex to its final configuration using either a routing fixture or steel rule dies.

10.4.2 Double-Sided Flexible Printed Circuit with Plated-Through Holes

10.4.2.1 Fabrication process. A flow diagram for the fabrication of a double side flexible printed wiring board with plated-through holes is shown in Fig. 10.9. Details of these operations are given below.

Select material. Select the materials (coverlays and copper clad laminates) that are required for the fabrication and using a sharp cutter, cut them into the required panel size. Store all cut materials in a temperature (40 to 80°F) and humidity controlled (below 70 percent) chamber (i.e., nitrogen cabinets, dry box, etc.) until ready for use.

Drilling. Prepare the copper clad laminates for drilling as follows. Stack the copper clad laminates to be drilled between similarly sized pieces of entry (this is the top sheet and is usually 0.005 to 0.015 in. thick aluminum) and backup material (this is the bottom sheet and is usually 0.062 to 0.093 in. aluminum clad particle board). The number of pieces of copper clad laminates in the stack should not exceed ten. Attach the stack to the drill table, making sure it is flat and firmly secured to the table. Using carbide drills, drill the stack using the drilling parameters (chip load, drill speed, and feed rates) per the material or drill manufacturer's recommendation.

Deburring and hole cleaning. The drilled holes must be deburred prior to further processing. The most commonly used method is hand deburring using a hand sander with fine wet and dry paper. Apply light pressure so as not to distort the laminate. After deburring, rinse the laminate thoroughly in running water and dry.

Electroless plating. Following the deburring process, the drilled holes in the panel must be made conductive. This is done using an electrolysis plating process that is basically the same process used for rigid boards. The sequence is as follows:

- Clean the laminates in a mildly alkaline cleaner/conditioner
- Microetch

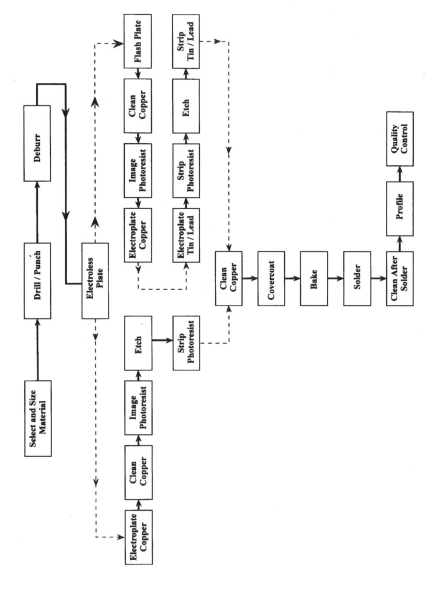

Figure 10.9 Flow diagram for fabricating a double-sided flex circuit.

- Catalyst dip
- Accelerator
- Electroless copper plate

Copper Plating

Flash copper plating. After the laminate has been electroless copper coated, plate the panel in a copper electroplating bath to 0.2 to 0.3 mils. Rinse thoroughly in running water and dry.

Panel plating. After the laminate has been electroless copper coated, plate the laminate in a copper electroplating solution (preferably acid) and plate it up to the full required thickness(1–2 mils). The laminates are then rinsed in running water and thoroughly dried prior to further processing.

Cleaning before applying photoresist.

Prior to applying the photoresist, the laminates should be recleaned. This can be accomplished by using either a mechanical or chemical cleaning process.

Mechanical cleaning. Hand scrubbing using a rotary brush with a 3F grit pumice slurry is generally used to clean the copper surfaces. Scrub both surfaces (top and bottom), then thoroughly rinse the copper laminate in running water to ensure that all the pumice is removed and that you have a water-break free surface. After scrubbing and rinsing, completely dry the copper so as to prevent the surfaces from reoxidizing.

Chemical cleaning. Chemical cleaning using a combination of cleaners and microetchants is probably the better of the two cleaning methods and causing the least distortion to the copper clad laminate. The following sequence is recommended:

- Soak clean in either an acid (preferred) or alkaline cleaner.
- Rinse thoroughly.
- Soak in a heated (120°F) 10 percent H_2SO_4 (sulfuric acid) solution for 1 minute.
- Rinse thoroughly and dry.

Applying and developing the photoresist.

After the laminates have been plated, a photoresist is then applied. Using the proper equipment (air pressure, roll speed, roll pressure, air and roll temperature), apply the dry film photoresist to the cleaned copper surfaces as soon as possible (but within 4 hr) after the surfaces have been cleaned and dried. Place the artwork on top of the laminate and expose it in a vacuum frame with the proper vacuum, pressure, and light source as recommended by the material supplier. After the laminate has been exposed, it is then developed in a solution suitable for that photoresist. *Note:* the photoresist is applied, exposed, and developed on the copper surfaces that we want to retain.

Stripping the photoresist. Remove the unwanted photoresist by immersing the laminate in a solution that is recommended for the specific type of photoresist being used. After stripping the photoresist, the exposed copper will be that copper that we want etched away. After the resist is removed, rinse the laminates thoroughly in warm water and dry.

Etching excess copper. Etch away all exposed copper on the laminate using an ammoniacal (preferred) etching solution. After etching, thoroughly clean the exposed copper surfaces. Test the cleaned copper for water breaks free surface then dry the panel thoroughly. *Note:* to minimize shrinkage of the laminate as a result of etching, it is recommended that copper borders be placed around the perimeter of the laminate prior to etching. Typical borders should be 0.5 to 2 in. wide. Some borders designs are shown in Fig. 10.7.

10.4.2.2 Covercoat. A flow diagram for applying the covercoat to the etched laminate is shown in Fig. 10.8. Prior to covercoat lamination, the laminate should again be cleaned using a chemical cleaning process rather than mechanical. The cleaning process detailed in Sec. 10.4.1.4 should be followed. Process the covercoat as described below.

Drilling access holes. Stack the coverlay material with the adhesive side up, between sheets of entry material (this is the top sheet and is usually 0.005 to 0.015 in. thick aluminum) and back-up (this is the bottom sheet and is usually 0.062 to 0.093 in. thick aluminum clad particle board) material. The maximum number of sheets in the stack should not exceed 15. Attach the stack to the drill table, making sure it is flat and firmly secured. Drilling parameters should be established (chip load, drill speed, feed rates, etc.) prior to drilling the covercoat material. Access holes should be 0.005 to 0.010 in. larger than the holes drilled in the laminate.

Cleaning the copper clad laminate. The preferred method for cleaning the clad laminate is to use the chemical procedure rather than the mechanical process. The chemical procedure will cause the least distortion to the circuitry. Chemically clean the copper clad laminate using a combination of cleaners and microetchants. The following sequence should be used:

- Soak in an alkaline or acid (preferred) cleaner.
- Rinse thoroughly in running water.

- Soak in a 10 percent H_2SO_4 solution (sulfuric acid) for 1 min at 125°F.
- Rinse thoroughly in running water and dry.

Laying up the covercoat. Remove the polyester protective film from the covercoat material and place the covercoat on top of the copper clad laminate, adhesive side down. Usually, tooling pins will assist in registering the covercoat film to the etched laminate. If registration is not a problem, then, if you heat tack the four corners of the laminate using a soldering iron, the covercoat film will hold in place.

Laminating the covercoat to the laminate. Any one of several press types can be used. The original laminating presses were hydraulic presses, which are quickly being replaced by vacuum-type presses.

Hydraulic press. Typically, a covercoat stack can be laminated using either a cold or hot press. When a cold press is used, place the stack in the press (temperature of press should be less than 150°F) and then heat it up to laminating temperature (360 to 390°F). Once laminating temperature is reached, full pressure (300 to 350 psi) is applied immediately and held for 60 min. The stack is cooled under pressure. When a hot cycle is used, the stack is loaded into the hot press and full pressure (300 to 350 psi) is applied. The stack should not reach full temperature (375°F) in less than 20 to 30 min. After curing, the stack is removed from the hot press and placed in another press to cool down under pressure.

Vacuum frames and bags. Vacuum frames and bags are used with hydraulic presses. The stack is placed in either a vacuum bag or frame, and a full vacuum is applied (29 in. Hg) at room temperature. Once full vacuum is achieved, the stack should be exposed to full laminating temperature and pressure. Once the lamination cycle is complete, cool the package under pressure. Apply a solder coating using the hot air leveling process (HASL) or an organic preservative (OSP). After solder coating the board, rout or die stamp parts to final configuration.

10.4.3 Multilayer Flexible Circuits

10.4.3.1 Fabrication process. A flow diagram for the fabrication of a multilayer flexible printed wiring board is shown in Fig. 10.10. Details of the fabrication process for these circuits are given below.

10.4.3.2 Selecting material. Select the thinnest materials (i.e., copper foil, laminate material, adhesive etc.) that can be used and still meet

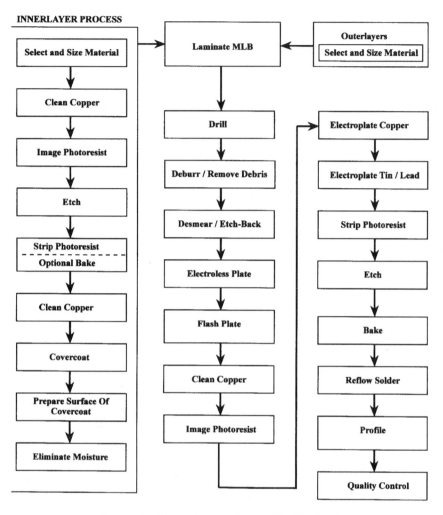

Figure 10.10 Flow diagram for fabricating a multilayer flexible circuit.

the electrical and overall board thickness requirements. Cut these materials into the required size using a sharp cutter. Store all cut material in a temperature (40°–80°F) and humidity controlled (below 70 percent) chamber (i.e. nitrogen cabinet, dry box etc.) until ready for use.

10.4.3.3 Inner layer processing

Cleaning before applying photoresist. All copper clad laminates have an anti-tarnish coating applied to the exposed copper surfaces. This coating must be removed prior to applying the photoresist. Cleaning the anti-tarnishing agent from the copper surfaces can be accomplished using either a mechanical or chemical cleaning process.

Mechanical cleaning. Hand scrubbing using a rotary brush with 3F grit pumice slurry is generally used to clean the copper surfaces. After scrubbing, rinse the clad laminate thoroughly in running water to ensure the complete removal of the pumice. These surfaces should be water-break free. After scrubbing and rinsing, completely dry the copper so as to prevent them from reoxidizing.

Chemical cleaning. Chemical cleaning using a combination of cleaners and microetchants is probably the better of the two cleaning processes, causing the least distortion to the copper surface. The following sequence is recommended:

- Soak the laminate in either an acid (preferred) or alkaline cleaner.
- Rinse thoroughly.
- Soak the laminate for 1 min in a heated (120°F) 10 percent H_2SO_4 (sulfuric acid) solution.
- Rinse thoroughly and dry.

Applying and developing the photoresist. Using the proper equipment (air pressure roll pressure, air and roll temperature), apply the dry film photoresist to the clean copper surfaces as soon as possible (but within 4 hr) after the surfaces have been cleaned and dried. Place the artwork on top of the laminate and expose it in a vacuum chamber with the proper vacuum, pressure, and light source as recommended by the material supplier. After the laminate has been exposed, it is then developed in a solution suitable for that photoresist.

Etching the pattern. Using any of the three basic etching solutions, (ammoniacal, cupric, or ferric chloride) etch off all exposed copper surfaces. After etching, rinse the panel thoroughly in warm (65 to 80°F) water and dry. *Note:* to minimize shrinkage of the etched laminate, it is recommended that copper borders be placed around the edge of the laminate prior to etching. Typical borders should be 0.5 to 2 in. wide all around the perimeter of the laminate. Typical borders are shown in Fig. 10.7.

Stripping the photoresist. Strip the photoresist by immersing the laminate in a solution that is recommended for the specific type of resist being used. After the resist has been removed, rinse the laminate thoroughly in a running water rinse (65 to 85°F) and then bake the laminate at 250°F for 20 to 30 min. Do not stack the laminates on top of each other. Hang each laminate separately in the oven.

10.4.3.4 Cleaning copper laminate for covercoat. Prior to covercoat lamination, the laminates should again be cleaned, using either the mechanical or chemical cleaning processes.

Covercoat lamination. A flow diagram for applying the covercoat to the etched laminate is shown in Fig. 10.8. The covercoating process should be performed prior to bonding all the inner and outer layers together. Remove the protective polyester film from the covercoat material and place this material on top of the etched clad laminate. Usually, tooling pins will assist in registering the covercoat to the etched laminate. If registration is not a problem, then, if you heat tack all four corners of the covercoat to the laminate, using a soldering iron, the covercoat film will stay in place.

Laminating covercoat to the laminate and curing

Laminating covercoat process. Both hot and cold cycles are used in covercoat lamination. However, a cold start (less than 150°F) is the preferred method, since the temperature ramp-up rate can be better controlled. If a vacuum is used, the vacuum draw down should be performed while the press is cool. Once the vacuum draw down is complete, the parts should be exposed to full lamination pressure as the temperature starts to rise. If no vacuum is used, the parts should be placed under full pressure immediately. Once the lamination cycle is complete, cool the package under pressure.

10.4.3.5 Preparing covercoated laminates for lamination. Preparing the covercoat laminate for lamination into a multilayer board requires that the covercoat be pretreated so as to enhance the bond strength between the adhesive and the covercoat. This is generally performed using a plasma etching process. Prior to plasma etching, the covercoat laminate should be baked out at 250°F for approximately 10 to 15 min. The plasma etching (cleaning) cycle consists of applying 100 percent oxygen gas for 10 to 15 min. Other parameters such as power, pressure, and temperature will depend on the complexity of the laminate and the type of unit being used.

10.4.3.6 Preparing the outer layers

Drilling. Prepare the copper clad laminate for drilling as follows. Stack the copper clad laminates to be drilled between similarly sizes pieces of entry (this is the top sheet and is usually 0.005 to 0.010 in. thick aluminum foil) and back up (this is the bottom sheet and is usually 0.062 to 0.093 in. thick aluminum clad particle board) material.

The number of copper clad laminates in a stack should not exceed 10. Place the package on the drill table and firmly secure all the outer edges of the stack by taping the edges firmly down on the drill table. Using new carbide drills, drill the stack using drilling parameters established by either the drill manufacturer or the material producer.

Deburring and hole cleaning. The drilled holes must be deburred and cleaned prior to further processing. The most commonly used process is to hand deburr using a hand or orbital sander with fine wet and dry paper, followed by pressure cleaning using a jet spray and then thoroughly drying.

Desmear. Wet chemical desmearing processes normally used to remove drill smear from drilled holes in rigid boards do not perform well with rigid flex materials. The most widely used and proven process is to use plasma to etch the drilled holes. Prior to plasma etching the laminates, they should be prebaked at 250°F for 30 min minimum. Thicker boards may require longer times. The preferred gas is a mixture of oxygen/Freon in a 80/20 percent mixture. Time, temperature, and pressure will depend on the complexity of the laminate and the type of plasma machine being used. After the board has been plasma etched, it is a good practice to subject the panel to a short plasma cycle of 100 percent oxygen for approximately 10 min.

10.4.3.7 Electroless copper plate.
The following process can be used to plate electroless copper in the drilled hole:

- Immerse the panel in cleaner/conditioner for approximately 5 min.
- Temperature of the bath should not exceed 140°F.
- Microetch for approximately 1 min in a solution at room temperature.
- Acid dip in a 10 percent solution of H_2SO_4 (sulfuric acid) at room temperature for about 3 min.
- Immerse in the catalyst at 95 to 105°F for about 5 min.
- Immerse the panel in the electroless copper bath (110 to 115°F) for 30 min.
- Acid dip for 1 min and then dry.

Flash plate. After the laminate has been electroless copper coated, plate the laminate in a copper electroplating solution (preferably acid copper) to approximately 0.3 mils. Remove the laminate, rinse thoroughly in running water, and dry. Board should be thoroughly dried before applying the photoresist.

Clean copper before applying photoresist. Prior to applying the photoresist, the panel should be recleaned using either a chemical or mechanical method.

The mechanical method provides for hand scrubbing the panel with a rotary brush using a 3F grit pumice slurry. Both the top and bottom surfaces must be cleaned. After cleaning, thoroughly rinse the laminate with running water to ensure that all the pumice has been removed, then dry the laminate. If the chemical process is used, it is generally a combination of cleaned/conditioner. This method causes the least distortion to the copper. The process is as follows:

1. Soak the panel in either an acid (preferred) or alkaline cleaner.
2. Rinse thoroughly.
3. Soak in a heated (120°F) 10 percent H_2SO_4 (sulfuric acid) solution for 1 min.
4. Rinse and dry.

10.4.3.8 Applying photoresist

Applying photoresist to flash copper plated laminates. After the laminate has been flashed copper plated (0.3 mils), the dry film photoresist is applied to the laminate, exposed and developed. The panel is then plated to the full copper thickness (1–2 mils) followed by tin/lead plating (0.2–0.3 mils).

Applying photoresist to copper panel plated (1–2 mil)s laminate. After the laminate has been fully copper plated (1–2 mils), a photoresist is applied to the laminate, exposed, and developed followed by solder plating to a thickness of 0.2 to 0.3 mils.

Strip photoresist and etch. Prior to etching off the unwanted copper, strip the photoresist with a solution recommended by the material supplier. The photoresist can be stripped from the panel using either a conveyorize photoresist stripping machine or by immersing the panel in the solution until the resist is completely removed. Etch the exposed copper with an alkaline etchant. Do not use cupric or ferric chloride etchants, as these may attack the solder plating.

10.4.3.9 Bake.
Bake out the board at 250°F for about 2 hr minimum. Longer times (up to 10 hr) may be required due to board thickness. Some boards reabsorb moisture quite readily, and therefore the reflow process should be performed as quickly as possible after the board has

been removed from the oven. Vacuum oven baking under full vacuum (29 in. Hg) can also be used and with a lower temperature (150 to 175°F).

10.4.3.10 Reflow of solder. After the laminate has been baked dry, and you are not ready to fuse the tin/lead plating, store the laminates in a temperature (40° to 80°F) and humidity (below 70 percent) controlled chamber. Fusing the tin/lead plating is performed as follows:

- Remove the laminate from the chamber and immerse it in hot oil solution (300°F) for 15 to 30 s (time is dependent on laminate thickness).

- Remove it from the first hot oil bath and quickly immerse it in a second hot oil bath (350°F) for 20 to 30 s (time is dependent on laminate thickness).

- Remove it from the second tank and resubmerge it in the first hot oil bath for 15 to 30 s. Remove it from the last hot oil bath and allow it to cool to room temperature before washing the laminate in running warm water to remove the hot oil residue. Following the water washing, the multilayer laminate should be baked dry.

10.4.3.11 Route. The panel is routed using standard routing processing similar to that used for rigid multilayer laminates.

10.4.4 Rigid Flex Circuits

Details of the fabrication process for the rigid flexible printed wiring are given below.

10.4.4.1 Selecting the material. Select materials (coverlay, flexible and rigid copper clad laminates, prepreg, adhesives, etc.). Cut these materials into the required sizes using a sharp cutter.After cutting, store the materials in a temperature (40 to 80°F) and humidity controlled (below 70 percent) chamber (i.e., nitrogen cabinet, dry box, etc.) until ready for use.

10.4.4.2 Inner layer processing

Cleaning before applying photoresist. All copper clad laminates have an anti-tarnish coating applied to the exposed copper surfaces. This coating must be removed prior to applying the photoresist. This can be accomplished by using either a mechanical or chemical cleaning process.

Mechanical cleaning process. Hand scrub the copper surface using a rotary brush with a 3F grit pumice slurry. After scrubbing, rinse the clad laminate thoroughly with running water to ensure complete removal of the pumice. The cleaned copper surfaces should be water break-free. After scrubbing and rinsing, completely dry the copper surfaces so as to prevent them from reoxidizing.

Chemical cleaning process. Chemical cleaning with a combination of cleaners and microetchants is probably the better of the two cleaning processes and causes the least amount of distortion to the copper foil. The following sequence is recommended:

- Soak the laminate in either an acid (preferred) or alkaline cleaner.
- Rinse thoroughly.
- Soak the laminate for 1 min in a heated (120°F), 10 percent H_2SO_4 (sulfuric acid) solution.
- Rinse thoroughly and dry.

Applying and developing the photoresist. Using the proper equipment (air pressure roll pressure, air and roll temperature), apply the dry film photoresist to the cleaned copper surfaces as soon as possible (but within 4 hr) after the surfaces have been cleaned and dried. Place the artwork on top of the laminate and expose it in a vacuum frame with the proper vacuum, pressure, and light source as recommended by the material supplier. After the laminate has been exposed, it is then developed in a solution suitable for that photoresist.

Tooling holes. To minimize layer misregistration, tooling holes should be punched or drilled in the clad laminate either before or after covercoat lamination.

Etching the pattern. Using any of the three basic etching solutions (ammoniacal, cupric or ferric chloride) etch off all the exposed copper. After etching, rinse the panel thoroughly in warm running water (65 to 80°F) and dry. *Note:* to minimize shrinkage of the etched patterns, it is recommended that copper borders be placed around the edges of the copper panel prior to photoresist and etching. Typical borders are shown in Fig. 10.7.

Stripping the photoresist. Strip the photoresist in immersion tanks or conveyorized equipment with the appropriate solution for the type of photoresist being used. After stripping, rinse the panel in warm water (65 to 80°F) and then bake dry at 250°F for 20 to 30 min. Do not stack the laminates on top of each other. Hang each laminate separately in the oven.

10.4.4.3 Covercoat lamination. The covercoat lamination process should be performed prior to bonding all the inner flexible layers and outer rigid layers together. It is not advisable to try and bond all the layers together in one lamination cycle. Also the covercoat can be applied in one of two ways:

1. Apply the covercoat to the entire flex circuit (Fig. 10.11).
2. Apply the covercoat only to those areas that will be exposed after lamination (Fig. 10.12).

Prior to adhering the covercoat to the etched laminate, remove the polyester protective film from the covercoat material and place the coverlay material (adhesive side down) on top of the etched clad laminate. Usually, tooling pins will assist in registering the coverlay to the etched clad laminate. If registration is not a problem, then, if you heat tack all four corners of the laminate using a soldering iron, the coverlay film will stay in place.

Lamination cycles. Prior to applying any pressure or temperature, it is advisable to place the stack in a vacuum bag/frame and pull a vacuum of 29 in. Hg for 30 to 120 min to remove any trapped air. The stack, with the vacuum bag/frame, is then placed in the laminating press. Both hot and cold cycles can be used in rigid flex lamination. However a cold start (less than 150°F) is preferred, since the temperature ramp-up rate can be better controlled. The stack is placed in the press and then heated to laminating temperature. Full pressure is applied imme-

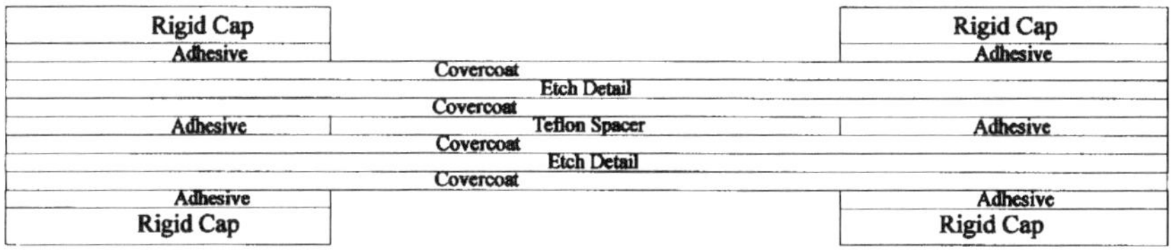

Figure 10.11 Covercoat material covering entire flex area.

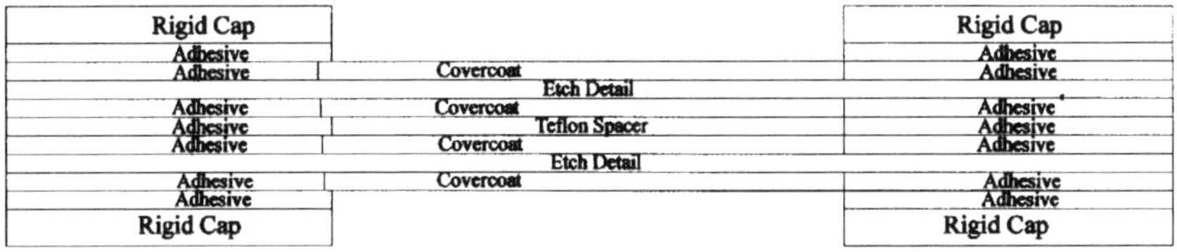

Figure 10.12 Covercoat material only in exposed area.

diately. After the laminate is cured, it is left in the press to cool down under full pressure. If a hot cycle is used, then the stack is placed into the laminating press that is at full laminating temperature. Pressure is applied immediately. Normal cycle for a hydraulic presses is

- Temperature: 360 to 390°F
- Time: 60 min
- Pressure: 300 to 350 psi for a hydraulic press and 250 to 300 psi for a vacuum press

10.4.4.4 Rigid flex final lamination cycle. After the covercoat has been laminated to the etched details, it now ready for final lay-up. Normally, the outer layers (first and last layers) are rigid materials. Preparing the covercoated laminate for lamination into a rigid flex requires that the laminate be pretreated so as to enhance the bond strength between the adhesive and the covercoat. This is generally performed using a plasma etching process. Prior to plasma etching, the covercoated laminate must be baked at 250°F for approximately 10 to 15 min. Place the laminate while still hot in the plasma etcher. Using 100 percent oxygen, clean/etch for about 10 to 15 min. Time, pressure, and temperature will depend on the type of equipment being used. During the lay-up process, place Teflon inserts between flex layers that you do not want bonded together during the lamination process (Fig. 10.13). After the laminate has been plasma prepared, lay up the stack as shown in Fig. 10.14.

After the stack has been completely assembled, place it in either a vacuum bag or frame and pull a full vacuum (29 in. Hg) for about 30 to 120 min. The time depends on complexity of the rigid flex construction. If a cold cycle is used, then the press temperature should be less than 150°F. The stack is placed in the press, encased in either the vacuum frame or bag, and then the press is heated to laminating temperature. Full pressure is applied immediately. After the laminate is cured, it is left in the press to cool down under pressure. If a hot cycle is used, then the stack is placed in the press at full laminating temperature. Pressure is applied immediately. When using vacuum bags or frames, apply a vacuum of 29 in. Hg for 60 to 120 min before applying any temperature or pressure. Normal laminating cycles using a hydraulic press are

- Temperature: 360 to 390°F
- Time: 60 min minimum
- Pressure: 300–350 psi

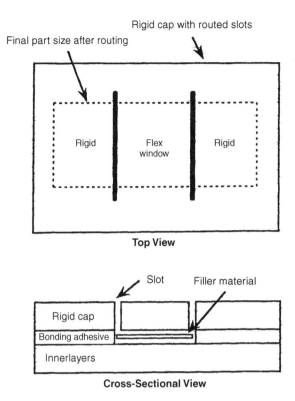

Figure 10.13 Teflon spacers (filler material).

10.4.4.5 Verifying registration. Before drilling the rigid flex, check the layer to layer registration using an x-ray machine. Adjust the drill tape to accommodate any misregistration.

10.4.4.6 Drilling. Drilling rigid flexes (in most cases) should be performed one laminate at a time using new carbide drill bits. Resharpened drills should not be used. The laminated panel must be flat and tightly secured to the drill table. This is usually accomplished by taping all four edges of the laminate to the drill table. Prior to drilling, place the top entry material (0.010 to 0.015 in. thick aluminum) and bottom back-up material (0.062 to 0.093 in. thick aluminum clad particle board) on top and bottom of the stack respectively. Drilling parameters such as speed, feed, and chip loads must be determined prior to drilling the panel. Each machine and each rigid flex design has its own specific drilling parameters.

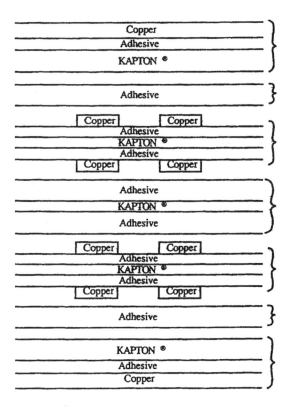

Figure 10.14 Lay-up for lamination.

Deburring. Deburr the drilled panel using either a rotary hand sander or a conveyorized debarring machine.

Etchback/plasma desmearing. The only successful process for desmearing rigid flex boards has been to use a plasma etcher. The rigid flex laminate is made up of three different materials.

1. A rigid material (usually a copper clad rigid polyimide or epoxy laminate)
2. A flexible material (usually a copper clad flexible polyimide film)
3. A flexible film material with an adhesive

All three of these materials etch at different rates.

Generally, the adhesives etch three times faster than the rigid material, which etches twice as fast as the flexible material. Before etching, parameters such as time, temperature, pressure, gas mixture, and gas flow rates must be established for each board design and each plasma etcher. Generally, the etch-back cycle consists of placing the drilled laminate in the chamber, applying the vacuum, and starting the gas flow. The etch back gasses are a mixture of oxygen/Freon (O_2/CF_4) in an 80/20 mixture. Following this, purge the chamber of the oxygen/Freon mixture and replace it with 100 percent oxygen. Apply this gas flow for about 10 min.

10.4.4.7 Electroless copper plate. The following process can be used to plate electroless copper in the drilled holes:

1. Immerse the panel in a cleaner/conditioner for approximately 5 min.
2. Temperature of the bath should not exceed 140°F.
3. Rinse thoroughly in running water.
4. Micro-etch for approximately 1 min at 70 to 80°F.
5. Rinse thoroughly in running water.
6. Acid dip in a 10 percent solution of H_2SO_4 (sulfuric acid) at room temperature for about 3 min.
7. Rinse thoroughly in running water.
8. Immerse in a catalyst at 95 to 105°F for about 5 min.
9. Rinse in running water.
10. Immerse the panel in the electroless copper bath (110 to 115°F) for 30 min.
11. Rinse in running water.
12. Acid dip for 1 min.
13. Rinse in water and dry.

10.4.4.8 Electroplating Copper

Flash plate. After the panel has been electroless copper coated, plate the panel in a copper electroplating solution (preferably acid copper) and plate approximately 0.3 mils of copper. Remove the panel, rinse thoroughly, and dry.

10.4.4.9 Applying photoresist, copper, and solder plating. Apply the dry film photoresist to the cleaned, dried laminate. After exposing and developing the resist, the laminate is then plated to the full required copper thickness (usually 2 mils). After copper plating, rinse thoroughly, and then electroplate the laminate with tin/lead (0.2 to 0.3 mils).

Strip photoresist and etch. Prior to etching off the unwanted copper, strip off the photoresist with a solution recommended by the material supplier. The photoresist can be stripped from the laminate either by a conveyorized photoresist stripping machine or by immersing the panel in a solution until the resist is completely removed.

Etch the exposed copper with an alkaline etchant. Do not use cupric or ferric chloride etchants as these may attack the solder plating.

Bake. Bake the board at 250°F for 2 hr minimum. Longer times (up to 10 hr) may be required due to board thickness. Some boards reabsorb moisture quite readily; therefore, the solder reflow process should be performed as quickly as possible after the board has been removed from the oven. Vacuum oven baking under full vacuum (29" in. Hg) can also be used and with a lower bake-out time and temperature.

10.4.4.10 Reflow solder. The rigid flex board should be reflowed prior to routing it into its final configuration. After the board has been baked out, flux the panel with a water soluble flux. Immerse the board in hot oil at 300°F for about 15 to 30 s. Remove the laminate and quickly immerse it in hot oil at 350°F for 20 to 40 s. Remove the laminate and again quickly immerse it in hot oil at 300°F for 15 to 30 s. Remove the laminate and allow it to cool to room temperature before cleaning off the flux residue. The following procedure can be used to clean the flux residue from the reflowed solder panel:

- Perform hot alkaline soak.
- Rinse in running water.
- Bake dry.

10.4.4.11 Profiling. Using an auto-router, rout the rigid flex from the panel. Remove all burrs by lightly sanding all edges if necessary.

10.5 References

1. Association Connecting Electronic Assemblies (IPC), 2215 Sanders Road Northbrook, Illinois. IPC-FC-231; IPC-FC-232; IPC-FC-241; IPC-MF-150.
2. E.I. du Pont de Nemours & Co. Inc. *Pyralux Technical Manual.*
3. Litton Industries, GKS Division, Shop Traveler.
4. Graphic Research Ins., Div. Method Inc. Shop Traveler.

Chapter

11

Fabrication and Properties of Electronic Ceramics and Composites

Jerry E. Sergent
Sergent & Sergent Consulting
Corbin, Kentucky

11.1 Introduction

Ceramic materials play an important role in the electronics industry. They have a high electrical resistivity and excellent high-frequency characteristics; they have a higher thermal conductivity than printed circuit boards; they are very stable, both chemically and thermally; and they have a high melting point. These properties are highly desirable in the design and manufacture of electronic circuits. Ceramics are used extensively as a foundation for thick and thin film hybrid circuits in a variety of applications, as insulators, as heat paths in circuits where thermal management is particularly important, and to fabricate electronic components, such as capacitors, for use in manufacturing electronic circuits.

A *composite* is a mixture of two or more materials that retain their original properties but, in concert, offer parameters that are superior to either. Composites in various forms have been used for centuries. Ancient peoples, for example, used straw and rocks in bricks to increase their strength. Modern-day structures use steel rods to reinforce concrete. The resulting composite structure combines the strength of steel with the lower cost and weight of concrete.

Ceramics are commonly used in conjunction with metals to form composites for electronic applications, especially in thermal management. Ceramic-metal (cermet) composites typically have a lower temperature coefficient of expansion (TCE) than metals, offer a higher thermal conductivity than ceramics, and are more ductile and more resistant to stress than ceramics. These properties combine to make cermet composites ideal for use in high-power applications where device temperature must be maintained at the lowest level possible.

Ceramics are crystalline in nature, with a dearth of free electrons. They are formed by the bonding of a metal and a nonmetal and may exist as oxides, nitrides, carbides, or silicides. An exception is diamond, which consists of pure carbon subjected to high temperature and pressure. Diamond substrates meet the criteria for ceramics and may be considered as such in this context.

The primary bonding mechanism in ceramics is ionic bonding, formed by the electrostatic attraction between positive and negative ions. Atoms are most stable when they have eight electrons in the outer shell. Metals have a surplus of electrons in the outer shell that are loosely bound to the nucleus and readily become free, even at moderately low temperatures, creating positive ions. Similarly, nonmetals have a deficit of electrons in the outer shell, and readily accept free electrons, creating negative ions. Figure 11.1 illustrates an ionic bond between a magnesium ion with a charge of +2 and an oxygen ion with a charge of –2, forming magnesium oxide (MgO). Ionically bonded materials are crystalline in nature and have both a high electrical resistance and a high relative dielectric constant. Due to the strong nature of the

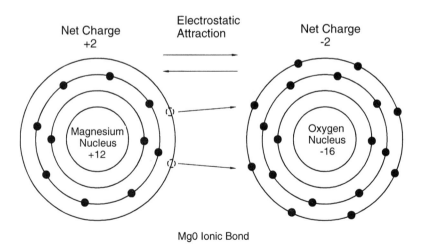

Figure 11.1 Magnesium oxide ionic bond.

bond, they have a high melting point and do not readily break down at elevated temperatures. By the same token, they are very stable chemically and are not attacked by ordinary solvents and most acids.

A degree of covalent bonding may also be present, particularly in some of the silicon and carbon-based ceramics. The sharing of electrons in the outer shell forms a covalent bond. A covalent bond is depicted in Fig. 11.2, illustrating the bond between oxygen and hydrogen to form water. A covalent bond is also a very strong bond and may be present in liquids, solids, or gases.

A knowledge of the properties of ceramics, diamond, and composite materials and how they are manufactured is critical to the overall understanding of the electronics industry. This chapter considers the manufacturing processes and properties of certain ceramics used in electronic applications, including aluminum oxide (alumina, Al_2O_3), beryllium oxide (beryllia, BeO), aluminum nitride (AlN), boron nitride (BN), diamond (C), and silicon carbide (SiC). A discussion of the basic material parameters is provided followed by the methods of fabrication and how they relate to the finished product. Two composite materials, aluminum silicon carbide (AlSiC) and Dymalloy®, a diamond/copper structure, are also described. Particular attention is given to these materials when intended for use as a so-called *substrate,* a thin, flat structure intended for use as a foundation for electronic circuitry. Although the conductive nature of composite materials prevents them from being used as a conventional substrate, they have a high thermal conductivity and may be used in applications where the relatively low electrical resistance is not a consideration.

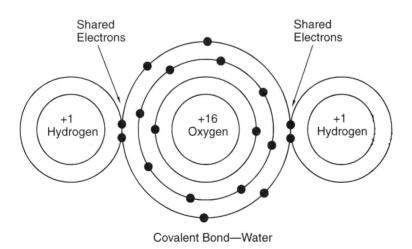

Figure 11.2 Covalent bond between oxygen and hydrogen to form water.

11.2 Surface Properties of Ceramics

The surface properties of interest, *surface roughness* and *camber*, are highly dependent on the particle size and method of processing. Surface roughness is a measure of the surface microstructure, and camber is a measure of the deviation from flatness. In general, the smaller the particle size, the smoother will be the surface.

Surface roughness may be measured by electrical or optical means. Electrically, surface roughness is measured by moving a fine-tipped stylus across the surface. The stylus may be attached to a piezoelectric crystal or to a small magnet that moves inside a coil, inducing a voltage proportional to the magnitude of the substrate variations. The stylus must have a resolution of 25.4 nm (1 μin) to read accurately in the most common ranges. Optically, a coherent light beam from a laser diode or other source is directed onto the surface. The deviations in the substrate surface create interference patterns that are used to calculate the roughness. Optical profilometers have a higher resolution than the electrical versions and are used primarily for very smooth surfaces. For ordinary use, the electrical profilometer is adequate and is widely used to characterize substrates in both manufacturing and laboratory environments.

The output of an electrical profilometer is plotted as shown in schematic form in Fig. 11.3, and in actual form in Fig. 11.4. A quantitative interpretation of surface roughness can be obtained from this plot in one of two ways: by the rms value and by the arithmetic average.

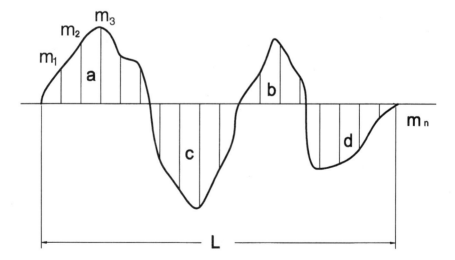

Figure 11.3 Schematic of a surface trace.

Fabrication and Properties of Electronic Ceramics and Composites 487

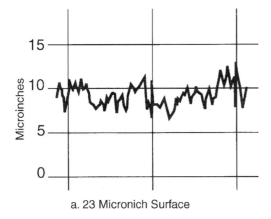

a. 23 Micronich Surface

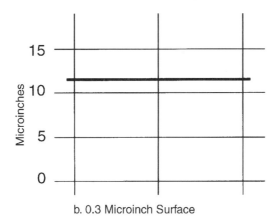

b. 0.3 Microinch Surface

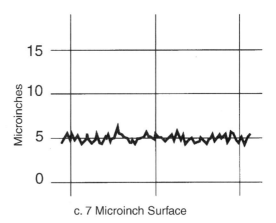

c. 7 Microinch Surface

Figure 11.4 Surface trace of three substrate surfaces.

The rms value is obtained by dividing the plot into n small, even increments of distance and measuring the height, m, at each point, as shown in Fig. 11.3. The rms value is calculated by

$$\text{rms} = \sqrt{\frac{m_1^2 + m_2^2 + \ldots + m_n^2}{n}} \qquad (11.1)$$

and the average value (usually referred to as the center line average, CLA) is calculated by

$$\text{CLA} = \frac{a_1 + a_2 + a_3 + \ldots + a_n}{L} \qquad (11.2)$$

where $a_1, a_2, a_3, \ldots$ = areas under the trace segments (Fig. 11.3)
L = length of travel

For systems where the trace is magnified by a factor M, Eq. (11.2) must be divided by the same factor.

For a sinewave, the average value is 0.636 × peak, and the rms value is 0.707 × peak, which is 11.2% larger than the average. The profilometer trace is not quite sinusoidal in nature. The rms value may be greater than the CLA value by 10–30 percent.

Of the two techniques, the CLA is the preferred method of use, because the calculation is more directly related to the surface roughness; but it also has several shortcomings.

- The method does not consider surface waviness or camber as shown in Fig. 11.5.[1]
- Surface profiles with different periodicities and the same amplitudes yield the same results, although the effect in use may be somewhat different.
- The value obtained is a function of the tip radius.

Surface roughness has a significant effect on the adhesion and performance of thick and thin film depositions. For adhesion purposes, it is desirable to have a high surface roughness to increase the effective interface area between the film and the substrate. For stability and repeatability, the thickness of the deposited film should be much greater than the variations in the surface. For thick films, with a typical thickness of 10–12 µ, surface roughness is not a consideration, and a value of 25 µin (625 nm) is desirable. For thin films, however, with a thickness measured in angstroms, a much smoother surface is required.[2] Figure 11.6 illustrates the difference in a thin film of tanta-

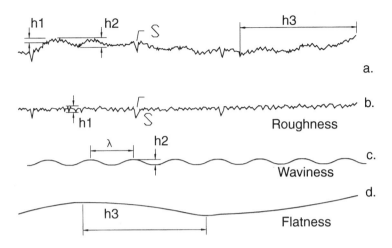

Figure 11.5 Surface characteristics.

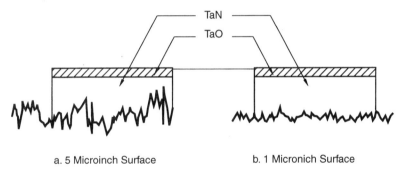

Figure 11.6 TaN resistor with TaO passivation on substrates with different surface roughness.

lum nitride (TaN) deposited on both a 1 µin surface and a 5 µin surface. Tantalum nitride is commonly used to fabricate resistors in thin film circuits and is stabilized by growing a layer of tantalum oxide, which is nonconductive, over the surface by baking the resistors in air. Note that the oxide layer in the rougher surface represents a more significant percentage of the overall thickness of the film in areas where the surface deviation is the greatest. The result is a wider variation in both the initial and post-stabilization resistor values and a larger drift in value with time.

Camber and waviness are similar in form in that they are variations in flatness over the substrate surface. Referring to Fig. 11.5, camber can be considered as an overall warpage of the substrate, while waviness is more periodic in nature. Both of these factors may occur as a result of uneven shrinkage during the organic removal/sintering process or as a result of nonuniform composition. Waviness may also occur as a result of a *flat spot* in the rollers used to form the green sheets.

Camber is measured in units of length/length, interpreted as the deviation from flatness per unit length, and is measured with reference to the longest dimension by placing the substrate through parallel plates set a specific distance apart. Thus, a rectangular substrate would be measured along the diagonal. A typical value of camber is 0.003 in/in (also 0.003 mm/mm), which for a 2×2 in. substrate, represents a total deviation of $0.003 \times 2 \times 1.414 = 0.0085$ in. For a substrate that is 0.025 in. thick, a common value, the total deviation represents a third of the overall thickness!

The nonplanar surface created by camber adversely affects subsequent metallization and assembly processes. In particular, screen printing is made more difficult due to the variable snap-off distance. Torsion bar printing heads on modern screen printers can compensate to a certain extent, but not entirely. A vacuum hold-down on the screen printer platen also helps, but it only flattens the substrate temporarily during the actual printing process. Camber can also create excessive stresses and a nonuniform temperature coefficient of expansion. At temperature extremes, these factors can cause cracking, breaking, or even shattering of the substrate.

Camber is measured by first measuring the thickness of the substrate and then placing the substrate between a series of pairs of parallel plates set specific distances apart. Camber is calculated by subtracting the substrate thickness from the smallest distance that the substrate will pass through and dividing by the longest substrate dimension. A few generalizations can be made about camber.

- Thicker substrates will have less camber than thinner.
- Square shapes will have less camber than rectangular.
- The pressed methods of forming will produce substrates with less camber than the sheet methods.

11.3 Thermal Properties of Ceramic Materials

11.3.1 Thermal Conductivity

Referring to Fig. 11.7, the thermal conductivity of a material is a measure of the ability to carry heat and is defined in one dimension as

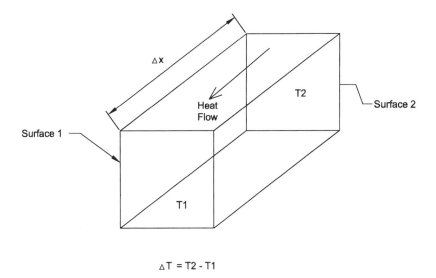

$$\Delta T = T2 - T1$$

Figure 11.7 Thermal conductivity of alumina vs. concentration, theoretical and actual.

$$q = -k\frac{\Delta T}{\Delta x} \qquad (11.3)$$

where k = thermal conductivity in W/m-°C
 q = heat flux in w/m^2
 ΔT = change in temperature from one surface to another in °C
 Δx = distance between the surfaces in meters

The negative sign denotes that heat flows from areas of higher temperature to areas of lower temperature.

The thermal conductivity of most materials decreases with temperature. A plot of the thermal conductivity vs. temperature for several materials is shown in Fig. 11.8.[3] One material not plotted in this graph is diamond. The thermal conductivity of diamond varies widely with composition and the method of preparation, and it is much higher than those materials listed. Diamond will be discussed in detail in a later section.

11.3.2 Specific Heat

The specific heat of a material is defined as

$$c = \frac{\Delta Q}{\Delta T} \qquad (11.4)$$

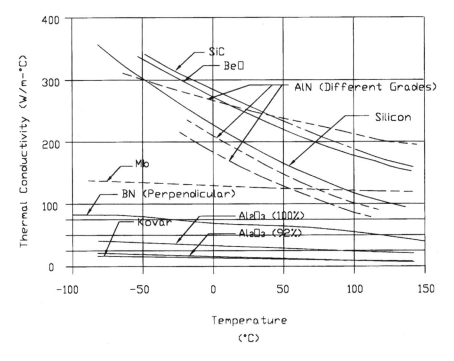

Figure 11.8 Thermal conductivity vs. temperature for selected materials.

where c = specific heat in W-s/g-°C
Q = energy in watt-s
T = temperature in K

The specific heat, c, is defined in a similar manner and is the amount of heat required to raise the temperature of one gram of material by one degree, with units of watt-s/g-°C. The quantity *specific heat* in this context refers to the quantity, c_V, which is the specific heat measured with the volume constant, as opposed to c_P, which is measured with the pressure constant. At the temperatures of interest, these numbers are nearly the same for most solid materials. The specific heat is primarily the result of an increase in the vibrational energy of the atoms when heated, and the specific heat of most materials increases with temperature up to a temperature called the Debye temperature, at which point it becomes essentially independent of temperature. The specific heat of several common ceramic materials as a function of temperature is shown in Fig. 11.9.

The heat capacity, C, is similar in form, except that it is defined in terms of the amount of heat required to raise the temperature of a mole of material by one degree and has the units of watt-s/mol-°C.

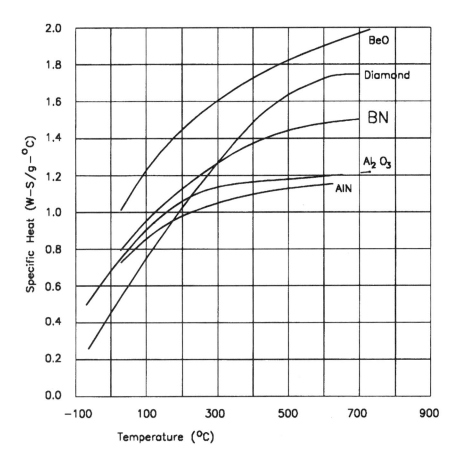

Figure 11.9 Specific heat vs. temperature for selected materials.

11.3.3 Temperature Coefficient of Expansion

The temperature coefficient of expansion (TCE) arises from the asymmetrical increase in the interatomic spacing of atoms as a result of increased heat. Most metals and ceramics exhibit a linear, isotropic relationship in the temperature range of interest, while certain plastics may be anisotropic in nature. The TCE is defined as

$$\alpha = \frac{\ell(T_2) - \ell(T_1)}{\ell(T_1)(T_2 - T_1)} \tag{11.5}$$

where α = temperature coefficient of expansion in ppm/°C^{-1}
T_1 = initial temperature
T_2 = final temperature

$\ell(T_1)$ = length at initial temperature
$\ell(T_2)$ = length at final temperature

The TCE of most ceramics is isotropic. For certain crystalline or single-crystal ceramics, the TCE may be anisotropic, and some may even contract in one direction and expand in the other. Ceramics used for substrates do not generally fall into this category, as most are mixed with glasses in the preparation stage and do not exhibit anisotropic properties as a result. The temperature coefficient of expansion of several ceramic materials is shown in Table 11.1.

Table 11.1 Temperature Coefficient of Expansion of Selected Ceramic Substrate Materials

Material	TCE (ppm/°C)
Alumina (96%)	6.5
Alumina (99%)	6.8
BeO (99.5%)	7.5
BN parallel	0.57
perpendicular	-0.46
Silicon carbide	3.7
Aluminum nitride	4.4
Diamond, type IIA	1.02
AlSiC (70% SiC loading)	6.3

11.4 Mechanical Properties of Ceramic Substrates

The mechanical properties of ceramic materials are strongly influenced by the strong interatomic bonds that prevail. Dislocation mechanisms, which create slip mechanisms in softer metals, are relatively scarce in ceramics, and failure may occur with very little plastic deformation. Ceramics also tend to fracture with little resistance.

11.4.1 Modulus of Elasticity

The temperature coefficient of expansion (TCE) phenomenon has serious implications in the applications of ceramic substrates. When a sample of material has one end fixed, which may be considered to be a result of bonding to another material that has a much smaller TCE, the net elongation of the hotter end per unit length, or *strain* (*E*), of the material is calculated by

$$E = TCE \times \Delta T \qquad (11.6)$$

where E = strain in length/length
 ΔT = temperature differential across the sample

Elongation develops a stress (*S*) per unit length in the sample as given by Hooke's law.

$$S = EY \qquad (11.7)$$

where S = stress in psi/in (N/m^2/m)
 Y = modulus of elasticity in lb/in^2 (N/m^2)

When the total stress, as calculated by multiplying the stress/unit length by the maximum dimension of the sample, exceeds the strength of the material, mechanical cracks will form in the sample that may even propagate to the point of separation. The small elongation that occurs before failure is referred to as *plastic deformation*. This analysis is somewhat simplistic in nature but serves as a basic understanding of the mechanical considerations. The modulus of elasticity of selected ceramics is summarized in Table 11.2, along with other mechanical properties.

Table 11.2 Mechanical Properties of Selected Ceramics

Material	Modulus of elasticity (GPa)	Tensile strength (MPa)	Compressive strength (MPa)	Modulus of rupture (MPa)	Flexural strength (MPa)	Density (g/cm^3)
Alumina (99%)	370	500	2600	386	352	3.98
Alumina (96%)	344	172	2260	341	331	3.92
Beryllia (99.5%)	345	138	1550	233	235	2.87
Boron nitride (normal)	43	2410	3525	800	53.1	1.92
Aluminum nitride	300	310	2000	300	269	3.27
Silicon carbide	407	197	4400	470	518	3.10
Diamond (type IIA)	1000	1200	11000	940	1000	3.52

11.4.2 Modulus of Rupture

Ordinary stress-strain testing is not generally used to test ceramic substrates since they do not exhibit elastic behavior to a great degree. An alternative test, the modulus of rupture (bend strength) test, is preferred. A sample of ceramic, either circular or rectangular, is suspended between two points, a force is applied in the center, and the elongation of the sample is measured. The modulus of rupture is the stress required to produce fracture and is shown in Table 11.2 for selected ceramics.

11.4.3 Tensile and Compressive Strength

A force applied to a ceramic substrate in a tangential direction may product tensile or compressive forces. If the force is tensile, in a direction such that the material is pulled apart, the stress produces plastic deformation as defined in Eq. (11.7). As the force increases past a value referred to as the *tensile strength,* breakage occurs. Conversely, a force applied in the opposite direction creates compressive forces until a value referred to as the *compressive strength* is reached, at which point breakage also occurs. The compressive strength of ceramics is, in general, much larger than the tensile strength. The tensile and compressive strength of selected ceramic materials is shown in Table 11.2.

In practice, the force required to fracture a ceramic substrate is much lower than predicted by theory. The discrepancy is due to small flaws or cracks residing within these materials as a result of processing. For example, when a substrate is sawed, small edge cracks may be created. Similarly, when a substrate is fired, trapped organic material may outgas during firing, leaving a microscopic void in the bulk. The result is an amplification of the applied stress in the vicinity of the void that may exceed the tensile strength of the material and create a fracture. If the microcrack is assumed to be elliptical with the major axis perpendicular to the applied stress, the maximum stress at the tip of the crack may be approximated by[4]

$$S_M = 2S_o \leq \left(\frac{a}{\rho_t}\right)^{1/2} \tag{11.8}$$

where S_M = maximum stress at the tip of the crack
S_O = nominal applied stress
a = length of the crack as defined in Fig. 11.10
ρ_t = radius of the crack tip

The ratio of the maximum stress to the applied stress may be defined as

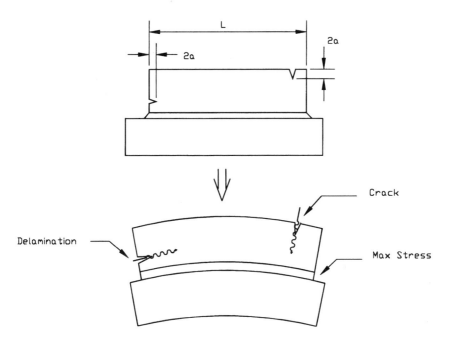

Figure 11.10 Cracks and chip-outs in substrates (from Ref. 4).

$$K_t = \frac{S_M}{S_o} = 2\left(\frac{a}{p_t}\right)^{1/2} \quad (11.9)$$

where K_t = stress concentration factor

For certain geometries, such as a long crack with a small tip radius, K_t may be much larger than 1, and the force at the tip may be substantially larger than the applied force.

Based on this analysis, a material parameter called the *plain strain fracture toughness*, a measure of the ability of the material to resist fracture, can be defined as

$$K_{IC} = ZS_c\sqrt{\pi a} \quad (11.10)$$

where K_{IC} = plain strain fracture toughness in psi-in$^{1/2}$ or Mpa-m$^{1/2}$
  Z = dimensionless constant, typically 1.2[4]
  S_c = critical force required to cause breakage

From Eq. (11.10), the expression for the critical force can be defined as

$$S_c = Z\frac{K_{IC}}{\sqrt{\pi a}} \tag{11.11}$$

When the applied force on the die due to TCE or thermal differences exceeds this figure, fracture is likely. The plain strain fracture toughness for selected materials is presented in Table 11.3. It should be noted that Eq. (11.11) is a function of thickness up to a point, but is approximately constant for the area to thickness ratio normally found in substrates.

Table 11.3 Fracture Toughness for Selected Materials

Material	Fracture toughness (MPa-m$^{1/2}$)
Silicon	0.8
Alumina (96%)	3.7
Alumina (99%)	4.6
Silicon carbide	7.0
Molding compound	2.0

11.4.4 Hardness

Ceramics are among the hardest substances known, and the hardness is correspondingly difficult to measure. Most methods rely on the ability of one material to scratch another, and the measurement is presented on a relative scale. Of the available methods, the Knoop method is the most frequently used. In this approach, the surface is highly polished, and a pointed diamond stylus under a light load is allowed to impact on the material. The depth of the indentation formed by the stylus is measured and converted to a qualitative scale called the *Knoop* or *HK* scale. The Knoop hardness of selected ceramics is given in Table 11.4.

11.4.5 Thermal Shock

Thermal shock occurs when a substrate is exposed to temperature extremes in a short period of time. Under these conditions, the substrate is not in thermal equilibrium, and internal stresses may be sufficient to cause fracture. Thermal shock can be liquid to liquid or air to air, with the most extreme exposure occurring when the substrate is transferred directly from one liquid bath to another. The heat is more rapidly absorbed or transmitted, depending on the relative tempera-

Table 11.4 Knoop Hardness for Selected Ceramics

Material	Knoop hardness (100 g)
Diamond	7000
Aluminum oxide	2100
Aluminum nitride	1200
Beryllium oxide	1200
Boron nitride	5000
Silicon carbide	2500

ture of the bath, due to the higher specific of the liquid as opposed to air.

The ability of a substrate to withstand thermal shock is a function of several variables, including the thermal conductivity, the coefficient of thermal expansion, and the specific heat. Winkleman and Schott[5] developed a parameter, called the *coefficient of thermal endurance,* that qualitatively measures the ability of a substrate to withstand thermal stress.

The coefficient of thermal endurance for selected materials is shown in Table 11.5. The phenomenally high coefficient of thermal endurance for BN is primarily a result of the high tensile strength to modulus of elasticity ratio as compared to other materials. Diamond is also high, primarily due to the high tensile strength, the high thermal conductivity, and the low TCE.

Table 11.5 Thermal Endurance Factor for Selected Materials at 25°C

Material	Thermal endurance factor
Alumina (99%)	0.640
Alumina (96%)	0.234
Beryllia (99.5%)	0.225
Boron nitride (a-axis)	648
Aluminum nitride	2.325
Silicon carbide	1.40
Diamond (type IIa)	30.29

The thermal endurance factor is a function of temperature in that several of the variables, particularly the thermal conductivity and the specific heat, are functions of temperature. From Table 11.5, it is also noted that the thermal endurance factor may drop rapidly as the alumina to glass ratio drops. This is due to the difference in the thermal conductivity and TCE of the alumina and glass constituents that increase the internal stresses. This is true of other materials as well.

11.5 Electrical Properties of Ceramics

The electrical properties of ceramic substrates perform an important task in the operation of electronic circuits. Depending on the applications, the electrical parameters may be advantageous or detrimental to circuit function. Of most interest are the resistivity, the breakdown voltage, or dielectric strength, and the dielectric properties, including the dielectric constant and the loss tangent.

11.5.1 Resistivity

The electrical resistivity of a material is a measure of the ability of that material to transport charge under the influence of an applied electric field. More often, this ability is presented in the form of the electrical conductivity, the reciprocal of the resistivity as defined in Eq. (11.12).

$$\sigma = \frac{1}{\rho} \tag{11.12}$$

where σ = conductivity in siemens/unit length
ρ = resistivity in ohm-unit length

Typical values of the resistivity of selected ceramic materials are presented in Table 11.6.

11.5.2 Breakdown Voltage

The term *breakdown voltage* is very descriptive. While ceramics are normally very good insulators, the application of excessively high potentials can dislodge electrons from orbit with sufficient energy to allow them to dislodge other electrons from orbit, creating an *avalanche effect*. The result is a breakdown of the insulation properties of the material, allowing current to flow. This phenomenon is accelerated by elevated temperature, particularly when mobile ionic impurities are present.

The breakdown voltage is a function of numerous variables, including the concentration of mobile ionic impurities, grain boundaries, and

Table 11.6 Electrical Properties of Selected Ceramic Substrates

Material	Electrical resistivity (Ω–cm)	Breakdown voltage (ac kV/mm)	Dielectric constant	Loss tangent (@ 1 MHz)
Alumina(96%)				
25°C	$> 10^{14}$			
500°C	4×10^{9}		9.0	
1000°C	1×10^{6}	8.3	10.8	0.0002
Alumina(95%)				
25°C	$> 10^{14}$			
500°C	2×10^{10}	8.7	9.4	0.0001
1000°C	2×10^{6}		10.1	
Beryllia				
25°C	$> 10^{14}$	6.6	6.4	0.0001
500°C	2×10^{10}		6.9	0.0004
Aluminum nitride	$>10^{13}$	14	8.9	0.0004
Boron nitride	$>10^{14}$	61	4.1	0.0003
Silicon carbide*	$> 10^{13}$	0.7	40	0.05
Diamond (type II)	$>10^{14}$	1000	5.7	0.0006

*Depends on method of preparation; may be substantially lower.

the degree of stoichiometry. In most applications, the breakdown voltage is sufficiently high as to not be an issue. However, there are two cases where it must be a consideration:

1. At elevated temperatures created by localized power dissipation or high ambient temperature, the breakdown voltage may drop by orders of magnitude. Combined with a high potential gradient, this condition may be susceptible to breakdown.
2. The surface of most ceramics is highly "wettable" in that moisture tends to spread rapidly. Under conditions of high humidity, coupled with surface contamination, the effective breakdown voltage is much lower than the intrinsic value.

11.5.3 Dielectric Properties

Two conductors in proximity with a difference in potential have the ability to attract and store electric charge. Placing a material with dielectric properties between them enhances this effect. A dielectric material has the capability of forming electric dipoles, displacements of electric charge, internally. At the surface of the dielectric, the dipoles attract more electric charge, thus enhancing the charge storage capability, or capacitance, of the system. The relative ability of a material to attract electric charge in this manner is called the *relative dielectric constant,* or *relative permittivity,* and is usually given the symbol K. The relative permittivity of free space is 1.0 by definition, and the absolute permittivity is

$$\varepsilon_o = \text{permittivity of free space}$$

$$= \frac{1}{36\pi} \times 10^{-9} \; \frac{\text{farads}}{\text{meter}} \quad (11.13)$$

The dielectric constant of a dielectric material is therefore

$$\varepsilon = \varepsilon_o K \quad (11.14)$$

In the presence of an electric field that is changing at a high frequency, the polarity of the dipoles must change at the same rate as the polarity of the signal so as to maintain the dielectric constant at the same level. Some materials are excellent dielectrics at low frequencies, but the dielectric qualities drop off rapidly as the frequency increases.

Changing the polarity of the dipoles requires a finite amount of energy and time. The energy is dissipated as internal heat, quantified by a parameter called the *loss tangent* or *dissipation factor.* Furthermore, dielectric materials are not perfect insulators. These phenomena may be modeled as a resistor in parallel with a capacitor. The loss tangent, as expected, is a strong function of the applied frequency, increasing as the frequency increases.

In alternating-current applications, the current and voltage across an ideal capacitor are exactly 90° out of phase, with the current leading the voltage. In actuality, the resistive component causes the current to lead the voltage by an angle less than 90°. The loss tangent is a measure of the real or resistive component of the capacitor and is the tangent of the difference between 90° and the actual phase angle.

$$\text{Loss tangent} = \tan(90° - \delta) \quad (11.15)$$

where δ = phase angle between voltage and current

The loss tangent is also referred to as the *dissipation factor (DF).*

The loss tangent may also be considered as a measure of the time required for polarization. It requires a finite amount of time to change the polarity of the dipole after an alternating field is applied. The resulting phase retardation is equivalent to the time indicated by the difference in phase angles.

11.6 Ceramic Fabrication

The manner in which ceramic substrates are fabricated will dramatically affect the properties of the substrate, including surface finish, density, porosity, tolerance, camber, and cost. Certain methods offer superior qualities for one or more of these parameters at the expense of another.

It is difficult to manufacture ceramic structures as a pure material. The melting point of most ceramics is very high, as shown in Table 11.7, limiting the ability to fabricate ceramic substrates in the pure form. Most are also very hard, making them difficult to machine except by laser. For these reasons, ceramic substrates are typically mixed with fluxing and binding glasses that melt at a lower temperature and make the finished product easier to machine.

Table 11.7 Melting Points of Selected Ceramics

Material	Melting point (°C)
SiC	2700
BN	2732
AlN	2232
BeO	2570
Al_2O_3	2000

The manufacturing processes for Al_2O_3, BeO, and AlN substrates are very similar. The base material is ground into a fine powder, several microns in diameter, and mixed with various fluxing and binding glasses, including magnesia and calcia, also in the form of powders. An organic binder, along with various plasticizers, is added to the mixture, and the resultant slurry is ball-milled to remove agglomerates and to make the composition uniform.

The slurry is formed into a sheet, the so-called "green state," by one of several processes and sintered at an elevated temperature to remove the organics and to form a solid structure.

11.6.1 Tape Casting

Referring to Fig. 11.11, the slurry is dispensed onto a sheet of mylar moving on a flat surface under a knife-edge, or "doctor blade." The doctor blade is adjusted to be parallel to the mylar at a specified height to form the sheet to the desired thickness. The viscosity and composition of the material are both critical parameters. The organic binder will typically contain a dispersant to equalize colloidal forces and a plasticizer to minimize slumping after the casting process. Control of particle size and uniformity are especially important. If the particle size is too large or is nonuniform, particles may be caught under the blade, leaving tracks or voids in the sheet. If the particles are nonuniform, the density will not be uniform. The percentage of ceramic in the overall mix (percent solids) is also very critical. If the percentage is too high, the material may be too thick to dispense properly, leaving voids and gaps in the film. If the percentage is too low, the fired film may be somewhat porous.

In the tape casting process, the organic binder consists of two parts: a non-volatile organic material that acts as the main binder, and a volatile organic material, or solvent, that thins the material down to the point where it may be properly dispensed. The volatile organic material is evaporated by heat lamps or by resistive heating elements after dispensing, leaving a thin sheet with approximately the consistency of putty. The solvent vapor is normally captured and condensed for later use. In this state, the sheet can be readily cut or stamped with a stainless steel die to a particular size and shape. It is often rolled up for later use. The tape casting process produces substrates ranging in thickness from about 12 μ to over 3 mm.

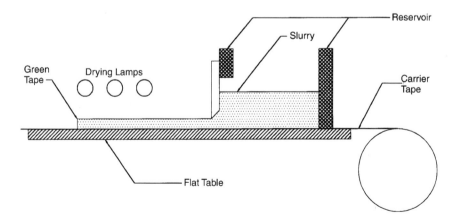

Figure 11.11 Tape casting process.

For added precision, two doctor blades may be used in sequence. The second doctor blade is set with a slightly smaller gap than the first. This operation will smooth out any irregularities due to improper dispensing from the reservoir.

11.6.2 Roll Compaction

There are two methods that use roll compaction. In the first, the slurry in the liquid form, thinner than for tape casting, is sprayed onto a flat surface and dried to remove the solvent, forming a thin sheet. The sheet is fed between a pair of large parallel rollers to form a sheet of uniform thickness. In the second, the slurry is much thicker, with only a small amount of water added to the organic binder, and is fed directly between the rollers. The resulting sheet needs very little if any drying. The tape width is typically 200–300 mm, and the thickness ranges from 0.1–3 mm. Roll compaction produces a very smooth surface but requires some maintenance on the rollers to keep them clean.

11.6.3 Powder Pressing

The powder is forced into a hard die cavity and subjected to very high pressure (up to 20,000 psi) throughout the sintering process. This produces a very dense part with tighter as-fired tolerances than other methods, although pressure variations may produce excessive warpage. This process produces substrates that have less camber than tape casting or roll compaction.

11.6.4 Hot Isostatic Powder Pressing (HIP)

This process utilizes a flexible die surrounded with gas, water, or glycerin and compressed with up to 45,000 psi during sintering. The pressure is more uniform and produces a part with less warpage. Excellent dimensional control and minimum post-processing are obtained with this process.

11.6.5 Extrusion

The slurry, less viscous than for other processes, is forced through a die. Tight tolerances and complex geometries are hard to obtain, but the process is very economical and produces a thinner part than are attainable by other methods.

11.6.6 Chemical Vapor Deposition/Infiltration

Chemical vapor deposition (CVD) may be used where thin coatings or thin substrates may be required. A substrate is placed in a vacuum chamber and heated. A gas of the proper composition is passed over the substrate and reacts to form a thin coating. Diamond, boron nitride (BN), silicon dioxide (SiO_2), silicon carbide (SiC), and silicon nitride (Si_3N_4) may be formed by this process.

Chemical vapor infiltration (CVI) is used to penetrate a preform to generate a composite material. Carbon/carbon composites are the most common applications of this process.

11.6.7 Sintering

Once the part is formed and punched, it is sintered at a temperature above the glass melting point to produce a continuous structure. The temperature profile is very critical, and the process may actually be performed in two stages; one stage to remove the nonvolatile organic materials and a second stage to sinter the glass/ceramic structure. The peak temperature may be as high as several thousand degrees centigrade and may be held for several hours, depending on the material and the type and amount of binding glasses. For example, pure alumina substrates formed by powder processing with no glasses are sintered at 1930°C.

It is essential that all the organic material be removed prior to sintering. Otherwise, the gases formed by the organic decomposition may leave serious voids in the ceramic structure and cause serious weakening. The oxide ceramics may be sintered in air. In fact, it is desirable to have an oxidizing atmosphere to aid in removing the organic materials by allowing them to react with the oxygen to form CO_2. The nitride ceramics must be sintered in the presence of nitrogen to prevent oxides of the metal from being formed. In this case, no reaction of the organics takes place; they are evaporated and carried away by the nitrogen flow.

During sintering, a degree of shrinkage takes place as the organic is removed and the fluxing glasses activate. Shrinkage may be as low as 10 percent for powder processing to as high as 22 percent for sheet casting. The degree of shrinkage is highly predictable and may be considered during design.

Boron nitride substrates are generally formed by powder pressing. Various silica and/or calcium compounds may be added to lower the processing temperature and improve machinability. Diamond substrates are typically formed by chemical vapor deposition (CVD). Composite substrates, such as AlSiC, are fabricated by creating a spongy structure of SiC and forcing molten aluminum into the crevices.

11.7 Ceramic Materials

The characteristics of various substrate materials have been summarized in previous sections. However, there may be substantial variations in the parameters due to processing, composition, stoichiometry, or other factors. This section covers the materials in more detail and also describes some common applications.

11.7.1 Aluminum Oxide

Aluminum oxide, Al_2O_3, commonly referred to as *alumina*, is by far the most common substrate material used in the microelectronics industry, because it is superior to most other oxide ceramics in mechanical, thermal, and electrical properties. The raw materials are plentiful, low in cost, and amenable to fabrication by a wide variety of techniques into a wide variety of shapes.

Alumina is hexagonal close-packed with a corundum structure. Several metastable structures exist, but they all ultimately irreversibly transform to the hexagonal alpha phase.

Alumina is stable in both oxidizing and reducing atmospheres up to 1925°C.

Weight loss in vacuum over the temperature range 1700 to 2000°C ranges from 10^{-7} to 10^{-6} g/cm^2-s. It is resistant to attack by all gases except wet fluorine to at least 1700°C. Alumina is attacked at elevated temperatures by alkali metal vapors and halogen acids, especially the lower-purity alumina compositions that may contain a percentage of glasses.

Alumina is used extensively in the microelectronics industry as a substrate material for thick and thin film circuits, for circuit packages, and as multilayer structures for multichip modules. Compositions exist for both high- and low-temperature processing. High-temperature cofired ceramics (HTCC) use a refractory metal, such as tungsten or molybdenum/manganese, as a conductor and fire at about 1800°C. The circuits are formed as separate layers, laminated together, and fired as a unit. Low-temperature cofired ceramics (LTCCs) use conventional gold or palladium silver as conductors and fire as low as 850°C. Certain power MOSFETs and bipolar transistors are mounted on alumina substrates to act as electrical insulators and thermal conductors.

The parameters of alumina are summarized in Table 11.8.

11.7.2 Beryllium Oxide

Beryllium oxide (BeO, beryllia) is cubic close-packed and has a zinc blende structure. The alpha form of BeO is stable to above 2050°C. BeO is stable in dry atmospheres and is inert to most materials. It hydrolyzes at temperatures greater than 1100°C with the formation and volatilization of beryllium hydroxide. BeO reacts with graphite at high temperature, forming beryllium carbide.

Table 11.8 Typical Parameters of Aluminum Oxide

Parameter	Units	Test	Percentage (%)			
			85	90	96	99.5
Density	g/cm^3	ASTM C20	3.40	3.60	3.92	3.98
Elastic modulus	GPa	ASTM C848	220	275	344	370
Poisson's ratio		ASTM C773	0.22	0.22	0.22	0.22
Compressive strength	MPa	ASTM C773	1930	2150	2260	2600
Fracture toughness	Mpa-m$^{0.5}$	Notched beam	3.1	3.3	3.7	4.6
Thermal conductivity	W/m-°C	ASTM C408	16	16.7	24.7	31.0
TCE	10^{-6}/°C	ASTM C372	5.9	6.2	6.5	6.8
Specific heat	W-s/g-°C	ASTM E1269	920	920	880	880
Dielectric strength	ac kV/mm	ASTMD116	8.3	8.3	8.3	8.7
Loss tangent (1 MHz)		ASTM D2520	0.0009	0.0004	0.0002	0.0001
Volume resistivity	25°C 500°C 1000°C Ω-cm	ASTM D1829	>10^{14} 4 × 10^8	>10^{14} 4 × 10^8 5 × 10^5	>10^{14} 4 × 10^9 5 × 10^6	>10^{14} 2 × 10^{10} 2 × 10^6

Beryllia has an extremely high thermal conductivity, higher than aluminum metal, and is widely used in applications where this parameter is critical. The thermal conductivity drops rapidly above 300°C but is suitable for most practical applications.

Beryllia is available in a wide variety of geometries formed using a variety of fabrication techniques. While beryllia in the pure form is perfectly safe, care must be taken when machining BeO, however, as the dust is toxic if inhaled.

Beryllia may be metallized with thick film, thin film, or by one of the copper processes. However, thick film pastes must be specially formulated to be compatible. Laser or abrasive trimming of BeO must be performed in the presence of a vacuum to remove the dust.

The properties of 99.5 percent beryllia are summarized in Table 11.9.

11.7.3 Aluminum Nitride

Aluminum nitride is covalently bonded with a wurtzite structure and decomposes at 2300°C under 1 atm of argon. In a nitrogen atmosphere of 1500 psi, melting may occur in excess of 2700°C. Oxidation of AlN in

Table 11.9 Typical Parameters for 99.5% Beryllium Oxide

Parameter	Units	Value
Density	g/cm^3	2.87
Hardness	Knoop 100 g	1200
Melting point	°C	2570
Modulus of elasticity	GPa	345
Compressive strength	MPa	1550
Poisson's ratio		0.26
Thermal conductivity 25°C 500°C	W/m-K	 250 55
Specific heat 25°C 500°C	W-s/g-K	 1.05 1.85
TCE	10^{-6}/K	7.5
Dielectric constant 1 MHz 10 GHz		 6.5 6.6
Loss tangent 1 MHz 10 GHz		 0.0004 0.0004
Volume resistivity 25°C 500°C	Ω-cm	 >10^{14} 2 × 10^{10}

even a low concentration of oxygen (<0.1 percent) occurs at temperatures above 700°C. A layer of aluminum oxide protects the nitride to a temperature of 1370°C, above which the protective layer cracks, allowing oxidation to continue. Aluminum nitride is not appreciably affected by hydrogen, steam, or oxides of carbon to 980°C. It dissolves slowly in mineral acids and decomposes slowly in water, It is compatible with aluminum to 1980°C, gallium to 1300°C, iron or nickel to 1400°C, and molybdenum to 1200°C.

Aluminum nitride substrates are fabricated by mixing AlN powder with compatible glass powders containing additives such as CaO and Y_2O_3, along with organic binders and casting the mixture into the desired shape. Densification of the AlN requires very tight control of both atmosphere and temperature. The solvents used in the preparation of substrates must be anhydrous to minimize oxidation of the AlN powder and prevent the generation of ammonia during firing.[6] For maximum densification and maximum thermal conductivity, the substrates must be sintered in a dry reducing atmosphere to minimize oxidation.

Aluminum nitride is primarily noted for two very important properties: a high thermal conductivity and a TCE closely matching that of silicon. There are several grades of aluminum nitride available, with different thermal conductivities. The prime reason is the oxygen content of the material. It is important to note that even a thin surface layer of oxidation on a fraction of the particles can adversely affect the thermal conductivity. Only with a high degree of material and process control can AlN substrates be made consistent.

The thermal conductivity of AlN does not vary as widely with temperature as that of BeO. Considering the highest grade of AlN, the crossover temperature is about 20°C. Above this temperature, the thermal conductivity of AlN is higher; below 20°C, BeO is higher.

The TCE of AlN closely matches that of silicon, an important consideration when mounting large power devices. The second level of packaging is also critical. If an aluminum nitride substrate is mounted directly to a package with a much higher TCE, such as copper, the result can be worse than if a substrate with an intermediate, although higher, TCE were used. The large difference in TCE builds up stresses during the mounting operation that can be sufficient to fracture the die and/or the substrate.

Thick film, thin film, and copper metallization processes are available for aluminum nitride. Certain of these processes, such as *direct bond copper (DBC)* require oxidation of the surface to promote adhesion. For maximum thermal conductivity, a metallization process should be selected that bonds directly to AlN to eliminate the relatively high thermal resistance of the oxide layer.

Thick film materials must be formulated to adhere to AlN. The lead oxides prevalent in thick film pastes that are designed for alumina and beryllia oxidize AlN rapidly, causing blistering and a loss of adhesion. Thick film resistor materials are primarily based on RuO_2 and MnO_2

Thin film processes available for AlN include NiCr/Ni/Au, Ti/Pt/Au, and Ti/Ni/Au.[7] Titanium in particular provides excellent adhesion by diffusing into the surface of the AlN. Platinum and nickel are transition layers to promote gold adhesion. Solders, such as Sn60/Pb40 and Au80/Sn20, can also be evaporated onto the substrate to facilitate soldering.

Multilayer circuits can be fabricated with W or MbMn conductors. The top layer is plated with nickel and gold to promote solderability and bondability. Ultrasonic milling may be used for cavities, blind vias, and through vias. Laser machining is suitable for through vias as well.

Direct bond copper may be attached to AlN by forming a layer of oxide over the substrate surface, which may require several hours at temperatures above 900°C. The DBC forms a eutectic with aluminum oxide at about 963°C. The layer of oxide, however, increases the thermal resistance by a significant amount, partially negating the high thermal conductivity of the aluminum nitride. Copper foil may also be brazed to AlN with one of the compatible braze compounds. Active metal brazing (AMB) does not generate an oxide layer. The copper may also be plated with nickel and gold.

The properties of aluminum nitride are summarized in Table 11.10.

Table 11.10 Typical Parameters for Aluminum Nitride (Highest Grade)

Parameter	Units	Value
Density	g/cm^3	3.27
Hardness	Knoop 100 g	1200
Melting point	°C	2232
Modulus of elasticity	GPa	300
Compressive strength	MPa	2000
Poisson's ratio		0.23
Thermal conductivity 25°C 150°C	W/m-K	 270 195
Specific heat 25°C 150°C	W-s/g-K	 0.76 0.94
TCE	10^{-6}	4.4
Dielectric constant 1 MHz 10 GHz		 8.9 9.0
Loss tangent 1 MHz 10 MHz		 0.0004 0.0004
Volume resistivity 25°C 500°C	Ω-cm	 >10^{12} 2×10^8

11.7.4 Diamond

Diamond substrates are primarily grown by chemical vapor deposition (CVD). In this process, a carbon-based gas is passed over a solid surface and activated by a plasma, a heated filament, or by a combustion flame. The surface must be maintained at a high temperature, above 700°C, to sustain the reaction. The gas is typically a mixture of methane (CH_4) and hydrogen (H_2) in a ratio of 1–2 percent CH_4 by volume.[8] The consistency of the film in terms of the ratio of diamond to graphite is inversely proportional to the growth rate of the film. Films produced by plasma have a growth rate of 0.1–10 μ/hr and are very high quality, while films produced by combustion methods have a growth rate of 100–1000 μ/hr and are of lesser quality.

The growth begins at nucleation sites and is columnar in nature, growing faster in the normal direction than in the lateral direction. Eventually, the columns grow together to form a polycrystalline structure with microcavities spread throughout the film. The resulting substrate is somewhat rough, with a 2–5 μ surface. This feature is detrimental to the effective thermal conductivity, and the surface must be polished for optimal results. An alternative method is to use an organic filler[9] on the surface for planarization. This process has been shown to have a negligible effect on the overall thermal conductivity from the bulk, and it dramatically improves heat transfer. Substrates as large as 10 cm^2 and as thick as 1000 μ have been fabricated.

Diamond can be deposited as a coating on refractory metals, oxides, nitrides, and carbides. For maximum adhesion, the surface should be a carbide-forming material with a low TCE.[10]

Diamond has an extremely high thermal conductivity, several times that of the next highest material. The primary application is, obviously, in packaging power devices. Diamond has a low specific heat, however, and works best as a heat spreader in conjunction with a heat sink. For maximum effectiveness,[10]

$$t_D = 0.5 \rightarrow 1 \times r_h$$
$$r_D = 3 \times r_h \qquad (11.16)$$

where t_D = thickness of diamond substrate
r_h = radius of heat source
r_D = radius of diamond substrate

Applications of diamond substrates include heat sinks for laser diodes and laser diode arrays. The low dielectric constant of diamond coupled with the high thermal conductivity makes it attractive for microwave circuits as well. As improved methods of fabrication lower the cost, the use of diamond substrates is expected to expand rapidly. The properties of diamond are summarized in Table 11.11.

Table 11.11 Typical Parameters for CVD Diamond

Parameter	Units	Value
Density	g/cm^3	3.52
Hardness	Knoop 100 g	7000
Modulus of elasticity	GPa	1000
Compressive strength	MPa	11000
Poisson's ratio		0.148
Thermal conductivity Normal Tangential	W/m-K	 2200 1610
Specific heat 25°C 150°C	W-s/g-K	 0.55 0.90
TCE	10^{-6}/K	1.02
Dielectric constant 1 MHz 10 GHz		 5.6 5.6
Loss tangent 1 MHz 10 MHz		 0.001 0.001
Volume resistivity 25°C 500°C	Ω-cm	 >10^{13} 2 × 10^{11}

11.7.5 Boron Nitride

There are two basic types of boron nitride (BN). Hexagonal (alpha) BN is soft and is structurally similar to graphite. It is white in color and is sometimes called *white graphite*. Cubic (beta) BN is formed by subjecting hexagonal BN to extreme heat and pressure, similar to the process used to fabricate synthetic industrial diamonds. Melting of either phase is possible only under nitrogen at high pressure.

Hot-pressed BN is very pure (>99 percent), with the major impurity being boric oxide (BO). Boric oxide tends to hydrolyze in water, degrading the dielectric and thermal shock properties. Calcium oxide (CaO) is frequently added to tie up the BO to minimize the water absorption. When exposed to temperatures above 1100°C, BO forms a thin coating on the surface, slowing further oxide growth.

Boron nitride in the hot-pressed state is easily machinable and may be formed into various shapes. The properties are highly anisotropic and vary considerable in the normal and tangential directions of the pressing force. The thermal conductivity in the normal direction is very high, and the TCE is very low, making BN an attractive possibility for a substrate material. However, it has not yet been proven possible to metallize BN,[11] thereby limiting the range of applications. It can be used in contact with various metals, including, copper, tin, and aluminum, and may be used as a thermally conductive electrical insulator. Applications of BN include microwave tubes and crucibles.

The properties of boron nitride are summarized in Table 11.12.

Table 11.12 Typical Parameters for Boron Nitride

Parameter	Units	Value
Density	g/cm^3	1.92
Hardness	Knoop 100g	5000
Modulus of elasticity	GPa	
Normal		43
Tangential		768
Compressive strength	MPa	
Normal		110
Tangential		793
Poisson's ratio		0.05
Thermal conductivity	W/m-K	
Normal		73
Tangential		161
Specific heat	W-s/g-K	
25°C		084
150°C		1.08
TCE	10^{-6}/K	
Normal		0.57
Tangential		−0.46
Dielectric constant		
1MHz		4.1
Loss tangent		0.0003
1 MHz		
Volume resistivity	Ω-cm	
25°C		1.6×10^{12}
500°C		2×10^{10}

11.7.6 Silicon Carbide

Silicon carbide (SiC) has a tetrahedral structure and is the only known alloy of silicon and carbon. Both elements have four electrons in the outer shell, with an atom of one bonded to four atoms of the other. The result is a very stable structure, not affected by hydrogen or nitrogen up to 1600°C. In air, SiC begins decomposing above 1000°C. As with other compounds, a protective oxide layer forms over the silicon, reducing the rate of decomposition. Silicon carbide is highly resistant to both acids and bases. Even the so-called "white etch" (hydrofluoric acid mixed with nitric and sulfuric acids) has no effect.

Silicon carbide structures are formed by hot pressing, dry and isostatic pressing (preferred), by CVD, or by slip casting. Isostatic pressing using gas as the fluid provides optimum mechanical properties.

Silicon carbide in pure form is a semiconductor, and the resistivity depends on the impurity concentration. In the intrinsic form, the resistivity is less than 1000 Ω-cm, which is unsuitable for ordinary use. The addition of a small percentage (<1 percent) of BeO during the fabrication process[12] increases the resistivity to as high as 10^{13} Ω-cm by creating carrier-depleted layers around the grain boundaries.

Both thick and thin films can be used to metallize SiC, although some machining of the surface to attain a higher degree of smoothness is necessary for optimum results. The two parameters that make SiC attractive as a substrate are the exceptionally high thermal conductivity, second only to diamond, and the low TCE, which matches that of silicon to a higher degree than any other ceramic. SiC is also less expensive than either BeO or AlN. A possible disadvantage is the high dielectric constant, 4 to 5 times higher than other substrate materials. This parameter can result in cross-coupling of electronic signals or in excessive transmission delay.

The parameters of SiC are summarized in Table 11.13.

11.8 Composite Materials

Composite materials have been in use for thousands of years. Ancient peoples used straw and rocks to strengthen adobe bricks for building shelters, and concrete reinforced with steel rods is used extensively in the construction industry. Only recently have composites been introduced into electronics applications. In particular, the demands for improved thermal management and higher packaging density have driven the development of composite materials. Today, several composite materials are available to meet the needs of electronic packaging engineers.

A two-component composite consists of a matrix and a filler, which may be in the form of long or short fibers, large or small particles, or

Table 11.13 Typical Parameters for Silicon Carbide

Parameter	Units	Value
Density	g/cm^3	3.10
Hardness	Knoop 100 g	500
Modulus of elasticity	GPa	407
Compressive strength	MPa	4400
Fracture toughness	Mpa-m$^{1/2}$	7.0
Poisson's ratio		0.14
Thermal conductivity 25°C 150°C	W/m-K	 290 160
Specific heat 25°C 150°C	W-s/g-K	 0.64 0.92
TCE	10^{-6}/K	3.70
Dielectric constant		40
Loss tangent		0.05
Volume resistivity 25°C 500°C	Ω–cm	 >10^{13} 2 × 10^9

as laminates. The theory of composite materials is well documented. For this discussion, it is sufficient to simply describe composites by the so-called "rule of mixtures," stated as

$$P_C = P_M V_{fM} + P_F V_{fF} \tag{11.17}$$

where P_C = parameter of the composite
P_M = parameter of the matrix
V_{fM} = volume fraction of the matrix
P_F = parameter of the filler
V_{fF} = volume fraction of the filler

The degree to which the composite will obey the rule of mixtures depends on a number of factors, including the size and orientation of the filler and the extent of the interaction between the matrix and the filler. Consider the structure in Fig. 11.12, consisting of continuous fibers uniformly dispersed in a matrix. Assuming that no slippage be-

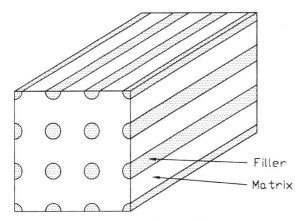

Figure 11.12 Matrix with continuous filler.

tween the two occurs, the strain, E, as defined in Eq. (11.6), is the same for both materials when a longitudinal force is applied. For this case,

$$E_C = E_M = E_F = \frac{\Delta L}{L_0} \tag{11.18}$$

where E_C = strain in composite
E_M = strain in matrix
E_F = strain in filler
ΔL = change in length
L_0 = initial length

A structure of this type will closely approximate the rule of mixtures. Where the filler is not uniformly distributed as in Fig. 11.12, a critical length may be defined as

$$L_C = \frac{D_F S_{MAX}}{2\sigma_M} \tag{11.19}$$

where L_C = critical length
D_F = diameter of filler
S_{MAX} = maximum shear strength of filler
σ_M = shear strength of matrix

To be effective, the filler must be greater than the critical length.

It is possible to tailor composite materials to obtain desirable properties superior to either component for specific applications. Composite materials may be divided into several categories based on the nature of the matrix and the filler.

11.8.1 Organic–Organic Composites

Composites composed of organic materials typically use an epoxy as the matrix and a variety of organic fibers as the filler. The epoxy may be in the liquid or powder form. For example, nylon reinforced plastics are commonly used to mold plastic parts for a number of applications.

11.8.2 Organic–Ceramic/Glass Composites

Organic–ceramic composites may use an epoxy as the matrix and glass or ceramic powder as the filler. A common example is the fiberglass-reinforced epoxy used as a printed circuit laminate. An epoxy substrate filled with alumina and carbon black has also been developed.[13] By weight, the composition is 10.8 percent epoxy resin, 89 percent alumina, and 0.2 percent carbon black. This material has a thermal conductivity of 3.0–4.0 W/m–K as compared to both glass epoxy printed circuit material (0.2 W/m–K) and glass-alumina low-temperature cofired substrates (2.5 W/m–K). The TCE (17 ppm) is substantially below that of PCB material (25–50 ppm). This composite can be utilized to make multilayer interconnection structures.

Organic–ceramic composites are also commonly used as materials for housings for electronic systems. Glass-reinforced plastics, for example, are very strong and are capable of withstanding considerable mechanical shock without breaking.

11.8.3 Ceramic–Ceramic Composites

Ceramic-based composites are somewhat difficult to fabricate due to their high melting point. Lawrence Berkeley National Laboratory[14] has developed a method of coating platelets of silicon carbide with alumina and forming them into a composite matrix that has exceptional fracture strength and fracture toughness. This approach was demonstrated by coating alpha SiC platelets with about 2 μ of alumina and combining them with a matrix of beta SiC. The coating prevents the alpha SiC from being assimilated into the beta SiC during processing and diverts cracks to the perimeter of the particles rather than through the body. The result is a SiC–SiC composite ceramic that is 2–3 times stronger than conventional SiC.

11.8.4 Ceramic–Glass Composites

In all but the simplest electronic circuits, it is necessary to have a method for fabricating multilayer interconnection structures to enable all the necessary points to be connected. The thick film technology is limited to three layers for all practical purposes, due to yield and planarity considerations, and thin film multilayer circuits are quite expensive to fabricate. The copper technologies are limited to a single layer due to processing limitations.

Ceramic–glass composite materials may be used to economically fabricate very complex multilayer interconnection structures. The materials in powder form are mixed with an organic binder, a plasticizer, and a solvent and formed into a slurry by ball or roll milling. The slurry is forced under a doctor blade and dried to form a thin sheet, referred to as *green tape* or *greensheet*. Further processing is dependent on the type of material. There are three basic classes of materials: high-temperature cofired ceramic (HTCC), low-temperature cofired ceramic (LTCC), and aluminum nitride.

High-temperature cofired ceramic (HTCC). HTCC multilayer circuits are primarily alumina based. The green tape is blanked into sheets of uniform size and holes are punched where vias and alignment holes are required. The metal patterns are printed and dried next. Despite their relatively high electrical resistance, refractory metals such as tungsten and molybdenum are used as conductors due to the high firing temperature. Via fills may be accomplished during conductor printing or during a separate printing operation. The process is repeated for each layer.

The individual layers are aligned and laminated under heat and pressure to form a monolithic structure in preparation for firing. The structure is heated to approximately 600°C to remove the organic materials. Carbon residue is removed by heating to approximately 1200°C in a wet hydrogen atmosphere. Sintering and densification take place at approximately 1600°C.

During firing, HTCC circuits shrink anywhere from 14 to 17 percent, depending on the organic content. With careful control of the material properties and processing parameters, the shrinkage can be controlled to within 0.1 percent. Shrinkage must be taken into consideration during the design, punching, and printing processes. The artwork enlargement must exactly match the shrinkage factor associated with a particular lot of green tape.

Processing of the substrate is completed by plating the outer layers with nickel and gold for component mounting and wire bonding. The gold is plated to a thickness of 25 μin for gold wire and 5 μin for aluminum wire. Gold wire bonds to the gold plating, while aluminum wire bonds to the nickel underneath. The gold plating in this instance is simply to protect the nickel surface from oxidation or corrosion.

The properties of HTCC materials are summarized in Table 11.14.

Table 11.14 Properties of Multilayer Ceramic Materials

	Low-temperature cofired ceramic (LTCC)	High-temperature cofired ceramic (HTCC)	Aluminum nitride
Material	Cordierite MgO, SiO_2, Al_2O_3 Glass-filled composites SiO_2, B_2O_3, Al_2O_3 PbO, SiO_2, CaO, Al_2O_3 Crystalline phase ceramics Al_2O_3, CaO, SiO_2, MgO, B_2O_3	88–92% alumina	AlN, yttria, CaO
Firing temperature	850–1050°C	1500–1600°C	1600–1800°C
Conductors	Au, Ag, Cu, PdAg	W, MoMn	W, MoMn
Conductor resistance	3–20 mΩ/☐	8–12 mΩ/☐	8–12 mΩ/☐
Dissipation factor	$15–30 \times 10^{-4}$	$5–15 \times 10^{-4}$	$20–30 \times 10^{-3}$
Relative dielectric constant	5–8	9–10	8–9
Resistor values	0.1 Ω–1 MΩ	n/a	n/a
Firing shrinkage X, Y, Z	12.0 ± 0.1% 17.0 ± 0.5%	12–18% 12–18%	15–20% 15–20%
Repeatability	0.3–1%	0.3–1%	0.3–1%
Line width	100 μm	100 μm	100 μm
Via diameter	125 μm	125 μm	125 μm
Number of metal layers	33	63	8
CTE	3–8 ppm/°C	6.5 ppm/°C	4.4 ppm/°C
Thermal conductivity	2–6 W/m·°C	15–20 W/m·°C	180–200 W/m·°C

Low-temperature cofired ceramic (LTCC). LTCC circuits consist of alumina mixed with glasses with the capability to simultaneously sinter and crystallize.[15] These structures are often referred to as *glass-ceramics*. Typical glasses are listed in Table 11.15. The processing steps are similar to those used to fabricate HTCC circuits with two exceptions;[15] the firing temperature is much lower, 850 to 1050°C, and the metalli-

Table 11.15 Typical Parameters for AlSiC (70% by Volume), SiC, and Aluminum

Parameter	Units	AlSiC	SiC	Al
Density	g/cm^3	3.02	3.10	2.70
Modulus of elasticity	GPa	224	407	69
Tensile strength	Mpa	192	*	55
Thermal conductivity (25°C)	W/m-K	218	290	237
TCE	10^{-6}/K	7.0	3.70	23
Volume resistivity (25°C)	µΩ-cm	34	>10^{13}	2.8

*Depends on method of preparation and number/size of defects.

zation is gold-based or silver-based thick film formulated to be compatible with the LTCC material. Frequently, silver-based materials are used in the inner layers with gold on the outside for economic reasons. Special via fill materials are used between the gold and the silver layers to prevent electrolytic reaction.

The shrinkage of LTCC circuits during firing is in the range of 12 to 18 percent. If the edges are restrained, the lateral shrinkage can be held to 0.1 percent with a corresponding increase in the vertical shrinkage.

One advantage that LTCC has over the other multilayer technologies is the ability to print and fire resistors. Where trimming is not required, the resistors can be buried in intermediate layers with a corresponding saving of space. It is also possible to bury printed capacitors of small value.

Aluminum nitride. Aluminum nitride multilayer circuits are formed by combining AlN powder with yttrium oxide (yttria) or calcium oxide (calcia).[16] Glass may also be added. Sintering may be accomplished in three ways;

1. Hot pressing during sintering

2. High-temperature (>1800°C) sintering without pressure

3. Low-temperature (<1650°C) sintering without pressure

Tungsten or molybdenum pastes are used to withstand the high firing temperatures. Control of the processing parameters during sintering is critical if optimal properties are to be attained. A carbon atmosphere helps in attaining a high thermal conductivity by preventing oxidation of the AlN particles. Shrinkage is in the range of 15 to 20 percent.

The properties of aluminum nitride multilayer materials are summarized in Table 11.14.

11.8.5 Metal–Ceramic Composites

Metal–ceramic composites are primarily formed by one of two methods. One is to form *in situ,* whereby the filler and the matrix are mixed together and fused together by a combination of heat and pressure. A more common approach is to form the filler into a porous shell and subsequently fill the shell with the matrix material in the molten state.

Ceramics typically have a low thermal conductivity and a low TCE, while metals have a high thermal conductivity and a high TCE. It is a logical step to combine these properties to obtain a material with a high thermal conductivity and a low TCE. The ceramic in the form of particles or continuous fibers is mixed with the metal to form a structure with the desirable properties of both. The resultant material is referred to as a *metal matrix composite (MMC)*. MMCs are primarily used in applications where thermal management is critical.

The most common metals used in this application are aluminum and copper, with aluminum being more common due to lower cost. Fillers include SiC, AlN, BeO, graphite, and diamond. Compatibility of the materials is a prime consideration. Graphite, for example, has an electrolytic reaction with aluminum but not with copper.[13]

Two examples will be described here: AlSiC, a composite made up of aluminum and silicon carbide, and Dymalloy®, a combination of copper and diamond.

Aluminum silicon carbide (AlSiC) is produced by forcing liquid aluminum into a porous SiC preform. The preform is made by any of the common ceramic processing technologies, including dry pressing, slip molding, and tape casting. The size and shape of the preform is selected to provide the desired volume fraction of SiC. The resulting combination has a thermal conductivity almost as high as that of pure aluminum, with a TCE as low as 6.1 ppm/°C. AlSiC is also electrically conductive, prohibiting its use as a conventional substrate.

The mechanical properties of the composite are determined by the ratio of SiC to aluminum as shown in Fig. 11.13.[18] A ratio of 70 to 73 percent of SiC by volume provides the most optimal properties for electronic packaging.[18] This ratio gives a TCE of about 6.5 ppm/°C, which closely matches that of alumina and beryllia. This allows AlSiC to be used as a baseplate for ceramic substrate, using its high thermal conductivity to maximum advantage.

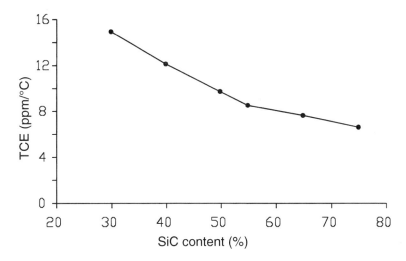

Figure 11.13 TCE vs. SiC content for AlSiC.

AlSiC, being electrically conductive, may be readily plated with aluminum to provide a surface for further processing. The aluminum coating may be plated with nickel and gold to permit soldering or may be anodized where an insulating surface is required.[17] An alternate approach is to flame-spray the AlSiC with various silver alloys for solderability.

Two other advantages of AlSiC are strength and weight. The aluminum is somewhat softer than SiC and reduces the propagation of cracks. The density is only about one-third that of Kovar®, and the thermal conductivity is >12 times greater.

AlSiC has been used to advantage in the fabrication of hermetic single chip and multichip packages and as heat sinks for power devices and circuits. While difficult to drill and machine, AlSiC can be formed into a variety of shapes in the powder state. It has been successfully integrated with patterned AlN to form a power module package.[18]

The TCE linearly increases with temperature up to about 350°C and then begins to decrease. At this temperature, the aluminum matrix softens, and the SiC matrix dominates. This factor is an important feature for power packaging.

The parameters of AlSiC are summarized in Table 11.15, along with those of aluminum and silicon carbide. Table 11.15 provides an interesting comparison of how the properties of a composite compare to those of the constituents.

Dymalloy is a matrix of Type I diamond and Cu20/Ag80 alloy.[19] The diamond is ground into a powder in the 6 to 50 mm range. The powder is coated with W74/Rh26 to form a carbide layer approximately 100 Å thick followed by a 1000 Å coating of copper. The copper is plated to a thickness of several mm to permit brazing.

The powder is packed into a form and filled in a vacuum with Cu20/Ag80 alloy that melts at approximately 800°C. This material is selected over pure copper, which melts at a much higher temperature, to minimize graphization of the diamond. The diamond loading is approximately 55 percent by volume.

The parameters of Dymalloy are summarized in Table 11.16.

Table 11.16 Typical Parameters for Dymalloy (55% by Volume)

Parameter	Units	AlSiC
Density	g/cm^3	6.4
Tensile strength	Mpa	400
Specific heat*	W-s/g-°C	$0.3165 + 8.372 \times 10^{-4}$ T
Thermal conductivity	W/m-K	360
TCE†	10–6/K	$5.48 + 6.5 \times 10^{-3}$ T

*Temperature in °C from 25 to 75°C.
†Temperature in °C from 25 to 200°C.

11.9 Forming Ceramics and Composites to Shape

Ceramic substrate materials are typically hard and brittle, making them difficult to machine by conventional means. Sawing and drilling are virtually impossible to accomplish, while grinding often results in surface voids. Certain composites, on the other hand, are relatively easy to machine. The high metal content of these structures permits sawing and drilling to shape.

The most common methods of forming ceramics to shape are punching in the green state and laser scribing. The ceramic in the green state is formed to the proper thickness by one of the described methods and punched to shape with a carbide die in the desired shape or drilled with a carbide-tipped drill. During the firing process, the ceramic may shrink anywhere from 10 to 22 percent, depending primarily on the organic content, the ratio of glass to ceramic, the particle size distribution, and the particle shape distribution. The amount of shrinkage during firing is highly predictable and consistent, and may be compensated for during the design process. Typical tolerances after firing are typically 1 to 5 percent.

Ceramic substrates may also be formed to shape after firing with a CO_2 laser. Holes and patterns may be created to a high degree of tolerance, typically 1 percent. It is common to partially scribe a groove in the substrate so that a single process may create several patterns. After processing, the circuits may readily be separated by a gentle force applied along the scribe line. The surface finish of laser scribe lines may vary greatly from the bulk of the ceramic, tending to have a higher glass content due to the nature of the laser machining process. Annealing at an elevated temperature will return the scribed surface of the substrate to the original state.[1] A diagram of a scribed substrate and typical examples are found in Figs. 11.14 and 11.15, respectively.

The surface finish of ceramic substrates may be made smoother by grinding, honing, ultrasonic machining, lapping, or polishing.[20] Grinding is accomplished by exposing the substrate surface to abrasive particles bonded to a wheel rotating at high speed. The orientation of the wheel to the substrate is obviously critical as is the particle size of the abrasive particles. The surface finish of the substrate is limited to a large extent by the size of the particles in the substrate itself. When a particle is dislodged from the substrate, as frequently happens, a pit equal to the size of the particle will be present on the substrate surface.

Honing utilizes a cylinder or other surface formed from the abrasive particles. The cylinder is moved rapidly back and forth across the surface of the substrate to form the finish. Honing provides a smoother, more accurate surface than grinding.

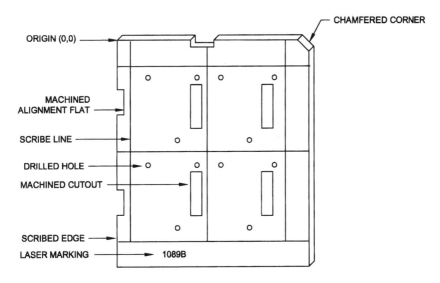

Figure 11.14 Laser scribing. *(Courtesy of Applied Laser Technology)*

Figure 11.15 Examples of laser machined substrates. *(Courtesy of Applied Laser Technology)*

In ultrasonic machining, a tool in the desired shape is formed from a soft, malleable metal. The abrasive in the form of a slurry is applied to the substrate and ultrasonic energy is applied to the tool. This process is somewhat slow, but permits round holes and other shapes to be formed in fired ceramic.

Lapping utilizes a fine abrasive powder, such as aluminum oxide or silicon carbide, suspended in a vehicle. The substrate surface is placed in contact with the powder and a weight is placed on the back of the substrate. The apparatus is placed on a rotating wheel made of cast iron that is rotated to produce a planetary rotation of the substrate/weight.

Polishing is similar to lapping in principle, except that the abrasive powder is placed in a soft material, such as felt, and the substrate is rotated against the powder. Typical powders used in this application are cerium oxide and ferrous oxide. These materials can be made into very fine powders, producing a very smooth surface.

11.10 References

1. Sergent, J., and Harper, C., *Hybrid Microelectronics Handbook,* 2nd ed., McGraw-Hill, 1995.
2. Richard Brown, Thin Film Substrates, in *Handbook of Thin Film Technology,* Leon Maissel and Reinhard Glang, Eds., McGraw-Hill, 1971.
3. Philip Garrou and Arne Knudsen, Aluminum Nitride for Microelectronic Packaging, *Advancing Microelectronics,* 21, 1, January/February 1994.

4. C. G. M. Van Kessel, S. A. Gee, and J. J. Murphy, The Quality of Die Attachment and Its Relationship to Stresses and Vertical Die-Cracking, *Proc. IEEE Components Conf., 1983.*
5. A. Winkleman and O. Schott, *Ann. Phys. Chem.*, vol. 51, 1984.
6. Ellice Y. Yuh, John W. Lau, Debra S. Horn, and William t. Minehan, Current Processing Capabilities for Multilayer Aluminum Nitride, *International Journal of Microelectronics and Electronic Packaging,* 16, 2, second quarter, 1993.
7. Nobuyiki Karamoto, Thin Film and Co-Fired Metallization on Shapal Aluminum Nitride, *Advancing Microelectronics,* 21, 1, January/February, 1994.
8. Paul W. May, CVD Diamond—A New Technology for the Future? *Endeavor Magazine,* 19, 3, 1995.
9. Ajay P. Malshe, S. Jamil, M. H. Gordon, H. A. Naseem, W. D. Brown, and L. W. Schaper, Diamond for MCMs, *Advanced Packaging,* September/October, 1995.
10. Thomas Moravec and Arjun Partha, Diamond Takes the Heat, *Advanced Packaging,* special issue, October, 1993.
11. S. Fuchs and P. Barnwell, A Review of Substrate Materials for Power Hybrid Circuits, *The IMAPS Journal of Microcircuits and Electronic Packaging,* 20, 1, first quarter, 1997.
12. Mitsuru Ura and Osami Asai, Internal Report, Hitachi Research Laboratory, Hitachi Industries, Ltd.
13. Koichi Hirano, Seiichi Nakatani, and Jun'ichi Kato, A Novel Composite Substrate with High Thermal Conductivity for CSP, MCM, and Power Modules, *Proceedings, International Microelectronic and Packaging Society,* 1998.
14. C. De Johghe, T. Mitchell, W.J. Moberly-Chan and R.O. Ritchie, Silicon Carbide Platelet/Silicon Carbide Composites, *J. Am. Cer. Soc.,* 1994.
15. Jerry E. Sergent, Materials for Multichip Modules, *Electronic Packaging and Production,* December, 1996.
16. Philip E. Garrou and Iwona Turlik, *Multichip Module Technology Handbook,* McGraw-Hill, 1998.
17. M. K. Premkumar and R. R. Sawtell, Alcoa's AlSiC Cermet Technology for Microelectronics Packaging, *Advancing Microelectronics,* July/August, 1995.
18. M. K. Premkumar and R. R. Sawtell, Aluminum-Silicon Carbide, *Advanced Packaging,* September/October 1996.
19. J. A. Kerns, N. J. Colella, D. Makowiecki, and H. L. Davidson, Dymalloy: A Composite Substrate for High Power Density Electronic Components, *International Journal of Microcircuits and Electronic Packaging,* 19, 3, third quarter, 1996.
20. Ioan D. Marinescu, Hans K. Tonshoff, and Ichiro Inasaki, *Handbook of Ceramic Grinding and Polishing,* Noyes Publications, 1998.

Chapter

12

Hybrid Microelectronics and Multichip Module Technologies

Jerry E. Sergent
Sergent & Sergent Consulting
Corbin, Kentucky

12.1 Introduction

The hybrid microelectronics technology is one branch of the electronics packaging technology and is primarily differentiated from other branches by the way that the interconnection scheme is generated. The foundation for the hybrid circuit is a substrate fabricated on one of the refractory ceramics discussed in Chap. 11. A metallization pattern is created on the substrate by one of the film technologies, forming the mounting pads and circuit traces to bond and interconnect additional active and passive devices as necessary. Another characteristic of the hybrid technology is the ability to fabricate passive components. The thick and thin film technologies, for example, can be used to manufacture resistors with parameters superior to the carbon resistors commonly used in conjunction with printed circuit boards.

The most commonly accepted definition of a hybrid circuit is a ceramic substrate metallized by one of the methods shown in Fig. 12.1, containing at least two components, one of which must be active. This definition is intended to exclude single-chip packages and circuits that contain only passive components, such as resistor networks. By this definition, a hybrid circuit may range from a simple diode-resistor logic gate to a circuit containing in excess of 100 integrated circuits (ICs).

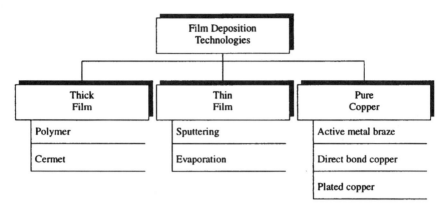

Figure 12.1 Film deposition methods.

Multichip modules (MCMs) are close relatives of hybrid circuits. While many of the assembly processes are common to those used to fabricate hybrid circuits, MCMs employ a wider range of substrate materials and metallization processes that provide a much higher packaging density.

This chapter describes the methods and properties of the various materials, metallization processes, and assembly methods used to manufacture hybrid circuits and MCMs. Also included are design guidelines, a discussion of reliability considerations, and applications of hybrid circuits and MCMs.

12.2 Thick Film Technology

Thick film circuits as depicted in Fig. 12.2 are fabricated by screen printing conductive, resistive, and insulating materials in the form of a viscous paste onto a ceramic substrate as required. The printed film is dried to remove volatile components and exposed to an elevated temperature to activate the adhesion mechanism that adheres the film to the substrate. In this manner, by depositing successive layers, as shown in Fig. 12.3, multilayer interconnection structures can be formed that may contain integrated resistors, capacitors, or inductors.

All thick film pastes have two general characteristics in common.

1. They are viscous fluids with a non-Newtonian rheology suitable for screen printing.

2. They are composed of two different multicomponent phases; a functional phase that imparts the electrical and mechanical properties to the finished film and a vehicle phase that imparts the proper rheology.

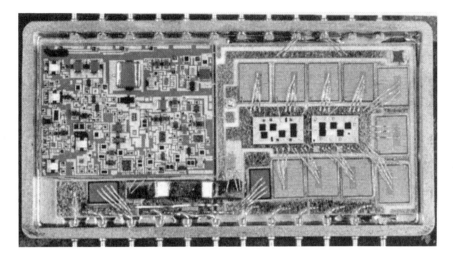

Figure 12.2 Power hybrid circuit.

There are numerous ways of categorizing thick film pastes. One such way is shown in Fig. 12.4, which depicts three basic categories: polymer thick films, refractory thick films, and cermet thick films. Refractory thick films are a special class of cermet thick films and are frequently categorized separately. These materials are designed to be fired at much higher temperatures (1500–1600°C) than conventional cermet materials and are also fired in a reducing atmosphere.

Polymer thick films consist of a mixture of polymer materials with conductor, resistor, or insulating particles, and cure at temperatures ranging from 85–300°C. Polymer conductors are primarily silver, with carbon being the most common resistor material. Polymer thick film materials are more commonly associated with organic substrate materials as opposed to ceramic and will not be considered in further detail.

Cermet thick film materials in the fired state are a combination of glass ceramic and metal and are designed to be fired in the range 850–1000°C.

A conventional cermet thick film paste has four major constituents: an active element, which establishes the function of the film, an adhesion element, which provides the adhesion to the substrate and a matrix that holds the active particles in suspension, an organic binder, which provides the proper fluid properties for screen printing, and a solvent or thinner, which establishes the viscosity of the vehicle phase.

12.2.1 The Active Element

The active element within the paste determines the electrical properties of the fired film. If the active element is a metal, the fired film will

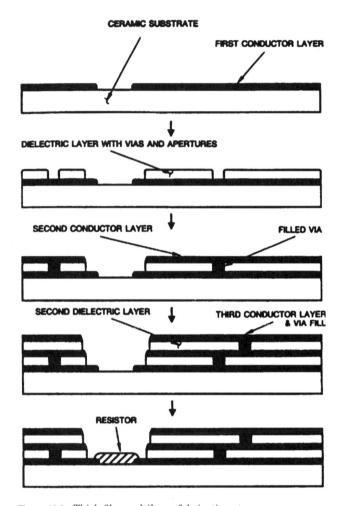

Figure 12.3 Thick film multilayer fabrication steps.

be a conductor; if it is a conductive metal oxide, a resistor; and if it is an insulator, a dielectric. The active element is most commonly found in powder form ranging from 1 to 10 µ in size, with a mean diameter of about 5 µ. Particle morphology can vary greatly depending on the method used to produce the metallic particles. Spherical, flaked, or circular shapes (both amorphous and crystalline) are available from powder manufacturing processes. Structural shape and particle morphology are critical to the development of the desired electrical performance, and extreme control over the particle shape, size, and distribution must be maintained to ensure uniformity of the properties of the fired film.

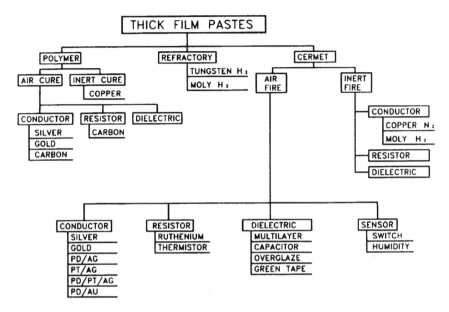

Figure 12.4 Thick film paste matrix.

12.2.2 The Adhesion Element

There are two primary constituents used to bond the film to the substrate: glass and metal oxides, which may be used singly or in combination. Films that use a glass, or frit, are referred to as *fritted* materials and have a relatively low melting point (500–600°C). There are two adhesion mechanisms associated with the fritted materials, a chemical reaction and a physical reaction. In the chemical reaction, the molten glass chemically reacts with the glass in the substrate to a degree. In the physical reaction, the glass flows into and around the irregularities in the substrate surface, flowing into holes or voids, and clinging to the small outcroppings of ceramic. The total adhesion is the sum of the two factors. The physical bonds are more susceptible to degradation by thermal cycling or thermal storage than the chemical bonds and are generally the first to fracture under stress. The glass also creates a matrix for the active particles, holding them in contact with each other to promote sintering and to provide a series of three-dimensional continuous paths from one end of the film to the other. Principal thick film glasses are based on B_2O_3-SiO_2 network formers with modifiers such as PbO, Al_2O_3, Bi_2O_3, ZnO, BaO, and CdO added to change the physical characteristics of the film, such as the melting point, the viscosity, and the coefficient of thermal expansion. Bi_2O_3 also has excellent wetting properties, both to the active element and to the substrate and is frequently used as a fluxing agent. The glass phase may be introduced as a prereacted particle or formed *in situ* by using glass precursors such as boric oxide, lead oxide, and silicon. Fritted conductor materials tend to have glass on the surface, making subsequent component assembly processes more difficult.

A second class of materials utilizes metal oxides to provide the adhesion to the substrate. In this case, a pure metal, such as copper or cadmium, is mixed with the paste and reacts with oxygen atoms on the surface of the substrate to form an oxide. The conductor adheres to the oxide and to itself by sintering, which takes place during firing. During firing, the oxides react with broken oxygen bonds on the surface of the substrate, forming a Cu or Cd spinel structure. Pastes of this type offer improved adhesion over fritted materials and are referred to as *fritless, oxide-bonded,* or *molecular-bonded* materials. Fritless materials typically fire at 950–1000°C, which is undesirable from a manufacturing aspect. Ovens used for thick film firing degrade more rapidly and need more maintenance when operated at these temperatures for long periods of time.

A third class of materials utilizes both reactive oxides and glasses. The oxides in these materials, typically Zn or Ca, react at lower temperatures but are not as strong as copper. A lesser concentration of glass than found in fritted materials is added to supplement the adhesion. These materials, referred to as *mixed bonded systems,* incorporate the advantages of both technologies and fire at a lower temperature.

12.2.3 The Organic Binder

The organic binder is generally a thixotropic fluid and serves two purposes: it holds the active and adhesion elements in suspension until the film is fired, and it gives the paste the proper fluid characteristics for screen printing. The organic binder is usually referred to as the *nonvolatile* organic, since it does not evaporate but begins to burn off at about 350°C. The binder must oxidize cleanly during firing, with no residual carbon that could contaminate the film. Typical materials used in this application are ethyl cellulose and various acrylics.

For nitrogen-fireable films, where the firing atmosphere can contain only a few ppm of oxygen, the organic vehicle must decompose and thermally depolymerize, departing as a highly volatile organic vapor in the nitrogen blanket provided as the firing atmosphere, since oxidation into CO_2 or H_2O is not feasible due to the oxidation of the copper film.

12.2.4 The Solvent or Thinner

The organic binder in the natural form is too thick to permit screen printing, which requires the use of a solvent or thinner. The thinner is somewhat more volatile than the binder, evaporating rapidly above about 100°C. Typical materials used for this application are terpineol, butyl carbitol, and certain of the complex alcohols into which the nonvolatile phase can dissolve. The low vapor pressure at room temperature is desirable to minimize drying of the pastes and to maintain a constant viscosity during printing. Additionally, plasticizers, surfactants, and agents that modify the thixotropic nature of the paste are added to the solvent to improve paste characteristics and printing performance.

To complete the formulation process, the ingredients of the thick film paste are mixed together in proper proportions and milled on a three-roller mill for a sufficient period of time to ensure that they are thoroughly mixed and that no agglomeration exists. There are three important parameters that may be used to characterize a thick film paste: fineness of grind, percent solids, and viscosity.

12.2.5 Parameters of Thick Film Paste

Fineness of grind is a measure of the particle size distribution and dispersion within the paste. A fineness-of-grind (FOG) gauge is a hard steel block with a tapered slot ground into one surface to a maximized depth, typically 50 μ, with a micrometer scale marked along the groove. The paste is placed in the deep end and drawn down the block toward the shallow end by a tapered doctor blade. At the point where the largest particles cannot pass under the gap between the groove and the doctor blade, the film will begin to form streaks, or areas with no paste. The location of the first streak with respect to the scale denotes the largest particle, and the point where approximately half of the width of the groove is composed of streaks is the mean value of the particle size. At some point, essentially all of the particles will be trapped, which represents the smallest particle.

The percent solids parameter measures the ratio of the weight of the active and adhesion elements to the total weight of the paste. This test is performed by weighing a small sample of paste, placing it in an oven at about 400°C until all the organic material is burned away, and reweighing the sample. The percent solids parameter must be tightly controlled so as to achieve the optimum balance between printability and the density of the fired film. If the percent solids content is too high, the material will not have the proper fluid characteristics to print properly. If it is too low, the material will print well, but the fired print may be somewhat porous or may lack in definition. A typical range for percent solids is 85–92 percent by weight. By volume, of course, the ratio is somewhat lower due to the lower density of the vehicle, as compared with the active and adhesion elements.

The viscosity of a fluid is a measure of the tendency of the fluid to flow and is the ratio of the shear rate of the fluid in s^{-1} to the shear stress in force/unit area. The unit of viscosity is the poise, measured in dynes/cm^2-s. Thick film pastes typically have the viscosity expressed in centipoise (cp), although the actual viscosity may be in the thousands of poise. An alternate unit of viscosity is the Pascal-s. One Pascal-s is equivalent to 0.001 centipoise.

In an ideal or "Newtonian" fluid, the relationship between shear rate and shear stress is linear, and the graph passes through the origin. Newtonian fluids are not suitable for screen printing, since the force of gravity assures that some degree of flow will always be present. As a basis for comparison, the flow properties of water approach those of Newtonian fluids.

To be suitable for screen printing, a fluid must have certain characteristics.

- The fluid must have a yield point, or minimum pressure required to produce flow, which must obviously be above the force of gravity. With a finite yield point, the paste will not flow through the screen at rest and will not flow on the substrate after printing.

- The fluid should be somewhat *thixotropic* in nature. A thixotropic fluid is one in which the shear rate/shear stress ratio is nonlinear. As the shear rate (which translates to the combination of squeegee pressure, velocity, and screen tension) is increased, the paste becomes substantially thinner, causing it to flow more readily. The corollary to this term is *pseudoplastic*. A pseudoplastic fluid is one in which the shear rate does not increase appreciably as the force in increased.

- The fluid should have some degree of hysteresis so that the viscosity at a given pressure depends on whether or not the pressure is increasing or decreasing. Preferably, the viscosity should be higher with decreasing pressure, as the paste will be on the substrate at the time and will have a lesser tendency to flow and lose definition.

The shear rate versus shear stress curve for a thixotropic paste with these characteristics is shown in Fig. 12.5. A third variable (time) should also be considered in this figure. In practice, a finite and significant amount of time elapses between the application of the force and the time when the steady-state viscosity is attained. During the printing process, the squeegee velocity must be sufficiently slow to allow the viscosity of the paste to lower to the point where printing is opti-

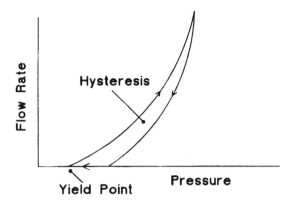

Figure 12.5 Viscosity curve for thick film paste.

mized. After the print, sufficient time must be allowed for the paste to increase to nearly the rest viscosity (leveling). If the paste is placed in the drying cycle prior to leveling, the paste will become still thinner due to the increased temperature, and the printed film will lose line definition.

Under laboratory conditions, the viscosity of the paste may be measured with a cone-and-plate or spindle viscometer. The cone-and-plate viscometer consists of a rotating cylinder with the end ground to a specific angle. A sample of paste is placed on a flat plate, and the cylinder is inserted into the paste parallel to the plate. The cylinder is rotated at a constant velocity, and the torque required to maintain this velocity is converted to viscosity. The spindle method utilizes a cylinder of known volume filled with paste. A spindle of known size is rotated at a constant speed inside the cylinder, and the torque measurement is converted to viscosity, as with the cone-and-plate. The spindle method provides a more consistent reading, because the boundary conditions can be more tightly controlled, and it is the most common method of characterizing thick film pastes in a development or manufacturing facility.

It is important to understand the limitations of viscometers for laboratory use. These are not analytical tools in that a viscometer reading may be directly translated to the settings on the screen printer. Most viscometers of this type are designed to operate at two speeds, which would yield only two points on the curve in Fig. 12.5. This limits the utilization of the viscometer to simply comparing one paste with another. When used for this purpose, the measurement conditions must be identical; the same type spindle in the same volume of paste must be used, and the sample must be at the same temperature if the correlation is to be meaningful.

Viscosity can be readily lowered by addition of the appropriate solvent. This is frequently required when the paste jar has been opened a number of times or if paste has been returned to the jar from the screen. If the original viscosity is recorded, the paste can be returned to the original viscosity with the aid of the viscometer. Increasing the viscosity is more difficult, requiring the addition of more nonvolatile vehicle, followed by remilling of the paste.

Cermet thick film pastes are divided into three broad categories: conductors, resistors, and dielectrics. Each of these categories may have several subcategories that describe materials for a specific application.

12.3 Screen Printing

Thick film paste is applied to the substrate by screen printing through a stainless steel mesh screen. During the design process, 1:1 artwork corresponding to each of the individual layers to is generated. These artworks are used to expose a screen coated with a photosensitive material or *emulsion* to generate the pattern. The emulsion not protected by the dark areas of the film is hardened by the UV light, and the protected portions may be simply washed away with water, leaving openings in the emulsion corresponding to the dark areas in the film.

Commercial screen printers are designed to hold the screen parallel to and in proximity to the substrate. A squeegee is used to provide the force necessary to force the paste through the openings onto the substrate. The process sequence is as follows.

1. The screen is placed in the screen printer.
2. The substrate is placed directly under the screen.
3. Paste is applied to the top of the screen.
4. The squeegee is activated and travels across the surface of the screen, forcing the paste through the openings onto the substrate.

In this manner, the paste can be applied in very precise geometries, allowing complex interconnection patterns to be generated.

Screen printing has been used for thousands of years to generate designs. In ancient China, silk was one of the first materials used as a mesh. A pattern was created in the silk using pitch or similar materials to block out unwanted areas, and dye was forced through the pattern by hand onto cloth or other surfaces to create colored patterns. By performing several sequential screenings with different colors and patterns, complex decorative patterns could be formed.

Silk continued to be one of the most common materials used until the development of synthetic materials, and the term *silk screening* is still commonly used to describe the screen printing process. The development of synthetic fibers, such as nylon, made possible greater control of the mesh materials and the added development of photosensitive materials used for creating the patterns allowed screen printing to become much more precise, repeatable, and controllable.

Today, in the hybrid microelectronics industry, the primary mesh material is stainless steel, which adds an additional degree of control and precision over nylon in addition to added resistance to wear and stretching. The crude, hand methods of printing have evolved into sophisticated, microprocessor-controlled machines that are self-aligning, have the ability to measure the thickness of the film, and have the ability to adjust the printing parameters to compensate for variations in the properties of the thick film paste.

Still, of all the processes used to manufacture thick film hybrid microcircuits, the screen printing process is the least analytical. It is not possible to measure the parameters of the paste and convert them to the proper printer settings needed to produce the desired results, due to the large number of variables involved. Many of the variables are not in the direct control of the process engineer and may change as the printing process proceeds. For example, the viscosity of the paste may change during printing due to solvent evaporation.

12.3.1 The Screen

The screen mesh is manufactured by weaving stainless steel wires to form a long sheet. The direction along the length of the sheet is referred to as the "warp" direction, while the direction across the width of the sheet is referred to as the "weft" direction. The vast majority of meshes used in thick film screen printing are woven in the so-called "plain" weave pattern, as shown in Fig. 12.6, in which each wire simply goes over and under only one wire at a time. In the "twilled" weave pattern, each wire goes over and under two wires at a time. The plain weave has more open area for a given mesh count and wire size, while the twilled weave is stiffer and is less likely to stretch.

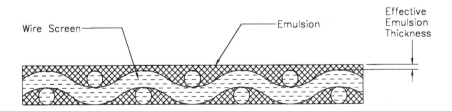

Figure 12.6 Cross section of screen and emulsion.

One of the most important parameters of the screen is the "mesh count," or the number of wires/unit length. In general, the mesh count is the same for both the warp and weft directions, as is the wire size. In practice, the mesh count may vary from 80 wires/in. for coarse screening, such as solder paste, to 400 wires/in. for fine line printing.

Another important parameter is the size of the opening in the screen, which strongly influences the amount of paste that can be transferred during the printing process and limits the maximum particle size of the material used to manufacture the paste. The opening is dependent on both the mesh size and the wire count, as shown in Fig. 12.7, and may be calculated by Eq. (12.1).

$$O = \frac{1}{M} - D \qquad (12.1)$$

where O = dimension of the opening
  M = mesh count
  D = diameter of the wire

A graph of the size of the opening as a function of mesh count and wire diameter is shown in Fig. 12.8.

The overall thickness of the screen mesh will be approximately two times that of the wire diameter but may be slightly smaller or larger than 2×, depending on the technique used in the weaving process. In weaving processes where a great deal of pressure is applied to the

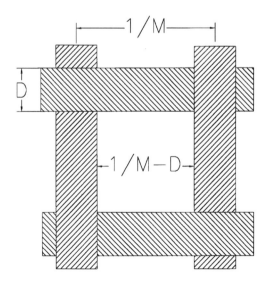

Figure 12.7 Screen opening.

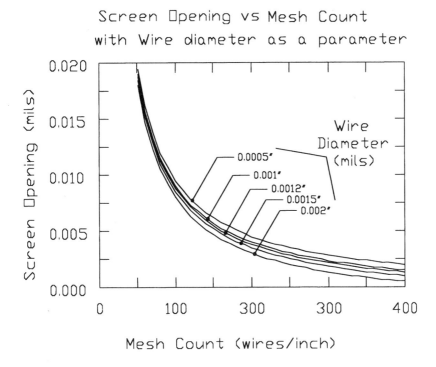

Figure 12.8 Screen opening vs. mesh count.

mesh, the mesh thickness is slightly less than 2×, and the mesh is referred to as a "hard mesh." A hard mesh is not as pliable as a "soft mesh" and is less desirable in most applications.

The amount of paste transferred during the printing process is dependent on the volume of the opening in the screen. Assuming that the thickness of the mesh is 2 × the wire diameter, the volume of the opening may be approximated by Eq. (12.2).

$$V = (2D)\left(\frac{1}{M} - D\right)^2 \tag{12.2}$$

Empirically, from Eq. (12.2), the volume of the opening is zero when either $D = 1/M$ or when $D = 0$. At some value of D between 0 and $1/M$, therefore, the volume must be at a maximum. Expanding Eq. (12.2) and taking the derivative with respect to D, the volume is maximized when the relationship defined in Eq. (12.3) is met.

$$D = \frac{1}{6M} \tag{12.3}$$

In most cases, this results in a wire diameter that is too small to be readily woven. In practice, however, this is not a severe problem, since the amount of paste transferred is adequate for most purposes.

Since most of the traces in a typical thick film circuit are parallel or at right angles to the screen frame, it is preferable that the mesh be oriented at an angle of 22 to 45° to the frame to prevent the partial blocking of the opening on one side of the trace by the screen wire. On long traces, this can lead to narrowing of the print at periodic intervals, and, on fine line prints of less than 0.010 in. in width, can lead to a discontinuity in the trace. Due to the manner in which the screen is attached to the frame, a screen attached at 45° is more expensive than one at 90°, since there is more waste of the screen material.

The screen frame is usually made from cast aluminum, with the bottom of the frame machined to be parallel to and a fixed distance from the top. In this manner, the screen will be parallel to the printer and will have the same reference point with respect to the substrate. This precaution will greatly improve the quality and reproducibility of the print as well as minimizing the setup time.

The screen is prepared for use by stretching the mesh by pneumatic or mechanical methods over a large frame capable of accommodating several smaller screen frames. The tension may be measured by an electronic tensiometer capable of measuring the tension in either the warp or the weft direction or by simply measuring the deflection in the center of the screen produced by a one-pound weight. The deflection method is the most common, but the tensiometer allows much greater control over the process. The mesh is attached to the small frames with epoxy that cures at room temperature. After curing, the mesh is trimmed away around the periphery of the epoxy, simultaneously separating the individual screen frames. A screen manufactured in this manner can be expected to last for thousands of prints without losing tension when handled and treated properly.

The final step in preparing the screen for use is to coat it with a photosensitive emulsion. This can be one of two types. The so-called "indirect" emulsion utilizes a photosensitive film of predetermined thickness with a Mylar backing for ease of handling. The film is placed on a flat surface with the emulsion side up, and the bottom of the screen mesh is placed on top. A photosensitive gel is poured onto the top of the screen and forced through the screen by a squeegee so that it fills the screen mesh and contacts the film below. The gel is allowed to dry, and the Mylar backing is removed, allowing the screen to be exposed. This process is very simple, except that excessive pressure on the squeegee can force the mesh to penetrate the film, making the overall thickness less than that expected. In addition, screens made with this process have a shelf life of only a few hours before exposure, which means that they must be sensitized just prior to use.

The so-called "direct" emulsion is initially in liquid form. To sensitize the screens, a dam or mold is formed around the periphery of the screen with cellophane tape or similar material to control the thickness. The top of the screen mesh is placed on a flat surface exactly the size of the inside of the frame. The emulsion is poured onto the mesh and smoothed with a straightedge to coat the screen evenly and fill the mesh. The thickness may be built up if desired by allowing the initial coating to dry and repeating the process with a second dam. If stored in a black, light-free plastic bag in a cool environment, screens coated with a direct emulsion may be stored for several months prior to exposure. All commercial screen makers use direct emulsions for this reason.

The screen is exposed by placing the emulsion side of the artwork in contact with the emulsion (bottom) side of the screen and exposing the screen to ultraviolet (UV) light, preferably collimated. The period of exposure varies with the strength of the source, the distance from the source, the type of emulsion, the quality of the artwork, and the thickness of the emulsion. The unexposed emulsion protected by the artwork may be removed by gently washing the screen with a spray of warm water. After drying at room or slightly elevated temperature, the screen is again exposed to UV light to further harden the remaining emulsion.

The quality of the screen is critical to the screen printing process. The wire mesh should initially be inspected for uniformity of wire size and the size of the opening. The screen must be cleaned with detergent to remove any oils and dirt prior to sensitization, and all photoprocesses must be performed under yellow light. The emulsions used for screen printing are not ultrasensitive to light, but the exposure time even to yellow light should be minimized.

12.3.2 Stencils

For applications in which it is required to print a viscous material with large particles, such as solder paste, a mesh screen may be inadequate. The screen wires interfere with the transfer of the paste to the substrate, leaving voids in the printed film and blocking the openings in the screen. In this case, it is common to use a stencil—a thin foil of metal with openings corresponding to the print. To the present time, stencils have been formed by photoetching a pattern through a brass or stainless steel sheet from both sides of the metal. The opening created in this manner has a characteristic "hourglass" shape, narrower in the middle than at the top and bottom. There is also a limitation in the minimum size of the opening due to the fact that the etching process proceeds laterally at the same time it is etching vertically through the metal. This not only limits the pitch of the devices that can be mounted, it also necessitates complicated correction factors that vary with the size and thickness of the metal. While small openings can be created with electrical discharge machining (EDM) or with a laser, these are relatively expensive processes.

The electroforming process grows the stencil around a pattern exposed on a thick photoresistive film placed on a flat conducting surface called a *mandrel*. The mandrel, usually copper, allows nickel to be electroplated in the openings in the photoresist to generate a stencil with extremely fine pitch and dimensional control. Stencils grown in this manner have been used to print epoxy and solder in geometries as small as 0.002 in.

12.3.3 The Printing Process

There are two basic methods of screen printing: the contact process and the off-contact process. In the contact process, the screen remains in contact with the substrate during the print cycle and then is separated abruptly by either lowering the substrate or raising the screen. In the off-contact process, the screen is separated from the substrate by a small distance and is stretched by the squeegee until it contacts the substrate only at a point directly under the squeegee. Once the squeegee passes, the screen "snaps back," leaving the paste on the substrate. In general, the best line definition is obtained with the off-contact process, and most printing of thick film pastes is performed in this manner. The contact process is generally used when a stencil is utilized to print solder paste. The stencil, being solid metal, cannot be continually stretched in the same manner as a screen without permanent deformation.

In the printing process, the paste is applied to the screen, and the squeegee is activated, sweeping across the screen. The pressure from the squeegee forces the paste through the openings in the screen onto the substrate. The rough substrate surface creates somewhat more surface tension than does the smooth wires of the screen mesh, causing the paste to stay on the substrate when the squeegee passes. The process is facilitated by the thixotropic nature of the paste. As the squeegee applies force to the paste, it becomes thinner and flows more readily. As the squeegee passes, the paste becomes thicker again and retains the line definition on the substrate.

12.3.4 Screen Printing Parameters

More than 100 variables have been identified that affect the screen printing process, ranging from the paste properties to the printer setup to the screen properties. Referring to Fig. 12.9, the screen printer adjustments that affect the print the most are as follows.

1. *Screen-to-substrate spacing (the "snap-off" distance).* This is arguably the most important parameter in the screen printer setup. If it is too large, the screen will rapidly lose tension and the print will lose definition. If it is too small, the tension on the screen during the print will not be sufficient to transfer the paste to the substrate. The amount of snap-off is dependent on the size of the screen, as shown in Table 12.1.

Table 12.1 Snap-Off Distance vs. Screen Size

Screen size	Snap-off distance
5 × 5 in.	0.025 in.
5 × 7 in.	0.035 in.
8 × 10 in.	0.050 in.

2. *Screen-to-substrate parallelism.* If the screen is not exactly parallel to the substrate, the snap-off distance will change across the print, causing variability in the print definition and thickness.

3. *Squeegee velocity.* The squeegee velocity applies pressure primarily in the tangential direction, but also in the normal direction. If the squeegee speed is too fast, the print definition may be poor due to voiding. If the viscosity of the paste is not allowed sufficient time to drop to the proper value, insufficient paste may be transferred, and the print will be thinner than normal. If the velocity is increased past this point, the squeegee will begin to "plane" over the paste instead of sweeping it along the surface of the screen, and the print will become thicker. If the speed is too slow, the paste may not be properly sheared, and the print may be too thick. In addition, the process time increases with an increase in cost due to the subsequent drop in throughput.

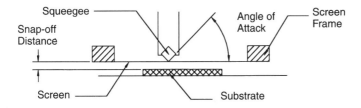

Figure 12.9 Screen setup parameters.

4. *Squeegee position.* The squeegee position with respect to the substrate is also an important variable. If the squeegee is too high, the print will be thin and/or contain voids. If the squeegee is too low, excessive pressure will be applied to the paste, causing the print to be too thin and possibly forcing paste between the screen and the substrate with a corresponding loss of print definition. The deflection of the squeegee during the printing cycle creates a downward pressure on the paste.

5. *Squeegee pressure.* Squeegee pressure is applied by a spring force that pushes the squeegee downward toward the substrate and is set by adding tension to the spring. The squeegee pressure is most significant when printing with a highly viscous paste. If the pressure is too low in this instance, the squeegee may plane on the paste, resulting in a thick print with poor definition.

6. *Attack angle.* The attack angle is a measure of the degree to which the squeegee is tilted with respect to the normal. A high degree of tilt will have the same effect as increased squeegee pressure or having the squeegee set too low.

12.3.5 Screen Printer Setup

This generalized procedure for setting up a screen printer described is applicable to most printers and should be followed in sequence.

1. With the screen removed, install the squeegee and place a substrate on the platen. Lower the squeegee to the point where it just touches the substrate and adjust it so that it is parallel to the substrate. With the squeegee just touching the substrate, lower it another 3–5 mils.

2. Set a reference level on the platen using a three-point position indicator. Install the screen and check the parallelism of the screen using the position indicator. If the screen is not parallel, make the appropriate adjustments.

3. Lower the screen to the point where it is just touching the substrate, and set a reference level of "0" on the screen position indicator. Set the snap-off distance as determined from Table 12.1.

4. Place a substrate on the platen and visually align the substrate to the pattern on the screen (if the printer does not have a vision system).

5. Apply paste to the screen and adjust the velocity and alignment of the print as necessary to optimize the definition and thickness of

the print. The squeegee pressure adjustment should be used as a fine control.

This procedure should result in a print that is nearly optimal with a minimum of time. The squeegee speed and squeegee pressure can be used to fine-tune the process if necessary.

12.3.6 Geometric Effects on Print Thickness

In theory, the wet thickness of a print may be increased indefinitely by simply increasing the emulsion thickness. For smaller openings, this is largely true. The volume created by the screen mesh and the squeegee in tandem fills with paste, and the shape of the print is somewhat flat. As the size of the opening increases, however, the screen can be deflected as a result of squeegee pressure, and the profile of the printed film becomes concave. If the size of the opening is sufficiently large that the screen can be deflected to the point where it is able to touch the substrate, the emulsion thickness becomes less of a factor, and the print thickness is largely determined by the screen mesh. A thicker emulsion in this case simply results in a more pronounced concave effect. The size of the opening where this phenomenon occurs depends on the mesh count and the size of the wire.

12.3.7 Measurement of Print Thickness

There are two basic approaches to measuring print thickness; off-contact and contact. Off-contact systems may be used to measure wet, dry, or fired films, while contact systems may be used only on dry or fired films. The simplest noncontact method is to focus a high-powered metallurgical microscope with a narrow depth of field on the substrate, mark a reference, and refocus the microscope on the top of the film. The print thickness is the distance the object lens must be moved. This method is somewhat inexact, since the profile of the print is nonuniform, allowing the thickness to vary considerably across the print.

Another method of off-contact thickness measurement utilizes light to determine the profile. The light-section microscope employs a split coherent light beam shined directly on a print. The interference pattern of the beam outlines the print, and the thickness is measured by noting the distance between the lines marking the substrate and the top of the print. While substantially more accurate than the metallurgical microscope, the light-section microscope is highly operator dependent and does not produce a written record of the measurement.

More sophisticated off-contact systems have a laser that sweeps across the film, with the reflection picked up by a light detector. Either the exact profile or the mean thickness of the print may be used as a basis of comparison with other prints.

Contact systems use a stylus that moves across the film at a selected rate of speed. The position of the stylus is detected, and the profile may be plotted as the stylus moves. These can be quite sensitive if the profilometer is placed on a steady table and shielded from air currents. Contact systems may be used only on dried and fired prints.

12.4 Drying

There are two organic components that compose the vehicle in the printed film: the volatile component and the nonvolatile component. Immediately after printing, thick film materials are simply discrete particles of glass or metal suspended in a thickened vehicle and are somewhat tacky and fragile. The volatile component must be removed at a low temperature prior to firing. The volatile solvents evaporate rapidly at temperatures above 100°C and may cause extreme voiding of the fired film if exposed to temperatures in excess of 150°C.

After printing, parts are usually allowed to "level" in air for a period of time (usually, 5 to 15 min). The leveling process permits screen mesh marks to fill in and some of the more volatile solvents to evaporate slowly at room temperature. The leveling process is critical to the precision of the fired film. The viscosity drops considerably during the printing process due to the thixotropic nature of the paste. Immediately after printing, the viscosity is still quite low and requires a period of time to return to a higher viscosity before drying. If the film is exposed to elevated temperatures immediately after printing, the viscosity will drop still lower, and the paste may spread across the substrate, degrading the definition of the printed film.

After leveling, the parts are force dried at temperatures ranging from 70 to 150°C for about 15 min. Drying is usually accomplished in a low-temperature, moving-belt dryer. For smaller or laboratory operations, drying may be accomplished in batch, forced-air dryers, or even by placing the substrates on a hotplate. In a production environment, it is important to have an exhaust system to remove the solvent vapor from the environment. Certain of the solvents have a strong odor and may also adversely affect the atmosphere in a firing furnace if allowed to remain in the immediate area.

There are two important considerations in drying: the cleanliness of the atmosphere and the drying rate. Drying must be accomplished in a clean room (<Class 100,000) to prevent dust and/or lint particles from accumulating in the dried film. During firing, the particles will burn away, leaving voids in the film. The rate of temperature rise dur-

ing the drying process must be controlled to prevent cracking of the film as a result of rapid solvent evaporation.

Drying removes the most volatile fraction of the paste. Perhaps 90 percent of the solvents and gums are removed in the drying step. Such solvents may be terpineol, butyl carbitol, higher alcohols such as decanol and octanol, or xylene. Because of the potentially toxic nature of these solvents, drying must be carried out under a hood or other extraction device. Since each paste system has its own solvents, gums, and wetting agents, the paste manufacturer will recommend the exact drying schedule for its materials.

12.5 Firing

After drying, the parts are placed on a moving belt or conveyor furnace. As with the drying profile, each paste manufacturer develops precise profiles for its products and should be consulted for the most current information.

A thick film furnace must meet the following criteria:

1. A clean furnace environment

2. A uniform and controllable temperature profile

3. A uniform and controllable atmosphere

To provide for both a clean environment and a controllable atmosphere, all modern thick film furnaces have impervious muffles. Both metal and quartz muffles are used, and both are satisfactory when properly designed. Larger production furnaces and multi-atmosphere furnaces all must use metal muffles (usually Inconel), since a large cross section of impervious quartz is too expensive to fabricate. Thick film furnaces are designed to operate under 1000°C, and resistive heated furnaces use wound nichrome heating elements.

In some designs, traditional firebrick insulation has been replaced with lightweight foam insulation. Lightweight insulation has many advantages over heavy brick. Since lightweight insulation does not absorb moisture to the same extent as brick, the furnace may be turned off if not being used. This is not recommended with brickwork furnaces, because the evaporating steam will damage the bricks. Given the high cost of electricity in many parts of the country today, the ability to shut a furnace down if it is not being used offers a tremendous cost saving. Conversely, in the case of brickwork furnaces, the furnace may be "banked" to a somewhat lower temperature if not being used, but it must not be turned off.

Lightweight insulation, by its nature, has lower thermal mass and responds to temperature changes more rapidly than brick. In fact, it can be integrated with the heating elements, making it is possible to use a single furnace for two or more profiles. A traditional furnace could take up to 12 hr to stabilize from an 850 to a 600°C profile. A furnace made with foam or fiber insulation can stabilize in one or two hours.

12.5.1 Temperature Control

Temperature control is accomplished by dividing the furnace into several "zones" of control and controlling each zone independently by means of thermocouples and closed-loop feedback controllers. On furnaces with wide belts, some cross-belt temperature control is also employed. Generally, temperature uniformity of ±3°C is possible across the belt and over the length of the profile, which is adequate for thick film firing. Type K or Platinel® thermocouples are used for control at the furnace operating temperatures of approximately 850°C.

Good practice requires that the actual furnace profile must be periodically checked independently of the furnace controllers. A long, Type K thermocouple is attached to the belt and run through the furnace with several substrates added as a thermal load. The actual profile is printed on a strip chart recorder. Typical profiles are shown in Figs. 12.10 and Fig. 12.11.

Figure 12.10 illustrates the profile for the so-called "long-profile" materials. The length of the profile is approximately one hour. Referring to Fig. 12.10, the ratios

$$\text{Temperature Rise Rate} = \frac{\Delta T_r}{\Delta T_r} \qquad (12.4)$$

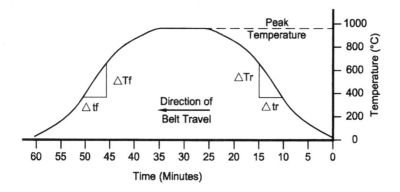

Figure 12.10 Typical profile for firing thick film paste.

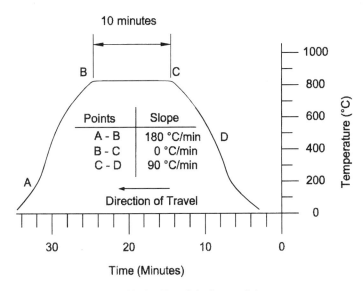

Figure 12.11 Firing profile for "fast-firing" materials.

$$\text{Temperature Fall Rate} = \frac{\Delta T_f}{\Delta T_r} \qquad (12.5)$$

are both approximately 50°C/min. The rise rate is a compromise between the throughput and the burnout rate of the nonvolatile organic binder. It is essential that the organic material be removed from the film before the temperature exceeds 500°C. Above this temperature, the heat will cause the organic material to break down, forming carbon monoxide (CO). The CO acts as a reducing agent, reacting with metal oxides in the film to form pure metal. This results in resistor values that are substantially lower than expected and also a loss of adhesion to the substrate. The fall rate is a compromise between the throughput and the cooling rate of the glass binder. If the film is cooled too fast, it may "quench" the glass and form small cracks throughout the structure. This will also result in a loss of adhesion and may cause instability in resistors.

Figure 12.11 represents the profile of "fast-firing" materials, with a time duration approximately half that of conventional "long-profile" materials. By contrast, the rise rate of fast-firing materials may be as high as 180°C/min and the fall rate as high as 90°C/min. These pastes contain glasses and organic materials substantially different from those used in conventional materials so as to withstand the higher rise and fall rates.

The use of microprocessors, highly accurate direct digital controls, and cross-belt heater trimming make control of furnace temperature a simple and repeatable operation. Modern furnaces have thermocouples installed at intervals and are equipped with microprocessors to allow more precise control.

Small variations in temperature or time at temperature can cause significant changes in the mean value and distribution of values. In general, the higher the ohmic value of the resistor, the more dramatic will be the change. As a rule, high ohmic values tend to decrease as temperature and/or time is increased, while very low values (<100 $\Omega/\square$) may tend to increase.

12.5.2 Atmosphere Control

Thick film resistors are very sensitive to the firing atmosphere. For resistor systems used with air-fireable conductors, it is critical to have a strong oxidizing atmosphere in the firing zone of the furnace. In a neutral or reducing atmosphere, the metallic oxides composing the active material will reduce to pure metal at the temperatures used to fire the resistors, dropping resistor values sometimes by more than an order of magnitude. Again, high-ohmic-value resistors are more sensitive than low-ohmic-value components. Atmospheric contaminants, such as vapors from hydrocarbons or halogenated hydrocarbons, will break down at firing temperatures, creating a strong reducing atmosphere. For example, one of the breakdown components of hydrocarbons is carbon monoxide, one of the strongest reducing agents known. Concentrations of fluorinated hydrocarbons in a firing furnace of only a few ppm can drop the value of a 100-kΩ resistor to below 10 kΩ. As a rule of thumb, no solvents, halogenated or carbon-based, should be permitted in the vicinity of a furnace used to fire thick film materials.

The adhesion element in thick film conductors may be partially reduced by a reducing atmosphere. As a result, blistering of the conductor along with a loss of adhesion may occur.

12.5.3 High-Temperature Furnaces

Refractory metallization, such as tungsten or "molybdenum/manganese" metallization, is fired at 1300 to 1550°C in pure hydrogen or disassociated ammonia (DA), composed of 75 percent hydrogen and 25 percent nitrogen. Tungsten metallization is commonly employed when processing high-temperature, 96 percent alumina tape for packages, because it can be cofired with the alumina tape. Molybdenum/manganese metallization is used for high-adhesion films on post-fired alumina ceramic bodies.

Since mesh belts and conventional heating elements fail at these temperatures, special furnace construction techniques are employed. Such furnaces are either "pusher" or "walking beam" designs with molybdenum or graphite elements. Safety interlocks and burn-off ports are needed to render the explosive H_2 gas safe to use. Because the heating elements degrade when exposed to air, common practice requires a continuous flow of hydrogen gas through the furnace. The excess hydrogen must be burned off to prevent accumulation of explosive hydrogen gas.

Such furnaces are difficult to maintain and expensive to operate; the use of such high-temperature metallizations is limited to a few specialty manufacturers.

12.5.4 Fast-Fire and Infrared (IR) Furnaces

A major expense of thick film manufacture is the cost of electricity to maintain a firing furnace. Traditional furnace design utilizes heavy firebrick insulation, which surrounds a quartz or metal muffle. To maintain a stable profile, the furnace must be kept at its operating temperature 24 hr a day, 7 days a week. For example, a change in profile from 850 to 600°C might take 8–12 hr to stabilize.

The "fast-fire" furnace uses high-efficiency, low-mass fiber or foam insulation. This low-mass design allows rapid profile changes and allows the furnace to be shut down when not in use. One manufacturer claims that a low-mass furnace can go from ambient temperature to 850–900°C in 10 min, with an additional 15 min required for profile stabilization. In addition, the "hot zone" occupies a greater portion of the total furnace length. This means that belt speed and, therefore, throughput is increased. By increasing the overall hot zone, ramp-up and cool-down slopes are much more severe. Belt speeds may be as high as 20 in./min with ramp rates of 500°C/min. However, this seems to have no deleterious effects on the materials. In fact, many years ago, parts with a heating and cooling rate of over 200°C/s were fired with no ill effects.

An alternative method of utilizing fast-fire technology is by employing IR heating elements in a furnace. Near IR energy (1.1–2.9 μm) is generated by banks of tungsten filament lamps encased in a clear fused quartz sheath. The sheath is hermetically sealed from the atmosphere and backfilled with an inert gas, or more recently with a halogen gas such as bromine. Such lamps can operate over a wide temperature range, usually from 1000–2500°C. One common application of IR furnaces is solder reflow.

While there are clearly many differences between conventional and IR firing, certain conclusions may be drawn.

1. Conventional pastes, when processed through an IR furnace using an appropriate schedule, yield stable and reproducible results.
2. It is possible to fire all thick film materials through an IR furnace at very short schedules. A 7-min total furnace cycle with 2 min in the hot zone has been reported.
3. To achieve optimal results, the exact profile utilized in an IR furnace must be somewhat different from the profile used in a conventional furnace.

Table 12.2 shows the relationship between conventional, fast-fire, and IR furnaces with respect to firing cycle.

Table 12.2 Comparison of Furnace Types

Furnace type	Firing temperature (°C)	Time at temperature (min)	Cycle time (min)
Conventional	850	15	60
Fast-fire	850	7	30
Infrared	900	2	7

There is some evidence that both resistor value and conductor adhesion could vary both with the furnace loading and coverage of paste on the substrate when an IR furnace is used. It is thought that the change in absorption of radiant energy (due to increased thermal mass or paste color) changes the effective temperature of the substrate and, therefore, the film properties.

In addition, exhaust protocols are very important with IR furnaces. Because the ramp-up is more severe than with conventional furnaces, the organic loads are generated more rapidly, and, in a smaller furnace volume, more organic burn-out by products per unit time are generated.

12.5.5 Inert Atmospheric Furnaces

Recent advances in materials technology have caused renewed interest in non-air atmosphere furnaces. Pure nitrogen and, more recently, nitrogen-NO_x blends have been used with copper-based conductor systems. Conductors and resistors using copper, nickel, or nickel-chrome alloys all require high-purity nitrogen atmospheres to prevent oxidation of the metal phase. In place of the air compressor used to deliver a stream of air, a liquid nitrogen bulk storage facility is used. Because

N$_2$ from liquid is always clean and dry, additional filtration is not needed. Therefore, in some ways, an inert atmosphere is easier than air to use and control.

However, the vehicles in most thick film inks are not cleanly removed in pure nitrogen within a time frame consistent with commercial practice. Therefore, small quantities of oxygen (1–5 ppm) are usually introduced to aid oxidation of the gums and to prevent reduction of glasses and oxides.

A modified GC known as an oxygen analyzer is required to make certain the proper oxygen level is maintained in the furnace. One particularly unforgiving paste system requires an oxygen content of 3–5 ppm. This is a very narrow range to maintain and would be impossible without an oxygen analyzer.

To maintain the low oxygen levels specified by most paste manufacturers, gas must flow through the furnace 24 hr a day. If the flow of N$_2$ is interrupted, oxygen will be adsorbed onto the belt and furnace walls, to be stripped off during processing. Thus, it is not practical to use a single furnace for both air and non-air atmospheres.

Typical gas settings for the furnace schematic are listed in Table 12.3. These settings are approximate and were developed for a 12-in. wide furnace. These settings will change as the load and furnace volume change.

Table 12.3 Typical Gas Settings for a Nitrogen Furnace

Flow meter control	Gas setting (SCFH)
1. Entry purge	60
2. Entry purge	60
3. Exhaust plenum	60
4. Exhaust plenum	60
5. Exhaust plenum	60
6. Exhaust plenum	60
7. Prefire section	60
8. Firing section	3
9. Exit purge	3

12.6 Cermet Thick Film Conductor Materials

Thick film conductors must perform a variety of functions in a hybrid circuit.

- The most fundamental function is to provide electrically conductive traces between the nodes of the circuit.
- They must provide a means to mechanically mount components by solder, by epoxy, or by direct eutectic bonding.
- They must provide a means for the electrical interconnection of components to the film traces and to the next higher assembly.
- They must provide a means of terminating thick film resistors.
- They must provide electrical connections between conductor layers in a multilayer circuit.

Thick film conductor materials are of three basic types: air-fireable, nitrogen-fireable, and those that must be fired in a reducing atmosphere. Air-fireable materials are made up of noble metals that do not readily form oxides. The basic metals are gold and silver, which may be used in the pure form or alloyed with palladium and/or platinum. Nitrogen-fireable materials include copper, nickel, and aluminum, with copper being the most common. The refractory materials, molybdenum, manganese, and tungsten, are intended to be fired in a reducing atmosphere consisting of a mixture of nitrogen and hydrogen.

12.6.1 Gold Conductors

Gold is most often used in applications where a high degree of reliability is required, such as military and medical applications, or where gold wire bonding is desirable for reasons of speed. The assembly processes (i.e., soldering, epoxy bonding, and wire bonding) used with gold thick films must be selected with care if reliability is to be maintained at a high level. For example, gold readily alloys with tin and will leach rapidly into certain of the tin-bearing solders, such as the lead-tin (Pb/Sn) alloys. Gold and tin also form brittle intermetallic compounds with a high electrical resistivity. Where Pb/Sn solders are to be used for component or lead attachment, gold must be alloyed with platinum or palladium to minimize leaching and intermetallic compound formation. Gold also forms intermetallic compounds with aluminum commonly used as the contact material on semiconductor devices and for wire bonding. The diffusion coefficient of aluminum into gold is much higher than that of gold into aluminum, with the diffusion rate increasing rapidly with temperature. Consequently, when an Au-Al interface occurs, as when an aluminum wire is bonded to a gold thick film conductor, the aluminum will diffuse into the gold wire, leaving voids in the interface (Kirkendall voids) that weaken the bond strength and increase the electrical resistance. This phenomenon is accelerated at temperatures above about 170°C and represents a reliability risk. The addition of palladium alloyed with the gold lowers the rate of aluminum diffusion significantly and improves the reliability of aluminum wire bonds.

12.6.2 Silver Conductors

Silver is often used in commercial applications where cost is a factor. Like gold, silver leaches into Pb/Sn solders, although at a slower rate. Pure silver may be used in applications where the exposure to Pb/Sn solder in the liquidus state is minimized and may also be nickel plated to further inhibit leaching.

Silver also has a tendency to migrate when an electrical potential is applied between two conductors in the presence of water in the liquid form. As shown in Fig. 12.12, positive silver ions dissolve into the water from the positive conductor. The electric field between the two conductors transports the Ag ions to the negative conductor, where they recombine with free electrons and precipitate out of the water onto the substrate as metallic silver. Over time, a continuous silver film grows between the two conductors, forming a conductive path. While other metals, including gold and lead, will migrate under the proper conditions, silver is the most notorious because of its high ionization potential.

Alloying silver with palladium and/or platinum slows down both the leaching rate and the migration rate, making it practical to use these alloys for soldering. Palladium/silver conductors are used in most commercial applications and are the most common materials found in hybrid circuits. However, the addition of palladium increases both the electrical resistance and the cost. A ratio of 4 parts silver to 1 part palladium is frequently used, providing a good compromise between performance and cost.

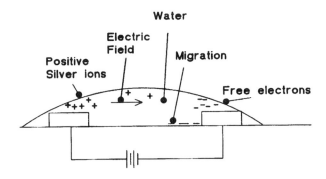

Figure 12.12 Silver migration.

12.6.3 Copper Conductors

Copper-based thick films were originally developed as a low-cost substitute for gold, but copper is now being selected when solderability, leach resistance, and low resistivity are required. These properties are particularly attractive for power hybrid circuits. The low resistivity allows the copper conductor traces to handle higher currents with a lower voltage drop, and the solderability allows power devices to be soldered directly to the metallization for better thermal transfer.

Copper thick film systems are known to exhibit the following problems:

- The requirement for a nitrogen atmosphere (<10 ppm oxygen) has created problems when scaling up from a prototype effort to high-volume production. Organic materials used in air-fireable systems combine with oxygen in the furnace atmosphere and "burn" off, while those used in copper paste systems "boil" off (or "unzip") and must be carried away by the nitrogen flow. It has proven difficult to maintain a consistent, uniform nitrogen blanket in the larger furnaces required for production, which has necessitated collaboration between furnace manufacturers, paste manufacturers, and users to come up with special, innovative furnace designs for copper firing. While some furnaces may be used for both nitrogen and air firing in prototype quantities, it is not practical to switch back and forth for production.

- Due to the large print areas normally required for dielectric materials, the problem of organic material removal is amplified when these materials are used to manufacture multilayer circuits. Consequently, multilayer dielectric materials designed to be used with copper are generally more porous than those designed for air-fireable materials, and, as a result, it has proven difficult to manufacture dielectric materials that are hermetic. This generally has required three layers of dielectric material between conductor layers to minimize short circuits and leakage, as opposed to the normal two required for air-fireable systems.

- Many resistor materials, particularly in the high ohmic range, have not proven to be as stable as air-fireable resistors when fired at temperatures below 980°C.

Refractory thick film materials, typically tungsten, molybdenum, and titanium, may also be alloyed with each other in various combinations. These materials are designed to be cofired with ceramic substrates at temperatures ranging up to 1600°C and are post-plated with nickel and gold to allow component mounting and wire bonding.

The capabilities of thick film conductors are summarized in Table 12.4.

Table 12.4 Thick Film Conductor Capabilities

	Au wire bonding	Al wire bonding	Eutectic bonding	Sn/Pb solder	Epoxy bonding
Au	Y	N	Y	N	Y
Pd/Au	N	Y	N	Y	Y
Pt/Au	N	Y	N	Y	Y
Ag	Y	N	N	Y	Y
Pd/Ag	N	Y	N	Y	Y
Pt/Ag	N	Y	N	Y	Y
Pt/Pd/Ag	N	Y	N	Y	Y
Cu	N	Y	N	Y	N

12.7 Thick Film Resistor Materials

Thick film resistors are formed by mixing metal oxide particles with glass particles and firing the mixture at a temperature/time combination sufficient to melt the glass and sinter the oxide particles together. The resulting structure consists of a series of three-dimensional chains of metal oxide particles embedded in a glass matrix. The higher the metal-oxide-to-glass ratio, the lower will be the resistivity of the fired film, and vice versa.

Referring to Fig. 12.13, the electrical resistance of a material in the shape of a rectangular solid is given by the classic formula

$$R = \frac{\rho_B L}{WT} \quad (12.6)$$

where
  R = electrical resistance in ohms
  ρ_B = bulk resistivity of the material in ohms-length
  L = length of the sample in the appropriate units
  W = width of the sample in the appropriate units
  T = thickness of the sample in the appropriate units

A "bulk" property of a material is one that is independent of the dimensions of the sample.

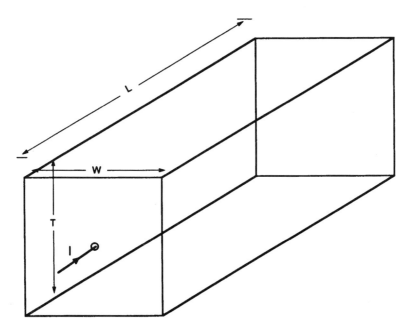

Figure 12.13 Resistance of a rectangular solid.

When the length and width of the sample are much greater than the thickness, a more convenient unit to use is the "sheet" resistivity, which is equal to the bulk resistivity divided by the thickness.

$$\rho_s = \frac{\rho_B}{T} \tag{12.7}$$

where ρ_S = sheet resistivity in ohms/square/unit thickness

The sheet resistivity, unlike the bulk resistivity, is a function of the dimensions of the sample. Thus, a sample of a material with twice the thickness as another sample will have half the sheet resistivity, although the bulk resistivity is the same. In terms of the sheet resistivity, the electrical resistance is given by

$$R = \frac{\rho_s L}{W} \tag{12.8}$$

For a sample of uniform thickness, if the length is equal to the width (i.e., the sample is a square), the electrical resistance is the same as the sheet resistivity, independent of the actual dimensions of the sample. This is the basis of the units of sheet resistivity, "ohms/

square/unit thickness." For thick film resistors, the standard adopted for unit thickness is 0.001 in. or 25 µ of dried thickness. The dried thickness is chosen as the standard, as opposed to the fired thickness, for convenience in process control. The dried thickness can be obtained in minutes, while obtaining the fired thickness can take as much as an hour. The specific units for thick film resistors are Ω/☐/0.001 in. of dried thickness, or simply Ω/☐.

A group of thick film materials with identical chemistries that are blendable is referred to as a "family" and will generally have a range of values from 10 Ω/☐ to 1 MΩ/☐ in decade values, although intermediate values are available as well. There are both high and low limits to the amount of material that may be added. As more and more material is added, a point is reached where there is not enough glass to maintain the structural integrity of the film. With conventional materials, the lower limit of resistivity is about 10 Ω/☐. Resistors with a sheet resistivity below this value must have a different chemistry and often are not blendable with the regular family of materials. At the other extreme, as less and less material is added, a point is reached where there are not enough particles to form continuous chains, and the sheet resistance rises very abruptly. Within most resistor families, the practical upper limit is about 2 MΩ/☐. Resistor materials are available to about 20 MΩ/☐, but these are not amenable to blending with lower value resistors.

The active phase for resistor formulation is the most complex of all thick film formulations due to the large number of electrical and performance characteristics required. The most common active material used in air-fireable resistor systems is ruthenium, which can appear as RuO_2 (ruthenium dioxide) or as $BiRu_2O_7$ (bismuth ruthenate). With the addition of TCR modifiers, these materials can be formulated to provide a temperature coefficient of resistance, as defined in Eq. (12.4), of ±50 ppm with a stability of better than 1 percent after 1000 hr at 150°C.

Certain properties of thick film resistors as a function of ohmic value are predictable with qualitative conduction models.[1]

- High-ohmic-value resistors tend to have a more negative TCR than low-ohmic-value resistors. This is not always the case in commercially available systems, due to the presence of TCR modifiers, but it always holds true in pure metal-oxide-glass systems.

- High-ohmic-value resistors exhibit substantially more current noise than low-ohmic-value resistors as defined in MIL-STD-202. Current noise is generated when a carrier makes an abrupt change in energy levels, as it must when it makes the transition from one metal oxide particle to another across the thin film of glass. When the metal oxide particles are directly sintered together, the transition is less abrupt, and little or no noise is generated.

- High-ohmic-value resistors are more susceptible to high-voltage pulses and static discharge than low-ohmic-value resistors. The high-voltage impulse breaks down the thin film of glass and forms a sintered contact, permanently lowering the value of the resistor. The effect of static discharge is highly dependent on the glass system used. Resistors from one manufacturer may drop by as much as half when exposed to static discharge, while others may be affected very little. This can be verified experimentally by heating a previously pulsed resistor to about 200°C. The value increases somewhat, indicating a healing of the glass oxide layer.

12.7.1 Electrical Properties of Thick Film Resistors

The electrical properties of resistors can be divided into two categories.

1. Time-zero (as-fired) properties

    a. Temperature coefficient of resistance (TCR)

    b. Voltage coefficient of resistance (VCR)

    c. Resistor noise

    d. High voltage discharge

2. Time-dependent (aged) properties

    a. High temperature drift

    b. Moisture stability

    c. Power handling capability

12.7.2 Time-Zero Properties

1. Temperature coefficient of resistance (TCR). Referring to Eq. (12.9), all real materials exhibit some change in resistance with temperature, and most are nonlinear to a greater or lesser degree. Figure 12.14 shows a graph of resistance versus temperature for a typical material. The TCR is a function of temperature and is defined as the slope of the curve at the test temperature, T.

$$\text{TCR}(T) = \frac{dR(T)}{dT} \qquad (12.9)$$

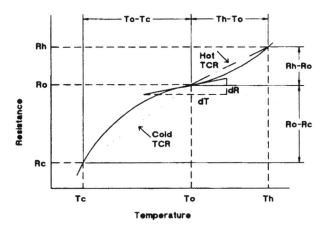

Figure 12.14 Resistance vs. temperature.

Referring again to Fig. 12.14, the TCR is often linearized over a range of temperatures as depicted in Eq. (12.10).

$$\text{TCR} = \frac{\Delta R}{\Delta T} \qquad (12.10)$$

In general, this result is a small number expressed as a decimal with several preceding zeroes. For convenience, Eq. (12.5) is typically normalized to the initial value of resistance and is multiplied by one million to produce a whole number as shown in Eq. (12.11).

$$\text{TCR} = \frac{R(T_2) - R(T_1)}{R(T_1)(T_2 - T_1)} \times 10^6 \text{ ppm/}°\text{C} \qquad (12.11)$$

where $R(T_2)$ = resistance at a temperature T_2
$R(T_1)$ = resistance at a temperature T_1

Most paste manufacturers present the TCR as two values:

- The average from 25 to 125°C (the "hot" TCR)
- The average from 25 to −55°C (the "cold" TCR)

An actual resistor paste carefully balances the metallic, nonmetallic, and semiconducting fractions to obtain a TCR as close to zero as possible. This is not a simple task, and the "hot" TCR may be quite different from the "cold" TCR. While linearization does not fully define the curve, it is adequate for most design applications.

It is important to note that the TCR of most materials is not linear, and the linearization process is at best an approximation. For example, the actual TCR of most thick film resistor materials at temperatures below −40°C tends to drop very rapidly and may be somewhat below the linearized value. The only completely accurate method of describing the temperature characteristics of a material is to examine the actual graph of temperature versus resistance. The TCR for a material may be positive or negative. By convention, if the resistance increases with increasing temperature, the TCR is positive. Likewise, if the resistance decreases with increasing temperature, the TCR is negative.

In general, metals exhibit positive TCRs, and nonmetals exhibit negative TCRs. In metals, the electron cloud is more disordered with increasing thermal energy, and resistance increases. Nonmetals (or semiconductors), which have electrons firmly bonded to crystal locations, become more mobile with energy and are better conductors as temperature is increased. They have a negative TCR.

While the absolute TCR of a resistor is important, in some cases, the ratio of resistance change between two resistors is more important. For example, assume that two resistors with a value of 1 kΩ have TCRs of +75 ppm and −40 ppm, respectively, at 0°C. After a change in temperature of +100°C, the respective values of resistance would be 1008 Ω and 996 Ω. The ratio of these two resistors at 100°C is

$$\frac{R_1}{R_2} = \frac{1008}{996} = 1.012$$

which translates to a net TCR of 115 ppm/°C, or the algebraic difference of the TCRs of the individual resistors.

Now consider that the TCRs of the two resistors are 200 ppm/°C and 175 ppm/°C, substantially higher in magnitude than the previous example. For this case, $R_1 = 1020$ Ω and $R_2 = 1018$ Ω at 100°C, and the ratio is

$$\frac{R_1}{R_2} = \frac{1020}{1018} = 1.002$$

which is equivalent to a TCR of only 24.6 ppm/°C for the ratio.

It is therefore possible to have low TCR tracking with high absolute TCR if both resistors move in the same direction with temperature variations. In many circuit designs, this parameter is more important than a low absolute TCR. Following these guidelines can enhance the TCR tracking of two resistors.

- Resistors made from the same value paste will track more closely than resistors made from different decades.
- Resistors of the same length will track more closely than resistors of different lengths.
- Resistors printed with the same thickness will track more closely than resistors of different thicknesses.

2. Voltage coefficient of resistance (VCR). Certain resistor materials also exhibit sensitivity to high voltages or, more specifically, to high electric fields as defined in Eq. (12.12). Note that the form of this equation is very similar in form to that of the TCR.

$$\text{VCR} = \frac{R(V_2) - R(V_1)}{R(V_1)(V_2 - V_1)} \times 10^6 \text{ ppm/V} \qquad (12.12)$$

where $R(V_1)$ = resistance at V_1
 $R(V_2)$ = resistance at V_2
 V_1 = voltage at which $R(V_1)$ is measured
 V_2 = voltage at which $R(V_2)$ is measured

Due to the semiconducting component in resistor pastes, the VCR is invariably negative. That is, as V_2 is increased, the resistance decreases. Also, because higher resistor decade values contain more glass and oxide constituents and are more semiconducting, higher paste values tend to have more negative VCRs than lower values.

Finally, VCR is dependent on resistor length. The voltage effect on a resistor is a gradient, and it is the volts/unit length rather than the absolute voltage that causes resistor shift. Therefore, long resistors show less voltage shift than short resistors, for similar compositions and voltage stress.

Resistor noise. On a fundamental level, noise occurs when an electron is moved to a higher or lower energy level. This change in energy of the electron is noise. The greater the potential difference between the energy levels, the greater the noise. Metals with many available electrons in the "electron cloud" have low noise, while semiconducting materials have fewer free electrons and exhibit higher noise.

There are two primary noise sources present in thick film resistors: thermal, or "white" noise, and current, or "pink" noise. The thermal noise is generated by the random transitions between energy levels as a result of heat and is present to a degree in all materials. Current noise occurs as a result of a transition between boundaries in a material, such as grain boundaries, where the energy levels may undergo abrupt changes from one region to another. In thick film resistors, the prime source of current noise is the thin layer of glass that may exist between the active particles.

The frequency spectrum of thermal noise is independent of frequency and is expressed in dB/Hz. The total noise is calculated by multiplying the noise figure by the bandwidth of the system. Current noise, on the other hand, has a frequency spectrum proportional to the inverse of the frequency, or $1/f$, and is expressed in units of dB/decade. The level of current noise in most applications is insignificant after the frequency exceeds 10–20 kHz.

In applications where the noise level is of significance, these guidelines may help to improve performance.

- High-value resistors exhibit a higher noise level than low value.
- Large area resistors exhibit a lower noise level.
- Thicker resistors exhibit a lower noise level.

Noise information is particularly important for low-signal applications as well as a quality check on processing. A shift in noise index, with constant resistor value, geometry and termination, indicates a process variation that must be investigated. For example, thin or underfired resistors generate higher noise than normal. The conductor/resistor interface can also be an important noise generator if it is glassy or otherwise imperfect. Finally, poor or incomplete resistor trimming also generates higher noise. A resistor noise test is an excellent method of measuring a resistor attribute not easily obtained by other methods.

12.7.3 Time-Dependent Properties

1. High-temperature drift. Thick film resistors in the untrimmed state exhibit a slight upward drift in value, primarily as a result of stress relaxation in the glasses that make up the body of the resistor. In properly processed resistors, the magnitude of the drift over the life of the resistor is measured in fractions of a percent and is not significant for most applications. At high temperatures, however, the drift is accelerated and may affect circuit performance in resistors that have not been properly fired or terminated or that are incompatible with the substrate.

To characterize the drift parameters, accelerated testing is frequently performed. A standard test condition is 125°C for 1000 hr at normal room humidity, corresponding to test condition S of MIL-STD-883C, method 1005.4. This test is considered to be equivalent to end-of-life conditions. More aggressive testing conditions are 150 °C for 1000 hr or 175°C for 40 hr.

2. Moisture stability. Resistance drift in the presence of moisture is a discriminating and important test. The most common test condition is 85 percent relative humidity and 85°C. Past studies indicate that this condition accelerates failure in thick film circuits by a factor of almost 500, compared with normally stressed circuits in the field. Humidity testing of resistors and circuits is more expensive than simple heat aging, but all the evidence indicates that it is a good predictor of reliability.

3. Power-handling capability. Drift due to high power is primarily due to internal resistor heating. It is different from thermal aging in that the heat is generated at the point-to-point metal contacts within the resistor film. When a resistor is exposed to elevated temperature, the entire bulk of the resistor is uniformly heated. Under power, local heating can be much greater than the surrounding area. Because lower-value resistors have more metal and, therefore, many more contacts, low-value resistors tend to drift less than higher-value resistors under similar loads. For most resistor systems, the shape of the power aging curve is a "rising exponential" as shown in Fig. 12.15.

The most generally accepted power rating of thick film resistors to achieve a drift of less than 0.5 percent over the resistor life is 50 W/in^2 of active resistor area. If more drift can be tolerated, the resistor can be rated at up to 200 W/in^2, as catastrophic failure will not occur until this rating is exceeded by a factor of several times. Typical properties of thick film resistors are shown in Table 12.5.

12.8 Thick Film Dielectric Materials

Thick film dielectric materials are used primarily as insulators between conductors, either as simple crossovers or in complex multilayer structures. Small openings, or vias, may be left in the dielectric layers so that adjacent conductor layers may interconnect. In complex structures, as many as several hundred vias per layer may be required. In this manner, complex interconnection structures may be created. Although the majority of thick film circuits can be fabricated with only three layers of metallization, others may require several more. If more than three layers are required, the yield begins dropping dramatically with a corresponding increase in cost.

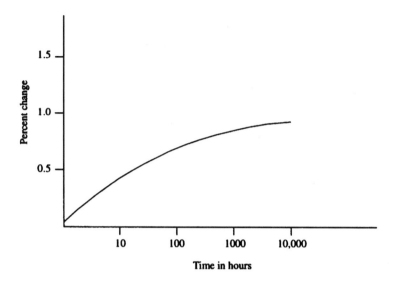

Figure 12.15 Resistance drift vs. time at 150°C.

Table 12.5 Typical Thick Film Resistor Characteristics

Parameter	Value
Tolerances, as fired	± 20%
Tolerances, laser trimmed	±0.5% to ±1%
TCRs 5–100 kΩ, −55 to + 125°C 100 kΩ–1MΩ, −55 to + 125°C	±100 ±150 ppm/°C ±150 ±750 ppm/°C
Resistance drift after 1000 hr at 150°C, no load	±0.3%
Resistance drift after 1000 hr at 85°C, with 25 W/in^2	0.3% max
Resistance drift, short-term overload (2.5 times rated voltage)	<0.5%
Voltage coefficient	−20 ppm/V
Noise (Quan-Tech®) 100 Ω/☐ 100 kΩ/☐	−30 to −20 db 0 to + 20 db
Power rating	40 to 50 W/in^2

Dielectric materials used in this application must be of the devitrifying or recrystallizable type. These materials in paste form are a mixture of glasses that melt at a relatively low temperature. During firing, when they are in the liquid state, they blend together to form a uniform composition with a higher melting point than the firing temperature. Consequently, on subsequent firings, they remain in the solid state, which maintains a stable foundation for firing sequential layers. By contrast, vitreous glasses always melt at the same temperature and would be unacceptable for layers to either "sink" and short to conductor layers underneath, or "swim" and form an open circuit. Additionally, secondary loading of ceramic particles is used to enhance devitrification and to modify the temperature coefficient of expansion (TCE).

Dielectric materials have two conflicting requirements in that they must form a continuous film to eliminate short circuits between layers and, at the same time, they must maintain openings as small as 0.010 in. In general, dielectric materials must be printed and fired twice per layer to eliminate pinholes and prevent short circuits between layers.

The TCE of thick film dielectric materials must be as close as possible to that of the substrate to avoid excessive bowing, or warpage, of the substrate after several layers. Excessive bowing can cause severe problems with subsequent processing, especially where the substrate must be held down with a vacuum or where it must be mounted on a heated stage. In addition, the stresses created by the bowing can cause the dielectric material to crack, especially when it is sealed within a package. Thick film material manufacturers have addressed this problem by developing dielectric materials that have an almost exact TCE match with alumina substrates. Where a serious mismatch exists, matching layers of dielectric must be printed on the bottom of the substrate to minimize bowing, which obviously increases the cost.

Dielectric materials with higher dielectric constants are also available for manufacturing thick film capacitors. These generally have a higher loss tangent than chip capacitors and utilize a great deal of space. While the initial tolerance is not good, thick film capacitors can be trimmed to a high degree of accuracy.

Another consideration for selecting a thick film dielectric is the compatibility with a resistor system. As circuits become more complex, the necessity for printing resistors on dielectric, as opposed to directly on the substrate, becomes greater. In addition, it is difficult to print resistors on the substrate in the proximity of dielectric several dielectric layers. The variations in the thickness affect the thickness of the substrate and, therefore, affect the spread of the resistor values.

12.9 Overglaze Materials

Dielectric overglaze materials are vitreous glasses designed to fire at a relatively low temperature, usually around 550°C. They are designed to provide mechanical protection to the circuit, to prevent contaminants and water from spanning the area between conductors, to create solder dams, and to improve the stability of thick film resistors after trimming.

Noble metals, such as gold and silver, are somewhat soft by nature. Gold, in particular, is the most malleable of all metals. When subjected to abrasion or to scraping by a sharp object, the net result is likely to be a bridging of the metal between conductors, resulting in a short circuit. A coating of overglaze can minimize the probability of damage and can also protect the substrates when they are stacked during the assembly process.

Contaminants that are ionic in nature, combined with water in the liquid form, can accelerate metal migration between two conductors. An overglaze material can help to limit the amount of contaminant actually contacting the surface of the ceramic and can help to prevent a film of water from forming between conductors. A ceramic substrate is a very "wettable" surface in that it is microscopically very rough, and the resulting capillary action causes the water to spread rapidly, creating a continuous film between conductors. A vitreous glass, being very smooth by nature, causes the water to "bead up," much like on a waxed surface, helping to prevent the water from forming a continuous film between conductors, thereby inhibiting the migration process.

When soldering a device with many leads, it is imperative that the volume of solder under each lead be the same. A well designed overglaze pattern can prevent the solder from wetting other circuit areas and flowing away from the pad, keeping the solder volume constant. In addition, overglaze can help to prevent solder bridging between conductors.

Overglaze material has long been used to stabilize thick film resistors after laser trim. In this application, a green or brown pigment is added to enhance the passage of an yttrium-aluminum-garnet (YAG) laser beam. Colors toward the shorter wavelength end of the spectrum, such as blue, tend to reflect a portion of the YAG laser beam and reduce the average power level at the resistor. There is some debate as to the effectiveness of overglaze in enhancing the resistor stability, particularly with high ohmic values. Several studies have shown that, while overglaze is undoubtedly helpful at lower values, it can actually increase the resistor drift of high-value resistors by a significant amount.

12.10 Thick Film—Conclusions

It is apparent that cermet thick film materials are complex structurally and electrically. More than 120 variables related to material properties and processes have been identified with thick film resistors. Clearly, it is impossible to control all these variables in a manufacturing environment. To achieve the best results, it is important that the user know how the various constituents interact and the role each of them plays in the final product. While the user need not know all the ingredients making up a particular paste, it is critical that the degree of compatibility with other pastes to be used in the circuit be assured. Very few hybrid circuits can be made with only one thick film paste, and if one or more of the pastes are incompatible with the others, unexpected reactions can occur that result in circuit failure. Usually, the only source of information, other than direct experimentation, is the paste manufacturer. A close working relationship with the manufacturer is a vital element toward creating a successful operation.

12.11 Thin Film Technology

The thin film technology is a subtractive technology in that the entire substrate is coated with several layers of metallization, and the unwanted material is etched away in a succession of photoetching processes. The use of photolithographic processes to form the patterns enables much narrower and better defined lines than can be formed by the thick film process. This feature promotes the use the thin film technology for high-density and high-frequency applications.

Thin film circuits typically consist of three layers of material deposited on a substrate. The bottom layer serves two purposes: it is the resistor material and also provides the adhesion to the substrate. The middle layer acts as an interface between the resistor layer and the conductor layer, either by improving the adhesion of the conductor or by preventing diffusion of the resistor material into the conductor. The top layer acts as the conductor layer.

The term *thin film* refers more to the manner in which the film is deposited onto the substrate as opposed to the actual thickness of the film. Thin films are typically deposited by one of the vacuum deposition techniques or by electroplating.

12.11.1 Sputtering

Sputtering is the prime method by which thin films are applied to substrates. In ordinary triode sputtering, as shown in Fig. 12.16, a current is established in a conducting plasma formed by striking an arc in

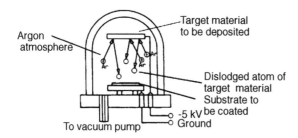

Figure 12.16 A dc sputtering chamber.

a partial vacuum of approximately 10 μ pressure. The gas used to establish the plasma is typically an inert gas, such as argon, which does not react with the target material. The substrate and a target material are placed in the plasma with the substrate at ground potential and the target at a high potential, which may be ac or dc. The high potential attracts the gas ions in the plasma to the point where they collide with the target with sufficient kinetic energy to dislodge microscopically sized particles with enough residual kinetic energy to travel the distance to the substrate and adhere.

The adhesion mechanism of the film to the substrate is an oxide layer that forms at the interface. The bottom layer (first layer sputtered) must therefore be a material that oxidizes readily. Gold and silver, for example, being noble metals, do not adhere well to ceramic surfaces. The adhesion is enhanced by presputtering the substrate surface by random bombardment of argon ions prior to applying the potential to the target. This process removes several atomic layers of the substrate surface, creating a large number of broken oxygen bonds and promoting the formation of the oxide interface layer. The oxide formation is further enhanced by the residual heating of the substrate that results from the transfer of the kinetic energy of the sputtered particles to the substrate when they collide.

Ordinary triode sputtering is a very slow process, requiring hours to produce usable films. By utilizing magnets at strategic points, the plasma can be concentrated in the vicinity of the target, greatly speeding up the deposition process. The potential that is applied to the target is typically RF energy at a frequency of 12.56 MHz. The RF energy may be generated by a conventional electronic oscillator or by a magnetron. The magnetron is capable of generating considerably more power with a correspondingly higher deposition rate.

By adding small amounts of other gases (such as oxygen and nitrogen) to the argon, it is possible to form oxides and nitrides of certain target materials on the substrate. It is this technique, called *reactive sputtering*, which is used to form tantalum nitride, a common resistor material.

12.11.2 Evaporation

The evaporation of a material into the surrounding area occurs when the vapor pressure of the material exceeds the ambient pressure and can take place from either the solid state or the liquid state. In the thin film process, the material to be evaporated is placed in the vicinity of the substrate and heated until the vapor pressure of the material is considerably above the ambient pressure. The evaporation rate is directly proportional to the difference between the vapor pressure of the material, and the ambient pressure and is highly dependent on the temperature of the material.

Evaporation must take place in a relatively high vacuum ($<10^{-6}$ torr) for three reasons.

1. To lower the vapor pressure required to produce an acceptable evaporation rate, thereby lowering the required temperature required to evaporate the material.
2. To increase the mean free path of the evaporated particles by reducing the scattering due to gas molecules in the chamber. As a further result, the particles tend to travel in more of a straight line, improving the uniformity of the deposition.
3. To remove atmospheric contaminants and components, such as oxygen and nitrogen, which may react with the evaporated film.

At an ambient pressure of 10^{-7} torr, a vapor pressure of 10^{-2} torr is required to produce an acceptable evaporation rate. A table of common materials, their melting points, and the temperature at which the vapor pressure is 10^{-2} torr is shown in Table 12.6.

The "refractory" metals, those with a high melting point such as tungsten, titanium, or molybdenum, are frequently used as carriers, or boats, to hold other metals during the evaporation process. To prevent reactions with the metals being evaporated, the boats may be coated with alumina or other ceramic materials.

If it is assumed that the evaporation takes place from a point source, the density of the evaporated particles assumes a cosine distribution from the normal. The distance of the substrate from the source then becomes a compromise between deposition uniformity and deposition rate; if the substrate is closer (farther away), the deposition is greater (lesser), and the deposition is less (more) uniform over the face of the substrate.

Table 12.6 Melting Points and $P_V = 10^{-2}$ Torr Temperatures of Some Common Metals Used in Thin Film Applications

Material	Melting point (°C)	Temperature at which $P_V = 10^{-2}$ Torr (°C)
Aluminum	659	1220
Chromium	1900	1400
Copper	1084	1260
Germanium	940	1400
Gold	1063	1400
Iron	1536	1480
Molybdenum	2620	1530
Nickel	1450	2530
Platinum	1770	2100
Silver	961	1030
Tantalum	3000	3060
Tin	232	1250
Titanium	1700	1750
Tungsten	3380	3230

In general, the kinetic energy of the evaporated particles is substantially less than that of sputtered particles. This requires that the substrate be heated to about 300°C to promote the growth of the oxide adhesion interface. This may be accomplished by direct heating of the substrate mounting platform or by radiant infrared heating.

There are several techniques by which evaporation can be accomplished. The two most common of these are resistance heating and electron-beam (E-beam) heating.

Evaporation by resistance heating, as depicted in Fig. 12.17, usually takes place from a boat made with a refractory metal, a ceramic crucible wrapped with a wire heater, or a wire filament coated with the evaporant. A current is passed through the element, and the heat generated heats the evaporant. It is somewhat difficult to monitor the temperature of the melt by optical means due to the propensity of the evaporant to coat the inside of the chamber, and control must be done by empirical means. There exist closed-loop systems that can control the deposition rate and the thickness, but these are quite expensive. In general, adequate results can be obtained from the empirical process if proper controls are used.

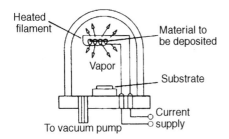

Figure 12.17 Thermal vacuum evaporation system.

The E-beam evaporation method takes advantage of the fact that a stream of electrons accelerated by an electric field tends to travel in a circle when entering a magnetic field. This phenomenon is utilized to direct a high-energy stream of electrons onto an evaporant source. The kinetic energy of the electrons is converted into heat when they strike the evaporant. E-beam evaporation is somewhat more controllable, since the resistance of the boat is not a factor, and the variables controlling the energy of the electrons are easier to measure and control. In addition, the heat is more localized and intense, making it possible to evaporate metals with higher 10^{-2} torr temperatures and lessening the reaction between the evaporant and the boat.

12.11.3 Comparison of Sputtering and Evaporation

While evaporation provides a more rapid deposition rate, there are certain disadvantages when compared with sputtering.

1. It is difficult to evaporate alloys such as NiCr due to the difference between the 10^{-2} torr temperatures. The element with the lower temperature tends to evaporate somewhat faster, causing the composition of the evaporated film to be different from the composition of the alloy. To achieve a particular film composition, the composition of the melt must contain a higher portion of the material with the higher 10^{-2} torr temperature, and the temperature of the melt must be tightly controlled. By contrast, the composition of a sputtered film is identical to that of the target.

2. Evaporation is limited to the metals with lower melting points. Refractory metals, ceramics, and other insulators are virtually impossible to deposit by evaporation.

3. Reactive deposition of nitrides and oxides is very difficult to control.

Comparisons between sputtering and evaporation are summarized in Table 12.7.

Table 12.7 **Comparison of Evaporation and Sputtering Processes for Nichrome**

Parameter	Vacuum evaporation	Sputtering
Mechanism	Thermal energy	Momentum transfer
Deposition rate	Up to 750,000 Å/min	Low (20 to 100 Å/min) except for some metals (e.g., Cu = 10,000 Å/min)
Control of deposition	Sometimes difficult	Reproducible and easy to control
Coverage for complex shapes	Poor, line of sight	Good, but nonuniform thickness
Coverage into small blind holes	Poor, line of sight	Poor
Metal deposition	Yes	Yes
Alloy deposition	Yes (flash evaporation)	Yes
Refractory metal deposition	Yes (by E-beam)	Yes
Plastics	No	Some
Inorganic compounds (oxides, nitrides)	Generally, no	Yes
Energy of deposited species	Low (0.1–0.5 eV)	High, 1 to >100 eV
Adhesion to substrate	Good	Excellent

12.11.4 Electroplating

Electroplating is accomplished by applying a potential between the substrate and the anode, which are suspended in a conductive solution of the material to be plated. The plating rate is a function of the potential and the concentration of the solution. In this manner, most metals can be plated to a metal surface.

In the thin film technology, it is a common practice to sputter a film of gold that is only a few angstroms thick and to build up the thickness of the gold film by electroplating. This is considerably more economical and results in much less target usage. For added savings, some companies apply photoresist to the substrate and electroplate gold only where actually required by the pattern.

12.11.5 Photolithographic Processes

In the photolithographic process, the substrate is coated with a photosensitive material, which is exposed with ultraviolet light through a pattern formed on a glass plate. The photoresist may be of the positive or negative type, with the positive type being prevalent due to its inherently higher resistance to the etchant materials. The unwanted material, which is not protected by the photoresist, may be removed by "wet," or chemical etching, or by "dry" or sputter etching.

In general, two masks are required, one corresponding to the conductor pattern and one corresponding to a combination of both the conductor and resistor patterns, generally referred to as the *composite* pattern. As an alternative to the composite mask, a mask that contains only the resistor pattern plus a slight overlap onto the conductor to allow for misalignment may be used. The composite mask is preferred, because it allows a second gold etch process to be performed to remove any bridges or extraneous gold to be removed that might have been left from the first etch.

Sputtering may also be used to etch thin films. In this technique, the substrate is coated with photoresist, and the pattern is exposed in exactly the same manner as with chemical etching. The substrate is then placed in a plasma and connected to a potential. In effect, the substrate acts as the target during the sputter etching process, with the unwanted material being removed by the impingement of the gas ions' on the exposed film. The photoresistive film, being considerably thicker than the sputtered film, is not affected. Sputter etching has two major advantages over chemical etching.

1. There is virtually no undercutting of the film. The gas ions strike the substrate in approximately a cosine distribution with respect to the normal of the substrate. This means that virtually no ions strike the film tangentially, leaving the edges intact. This results in more uniform line dimensions, which further results in better resistor uniformity. By contrast, the rate of the chemical etching process in the tangential is the same as in the normal direction, which results in the undercutting of the film by a distance equal to the thickness.

2. The potent chemicals used to etch thin films are no longer necessary, with less hazard to personnel and no disposal problems.

12.12 Thin Film Materials

The sputtering process may deposit virtually any inorganic material, but the outgassing of most organic materials is too extensive to allow sputtering to take place. A wide variety of substrate materials are also available, but these must contain an oxygen compound to permit adhesion of the film.

12.12.1 Thin Film Resistors

Materials used for thin film resistors must perform a dual role in that they must also provide the adhesion to the substrate, which narrows the choice to those materials that form oxides. The resistor film begins forming as single points on the substrate in the vicinity of substrate faults or other irregularities that might have an excess of broken oxygen bonds. The points expand into islands that, in turn, join to form continuous films. The regions where the islands meet are called *grain boundaries*, which are a source of collisions for the electrons. The more grain boundaries that are present, the more negative will be the TCR. Unlike thick film resistors, however, the boundaries do not contribute to the noise level. Furthermore, laser trimming does not create microcracks in the glass-free structure, and the inherent mechanisms for resistor drift are not present in thin films. As a result, thin film resistors have better stability, noise, and TCR characteristics than thick film resistors.

The most common types of resistor material are nichrome (NiCr) and tantalum nitride (TaN). Although NiCr has excellent stability and TCR characteristics, it is susceptible to corrosion by moisture if not passivated by sputtered quartz or by evaporated silicon monoxide (SiO). A TaN film, on the other hand, may be passivated by simply baking it in air for a few minutes. This feature has resulted in the increased use of TaN at the expense of NiCr, especially in military programs. The stability of passivated TaN is comparable to that of passivated NiCr, but the TCR is not as good unless annealed for several hours in a vacuum to minimize the effect of the grain boundaries. Both NiCr and TaN have a relatively low maximum sheet resistivity on alumina, about 400 $\Omega/\square$ for NiCr and 200 $\Omega/\square$ for TaN. This requires complex patterns to achieve a high value of resistance, resulting in a large required area and the potential for low yield. Chrome disilicide has a maximum sheet resistance of 1000 $\Omega/\square$ and overcomes this limitation to a large extent.

The TaN process is more often used due to the inherently high stability. In this process, N_2 is introduced into the argon gas during the sputtering process, forming TaN by reacting with pure Ta atoms on the surface of the substrate. By heating the film in air at about 425°C for 10 min, a film of TaO is formed over the TaN that is virtually impervious to further O_2 diffusion at moderately high temperatures, which helps to maintain the composition of the TaN film and stabilizes the value of the resistor. TaO is essentially a dielectric and, during the stabilization of the film, the resistor value is increased. The amount of increase for a given time and temperature is dependent on the sheet resistivity of the film. Films with a lower sheet resistivity increase proportionally less than those with a higher sheet resistivity. The resistance increases as the film is heated longer, making it possible to control the sheet resistivity to a reasonable accuracy on a substrate-by-substrate basis.

The properties of tantalum nitride and nichrome resistors are shown in Tables 12.8 and 12.9, respectively.

Table 12.8 Characteristics of Tantalum Nitride Resistors

Parameter	Value
Sheet resistance	20 to 50 $\Omega/\square$, 100 $\Omega/\square$, typical
Sheet resistance tolerance	±10% of nominal value
TCR	−75 ± 50 ppm/°C typical
	0 ± 25 ppm/°C with vacuum anneal
TCR tracking (−55 to + 125°C)	<2 ppm
Resistance drift (1000 hr @ 150°C in air)	<1000 ppm (0.1%)
Ratio tracking	5 ppm
Resistor tolerance after anneal and laser trim	±0.10% standard, ± 0.03% bridge trim
Noise (100 Hz to 1 MHz)	< −40 db

12.12.2 Barrier Materials

When Au is used as the conductor material, a barrier material between the Au and the resistor is required. When gold is deposited directly on NiCr, the Cr has a tendency to diffuse through the Au to the surface, which interferes with both wire bonding and eutectic die bonding. To alleviate this problem, a thin layer of pure Ni is deposited over the NiCr. In addition, the Ni improves the solderability of the surface considerably. The adhesion of Au to TaN is very poor. To provide the necessary adhesion, a thin layer of 90Ti/10W may be used between the Au and the TaN.

12.12.3 Conductor Materials

Gold is the most common conductor material used in thin film hybrid circuits because of the ease of wire and die bonding and the high resis-

Table 12.9 Characteristics of Nichrome Resistors

Parameter	Value
Sheet resistance	25 to 300 Ω/☐, 100 to 200 Ω/☐, typical
Sheet resistance tolerance	±10% of nominal value
TCR	0 ± 50 ppm/°C typical 0 ± 25 ppm/°C with special anneal
TCR tracking (–55 to +125°C)	2 ppm
Resistance drift (1000 hr @ 150°C in air)	<2000 ppm (0.2%) <1000 ppm with special anneal <200 ppm sputtered with 350°C anneal
Ratio tracking	5 ppm
Resistor tolerance after anneal and laser trim	± 0.10%
Noise (100 Hz to 1 MHz)	–35 db maximum

tance of the gold to tarnish and corrosion. Aluminum and copper are also frequently used in certain applications. It should be noted that copper and aluminum will adhere directly to ceramic substrates, but gold requires one or more intermediate layers, since it does not form the necessary oxides for adhesion.

12.13 Comparison of Thick and Thin Film

Although the thin film process provides better line definition, smaller line geometry, and better resistor properties, it has several disadvantages as compared with thick film.

1. Due to the added labor involved, the thin film process is almost always more expensive than the thick film process. Only in the case where a number of thin film circuits can be fabricated on a single substrate can thin film compete in price.

2. Multilayer structures are extremely difficult to fabricate. While they are possible with multiple deposition and etching processes, this is a very expensive and labor-intensive process and is limited to very few applications.

3. The designer is, in most cases, limited to a single sheet resistivity. This requires a large area to fabricate both large- and small-value resistors.

Design guidelines for the thick and thin film technologies are shown in Figs. 12.18 through 12.34.

12.14 Copper Metallization Technologies

The thick film and thin film technologies are limited in their ability to deposit films with a thickness greater than 1 mil (25 μ). This factor directly affects the ohmic resistance of the circuit traces and affects their ability to handle large currents or high frequencies. The copper metallization technologies provide conductors with greatly increased conductor thickness, and these offer improved circuit performance in many applications. There are three basic technologies available to the hybrid designer: direct bond copper (DBC), active metal braze (AMB), and the various methods of plating copper directly to ceramic.

12.14.1 Direct Bond Copper

Copper may be bonded to alumina ceramic by placing a film of copper in contact with the alumina and heating to about 1065°C, just below the melting point of copper, 1083°C. At this temperature, a combination of 0.39 percent O_2 and 99.61 percent Cu forms a liquid that can melt, wet, and bond tightly to the surfaces in contact with it when

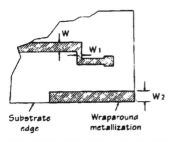

Dimension	Thick film, mils		Thin film, mils		Remarks
	Preferred	Minimum	Preferred	Minimum	
W	As required for current-carrying capability or electrical resistance				*
W_1	10	5	5	2	
W_2	20	15	15	10	†

*I_{MAX} may be calculated from Eq. (11.74).
†Wraparound metallization for thick film printing requires special tooling. Do NOT use laser-scribed substrates for either thick film or thin film fabrication with this process unless the substrates have been properly annealed, as the metallization will not adhere well to the initial laser-scribed surface.

Figure 12.18 Conductor line widths.

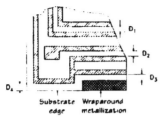

Dimension	Thick film, mils		Thin film, mils		Remarks
	Preferred	Minimum	Preferred	Minimum	
D_1	10	7.5	5	2	*
D_2	15	10	7.5	5	†
D_3	15	10	7.5	5	
D_4	10	10	7.5	7.5	‡

*Conductor lines ≤ mils in length.
†Conductor lines > 15 mils in length.
‡From minimum substrate size.

Figure 12.19 Conductor line spacing.

cooled to room temperature. In this process, the copper remains in the solid state during the bonding process, and a strong bond is formed between the copper and the alumina with no intermediate material required. The metallized substrate is slowly cooled to room temperature at a controlled rate to avoid quenching. To prevent excessive bowing of the substrate, copper must be bonded to both sides of the substrate to minimize stresses due to the difference in TCE between copper and alumina.

In this manner, a film of copper from 5 to 25 mils thick can be bonded to a substrate and a metallization pattern formed by photolithographic etching. For subsequent processing, the copper is usually plated with several hundred microinches of nickel to prevent oxidation. The nickel-plated surface is readily solderable, and aluminum wire bonds to nickel is one of the most reliable combinations.[2] Aluminum wire bonded directly to copper is not as reliable and may result in failure on exposure to heat and/or moisture.[3]

Multilayer structures of up to four layers have been formed by etching patterns on both sides of two substrates and bonding them to a common alumina substrate. Interconnections between layers are made by inserting oxidized copper pellets into holes drilled or formed in the substrates prior to firing. Vias may also be created by using one of the copper plating processes.

Bonding a copper sheet larger than the ceramic substrate and etching a lead frame at the same time as the pattern may create integral leads extending beyond the substrate edge.

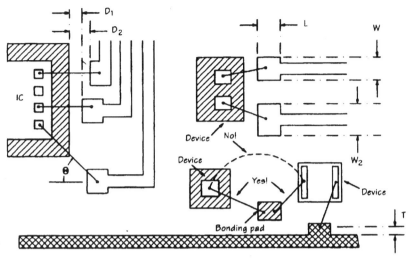

Figure 12.20 Bonding pad sizes.

The line and space resolutions of DBC are limited due to the difficulty of etching thick layers of metal without substantial undercutting. Special design guidelines must be followed to allow for this factor, as shown in Fig. 12.35. While the DBC technology does not have a resistor system, the thick film technology can be used in conjunction with DBC to produce integrated resistors and areas of high-density interconnections.

Aluminum nitride can also be used with copper, although the consistency of such factors as grain size and shape are not as good as aluminum oxide at this time. Additional preparation of the AlN surface is required to produce the requisite layer of oxide necessary to produce

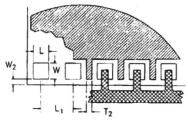

Dimension	Thick film, mils		Thin film, mils		Remarks
	Preferred	Minimum	Preferred	Minimum	
L	60	30	60	30	
W	60	50	60	50	
L_1	100	50	100	50	*
W_2	10	5	10	5	†
T_2	10	7.5	NA	NA	‡

*Pad spacing, center to center.
†Distance from edge of substrate.
‡Screened dielectric between pads.

Figure 12.21 Pads for lead frame attachment.

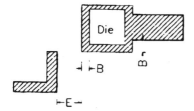

Dimension	Thick film, mils		Thin film, mils		Remarks
	Preferred	Minimum	Preferred	Minimum	
B	10	5	10	5	*,‡
B	10	10	10	10	†,‡
E	15	10	10	2	§

*Distance from edge of chip to edge of mounting pad; chip size≤100 mils.
†Chip size > 100 mils.
‡Distance=10 mils *maximum*.
§Pad distance to other metallization.

Figure 12.22 Mounting pads for semiconductor die.

the bond. This can be accomplished by heating the substrate to about 1250°C in the presence of oxygen.

Direct bond copper offers considerable advantages when packaging power circuits. The thick layer of copper can handle considerable current without excessive voltage drops and heat generation and allows

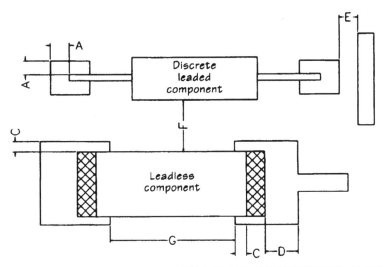

Dimension	Thick film, mils Preferred	Minimum	Thin film, mils Preferred	Minimum
A	10	10	10	10
C	10	5	10	5
D	20	10	20	10
E	15	15	15	15
F	40	40	40	40
G	AR	30	AR	30

1. Metallized pad overlap for leaded component.
2. Distance from edge of pad to chip metallization on leadless chip side and underneath.
3. Distance from edge of pad to chip metallization end.
4. Pad distance to other metallization.
5. Distance from leaded component.
6. Spacing between termination for leadless chip components.

Figure 12.23 Pads for discrete components.

the heat to spread rapidly outward from semiconductor devices, which lowers the thermal impedance of the system dramatically. The layer of copper on the bottom also contributes to heat spreading.

12.14.2 Plated Copper Technology

While a copper film can be mechanically bonded to a rough substrate surface, the adhesion is usually inadequate for most applications. The various methods of plating copper to a ceramic all begin with the formation of a conductive film on the surface. This film may be vacuum deposited by thin film methods, screen printed by thick film processes, or deposited with the aid of a catalyst. A layer of electroless copper

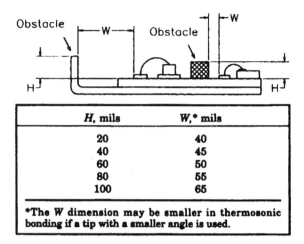

H, mils	W,* mils
20	40
40	45
60	50
80	55
100	65

*The W dimension may be smaller in thermosonic bonding if a tip with a smaller angle is used.

Figure 12.24 Pads for lead frame attachment.

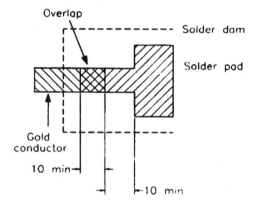

Figure 12.25 Wire bond distance from obstacle.

may be plated over the conductive surface, followed by a layer of electrolytic copper to increase the thickness.

A pattern may be generated in the plated surface by one of two methods. Conventional photolithographic methods may be used to etch the pattern, but this may result in undercutting and loss of resolution when used with thicker films. To produce more precise lines, a dry film photoresist may be utilized to generate a pattern on the electroless copper film that is the negative of the one required for etching. The traces may then be electroplated to the desired thickness using the photoresist pattern as a mold. Once the photoresist pattern is removed, the entire substrate may be immersed in an appropriate

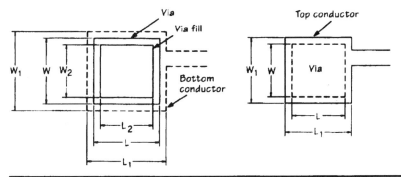

Dimension	Thick film, mils Preferred	Thick film, mils Minimum	Thin film, mils Preferred	Thin film, mils Minimum	Remarks
L	12.5	10	NA	NA	(1), (7), (8)
L_1	$L + 5$	L	NA	NA	(2)
L_2	L	L	NA	NA	(3), (9)
W	12.5	10	NA	NA	(4), (7), (8)
W_1	$L + 5$	L	NA	NA	(5)
W_2	W	W	NA	NA	(6), (9)

1. Via length.
2. Insert top and bottom.
3. Length of via fill.
4. Via width.
5. Conductor overlap, top and bottom.
6. Width of via fill.
7. It is preferred for inspection purposes that $L=W$.
8. It is preferred for inspection purposes that all vias on a substrate be the same size.
9. Via fills are required with more than two metallization layers. For more than two layers, the top via fill may be omitted.

Figure 12.26 Via and conductor pad sizes.

etchant to remove the unwanted material between the traces. Plated copper films created in this manner may be fired at an elevated temperature in a nitrogen atmosphere to improve the adhesion.

12.14.3 Active Metal Brazing Copper Technology

The active metal brazing (AMB) process utilizes one or more of the metals in the IV-B column of the periodic table, such as titanium, hafnium, or zirconium, to act as an activation agent with ceramic. These metals are typically alloyed with other metals to form a braze that can be used to bond copper to ceramic. One such example is an alloy of 70Ti/15Cu/15Ni, which melts at 960 to 1000°C. Numerous other alloys can also be used.[1]

The braze may be applied in the form of a paste, a powder, or a film. The combination is heated to the melting point of the selected braze in a vacuum to minimize oxidation of the copper. The active metal forms

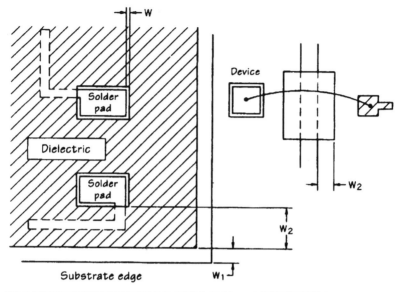

Figure 12.27 Thick film dielectric spacing and size.

a liquidus with the oxygen in the system that acts to bond the metal to the ceramic. After brazing, the copper film may be processed in much the same manner as DBC.

12.14.4 Comparison of Copper Metallization Technologies

The DBC and AMB processes are advantageous in applications where a thicker copper layer is beneficial. In particular, these technologies are very suitable for use in power circuits where their ability to handle high currents and to aid in dissipating heat can be utilized. Plated copper is most applicable to applications where fine lines and precise geometries are required, such as RF circuits. Design guidelines for each of these technologies are given in Table 12.10.

Table 12.10 Design Rules for Pure Copper Metallization

Technology	Line width (in.)		Etch factors* (in.)	Registration† (in.)	Copper thickness‡ (in)			Via diameters (in)		Integral leads from edge of ceramic	Camber§ (in/linear in)
	Min.	Typical			Min.	Typical	Max.	Min.	Typical		
Direct bond copper	0.015	0.020	0.004 to 0.008 with 0.020 pullback required from ceramic edge	±0.008	0.005	0.012	0.050	0.016	0.064	0.0100 in² required on ceramic for integral leads	±0.004
Plated copper	0.02	0.005	0.0005	±0.005	0.001	0.002	0.005	0.005	0.025	Leads not possible	±0.003
Active metal braze copper	0.015	0.020	0.004 to 0.008 with 0.020 pullback required from ceramic edge	±0.010	0.008	0.010	0.012	Vias not possible		0.100 in² required on ceramic for integral leads	±0.004

* The width of the artwork pattern determines final conductor width. Etching will reduce the original artwork width by several mils, depending upon copper thickness. Therefore, the artwork must have a wider dimension to compensate. This is called the *etch factor*.
† The artwork is registered to a feature on the ceramic substrate, usually a corner of the ceramic surface.
‡ Copper layers can be lapped to a lesser thickness after metallization.
§ Assume that both ceramic sides have copper metallization. In this case, camber depends on differences in percentage of ceramic surface coverage by copper.

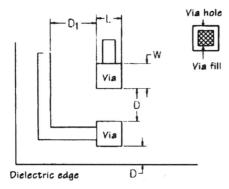

Dimension	Thick film, mils Preferred	Minimum	Remarks
L	20	10	*
W	20	10	*
D	20	10	†
D_1	15	10	‡
D_2	10	10	§

*It is preferred for inspection purposes that all vias on a substrate be the same size.
†This also applies to vias on adjacent layers. Vias must be staggered unless used for heat dissipation purposes.
‡Via distance to metallization.
§Conductor distance to metallization.

Figure 12.28 Dielectric via sizes and locations.

12.15 Summary of Substrate Metallization Technologies

The applications of the various substrate metallization technologies rarely overlap. Given a particular set of requirements, the choice is usually very apparent. Table 12.11 illustrates the properties of the different technologies.

12.16 Resistor Trimming

Printed and fired thick film resistors will have a spread of values distributed about some nominal value, normally about ±20 percent. The as-etched spread of thin film resistors is somewhat better, around ±10 percent. However, many designs call for 1 percent or even less for the initial tolerance. Therefore, further resistor adjustment is needed. The most common method of modifying the final resistance of the film is by

Table 12.11 Comparison of Hybrid Metallization Technologies

Technology	Ceramic selection	Metallization description	Adhesion mechanism	Geometry (typical)	Electrical and thermal	Hybrid assembly	Reliability	Cost
Thick film	Usually oxide, most types possible	Metal + glass	Chemical + mechanical	0.010 in (250 μm) lines and more than 0.0005 (12 μm) thickness	Poor	Good solder, rework can be a problem	Good, well understood	Variable, can use expensive precious metals
Thin film	All types	Pure metal, require adhesion layers	Chemical + mechanical	0.002 in (50 μm) lines and less than 0.0005 in (12 μm) thickness	Adequate	Adequate, not easily soldered	Good, well understood	High for equipment, process, and materials
Direct bond copper	More selective, oxide based	Pure copper	Chemical + mechanical	0.020 in (400 μm) lines and 0.0008 to 0.020 in (200 to 500 μm) thickness	Good	Good	Good, well understood	Reasonable
Plated copper	All types	Pure copper, usually thin adhesion layer	Chemical + mechanical	0.004 in (100 μm) lines and 0.0005 (12 to 125 μm) thickness	Good	Good	Less understood	Reasonable
Active metal braze copper	All types	Pure copper, with braze adhesion layer	Chemical + mechanical	0.020 in (400 μm) lines and 0.0008 to 0.020 in (200 to 500 μm) thickness	Good	Good	Adequate	Reasonable

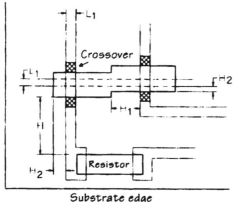

Dimension	Thick film, mils		Remarks
	Preferred	Minimum	
H	50	30	*
H_1	20	10	†
H_2	20	10	‡
L_1	15	10	§

*Distance to resistor.
†Conductor distance to dielectric step. Use staggered crossovers or H patterns when two or more adjacent lines must cross in the same direction. (See Fig. 11.40.)
‡Overlap distance on crossover.
§Conductor lines overlap.

Figure 12.29 Dielectric distance on crossovers.

mechanical "trimming," or removing a portion of the resistor to alter the effective number of squares. Equation (6.3) shows that the actual resistance depends on the ratio of L/W, the number of squares, while Fig. 12.36 shows how a two square resistor is effectively made into a four square resistor by cutting a notch into it.

As a first approximation, when a resistor is trimmed, only the untrimmed area is active, depicted by the shaded area in Fig. 12.37. If a cut is made a distance of 50 percent into the resistor, the width is decreased by 50 percent and, therefore, the number of squares is increased to 4. However, referring to Fig. 12.38, this is not a completely accurate picture of the electric field within a trimmed resistor. Part of the trimmed resistor is still active in the circuit. It is this phenomenon that allows us to make extremely accurate resistor trims.

The value of thick film resistors may be increased by removing part of the resistor either with a laser beam or with a stream of abrasive particles propelled by compressed air. Of the two methods, laser trim-

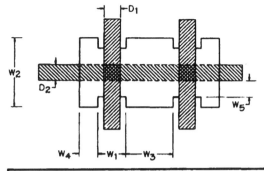

Dimension	Thick film, mils		Remarks
	Preferred	Minimum	
D_1	10	5	(1)
D_2	10	5	(2)
W_1	$D_1 + 10$	$D_1 + 10$	(3)
W_2	$D_1 + 10$	$D_1 + 10$	(4)
W_3	10	5	(5)
W_4	15	10	(6)
W_5	10	5	(7)

1. Top conductor widths.
2. Bottom crossover width.
3. Width of crossover.
4. Overlap of bottom conductor.
5. Width of dielectric print between crossovers.
6. Width of crossover arm.
7. Extension of crossover past conductor.

Figure 12.30 H-pattern crossovers.

ming is the most often used in production, although abrasive trimming has extensive use in trimming resistors used in power hybrid circuits.

12.16.1 Laser Trimming

Commercially available trimming systems are of the "closed-loop" type in which the resistor being trimmed is continually probed and the value fed back to the trimming system. When the value of the resistor falls within a prespecified range, the trimming process is terminated. A typical laser trim system is shown in Fig. 12.39.

More than any other development, laser trimming has contributed to the rapid growth of the hybrid microelectronics industry over the past few years. Once it became possible to trim resistors to precision values in less than a second, hybrid circuits became cost competitive with other forms of packaging, and the technology expanded rapidly.

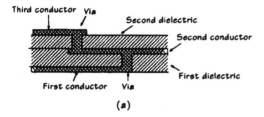

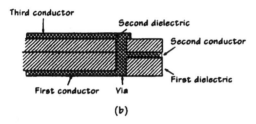

Figure 12.31 Via location on adjacent layers, (a) permitted and (b) not permitted. The only exception to this rule is when vias are used as heat sinks under components.

Lasers typically deliver enormous bursts of power in a short time, but this is not suitable for trimming thick film resistors, as the laser will penetrate too deeply into the substrate, causing structural damage to the resistor and/or the substrate, and further causing the resistor to be unstable. By the same token, lasers such as the CO_2 laser are too powerful, and the same type of damage results. Lasers used to trim thick film resistors are generally based on the yttrium-aluminum-garnet (YAG) crystal that has been doped with neodymium. YAG lasers are of lower power and are capable of trimming a thick film resistor without causing serious damage if the parameters are adjusted properly.

Even with the YAG laser, it is necessary to lower the maximum power and spread out the beam to avoid overtrimming and to increase the trimming speed. This is accomplished by a technique called "Q-switching," which decreases the peak power and widens the pulse width such that the overall average power is the same. One of the effects of this technique is that the first and last pulses of a given trim sequence have substantially greater energy than the intervening ones. The first pulse does not create a problem, but the last pulse occurs within the body of the resistor, penetrating deeper into the substrate and creating the potential for resistor drift by creating minute cracks (microcracks) in the resistor, which emanate from the point of trim termination.

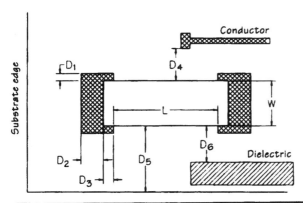

Dimension	Thick film, mils Preferred	Minimum	Remarks
L	40	20	(1)
W	See Table 11.4		(2)
D_1	10	5	(3)
D_2	10	5	(4)
D_3	10	7.5	(5)
D_4	20	15	(6)
D_5	30	20	(7)
D_6	20	20	(8)

1. Aspect ratio $0.5 \leq L/W \leq 5$ preferred, $0.3 \leq L/W \leq 10$ maximum.
2. Depends on the more restrictive of tolerance and power.
3. Minimum excess conductor width.
4. Minimum excess conductor length.
5. Resistor overlap onto conductor.
6. Conductor distance from resistance.
7. Distance from edge of substrate.
8. Distance to multilayer or crossover dielectric.

Figure 12.32 Thick film resistor dimensions.

The microcracks occur as a result of the large thermal gradient that exists between the area of the last pulse and the remainder of the resistor. During the life of the resistor, they will propagate through the body of the resistor until they reach a termination point, such as an active particle or even another microcrack. The propagation distance can be up to several mils in a high-value resistor with a high concentration of glass. The microcracks cause an increase in the value of the resistor and also increase the noise by increasing the current density in the vicinity of the trim. With proper design of the resistor, and with proper selection of the cut mode, the amount of drift can be held to less than 1 percent over the life of the resistor.

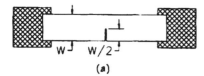

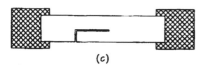

Figure 12.33 Laser trim cut modes. Do not penetrate more than halfway into the resistor. When trimming to less than 50% tolerance, use the double-plunge cut or L-cut mode. Trim to ≈5% or until $W/2$ is reached with a single-plunge cut, then switch to a double-plunge or L cut.

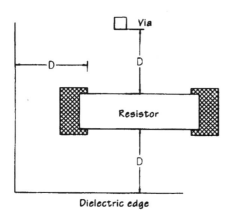

Dimension	Thick film, mils		Remarks
	Preferred	Minimum	
D	20	20	*, †, ‡

*Distance to edge of dielectric or via when resistors are printed on top of dielectric.
†The same dimensions for resistors in Fig. 11.43 also hold.
‡The use of resistors on dielectric requires an alternative set of resistor curves.

Figure 12.34 Dimensions of resistors on dielectric.

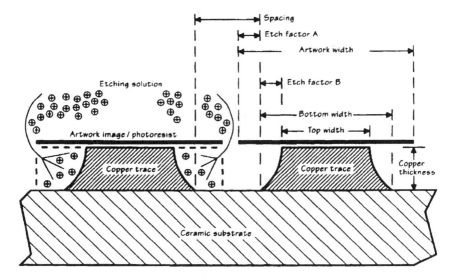

Figure 12.35 Design guidelines for photolithographic etch process.

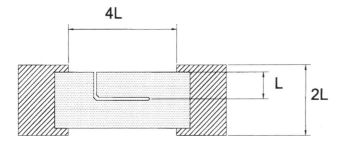

Figure 12.36 Resistor trimming.

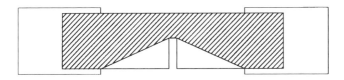

Figure 12.37 Active resistor area.

The propagation of the microcracks can be accelerated by the application of heat, which also has the effect of stabilizing the resistor. Where high-precision resistors are required, the resistors may be trimmed slightly low, exposed to heat for a period of time, and retrimmed to a value in a noncritical area.

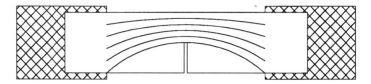

Figure 12.38 Field lines in a trimmed resistor.

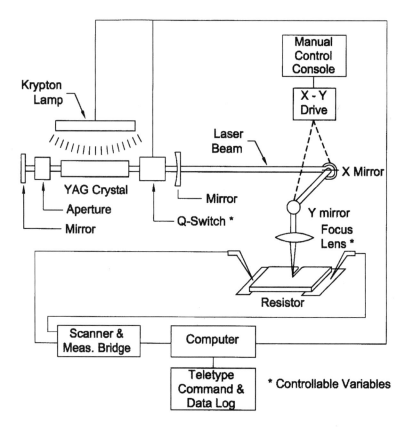

Figure 12.39 Block diagram of a laser resistor trimmer.

The amount of drift depends on the amount of propagation distance relative to the distance from the termination point of the trim to the far side of the resistor. This can be minimized by either making the resistor larger or by limiting the distance that the laser penetrates into the resistor. As a general rule of thumb, the value of the resistor should not be increased by more than a factor of two to minimize drift.

There are a variety of cut modes that can be used to trim resistors, as shown in Fig. 12.40. The selection of a cut mode is a compromise be-

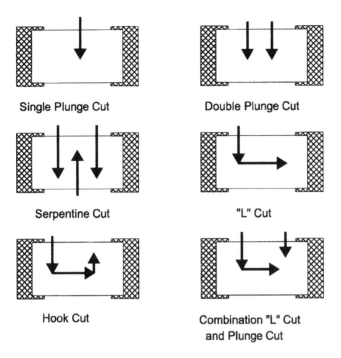

Figure 12.40 Laser trim cuts.

tween trim time, stability, and precision. Trim time can be a very significant factor in high-volume production, as a difference of only a fraction of a second in trim time can be significant where a large number of resistors are involved.

Single plunge cut. This is the simplest and most economical of all the trim methods, but it offers the least precision and the lowest stability. As the laser penetrates deeper into the resistor, the change in resistance per pulse becomes larger, causing a loss of sensitivity and resulting in a wider spread of final values and a higher tolerance. The single plunge cut should not be used to trim resistors where the desired tolerance is less than 5 percent. Furthermore, for a fixed change in resistance, a single plunge cut must penetrate deeper into the resistor than other cut modes. For a resistor with an aspect ratio of 2, the laser must penetrate approximately 70 percent of the resistor width to double the value. This creates instability by locating the last pulse in closer proximity to the edge of the resistor. As a general rule, the extent of penetration into the resistor should be limited to 50 percent of the resistor width. This limits the extent to which the value of the resistor can be increased.

Double plunge cut. The double plunge cut may be used either where greater precision is required or where the extent of penetration exceeds the limits. The laser may be programmed until the resistor reaches approximately 95 percent of the desired value or until the penetration reaches 50 percent of the resistor width, at which point the laser begins a second plunge cut. With the double plunge cut, resistors may be routinely trimmed to a tolerance of 1 percent or better. While the range of trim is increased by this method, it is still limited to about 1.6/1 if the rule limiting penetration to 50 percent is observed.

"L"-cut. The L-cut offers improvements in the trim time and the range of trim without sacrificing precision. However, for the L-cut to be completely effective, the resistor must have dimensions of greater than 0.030 in.

Hook cut. The hook cut is basically an L-cut in which the trim direction doubles back toward the point of initial penetration when the probing system senses that the value is within a few percent of the final value. In this manner, the final pulse is applied where most of the microcracks have less effect on the overall resistance. Resistors trimmed in this manner are inherently more stable than an equivalent trim using other cut modes. This mode requires added time over the other methods.

Serpentine cut. The serpentine cut utilizes plunge cuts from opposite sides of the resistor. In this manner, the value of the resistor can be increased by a factor of several times, but at the expense of greater instability, increased noise, and increased trim time. This mode is generally used only during the developmental phase of the hybrid circuit for these reasons.

The setup of the laser is very critical in achieving resistor precision and stability. The laser should be sharply focused with a beam size of less than 0.002 in. The beam power should be sufficient to provide a clean kerf without detritus and, at the same time, not be so high as to create an excess of microcracks. It has been shown that the best way to provide a given average power is to use a high peak power at a low pulse frequency. This combination results in rapid vaporization of the resistor material without excess damage to the region surrounding the kerf. The pulse frequency should be just high enough to obtain a continuous kerf at the desired speed. Typical ranges of trim parameters for a YAG laser are listed below.

- Average power   1.0 to 1.5 W
- Pulse frequency   2.5 to 3.5 kHz
- Trim speed   0.5 to 1.5 in/s

12.16.2 Abrasive Trimming

In the abrasive trimming process, a fine stream of sand propelled by compressed air is directed at the resistor, abrading away a portion of the resistor and increasing the value. Trimming resistors to a high precision is more difficult with abrasive trimming due to the size of the particle stream, although tolerances of 1 percent can be achieved with proper setup and slow trim speeds. It has also proven difficult to automate the abrasive trimming process to a high degree, and it remains a relatively slow process. In addition, substantially larger resistors are required.

Still, abrasive trimming plays an important role in the hybrid industry. The cost of an abrasive trimmer is substantially less than that of a laser trimmer, and the setup time is much less. In a developmental mode, where only a few prototype units are required, abrasive trimming is generally more economical and faster than laser trimming.

In terms of performance, abrasively trimmed resistors are more stable than laser-trimmed resistors and generate less noise due to the fact that no microcracks are generated during the abrasive trimming process. In the power hybrid industry, resistors that carry high currents or dissipate high power are frequently abrasively trimmed by trimming a groove in the middle of the resistor. This technique minimizes current-crowding and further enhances the stability.

12.17 Assembly of Hybrid Circuits

There are two basic methods of assembling hybrid circuits.

- The *chip-and-wire* approach, in which semiconductor devices in the unpackaged form are mechanically attached to the substrate metallization by epoxy, solder, or eutectic bonding, and electrically connected by wire bonding
- The *surface mount* approach, in which packaged devices are soldered to the substrate, making both the electrical and mechanical connections simultaneously

These methods are not mutually exclusive; it is very common to combine these approaches on a single substrate. This chapter is primarily concerned with the chip-and-wire technology.

12.17.1 Chip-and-Wire Technology

Semiconductor devices in the unpackaged state are very delicate. The metallization patterns are extremely thin and easily damaged, even with normal handling. The input/output (I/O) terminals are electrically open, and electrical charge has no path to ground, making the die extremely susceptible to damage by electrostatic discharge (ESD). ESD can be generated from a variety of sources, as shown in Table 12.12. The die must be handled with extreme care, using a vacuum pickup (as opposed to tweezers) that is designed to dissipate electrical charge. Operators must be properly grounded, and ionized air blowers may be necessary when especially susceptible devices are being mounted.

Table 12.12 Typical Electrostatic Voltages

	Relative humidity, %		
Event	10	40	55
Walking across carpet	35,000	15,000	7,500
Walking across vinyl floor	12,000	5,000	3,000
Motions of bench worker	6,000	800	400
Remove DIPs from plastic tubes	2,000	700	400
Remove DIPs from vinyl trays	11,500	4,000	2,000
Remove DIPs from Styrofoam®	14,500	5,000	3,500
Remove bubble pack from PWBs	26,000	20,000	7,000
Pack PWBs in foam-lined box	21,000	11,000	5,500

12.17.2 Direct Eutectic Bonding of Semiconductor Devices

Silicon forms a eutectic composition with gold in the ratio of 94 percent gold and 6 percent silicon by weight, which melts at 370°C. Gold is the only metal to which the eutectic temperature in combination with silicon is sufficiently low to be practical. By contrast, the eutectic temperature of silicon and silver is nearly 800°C.

In the eutectic bonding process, the substrate is preheated to about 200°C to minimize thermal shock and then transferred to a heated stage at about 400°C. The silicon die is picked up by a heated collet to further minimize thermal shock, which is sized to match the size of the die and is connected to a small vacuum pump for the purpose of holding the die in place. The collet is also connected to a motor capable of providing mechanical scrubbing to assist in making the bond. The die is picked up by the collet and transported to the desired metallization pattern on the substrate. Simply placing the die in contact with the heated substrate will not automatically form the bond; the materials must be in the proper proportion. This is accomplished by mechanically scrubbing the die into the gold metallization. During the scrubbing process, the eutectic alloy will be formed at some random point along the interface, which will then become molten. This, in return, will rapidly dissolve more material until the entire interface is liquid and the bond is formed. Devices larger than 0.020 × 0.020 in. tend to crack when mechanically scrubbed and require a gold-silicon preform to make the bond.

The eutectic process makes a very good bond (mechanically, electrically, and thermally), but the high bonding temperature is an extreme disadvantage for a variety of reasons, and this process is generally used only in single-chip applications. Where a metallurgical bond is required, solder is the preferred process. Semiconductor devices intended for solder bonding are typically metallized with a titanium-nickel-silver alloy on the bottom.

12.17.3 Organic Bonding Materials

The most common method of mechanically bonding active and passive components to a metallized substrate is with epoxy. Most epoxies for hybrid circuit applications have a filler added for electrical and/or thermal conductivity. The most common conductive filler is silver. Silver has a high electrical conductivity, and a smaller proportion of silver to epoxy is required to produce a given resistivity than other metals. Silver epoxy, therefore, has a higher mechanical strength, since more epoxy is present. Other conductive filler materials include gold, palladium-silver for reduced silver migration, and tin-plated copper. Nonconductive filler materials include aluminum oxide, beryllium oxide, and magnesium oxide for improved thermal conductivity.

Epoxy may be dispensed by screen printing, pneumatic dispensing through a nozzle, and by the die transfer method. Screen printing requires a planar surface, while pneumatic dispensing and die transfer may be used with irregular surfaces. Where many die are to be mounted on a single substrate, screen printing is the preferred method, because epoxy can be applied to all mounting pads in a single pass of the screen printer.

Where large surfaces are to be mounted, such as substrates, liquid epoxies are difficult to use. For this application, a large sheet of glass cloth can be impregnated with epoxy, which is then B-staged, or partially cured, to produce a solid structure of epoxy generally referred to as a *preform*. The sheet may then be cut into smaller pieces and used to mount large components and substrates. Where high thermal conductivity is required, the mesh may be made from silver wire and silver epoxy used.

Epoxy as a die attach material suffers from two major disadvantages.

1. The thermal and electrical resistances are quite high compared with direct methods of bonding such as soldering or eutectic bonding.
2. The operating temperature is limited to about 150°C, as many epoxies begin breaking down and outgassing at temperatures above this figure.

These factors inhibit the use of epoxy in mounting power devices, in applications where a high processing temperature is utilized, where performance at a high ambient temperature is required, or where a low bond resistance is needed. One phenomenon that can occur when the epoxy is operated at elevated temperatures near or above the glass transition temperature (T_g) is the settling of the filler away from the device interface, leaving a thin layer of resin at the bond line and increasing the electrical resistance of the bond to the point where circuit failure occurs. Another failure mechanism that can occur is silver migration, which can be accelerated by water absorption of a resin with a relatively high mobile ion content.

Other materials utilized for die bonding include polyimide and certain of the thermoplastic materials. Polyimide is stable out to around 350°C, unless filled with silver, which acts as a catalyst to promote adhesion loss. Polyimides are typically dissolved in a solvent, such as xylene, necessitating a two-step curing process. The first step, at a low temperature, evaporates the solvent, and the second step, at an elevated temperature, cross-links the polyimide. Thermoplastic materials are particularly advantageous in high-volume applications, as the cure cycle is much shorter than that of either epoxy or polyimide.

12.17.4 Solder Attachment

In the soldering process, an alloy of two or more metals is melted at the interface of two metal surfaces. The molten solder dissolves a portion of the two surfaces and, when the solder cools, a junction, or *solder joint,* is formed, joining the two metal surfaces.

For the solder joint to occur, both metal surfaces and the solder must be clean and free from surface oxides. Removal of the oxides is accomplished by the use of a *flux,* an organically based acid. Fluxes are categorized by their strength and by the requirement for cleaning. Fluxes requiring solvent cleaning are rapidly being phased out due to environmental regulations. The so-called "no-clean" fluxes, which remain on the circuit after the soldering process, are seeing extensive use in commercial applications. Water-soluble fluxes are still widely used in the printed circuit board industry. Surfaces that are heavily oxidized before soldering may require the use of a stronger flux such as an resin mildly activated (RMA) flux.

Solder materials are selected primarily for their compatibility with the surfaces to be soldered and their melting point. Solders are generally divided into two categories: "hard" solders, which have a melting point above about 500°C and are often referred to as *brazes,* and "soft" solders, which have a lower melting point. Soft solders are used primarily in the assembly process, while hard solders are used in lead attachment and package sealing.

Soft solders may also be divided into two categories: eutectic solders and non-eutectic solders. Eutectic solders have the lowest melting point and are typically more rigid in the solid state than other solders with the same constituents. This is due in part to the fact that eutectic solders go directly from the liquid to the solid state without going through a "plastic" region. Their lower melting point makes them very attractive in many applications, and their use is quite common. Some of the more common solders and their characteristics are listed in Table 12.13.

The compatibility of the solder with the material in the metal surfaces must be a prime consideration when selecting a solder, in particular the tendency of the metal to *leach* into the solder and the tendency to form intermetallic compounds, which might prove detrimental to reliability. Leaching is the process by which a material is absorbed into the molten solder to a high degree. While a certain degree of leaching must occur to produce the solder joint, excessive leaching can cause the metallization pattern to vanish into the molten solder, creating an open circuit. Tin-bearing solders used in conduction with gold or silver conductors are especially prone to this phenomenon, since these materials have a strong affinity with tin. A thick or thin film gold or silver conductor will dissolve into a tin-lead solder in a matter of seconds. The addition of platinum and/or palladium to the thick film conductor materials will greatly enhance the leach resistance to tin-bearing solders, but it must still be a consideration. The leach resistance is proportional to the amount of Pt and Pd added, but this has the adverse effect of adding both to the cost and the electrical resistance. Soldering at the lowest possible temperature for the shortest possible time consistent with forming the solder joint can minimize leaching. The addition of a small amount of silver to the solder lowers the melting point slightly, and, with proper control of the soldering temperature, the silver already present in the solder partially saturates the solution, further inhibiting leaching. Soldering directly to gold requires the use of AuSn, PbIn, or other solders that do not leach

Table 12.13 Processing Temperatures of Selected Organic and Metallic Attachment Materials

Organic attachment materials	
Material	Temperature (°C)
Polyimide*	250–350
Epoxy	150

Metallic attachment methods		
Alloy†	Liquidus‡ (°C)	Solidus (°C)
In 52 – Sn 48§	118	118
Sn 62.5 – Pb 36.1 – Ag 1.4§	179	179
Sn 63 – Pb 37§	183	183
In 60 – Pb 40	185	174
Sn 60 – Pb 40	188	183
Sn 96.5 – Ag 3.5§	221	221
Pb 60 – Sn 40	238	183
Pb 70 – Sn 27 – Ag 3	253	179
Pb 92.5 – Sn 5 – Ag 2.5	280	179
Sn 90 – Ag 10	295	221
Pb 90 – Sn 10	302	275
Au 88 – Ge 12§	356	356
Au 96.4 – Si 3.6§	370	370
Ag 72 – Cu 28§	780	780

* Polyimide materials may also require a precure step at 70°C to remove solvents
† Numerical values are percentages
# Eutectic composition
‡ The processing temperature of most alloys is ≥ 20°C above the liquidus

gold. The added Au in the AuSn combination inhibits leaching of the gold in the film.

Intermetallic compound formation is also a consideration. Certain compounds have a high electrical resistance and are susceptible to mechanical failure when exposed to temperature cycling or storage at temperature extremes. Tin forms intermetallic compounds with both gold and copper, and indium also forms intermetallic compounds with copper. The most commonly used solders are the tin-lead solders, which are used extensively on copper and the PdAg and PtAg alloys.

Lead-free solders, such as the tin-silver combinations, are becoming more widely used. Apart from environmental concerns, the tin-silver solders have proven to be more resilient to extensive temperature cycling conditions.

In the soldering process, it is desirable to minimize the exposure of the part to elevated temperatures. High temperatures accelerate the rate of chemical reactions that may be detrimental to the reliability of the circuit. Furthermore, excessive exposure of metal surfaces to liquid solder increases the rate of formation of intermetallic compounds and also increases leaching.

Solders for microelectronic applications are generally in the form of a paste, with the solder in powder form mixed with an appropriate flux and a dispensing vehicle. Solder in paste form may be screen printed or dispensed pneumatically. The part is placed in the wet solder paste prior to soldering by an automatic pick-and-place system or manually.

The most effective method of soldering is to place the part in conjunction with the solder paste into a tunnel furnace with several heated zones. By controlling the speed of the belt and the temperature of the individual zones, a time-temperature relationship, or *profile,* can be established that will optimize the soldering process. Heating of the part may be accomplished by resistance heating, by infrared (IR) heating, or by a combination of both.

The use of a nitrogen or forming gas blanket during the soldering process to prevent oxidation of the solder and/or the surfaces aids greatly in soldering, particularly to nickel, and improves the wetting of tin-bearing solders. The formation of a gas blanket can be easily accomplished by connecting a gas source to the furnace and by the use of baffles at the ends of the furnace to minimize the intrusion of air into the heated region. A typical profile has a duration of several minutes and has a plateau just below the melting point of the solder for a period of time followed by a rapid rise in temperature, or *spike,* above the melting point for a short duration, and a gradual decline down to room temperature. In general, the duration of the spike should be held as short as possible, consistent with good solder flow and wetting to minimize such effects as leaching of conductor materials and the effect of high temperatures on the components.

12.17.5 Wire Bonding

Ohmic contacts to semiconductor devices are typically made with aluminum, since that material diffuses well into the silicon structure at a moderate temperature. Wire bonding is used to make the electrical connections from the aluminum contacts to the substrate metallization or to a lead frame, from other components, such as chip resistors or chip capacitors, to substrate metallization, from package terminals to the substrate metallization, or from one point on the substrate metallization to another.

There are two basic methods of wire bonding: thermocompression wire bonding, which uses primarily gold wire, and ultrasonic wire bonding, which uses primarily aluminum wire. Thermocompression wire bonding, as the name implies, utilizes a combination of heat and pressure to form an intermetallic bond between the wire and a metal surface. In thermocompression bonding (Fig. 12.41), a gold wire is passed through a hollow capillary, generally made from a refractory metal such as tungsten, and a ball formed on the end by means of an electrical arc. The substrate is heated to about 300°C, and the ball is forced into contact with the bonding pad on the device with sufficient force to cause the two metals to bond. The capillary is then moved to the bond site on the substrate, feeding the wire as it goes, and as the wire is bonded to the substrate by the same process, except that the bond is in the form of a "stitch," as opposed to the "ball" on the device.

The wire is then clamped and pulled to break just above the stitch, and another ball formed as above. Thermocompression bonding is rarely used for a variety of reasons.

- The high substrate temperature precludes the use of epoxy for device mounting.

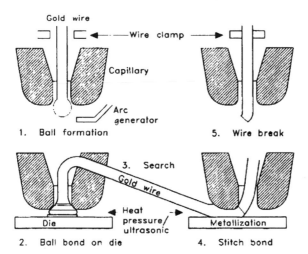

Figure 12.41 Typical ball bonding sequence.

- The temperature required for the bond is above the threshold temperature for gold-aluminum intermetallic compound formation. The diffusion rate for aluminum into gold is much greater than for gold into aluminum. The aluminum contact on a silicon device is very thin, and, when it diffuses into the gold, voids, called *Kirkendall voids,* are created in the bond area, increasing the electrical resistance of the bond and decreasing the mechanical strength.
- The thermocompression bonding action does not effectively remove trace surface contaminants that interfere with the bonding process.

The ultrasonic bonding process (Fig. 12.42) uses ultrasonic energy to vibrate the wire against the surface to combine the atomic lattices together at the surface. Localized heating at the bond interface caused by the scrubbing action, aided by the oxide on the aluminum wire, assists in forming the bond. The substrate itself is not heated. Intermetallic compound formation is not as critical as with the thermocompression bonding process, as both the wire and the device metallization are aluminum. Kirkendall voiding on an aluminum wire bonded to gold substrate metallization is not as critical, since there is substantially more aluminum available to diffuse than on device metallization. Ultrasonic bonding makes a stitch on both the first and second bonds. For this reason, ultrasonic bonding is somewhat slower than thermocompression, since the capillary must be aligned with the second bond site when the first bond is made. Ultrasonic bonding to package leads may be difficult if the leads are not tightly clamped, since the ultrasonic energy may be propagated down the leads instead of being coupled to the bond site.

The use of thermosonic bonding largely overcomes the difficulties noted with thermocompression bonding. In this process, used with gold

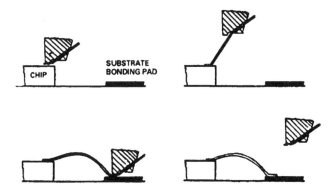

Figure 12.42 Aluminum ultrasonic wedge–wedge bonding.

wire, the substrate is heated to 150°C, and ultrasonic energy is coupled to the wire through the transducer action of the capillary, scrubbing the wire into the metal surface and forming a ball-stitch bond from the device to the substrate as in thermocompression bonding.

Thermosonic gold bonding is the most widely used bonding technique, primarily because it is faster than ultrasonic aluminum bonding. Once the ball bond is made on the device, the wire may be moved in any direction without stress on the wire, which greatly facilitates automatic wire bonding, as the movement need be only in the x and y directions.

By contrast, before the first ultrasonic stitch bond is made on the device, the circuit must be oriented so that the wire will move toward the second bond site only in the direction of the stitch. This requires rotational movement, which not only complicates the design of the bonder but increases the bonding time as well.

Attempts at designing an aluminum ball bonder have not proven successful to date because of the difficulty in making a ball on the end of the wire. Aluminum bonding is commonly used in hybrids or chip carriers where the package-sealing temperature may exceed the threshold temperature for the formation of intermetallic compounds, in power hybrids where the junction temperature is high, and in applications where large wires are required. Gold wire is difficult to obtain in diameters above 0.002 in., while aluminum wire is available up to 0.022 in.

Automatic thermosonic gold bonding equipment now exists that can bond up to four wires per second under optimal conditions. The actual rate depends on the configuration of the bonding pattern and the number of devices to be bonded. The utilization of an automatic bonder with pattern recognition is strongly dependent on the accuracy of the device placement and on the quality of the substrate metallization. If the devices are not placed within a few mils of the bond site and within a few degrees of rotation, the pattern recognition system will not be able to locate the device, resulting in lost production time while the operator feeds in the coordinates.

The substrates must be kept very clean and should be handled only with finger cots during the assembly process. In storage, the substrates should be kept in a clean, dry area. Thin film metallization makes the best bonding surface, since it is uniform in thickness and is almost pure gold. There exists a considerable variation in bondability among the various thick film golds, and bonding studies under a variety of conditions must be an integral part of the selection process of a thick film gold material. Thick film gold for use with aluminum wire must contain palladium to improve the strength of the bond under conditions of thermal aging. The same principle applies to aluminum bonding on palladium-silver materials. As the palladium-to-silver ratio is increased, the reliability of the bonds increases.

The integrity of the wire bonds may be tested by placing a small hook under the wire and pulling at a constant rate of speed, usually very slow, until the bond fails. The amount of force required is the *bond strength*. The mode of failure is just as important as the actual bond strength. There are five points at which a bond can fail.

1. The ball (or stitch) at the device may fail.
2. The wire may break just above the stitch at the device end.
3. The wire may break in the center.
4. The wire may break just above the stitch at the substrate end.
5. The stitch at the substrate may lift.

Of the five options, number three is the most desirable, with two and four following in turn. If an excessive number of lifts occur, particularly at the device end, a serious bonding problem may be indicated. Some of the more common bonding problems and their possible causes are listed below.

- Excessive bond lifts at the device. This problem may be a result of one of three factors: improper bond setup, contamination on the device, or incomplete etching of the glass passivation on the device.

- Excessive bond breaks just above the device. This is probably caused by either a worn bonding tool or by excessive bonding pressure, both of which can crimp the wire and create a weak point.

- Excessive bond breaks just above the substrate. This is probably caused by excessive bonding pressure. This may result from improper bonding setup, which in turn may be due to contamination or to poor substrate metallization. A worn bonding tool may result in the wire being nicked, creating a possible failure point.

- Excessive stitch lifts at the substrate. This may be due to a worn bonding tool, poor bonder setup, or poor device metallization.

The wire size is dependent on the amount of current that the wire is to carry, the size of the bonding pads, and the throughput requirements. Table 12.14 illustrates the wire size for a specified current for both gold and aluminum wire. For applications where current and bonding size is not critical, 0.001-in. wire is the most commonly used size. While 0.0007-in. wire is less expensive, it is difficult to thread through a capillary and bond without frequent breaking and consequent line stoppages.

Table 12.14 Maximum Current for Wire Size

Material	Diameter (in.)	Maximum current (A)	
		$L < 0.040$ in.	$L > 0.040$ in.
Gold	0.001	0.949	0.648
	0.002	2.683	1.834
Aluminum	0.001	0.696	0.481
	0.002	1.968	1.360
	0.005	7.778	5.374
	0.008	15.742	10.876
	0.012	28.920	19.981
	0.015	40.417	27.924
	0.022	71.789	49.600

12.18 Packaging

Packaging for individual devices or complete hybrid assemblies can be characterized as shown in Table 12.15. There are many driving forces and considerations in the choice of technology for packaging and the subsequent next level assembly technologies.

Table 12.15 Characterization of Package Types

Leads	Type
Through-hole	Hermetic
Plug-in	Metal
Dual in-line (DIP)	Ceramic
Pin grid array (PGA)	
TO styles	
Surface mount	Non-hermetic
Small outline (SOIC, SOT)	Plastic
Chip carrier (leadless, leaded)	
Thin small outline (TSOP)	
Ball grid array (BGA)	

12.18.1 Hermetic Packages

A true hermetically sealed package would prevent intrusion of contaminants (liquid, solid, or gas) for an indefinite period of time. In practice, however, this is not realistic. Even in a perfectly sealed structure, diffusion phenomena will occur over time, allowing the smaller molecules, such as helium or water vapor, to penetrate the barrier medium and ultimately reach equilibrium within the package interior. A hermetic package is defined as one in which the leak rate of helium after pressurization is below a specified rate with reference to the package size as shown in Table 12.16.

Table 12.16 Definition of Hermetic Package

Package volume (cm^3)	Bomb condition			Reject limit (atm–cm^3/s He)
	PSIG	Exposure time (hr)	Maximum dwell (hr)	
V < 0.40	60 ± 2	2 + 0.2, –0	1	5×10^{-8}
V > 0.40	60 ± 2	2 + 0.2, –0	1	2×10^{-7}
V > 0.40	30 ± 2	4 + 0.4, –0	1	1×10^{-7}

A hermetic package must either be metal, ceramic, or glass. Organic packages, or packages with an organic seal, may initially pass the pressurization test described above but will allow water vapor to pass back and forth from the atmosphere to the package interior. Therefore, they are not truly hermetic. Interconnections through a metal package may be insulated by glass-to-metal seals utilizing glass that matches the coefficient of expansion to the metal.

A hermetic package allows the circuit mounted inside to be sealed in a benign environment, generally nitrogen, which is obtained from a liquid nitrogen source. Nitrogen of this type is extremely dry, with a moisture content of less than 10 ppm. As a further precaution, the open package with the enclosed circuit mounted inside is subjected to an elevated temperature, usually 150°C, in a vacuum to remove absorbed and absorbed water vapor and other gases prior to sealing. For added reliability, the moisture content inside a package should not exceed 5000 ppm. This figure is below the dew point of 6000 ppm at 0°C, ensuring that any water that precipitates out will be in the form of ice, which is not as damaging as water in the liquid form.

A hermetic package adds considerably to the reliability of a circuit by guarding against contamination, particularly of the active devices. An active device is susceptible to a number of possible failure mechanisms, such as corrosion and inversion, and may be attacked by something as benign as distilled, deionized water, which can leach phosphorous out of the passivating oxide to form phosphoric acid which, in turn, can attack the aluminum bonding pads.

12.18.2 Metal Packages

The most common type of hermetic package is the metal package, which is fabricated primarily from ASTM F-15 alloy, Fe52Ni29Co18 (known also as Kovar®). The tub-type package is fabricated by forming a sheet of the F-15 alloy over a set of successive dies. Holes for the leads are then punched in the bottom for plug-in packages and in the side for flat packages. A layer of oxide is then grown over the package body. Beads of borosilicate glass, typically Corning 7052 glass, are placed over the leads and placed in the holes in the package body. Heating the structure above the melting point of the glass (approximately 500°C) forms a reactive glass-to-metal seal. The molten glass dissolves some of the oxides on the alloy (primarily iron oxide) which, upon cooling, provided the adhesion mechanism. The glass-to-metal seal formed in this manner has four distinctive layers.

1. Metal
2. Metal oxide
3. Metal oxide dissolved in glass
4. Glass

After the glass-to-metal seals have been formed, the oxide not covered by the glass must be removed and the metal surface plated to allow the package to be sealed and to allow the package leads to be soldered to the next higher assembly. The prime plating material is electrolytic nickel, although gold is frequently plated over the nickel to aid in sealing and to prevent corrosion. In either case, the package leads are plated with gold to allow wire bonding and to improve solderability. Although electroless nickel has better solderability, it tends to crack when the leads are flexed.

The glass-to-metal seal formed in this manner provides an excellent hermetic seal, and the close match in TCE between the glass and the F-15 alloy (approximately $5.0 \times 10^{-6}/°C$) maintains the hermeticity through temperature cycling and temperature storage.

There are three types of lids commonly used on metal packages: the domed lid, the flat lid, and the stepped lid. These are also fabricated with ASTM F-15 alloy with the same plating requirements as the packages. The domed lid is designed for use with platform packages and may be projection welded or soldered. The flat lid is designed for use with the tub package and is primarily soldered to the package.

The stepped lid is fabricated by photoetching a groove in a solid sheet of F-15 alloy such that the flange is about 0.004 in. thick. This lid is designed to be seam welded to a tub package. When lids are designed for soldering, a preform of the desired solder material is generally attached to the outer perimeter of the bottom of the lid.

12.18.3 Methods of Sealing Metal Packages

A flat lid or a stepped lid may be soldered to the package by hand, by the use of a heated platen, or in a furnace. While the platen is somewhat faster, the metal package acts as a heat sink, simultaneously drawing heat away from the seal area and raising the temperature inside the package, unless the glass beads used for insulating the leads extend entirely around the periphery of the package, an obviously impractical requirement. In addition, leaks through the solder, or "blow holes," caused by a differential in pressure between the inside and outside of the package, will occur unless the ambient pressure outside the package is increased at the same rate as the pressure inside created by heating the package. Because of the temperature rise inside the package, it is risky to use epoxy to mount components unless the glass beads extend around the periphery of the package as described above. Solder sealing may be accomplished in a conveyor-type furnace, which has a nitrogen atmosphere. The nitrogen prevents the oxidation of the solder and also provides a benign environment for the enclosed circuit. Furnace sealing requires a certain degree of fixturing to provide pressure on the lid.

Parallel seam welding is accomplished by the generation of a series of overlapping spot welds by passing a pair of electrodes along the edge of the lid. The alignment of the lid to the package is quite critical and is best accomplished outside the sealing chamber, with the lid being tacked to the package in two places by small spot welds. A stepped lid greatly facilitates the process and improves the yield, since it requires considerably less power than a flat lid of greater thickness. While the sealing process is relatively slow compared with other methods, a package sealed by parallel seam welding can be easily delidded by grinding the edge of the lid away. Since the lid is only about 0.004 in. thick in the seal area, this may be readily accomplished in a single pass of a grinding wheel. With minimal polishing of the seal area of the package, another lid may be reliably attached.

Certain classes of packages with a flange on the package may be sealed by a process called *projection*, or *one-shot*, welding. In this process, an electrode is placed around the flange on the package, and a large current pulse is passed through the lid and the package, creating a welded seam. Heavy-duty resistance welding equipment capable of supplying 500 lb of pressure and 12,000 A per linear inch of weld is required for these packages. The major advantages of one-shot welding are process time and a less expensive package. The major disadvantage is the difficulty of removing the lid and repairing the circuit inside. Delidding a projection-welded package is a destructive process, and the package must be replaced.

12.18.4 Ceramic Packages

Ceramic packages in this context are considered to be structures that permit a thick or thin film substrate to be mounted inside in much the same manner as a metal package. Ceramic structures that have metallization patterns, which allow direct mounting of components, are referred to as *multichip ceramic packages*. Ceramic packages for hybrid circuits generally consist of three layers of alumina. The bottom layer may or may not be metallized, depending on how the substrate is to be mounted. A ring of alumina is attached to the bottom layer with glass, and a lead frame is sandwiched between this ring and a top ring with a second glass seal. The top ring may be metallized to allow a solder seal of the lid or may be left bare to permit a glass seal.

12.18.5 Methods of Sealing Ceramic Packages

The most common method of sealing ceramic packages is solder sealing. During the manufacturing process, a coating of a refractory metal or combination of metals, such as tungsten or an alloy of molybdenum and manganese, is fired onto the ceramic surface around the periphery of the seal area. On completion, the surface area is successively nickel plated and gold plated. A lid made from ASTM F-15 alloy is plated in the same manner and soldered onto the package, usually with an alloy of Au80Sn20, in a furnace with a nitrogen atmosphere.

A less expensive, but also less reliable, method of sealing is to use a glass with a low melting point to seal a ceramic lid directly to a ceramic package. This avoids the use of gold altogether, lowering the material cost considerably. The glass requires a temperature of about 400°C for sealing, as opposed to about 300°C for the AuSn solder. The glass seal is somewhat susceptible to mechanical and thermal stress, particularly at the interface between the glass and the package.

These two techniques have a common problem; it is difficult to remove the lid for repair without rendering the package useless for further sealing. An alternative approach seeing increased use is to braze a ring of ASTM F-15 alloy, which has been nickel and gold plated as described above, onto the sealing surface of the ceramic package. It is then possible to use parallel seam welding with its inherent advantages for repair. This approach is also frequently used for ceramic multilayer packages designed for multichip packaging.

12.18.6 Non-hermetic Packaging Approaches

The term *non-hermetic* package encompasses a number of configurations and materials, all of which ultimately allow the penetration of moisture and/or other contaminants to the circuit elements. Most techniques involve encapsulation with one or more polymer materials, with the most common being the molding and fluidized bed approaches.

Both injection and transfer molding techniques utilize thermoplastic polymers, such as acrylics or styrenes, to coat the circuit. In transfer molding, the material is heated and transferred under pressure into a closed mold in which the circuit has been placed, whereas, in injection molding, the material is heated in a reservoir and forced into the mold by piston action.

The fluidized bed technique uses an epoxy powder kept in a constant state of agitation by a stream of air. The circuit to be coated is heated to a temperature above that of the melting point of the epoxy and is placed in the epoxy powder. The epoxy melts and clings to the circuit, with the thickness controlled by the time and the preheat temperature.

Both methods are used to encapsulate hybrids and individual devices and are amenable to mass production techniques. The overall process may be performed at a cost of only a few cents per circuit. The coatings are quite rugged mechanically, are resistant to many chemicals and have a smooth, hard surface suitable for marking.

12.18.7 Plug-In Packages

Plug-in packages have leads protruding from the bottom with a lead spacing of 0.100 in. A special case of plug-in packages is the DIP package (Fig. 12.43). Standard DIP packages are used to package individual die and have two rows of leads on 0.100-in. centers separated with each row being 0.300 in. apart. Other package types designed for hybrid use (platform or bathtub) often are found to have a lead configuration consistent with the DIP design for commonality with test mounting sockets.

Plug-in packages are designed for through-hole insertion into printed wiring boards. In addition to DIP packages, a single in-line (SIP) packaging technique was developed for resistor and capacitor components or networks. For all plug-in packages, the leads provide a convenient method for ensuring clearance on assembly and providing

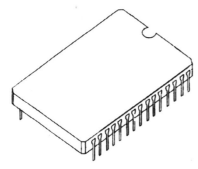

Figure 12.43 Dual in-line package (DIP).

a degree of compliance into the mechanical stress established by assembly or expansion coefficient mismatch between the package and the mounting substrate. DIP packages have been the mainstay in discrete device packaging while device complexity was low. However, with the advent of very large scale integration (VLSI) devices and high input/output (I/O) count, the 100-mil lead centers required development of very large packages. A standard 40-pin device required a length exceeding 2.00 in., and higher pin counts became increasingly difficult to package in DIP form.

12.18.8 Small Outline Package

The small outline (SO) package is shown in Fig. 12.44. The leads on the SO package are on 0.050-in. centers, as opposed to 0.100-in. centers for the DIP package. The SO has a low profile and occupies less than 50 percent of the area of the DIP. It weighs about one-tenth that of a DIP. The SO package family includes packaging of passive devices, packages that contain ICs, known as SOIC packages, and pack-

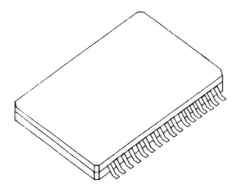

Figure 12.44 Small outline (SO) package.

ages that contain transistors, known as SOTs. Both plastic and ceramic SO packages are available.

12.18.9 Ceramic Chip Carriers

A special case of the ceramic package is the hermetic chip carrier as shown in Fig. 12.45. The wire bonding pads are routed to the outside between layers of ceramic and are connected to semicircular contacts called *castellations*. The most common material is alumina, which is metallized with a refractory metal during fabrication and then successively plated with nickel and gold. Most multilayer chip carriers are designed to be sealed with solder, usually Au80/Sn20.

Configurations of chip carriers for military applications have been standardized by JEDEC in terms of size, lead count, lead spacing, and lead orientation, although nonstandard carriers can be used for specialized applications. The most common lead spacing is 0.050 in., with high-lead-count packages having a spacing of 0.040 in.

The removal of heat from chip carriers in the standard "cavity-up" configuration has been a problem, because the only path for heat flow is along the bottom of the carrier out to the edge of the carrier, where it flows down to the substrate through the solder joints. This problem can be alleviated to a certain extent by printing pads on the bottom of the carrier, which are soldered directly to the substrate. This lowers the thermal impedance by a factor of several times. If this does not prove adequate, carriers with the cavity pointing down can be utilized. In this configuration, the chip is mounted on the top of the carrier in the upside-down position, the lid is mounted in a recess on the bottom of the carrier, and a heat sink is mounted to the top to enable the heat to be removed by convection. Beryllia and aluminum nitride chip carriers are being used to package high-power devices,

Devices mounted in chip carriers can be thoroughly tested and burned in before mounting on a substrate or printed circuit board. This process can be highly automated and is frequently done in mili-

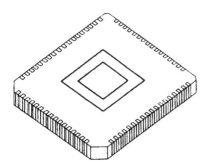

Figure 12.45 Leadless chip carrier (LLCC).

tary applications. Sockets exist for the standard sizes, which make contacts to the castellations without the necessity for solder.

Chip carriers have proven to be a viable approach for packaging hybrid circuits of minimum size. Although available with pin counts up to 128, chip carriers have proved to be a reliability risk when the pin count is greater than 84, because the net expansion of the carrier at temperature extremes, and, therefore, the stress on the solder joints is proportional to its size. Furthermore, the temperature at which chip carriers may be used on PC boards is limited due to the difference in coefficient of expansion (TCE) between the carrier and board material.

As the solder joint is made higher, the difference in TCE becomes less significant. The highest solder column that can be made by ordinary means is about 0.007 in. Above this height, a molten solder column begins collapsing of its own weight. Power cycling, in which the device in the carrier is powered on and off at periodic intervals, has proven to be a serious reliability risk, even more than temperature cycling, when power devices are mounted. While the device in the carrier is being power cycled, the carrier and the board are in a non-equilibrium state with respect to temperature. This causes considerable stress on the solder joints, ultimately resulting in failure due to metal fatigue.

12.18.10 Packages for Power Hybrid Circuits

As the power requirements of package materials have become more demanding, ASTM F-15 alloy and alumina become less attractive due to their relatively low thermal conductivity. Copper, molybdenum, copper-clad materials, aluminum nitride, and beryllia have all been used to manufacture packages. This requires some innovation on the part of the package manufacturer to develop methods of sealing and through-hole connections, as copper is not amenable to seam welding or glass-to-metal sealing.

One power package uses a cavity machined from a solid block of copper, plated with nickel and gold, which has a stainless steel seal ring brazed around the perimeter. The leads are made from copper-cored Alloy 52 material with a ceramic insulator. The insulator is generally metallized on the outside with an alloy of MbMn successively plated with Ni and Au. The pins are individually soldered to the package body with 80Au/20Sn solder. This package is compatible with conventional seam welding and offers the best thermal conductivity of the various configurations when oxygen-free high-conductivity (OFHC) copper is used. The copper must be processed such that the cold-worked mechanical properties are not destroyed to enable the package to withstand the constant acceleration test. This limits the temperature range that the package can see during processing and use.

An alternative method of attaching the leads to the package is to use a glass with a low melting point, such as potash soda barium glass, which has a different temperature coefficient of expansion (TCE) than the package material such that a compression seal is formed.

Another approach uses a copper base that has an ASTM F-15 lead frame brazed to it and is plated with Ni and Au. This package can use conventional glass-to-metal sealing in the manufacturing process and is compatible with parallel seam welding. This package must be less than 1×1 in. and must use a copper base less than 0.060 in. thick due to the large TCE mismatch between copper and ASTM F-15 alloy.

12.19 Design Of Hybrid Circuits

The successful design of a hybrid microcircuit involves not only the design of a circuit that meets the technical requirements but that also meets the cost, reliability, and schedule requirements. By using accepted design guidelines along with a systematic design approach, the hybrid designer will have the best chances for success. A procedure that meets these criteria is shown in Table 12.17. It will result in a drawing package as outlined in Table 12.18.

12.19.1 Sizing

The approximate size of a hybrid circuit can be determined by using one of the empirical methods based on component count and the weighting factors for each component. The weighting factors are different for thin film and thick film and are based on the standard line widths of 0.002 in. for thin film and 0.010 in. for thick film. For wider lines, the weighting factors must be adjusted proportionately. The weighting factors for different components are shown in Table 12.19.

The size of a hybrid is often determined by the pin count requirements. The maximum pin count is limited by the length around the periphery of the circuit and the pitch. If the size as determined by the empirical formula is not sufficient to accommodate the required pin count, a larger size must be used. For example, a 1.0×1.0-in. circuit with 0.1-in. pin spacing can accommodate 36 leads.

12.19.2 Selection of Technology

It is frequently possible for more than one substrate technology to be utilized to manufacture a given circuit. The selected technology must

Table 12.17 Hybrid Circuit Design Sequence

Operation	Output
Partitioning	Division of system into individual circuits
Initial concept review	Division of system into individual circuits
Circuit analysis	Verification of electrical design Design parameters centered Sensitivity analysis Voltage, current, and power levels in each component
Breadboard tests	Verification of circuit analysis
Component selection	Preliminary parts list
Preliminary thermal analysis	Indication of potential thermal problems
Sizing analysis	Approximate size of circuit
Technology selection	Determination of substrate technology
Process sequence	Selection of manufacturing processes
Material selection	Selection of attachment materials
Circuit layout	Location of components and interconnection traces Layer drawings Assembly drawings
Detailed thermal analysis	Temperature profile of circuit
Preliminary design review	Review prior to prototype build Exceptions to design guidelines Conformance to quality standards
Prototype build	Verification of performance Electrical performance Conformance to quality standards
Documentation release	Assembly drawings Process instructions Travelers
Detailed design review	Review of prototype build and documentation
Preproduction build	Verification of design and documentation
Production release review	Review of preproduction build prior to production
Production	Manufacturing of hybrid circuit

Table 12.18 The Design Package

Circuit schematic
Complete parts and materials list
Complete process sequence
Process specifications
Layout with individual layer drawings
Assembly diagram and instructions
Electrical test procedure
Environmental test procedure
Troubleshooting procedure
Qualification procedure
Handling and packaging instructions
Special instructions

Table 12.19 Weighting Factors for Components for Substrate Size

Component	Weighting factor	
	Thick film	Thin film
IC (chip-and-wire)	5	4
Transistor (small signal)	4	3
Diode (small signal)	4	3
Transistor (power)	5	5
Diode (power)	4	4
Resistor	3	3
Capacitor	2	2
Chip resistor	3	2
Inductor	3	2

first meet the technical requirements. After these are satisfied, cost becomes the prime consideration. The technology selection process begins by assuming that the circuit will be fabricated with the least expensive process, or with the "standard processes" that exist within a given company. This assumption remains in effect until it is proven that the standard processes cannot meet the technical requirements. In this manner, the most economical design will result.

12.19.3 Layout

The layout is a graphical representation of each of the individual layers on the substrate, the location of the individual components on the substrate, and the location of the substrate with respect to the input/output pins. An important part of the layout process is to perform a worst-case analysis on the dimensional tolerances to determine whether a potential interference fit exists.

Virtually all layouts are created on a CAD system with software designed specifically for the purpose. In addition to the drafting function, most CAD systems have the following features:

- *Automatic checking.* With the schematic entered into the system, the CAD system can perform a point-to-point continuity check of the layout.
- *Component boundaries.* The computer prevents the designer from placing a component within a predetermined distance from another component to prevent interference.
- *Parts list.* A complete parts list can be created. When thick film circuits are being laid out, the area of each print can be calculated and the amount of paste required can be determined from an algorithm.
- *Visual aids.* The visual aids needed for fabricating the circuit can be created by the CAD system directly from the layout.
- *Resistor design.* The data from the resistor characterizations can be fed into the computer, which will then calculate empirical design equations allowing for power dissipation, tolerance, and termination effects.

It is important to realize that no CAD system to date can perform the complete layout of a chip-and-wire hybrid circuit without operator intervention. There are simply too many variations to consider.

A successful hybrid design requires extreme attention to detail. Each aspect of the design, including thermal, electrical, and mechanical aspects, must be thoroughly analyzed and reviewed by the design team before the design proceeds. There is an old saying among hybrid engineers: "The sooner the layout begins, the longer it will take, and the more mistakes will be made."

12.20 Multichip Modules

Multichip modules are an extension of the hybrid technology and permit a higher packaging density than can be attained with other approaches, allowing a silicon-to-substrate area ratio of greater than 30 percent. There are three branches of the MCM technology as depicted in Table 12.20:[4] MCM-L, based on a laminated printed circuit board structure; MCM-C, based on cofired ceramic structures; and MCM-D, based on the thin film technology, which utilizes deposited conductors, resistor, and dielectrics.

Table 12.20 Types of Multichip Modules

MCM–L	Substrates formed by *laminating* layers of printed circuit.
MCM–C	Substrates formed by cofired *ceramic* or glass/ceramic structures, similar to the thick film process.
MCM–D	Interconnections formed by *depositing* alternate layers of conductors and dielectrics onto an underlying substrate, similar to the thin film process.

12.20.1 The MCM-L Technology

The MCM-L technology is based on printed circuit board technology. Multilayer structures are formed by etching patterns in copper foil laminated to both sides of resin-based organic panels ("cores") laminated along with one or more layers of the basic resin in between to act as an insulator. Interconnections between layers may be formed by "through" vias, which extend all the way through the board, "blind" vias, which extend from the surface part way through the board, or "buried" vias, which connect only certain of the inner layers and do not extend to the surface in either direction. Through vias may be drilled and plated after laminating, while blind and buried vias must be drilled and plated prior to laminating.

There are numerous materials that may be used to fabricate MCM-L structures, as shown in Table 12.21. The criteria for selection will vary with the application. Cyanate ester, for example, has a very low dielectric constant and excellent high-frequency characteristics.

Design guidelines for MCM-L substrates are presented in Table 12.22. Printed circuit boards for MCM applications must be suitable for wire bonding. This is accomplished by selectively plating nickel

Table 12.21 Key Properties of PWB Substrate Materials[6]

Reinforcement/resin	T_g (°C)	x–y TCE (ppm)	z CTE (ppm)	ε' @ 1 MHz	Dimensional stability	Water absorption (%)
E-glass/epoxy	125	14–18	90	4.7	0.04	0.15
E-glass/PI	250	12–16	60	4.5	0.05	0.35
Woven Kevlar™/epoxy	125	6–8	105	3.9	0.06	0.85
S-glass/cyanate ester	230	8–10	40	3.6	0.03	0.08
Quartz/PI	250	6–8	34	4.0	0.04	0.35
Thermount™/high-T_g epoxy	180	7–9	110	3.9	0.03	0.44
Thermount™/high-T_g epoxy	230	7–9	80	3.6	0.03	0.81

and gold on the copper traces as required. The difference in gold thickness is due to the different bonding mechanism between gold and aluminum wire. Gold wire bonding is accomplished by thermosonic bonding and is fundamentally a gold-to-gold bond, whereas aluminum wire bonding is accomplished by ultrasonic bonding and is an aluminum-nickel bond. Gold plating for aluminum bonding is thinner, since it acts only to keep the nickel from oxidizing and interfering with the bonding process.

12.20.2 The MCM-C Technology

MCM-C multilayer structures are fabricated from ceramic or ceramic/glass materials, with alumina being the primary base. There are two basic types of MCM-C substrates: high-temperature cofired ceramic (HTCC) and low-temperature cofired ceramic (LTCC). Both processes begin with thin sheets of unfired material approximately the consistency of putty, referred to as the *green tape* state. Green tape is created by mixing the base powder with an organic vehicle and forming it into sheets by doctor blading or other means as described in Chapter 11. Vias are punched in the green tape where interconnections between layers are needed and filled with thick film paste designed specifically for via filling. The individual layers are printed with thick film paste to create the metallization patterns, aligned with the other layers, and laminated at elevated temperature and pressure. The laminated structures are then subjected to a lengthy bake-out cycle to re-

Table 12.22 Design Guidelines for MCM–L Substrates

Parameter	Value
Maximum metal spacing	0.003 in
Minimum trace width/maximum metal thickness	0.003–0.0018 in
Minimum PTH pad diameter	0.015 in
Surface layer nickel thickness (μ in) Aluminum/gold wire bonding (μ in)	30 + 100–200
Surface layer gold thickness (μ in) Aluminum wire bonding (μ in) Gold wire bonding (μ in)	10 + 10–40
Minimum blind via diameter	50–100
Minimum finished hole diameter	0.010–0.000 in
Minimum buried via diameter	0.006 in
Minimum via pitch (no tracks)	0.015 in

move the organic material and cofired at an elevated temperature to form a monolithic structure.

The HTCC and the LTCC processes differ in two primary ways: the firing temperatures and the thick film materials.[5] HTCC ceramics are designed to be fired at approximately 1600°C and require the use of refractory metals such as tungsten and molybdenum/manganese alloys as the conductors, and the firing process must take place in a reducing atmosphere to avoid oxidation of the metals. The top and bottom metallization layers are plated with nickel and gold to permit die and wire bonding. LTCC materials primarily consist of glass/ceramics and are designed to be fired at much lower temperatures in the range 850 to 1050°C. This permits the use of standard thick film materials, such as silver and gold, which have much lower sheet resistivity than the refractory metals and do not require subsequent processing for assembly. A comparison of the HTCC and LTCC processes is shown in Table 12.23.

As a result of the organic burnout, the substrates shrink during firing. The shrinkage is very predictable, however, and may be accounted for during the design stage. This property is very critical when selecting a via fill material. The shrinkage of the via fill material must match that of the ceramic to prevent open circuits between layers.

Resistors and capacitors compatible with the LTCC green tape process may also be fabricated.[5] A distinction is made here between a

Table 12.23 Properties of Multilayer Ceramic Materials

	Low temperature cofired ceramic (LTCC)	High-temperature cofired ceramic (HTCC)	Aluminum nitride
Material	Cordierite MgO, SiO_2, Al_2O_3 Glass filled composites SiO_2, B_2O_3, Al_2O_3 PbO, SiO_2, CaO, Al_2O_3 Crystalline phase ceramics Al_2O_3, CaO, SiO_2, MgO, B_2O_3	88–92% alumina	AlN, yttria, CaO
Firing temperature	850–1050°C	1500–1600°C	1600–1800°C
Conductors	Au, Ag, Cu, PdAg	W, MoMn	W, MoMn
Conductor resistance	3–20 mΩ/☐	8–12 mΩ/☐	8–12 mΩ/☐
Dissipation factor	$15–30 \times 10^{-4}$	$5–15 \times 10^{-4}$	$20–30 \times 10^{-3}$
Relative dielectric constant	5–8	9–10	8–9
Resistor values	0.1 Ω–1 MΩ	N/A	N/A
Firing shrinkage x, y z	12.0 ± 0.1% 17.0 ± 0.5%	12–18% 12–18%	15–20% 15–20%
Repeatability	0.3–1%	0.3–1%	0.3–1%
Line width	100 µm	100 µm	100 µm
Via diameter	125 µm	125 µm	125 µm
Number of metal layers	33	63	8
CTE	3–8 ppm/°C	6.5 ppm/°C	4.4 ppm/°C
Thermal conductivity	2–6 W/m·°C	15–20 W/m·°C	180–200 W/m·°C

sandwiched resistor, which is formed between two layers of green tape, and a buried resistor, which is printed on a standard alumina substrate and covered with layers of green tape. The resistor pastes developed for this process show a high degree of stability after several refirings at high temperatures and exhibit TCRs comparable to those of standard materials (<100 ppm/°C). Although the accuracy of untrimmed resistors is adequate for many digital circuit applications, buried resistors printed and fired on the substrate may be laser trimmed prior to lamination with green tape. These show excellent stability under conditions of high-temperature storage, temperature cycling, and harsh environments, because they are covered with hermetic dielectric.

12.20.3 The MCM-D Technology

The MCM-D technology utilizes processes similar to those used to fabricate thin film hybrid circuits. The conductors are primarily sputtered or plated metals, gold, aluminum, or copper, deposited on a variety of substrate materials, including ceramic and silicon. The dielectric materials are primarily used in the liquid state and are applied by spinning. Vias may be opened in the dielectric film by applying photoresist and etching or by using photosensitive dielectric materials.

MCM-D structures can be made much denser than the other types as a result of the photoetching process. Line widths of 10 µ and via diameters of 15 µ are common.

12.20.4 Summary

Multichip modules are an important part of the repertoire of the packaging engineer for at least two reasons.

1. By utilizing the increased density available with this technology, more functionality can be incorporated into a smaller volume with all the advantages that this ability encompasses.
2. The variety of materials enables the substrate/interconnection structures to be more nearly tailored to a particular application.

A comparison of the various interconnection technologies is given in Table 12.24, and a representative MCM-C circuit is shown in Fig. 12.46.

12.21 References

1. *Handbook of Hybrid Microelectronics,* Jerry Sergent and Charles Harper, eds., McGraw-Hill, 1995.
2. George Harman, Wire Bond Reliability and Yield, *ISHM Monograph,* 1989.
3. Temperature Dependent Wear-Out Mechanism for Aluminum/Copper Wire Bonds, Craig Johnston, Robin A. Susko, John V. Siciliano, and Robert J. Murcko, *Proceedings of the ISHM Symposium, 1991.*
4. J. Sergent, Materials for Multichip Modules, *Semiconductor International,* June 1996.

Figure 12.46 Representative MCM-C circuit.

Table 12.24 Characteristics of MCM Technologies

Technology	Substrate material	Metallization	Characteristics
MCL–L	Organic FR-4 (epoxy glass) Cyanate ester Polyimide Polyester Kevlar Clad metal	Laminated copper	2.5-mil line width Low dielectric constant (3–5) Low T_g High TCE Up to 46 layers
MCM–C	Ceramic Alumina Ceramic glass	Thick film	3-mil line width Moderate dielectric constant (7–9) Low TCE (6–8 ppm) Up to 75 layers
MCM–D	Various Organic Ceramic Metal	Thin film	<1 mil line width Up to 8 layers Other parameters dependent on substrate

5. H. Kanda, R.C. Mason, C. Okabe, J. D. Smith, and R. Velasquez, Buried Resistors and Capacitors for Multilayer Hybrids, *Proceedings, ISHM Symposium, 1994.*
6. P. Garrou, I. Turlik, *Multichip Module Technology Handbook,* McGraw-Hill, 1998.

Chapter

13

Environmental Considerations in Electronic Assembly Fabrication

John W. Lott
DuPont iTechnologies
Research Triangle Park, North Carolina

13.1 Introduction

This chapter is designed to present an overview of the aspects of what constitutes an environmentally conscious approach to manufacturing and assembling printed wiring boards. This is merely an overview of concepts that are being addressed, both in a regulatory context and in terms of the electronic supply chain. It is important that PWB producers recognize these issues and needs.

The concept of environmentally conscious electronic assemblies encompasses a number of themes. Originally, companies that worked to make printed wiring boards with the minimum of pollution resulting from their processes were driven both by their own internal goals and regulatory requirements. This concept was broadened into the idea of *eco-efficiency*. Eco-efficiency includes the idea that processes are engineered to generate not only a minimum of waste products but also a maximum yield; that is, reduced waste in the form of "good" product. *Six sigma* methodology is very useful for realizing and implementing opportunities for improving product yield and process uptime (the time during which process equipment is actually used to make product).

One of the other industry drivers for environmentally conscious manufacturing, over and above regulatory concerns and requirements, is design for the environment. Design for the environment is used in two different contexts, one (DfE) by Environmental Protection Agency and one (DFE) by original equipment manufacturers (OEMs, the manufacturers at the top of the electronics supply chain). The EPA definition applies to a methodology for looking at individual process areas or groups of process steps, termed by EPA as *use clusters*. Furthermore, it involves determining if there are *cleaner technologies* that can be substituted for this *use cluster*. The OEM definition is much broader and actually fits in with the ISO 14000 concept of considering and assessing the environmental impacts of equipment design decisions/choices, even to the point of looking at the process impacts of individual components.

The OEM implementation of DFE means that original equipment manufacturers are requiring suppliers of parts, material, equipment, and processes to provide some means of assessing the environmental impact of the supplied element. Some electronics and automobile manufacturers have devised "black lists" of materials that must not be in the product or should not be used in processes supplied to these manufacturers. At least one industry effort for producing a common list of these materials has been completed by the Electronics Industry Alliance.[1] At this writing, another is also being developed by automakers globally.[2]

Several methods for obtaining this information or providing the information have been used and/or proposed. Many companies simply send lists of banned or restricted materials. Some companies also have a "prefer not to have" list as well. Volvo, Sony, Mitsubishi, Phillips, and the European Association of Consumer Electronic Manufacturers (EACEM) are examples. Other companies have various rating schemes that they use internally to rate the materials and subsystems that they employ. For example, IBM, Motorola, Hewlett-Packard, and Sun Microsystems all have rating systems.[3] The European Computer Manufacturers Association (ECMA) has proposed a product declaration attribute form for various electronic devices that lists not only materials that are not found in the products but also various attributes such as energy usage and physical emissions (VOCs).[4]

Others are attempting to develop various methodologies for weighing these factors for each item or process supplied. One of the more popular means of doing this is using a life cycle assessment or life cycle inventory.[5] This is one of the guidelines being developed as part of the ISO14000 series of environmental standards and guidelines (see below). This guidance is now adding additional focus on the recycling of equipment as well as materials being used to make products.

These concepts of assigning an environmental valuation to materials and processes is already being implemented by companies like IBM, Xerox, Motorola, and others.[6] They, in turn, are asking questions of their suppliers about environmental impacts. Initially these questions dealt with the ability of the supplied materials to be recycled, reused, and refurbished, but they are now requesting information about energy and resource usage in making the products supplied to them. This allows them to go even further into truly determining the environmental and resource impacts of the complete systems and products that they make and sell. It will further help them determine which are the most sustainable (i.e., those that use up resources vs. those that simply "borrow" resources and can be returned to the life cycle of the product at a later date as recycled, reused, or refurbished materials).

As noted above, the EPA definition of DfE is more closely associated with a highly detailed examination of a single process or *use cluster* within a manufacturing scheme.[7] The use cluster is chosen and then clearly defined, and all materials used in the processes that make up the use cluster are inventoried (along with any that have been or are being developed to have a lower environmental, health, safety, and resource impact). These processes and alternatives are then examined in a *cleaner technology substitute assessment (CTSA)* to determine cost, health and safety, energy usage, environmental, quality, and performance assessments. The results are then presented in a nonjudgmental format to allow individual users of the technology to determine which, if any, substitutes would make sense for them to use. Finally, after the processes have been tested in a demonstration process, implementation of the substitutes is followed up using actual practitioners of the new technologies to determine what was good and bad about the processes[8] (see Fig. 13.1).

The DFE process is more of a top-down, design-level evaluation that will eventually drive the move toward newer and more resource and environmentally "friendly" processes. The DfE process is an evaluation scheme for evaluating, implementing, and disseminating these processes.

13.1.1 ISO 14000—Continuous Improvement

Basically, ISO 14000 is an ISO 9000 approach to the management of systems used to ensure compliance and to continuously improve the systems that control the environmental impact of a plant. This includes clearly documenting the management systems, controls, tests, and procedures used to maintain these systems. It does not, however, require the entity or plant seeking certification under 14000 (actually 14001) to be in compliance with local regulations. That entity needs only to define these systems and the verification under which they are working. The Eco-Management and Audit Scheme or EMAS (the European equivalent of ISO 14000), BS 7750, and several other country

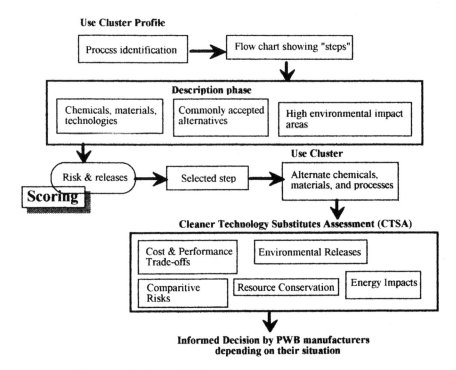

Figure 13.1 Cleaner technology substitute assessment.

specific EMS standards *do* require compliance with local regulations. They also require, as a first step, a compilation of all the regulatory and legal environmental standards that apply to the plant or entity seeking certification under the standard.

Many companies are using one of three approaches to ISO 14001:

1. *Full certification.* Implementation of the standard with auditing and certification by a third party. Several companies, such as Ford and GM, are also requiring first-tier suppliers to have certification.

2. *Evaluation.* Others have chosen to determine where their existing EMS stands, with regard to the standard, in case they decide to certify.

3. *Organization.* Others have found that the guidance document is useful for helping them to set up their environmental management system. This can be valuable because, in many companies, compliance is achieved only because one individual is knowledgeable and implements the process of compliance. The lack of documentation and procedure becomes evident when that individual is no longer

at the company. (This is also known as the "what happens if Joe gets hit by a truck?" scenario.)

Additional benefits of having a formalized environmental management system are the ability to

1. Audit their effectiveness
2. Review them and to improve their efficiency
3. Determine, with greater accuracy, the resources and costs associated with various environmental and regulatory requirements
4. Involve all those whose jobs are now recognized as part of environmental management in any improvement processes

An example of one aspect of management system scheme is shown in Fig. 13.2.

External benefits of an EMS can be recognition by the community of an environmentally committed company or plant, lower insurance rates due to better controls, and thus less susceptibility to environmental upsets and to fines and legal action. They may also promote more latitude in addressing problems by the regulatory and legal en-

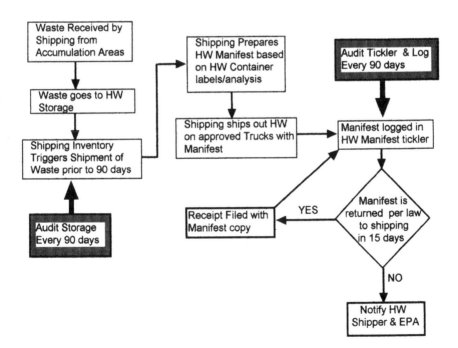

Figure 13.2   Example of an EMS for hazardous waste manifesting.

tities such as state and federal EPAs as well as the Department of Justice, better access to capital from financial markets, and better employee morale.

There are innumerable references on ISO 14000 in the literature and on the World Wide Web. The U.S. EPA Design for the Environment program has published two documents specific to the printed wiring board industry on this subject and the National Standards Foundation International (NSF) has also published a report[9] on a demonstration project with a number of case studies accessible from the Internet.[10] There are also numerous texts that deal with implementation of environmental management systems.[11]

13.1.2 Ecolabels

Many regions and countries have various environmental labels, *ecolabels*, that rely on the perceived environmental impact of the product labeled. There are also a number of other attempts to define the environmental impact of products for the benefit of whatever customer or user is buying the product. Original equipment manufacturers are particularly interested in the content of subsystems that they use in their equipment. They often need information to qualify their products for ecolabels but also often have to answer questions pertaining to the environmental impact and content of their product. This is becoming more of a need with the product take-back legislation for electronic equipment currently being decided in the European Union legislature[12] and discussed by the U.S. EPA and other stakeholders.

13.1.3 Financial Implications

In addition to looking at the properties of materials in use, several financial institutions are beginning to look at the environmental performance of companies and using this as a way to measure financial risk. This extends to the perceived ability of companies to survive in a world where resources, energy and materials will become more limited. The World Business Council on Sustainable Development is a resource of processes for examining the sustainability of a company and its value to its shareholders.[13] There are also a number of funds and organizations developing tools to be used when choosing how environmental impacts can effect financial investments.[14]

13.2 Material Considerations

As indicated in Sec. 13.1, there are a number of nonregulatory impacts that can affect material choices in PWB manufacture and assembly.

Some are very specific to a small number of materials, others are related to so-called "end-of-life" impacts, and, finally, others are driven both by performance needs and smaller environmental and human health impacts. The requirement for the ability to reuse, recycle, or refurbish the materials and components in electronic equipment means that designers and suppliers will have to consider more carefully the materials that are used. This means not only how they are used initially but also how they will be recycled. Brominated flame retardants used in polymers, for example, which have allegedly been implicated as sources of dioxin in incinerators, have found themselves being considered for elimination as materials of use in PWBs in Japan and Europe. Several U.S. states[15] have also begun to consider restricting the disposal of electronic equipment in municipal landfills. This would either cause the materials to be eliminated or require an improved infrastructure for capturing and reusing these materials.

13.2.1 Solder and Solder Coatings

Lead-based solder has been used for three main purposes in printed wiring board fabrication for many years.

1. Solder coating on copper contact areas is used to provide a solderable base upon which parts are soldered using additional lead-based solder.

2. It's used to protect the copper surface of connection pads from oxidation after the bare board has been manufactured and before parts are soldered on.

3. The third use is, of course, to provide the mechanical and electrical connection of components to the board.

Over the last few years the first two uses (coating the connecting pads with solder to maintain solderability and prevent oxidation/contamination) have been replaced in many shops. Either organic surface protectants or solder-compatible metals other than lead have been used. The U.S. EPA's Design for the Environment Program[16] has completed a study using the methodology described earlier in this chapter. The results, with respect to economic and environmental benefits, were mixed on the various alternatives versus the traditional lead solder process using hot air to level the solder to a thin coating (also known as hot air solder leveling or HASL).

Lead-based soldering for actually soldering components to circuit boards has come under increased scrutiny in the last two years, principally due to perceived take-back issues in Europe and Japan. The European Commission's Waste Electrical and Electronic Equipment (WEEE) directive on hazardous materials was intended to make the recycling of materials used in equipment less hazardous and pose less of a risk to the environment. Although lead has been recycled from board manufacturers as well as auto battery manufacturing for many years, the European commission felt that lead represented a hazard in recycling electrical and electronic equipment. The most recent draft would require substitution of other, less hazardous metals for lead no later than January 2008.

Several companies offer no-lead solders, and there has been a great deal of work to determine the technical issues around using no-lead solders. One of the biggest technical issues has been the increased temperatures required to use substitute solders. The higher temperature not only requires more energy to be used in the process, but also may have detrimental effects on the board laminate material and some of the components to be soldered on to the board. In addition, there has been *no* environmental assessment of the substitute solders. One of the main candidates contains silver, which is mined in conjunction with lead. Another prime candidate contained bismuth, which turned out to cause problems in the metal recovery process of most copper reclaimers. There have been some papers published that indicate that replacement of lead in solder is not necessarily a good thing.[17] So, the question of environmental benefit of replacing lead solder is still open.

13.2.2 Brominated Flame Retardants

At about the same time that concerns began to appear for lead-based solders, brominated flame retardants also began to be viewed as an environmental hazard. Some of this was related to concerns that incineration of brominated flame retardants were producing dioxins[18] in some incinerators. This concern showed up initially in Japan, where supposedly the incinerators are older and run at temperatures much lower than in Europe or North America. Brominated flame retardants are used in circuit board laminate materials such as glass epoxy to prevent fires in case of electrical problems, such as shorting or overheating.

With respect to studies done on brominated flame retardants, most have identified only polybrominated diphenyl ethers (PBDEs) as presenting a health hazard. The flame retardant in most circuit board laminates is called tetrabromobisphenol A (TBBPA). This flame retardant also differs from most others in that it reacts with the polymer in glass epoxy laminates to become part of the inert polymer. So, rather than being an unattached solid or liquid that exists as a separate material within the spaces of the matrix of the polymer, TBBPA, is part of the polymer matrix.[19]

In Japan, as well as Europe and North America, several companies have already developed "halogen-free" (meaning bromine- or chlorine-free) laminate materials. The substitution requires the use of a filler material such as a metal oxide or hydrate, addition of a separate material such as phosphorus compounds, or a change in the chemistry of the polymer, such as inclusion of significant amounts of nitrogen or a high level of carbon rings. The fillers tend to dramatically reduce the mechanical integrity of the polymer, particularly at the high levels required to achieve good flame retardance. Phosphorus can be extremely toxic, if the polymer is burned under the wrong set of conditions, and can also degrade the mechanical properties.[20] Changing the chemistry by adding nitrogen or carbon rings can be expensive in most cases, although polyimides, which normally have nitrogen as part of the polymer, are commercially available and have been in use for many years as flexible laminate.

As in the case of lead-free solders, no comprehensive environmental impact studies have been done on the replacements for bromine-containing polymers in circuit boards.[21] So, not only are the new candidates' environmental impacts untested, there appears to be significant evidence that the brominated materials used in circuit boards are not an environmental hazard.[22] The question of whether brominated flame retardants will be replaced may hinge on market forces rather than environmental benefits.

13.2.3 Recycling, Reuse, and Refurbishment

The electronic industry is focusing more attention on end-of-life product issues for its equipment. Many active components can be recycled and reused in lower-functionality applications or simply refurbished for further use in new equipment. PWB structures, however, are currently only being "recycled" by removing the metals through either thermal degradation of the polymers[23] or incineration to capture the energy value of the organic materials. Use of (a) organic structures that can be transformed back into reusable organic materials in the same or similar forms or (b) processes that can extract the organics in a more usable form are highly desired. Some plastics can be converted directly back to raw materials that were used to make them and thus recaptured.[24] There have been some reports on processes to reclaim glass epoxy materials and also polyester materials.[25] The opportunity exists for makers of polymers to develop sustainable closed-loop systems with their customers, particularly with those who have already initiated some level of take-back programs.[26]

13.2.4 Resources, Energy, and Waste

In Sec. 13.1, we indicated that eco-efficiency through process optimization is one way to improve and reduce the use of resources and energy by reducing waste. Recycling materials is another way to reduce the use of precious resources. Requirements of customers concerning material and energy usage in life cycle assessments constitute one way to drive PWB manufacturers and assemblers to more closely examine their own processes and systems. The following section looks at these approaches in greater depth.

13.3 Processes for Environmentally Conscious Manufacturing

13.3.1 Processes to Reduce Usage and Eliminate Materials of Concern

The following sections will discuss both the technical and management options that are available for developing an environmentally conscious manufacturing framework using process technologies. They are aimed principally at existing processes and materials. We will also reference some places to look for additional help. In general, environmentally conscious manufacturing can be accomplished by source reduction, recycling (internally or externally), or elimination/substitution.

13.3.1.1 Management of environmental and pollution systems. To improve any system or process, you must first understand it. There are many ways to do this. A simple diagram of the process steps, their layout, and how they are plumbed would be a first step that often can be done by simply checking with the facilities manager. Next, an analysis or inventory of the process steps with respect to what goes into each and what comes out of each is also useful. This would include not only product, but waste streams, recycle streams, and even lab analysis samples. Finally, the management systems used to control these various aspects through procedure documents, waste forms, analysis practices, audits, waste disposal methods and associated costs, etc. should be evaluated.

By taking these first steps, several things are accomplished.

1. You may discover some very obvious things to correct.

2. You will determine the source of most of your waste.

3. You will see a number of candidates for pollution prevention.

4. You will find some flaws in your environmental management system or that you may not have one at all!
5. You may discover that you are paying a lot more for pollution and/or waste than you should be!
6. You may also discover that some of the material or process choices that you have made may have significant unrecognized downstream costs associated with them.

Let's examine some of these possibilities.

1. Some obvious things

    Water overuse. Good rinsing between processes is essential, but a high rate of water rinsing is not necessarily the answer. Quick fixes such as timers and cascade rinse tanks are recognized as good practice.[27]

    Segregation of waste streams. Many mixed waste streams are treated with the wrong and/or more expensive technology than is required for separate streams.

    Material waste. Inadequate maintenance procedures cause chemistries changes more often than necessary.

2. Source of wastes

    Inventorying waste streams and assigning correct costs to them will allow you to correctly allocate time and resources. If you can tie a particular waste source to a particular material or process, you might substitute a process or material with less impact on your waste and costs. For example, many of the new direct-plating chemistries designed to replace electroless plating for plated-through holes don't require the use of high complexer concentration. The complexer makes the copper in the electroless plating bath rinses harder and more expensive to remove in waste treatment. Another example is photo resist stripping solution. If you can use a resist that breaks into filterable particles in your stripper and then doesn't easily redissolve, then you can treat most of the resist solution by simply filtering off the resist particles rather than having to chemically precipitate it later. In addition, your stripper solution lasts longer!

3. Sources of pollution prevention projects

    After diagramming your process systems, an inventory of materials used and processed will give you a clearer overall picture of the possibilities. Also, don't overlook your employees as sources of suggestions for pollution prevention. They will be more willing to participate in pollution prevention if they are the source of some of the ideas. In addition, they are the most knowledgeable people about the process and the actual situation around each process.

4. Environmental management system

    As noted above, an EMS offers you many advantages. It ensures that, no matter who is running any piece of equipment or process, the system will take care of any regulatory and environmental concerns. For the system to always be effective, the procedures, records, tests, etc. that make the system work should be documented. Those involved with each system should be trained to stick to the system. For example, suppose that Charlie, who runs your stripper in photoimage, knows to let Sharon in waste treat know when he's using a stripper with MEA rather than the usual plain caustic stripper. Sharon will then be aware that her system is going to receive a batch of water with a high concentration of copper and complexer in it, and she can treat it accordingly. If Charlie moves to another area and his replacement doesn't know to do this, then Sharon could have a waste treat upset. By having a written procedure for equipment bath dumps and by training people who come into the area, you have an ongoing system to handle potential problems. You can also check and see if they really work by auditing them from time to time and developing improvements that are *written into the procedure.*

5. Cost of waste and pollution

    This, of course, is very important, whether you are the operator or the owner. Significant savings can be captured by monitoring the costs of environmental and regulatory systems. Regulatory failure can cost you significant fines or even criminal prosecution; environmental failures can cost you money and affect the environment. Once you know your costs, you can assign priorities to what needs to be fixed.

13.3.1.2 Implementation of pollution prevention and cost estimates

Case 1. "Low-Hanging Fruit." Simple changes that can reduce costs of water (e.g., using cascade rinses and flow limiters on spray rinses to minimize water usage and its subsequent waste treatment) can be used to effect the same level of cleanliness.[28] There are several other examples of these simple, low-cost changes in work practices.[29]

Case 2. Multiple usage's of process waters. There are several examples of this. One company[30] utilized its microetchant from an electroless copper line preclean step. Copper concentration, controlled by additions on reaching a threshold, prevents frequent bath dumps and maintains the range where the microetch rate is more consistent. The overflow from this bath goes via gravity feed to a microetch used to prepare the imaged copper for pattern plating. This bath, in turn, flows into a bath used for electroless copper rack stripping. It is later pumped and used to clean extraneous copper from the electroless tank walls. Finally, the copper concentration is high enough to plate out copper economically for sale. The remainder of the solution is then waste treated.

Case 3. Water cleaned up for reuse in the same or similar process. Water reuse has become of significant interest to manufacturers in areas of high water cost or low availability. One recent paper dealt with a combination of technologies to clean up process water for reuse and capture some of the polluting materials.[31] The effective cleaning rates were as high as 100. Savings in water costs and improved yield due to better rinsing water quality depended on local water costs and production/water volumes.

Economic improvement to the bottom line can drive many simple environmentally conscious manufacturing processes. A few things to remember when determining economic value of such projects are as follows:

- For simple changes, simple calculations can be used to determine costs and payback times. For more complex changes, more accurate calculations on cost factors based on internal rates of return etc. should be used.

- Labor costs should be used with care, since the labor savings may not actually result in less labor costs overall, especially where employees serve many functions in working on processes.

- Agreements with material suppliers may be based on boards produced and not on amounts of chemicals actually used, so some considerations should be made in this case as to cost savings.

- All costs need to be included in making an assessment of the net impact of the pollution prevention project. These should include the following:

    Costs
    – Capital cost of the equipment
    – Installation costs of equipment, including plumbing changes for segregation of wastes
    – Start-up and operating costs, including consumables and energy costs

- Facilities costs if new construction is required, and any permitting costs
- Assignable labor costs requiring *new* labor

Financial benefits

- Reduced water and sewage costs (sometimes lumped together by municipalities)
- Reduced utility fees, including connection fees and/or time-of-day discounts
- Rebates from sewer authorities for water use reductions
- Reduced process chemical usage and waste treatment costs
- Increased capacity due to increased usage of same facilities

Other considerations

- Determination of impact on sewer use, waste water, and air permits
- Ensuring that changes do not negatively effect quality of products
- Improved relationship with local water and regulatory authorities

13.3.1.3 More complex solutions to eliminate waste and material usage. As was noted in the previous sections, once simpler source reduction, reuse, and recycling projects have been completed, pollution prevention must be more completely evaluated, since the cost will most likely be higher and the payback period longer. These processes can be classified as closed-loop systems for reuse, chemical recovery, chemical maintenance for reduced usage, and substitute technologies. A very comprehensive evaluation of these technologies as used in the plating industry can be found in *Pollution Prevention and Control Technologies for Plating Operations*.[32] A document focused specifically on the printed wiring board industry can be found in *Printed Wiring Board Pollution Prevention and Control: Analysis of Survey Results*[33] and several more recent publications of the U.S. EPA Printed Wiring Board Project on the U.S. EPA DFG web site.

13.3.1.4 Closed-loop systems. Closed-loop systems are generally of the type in which a chemical bath is plumbed into a regeneration system that regenerates the chemistry and sends it back to the process bath. Examples are etchant regeneration systems[34] in which either ammoniacal or cupric chloride or a combination of these etchants is regenerated using combinations of either solvent extraction plus electrowinning or electrodialysis and electrowinning (or electrolytic plating out of copper). Another example is regeneration of the solution used to strip solder plating from tab and connector fingers.[35] Even process solutions used for developing and stripping dry film photoresists can be reused with a minimum of added chemicals using closed-loop systems.[36]

13.3.1.5 Recovery systems. Chemical recovery is very similar to closed-loop system technology except that, rather than reusing the materials in the same process bath, the chemicals are recovered for recycling or reuse in another industry. This is usually the case because the recovery process cannot separate the impurities or by-products of the process sufficiently to allow reuse in the original process. Most typically, this is accomplished by one or more of the following technologies: ion exchange, electrowinning, reverse osmosis or ultrafiltration membranes, electrodialysis, filtration, or carbon treating.

13.3.1.6 Chemical maintenance for reduced usage. As indicated above, improved chemical maintenance can not only reduce the frequency of bath replacements (see Sec. 13.3.1.2, Case 2) but also improve the quality of the bath performance. *Chemical maintenance* is taken to mean maintaining a processing bath without application of additional chemicals. An example would be a feed-and-bleed process where used solution is replaced with new in a overflow type of system. Only if the overflow is "cleaned up" by one of the regeneration or recovery processes above would this be considered "maintenance." Sources of contamination are the chief reason for baths to require maintenance.

Proper analysis of the various baths is obviously a necessity for controlling the chemistry and thus the efficiency of the bath. Some analyses are also designed to measure the contamination levels. Many baths can be monitored on a continuous periodic basis by a number of techniques. This type of analysis, along with a continuous cleaning process, tends to keep the concentrations of all the ingredients at near the optimal level and removes contaminants in a continuous manner. Again, these operations would tend to provide a more consistent level of performance and reduce product scrapped due to low quality.

13.3.2 Processes that Improve Functionality and Environmental Impact.

As the feature size and demands on printed wiring boards have increased, the technologies to produce them have changed and are still changing. This, along with some regulatory drivers, has caused manufacturers to re-examine some of the processes used for making boards. Examples of substitute technologies are

- dry film photoresists to replace solvent coated liquid resists
- aqueous cleaners to replace solvent cleaners
- aqueous processable photoresist to replace solvent and semiaqueous processable resists and solder masks
- direct plating for through holes and blind vias to replace formaldehyde-based electroless plating baths

And, more recently, the following are being used or tested:

- high-density interconnect via processes
- direct imaging laser resists
- thinner resists for finer circuit density
- elimination of photoprinting of resists[37]

13.3.2.1 High-density interconnects. Most recently, a number of technologies have been developed to reduce the diameter of the via or hole interconnecting two layers. Several processes for doing that are based on the use of lasers to drill holes in either dielectric material or dielectric/copper sandwich. The copper may have holes the size of the vias etched into it. These may serve as a mask for the laser drilling process. Several other similar methods are used to removed very small diameter of dielectric to make vias.[38]

In addition to processes that create these "microvias," there are also processes to add either conductive or nonconductive material into the vias.[39] The purpose of this "filler" may be two-fold. It may be used to make the vias more reliably conductive, and it also provides the finished layer with a planar or flat surface with no holes present. This is important, because it permits the designers of boards to run lines over the vias with out fear of having a "dimple" in the dielectric copper laminate where it runs over the via. The "dimple" can be a problem from an imaging standpoint, and it also can modify the electrical characteristics at that point.

The environmental impact of the combination of microvias and via fillers is that the smaller vias or microvias take up less space, can be made layer to layer, and can have lines over them on the next layer(s). This translates to a much higher circuitry density and means that less material is used in achieving the same circuitry or function.

13.3.2.2 Laser-imaged photoresists. In the past, all photoresists have required the use of a photographic negative of the circuitry to image the photoresist. The photoresist is polymeric film or, in some cases, a liquid with photoactive ingredients. The combination is transformed by UV light into a polymer, which is insoluble in a slightly caustic developer. A negative of the circuit design is usually placed into intimate contact with the resist, the UV light passes through the clear areas (while it is blocked in black areas). The exposed resist then is treated with the developer, which removes the unpolymerized material that was blocked by the dark areas of the photo negative. If there is dirt or scratches on the negative, they can cause either unwanted holes in the resist or unwanted resist. The resist is used to either define where the copper is etched or plated. If there is unwanted resist or holes, then this translates to either open circuitry or shorts between lines, depending on whether the resist is used for plating or etching.

Newer versions of photoresist now can be exposed using a scanning laser beam.[40] Since there is no photo negative to be scratched or have dirt on it, these defects are effectively eliminated. The environmental impact is two fold in this case. First, all of the chemistry, materials, and energy formerly used to make the numerous photo negatives as well as the associated waste products and streams are eliminated. Second, since a major source of error—the photo negative—is no longer present to have scratches or dirt, the yield on the photoresist process is increased, with a resulting decrease in waste. In addition, the laser imaging can also image finer lines increasing the circuit density.

There have also been several papers on using ablative laser beams to directly pattern thin metal layers on flexible substrates. The process can be used to leave only metal on the flex substrate, which serves as a catalyst for electroless or additive plating to build up the thickness of the imaged lines.[41] This process would also minimize the resist processing solutions.

Photoresists, as noted above, are responsible for creating the circuit lines either through a plating or etching process. Most of the inner layers (layers inside) of a multilayer PWB are etched. The thinner the resist used to define these circuits, the finer the lines that can be etched, because the thinner resist can be imaged and developed to a smaller width. However, in the past, thinner resists could not adequately match the surface contours of the less-than-smooth copper surface of a copper/dielectric laminate. To overcome this problem, a water assist during the lamination or resist adhesion process has been used. The water acts as an agent to soften the dry film so that it flows slightly and adheres more effectively to the copper surface. Significant improvement in fine line yields have been documented.[42] Again, this leads to higher circuit density reducing the amount of energy and resources used in making the board itself. In addition, the image process uses less material to make the thinner resist itself, and also to process it and treat the resultant waste.

13.3.2.3 Embedded capacitor and resistor components.
One of the applications currently being developed for printed wiring board fabrication holds the promise of combining some of the functions of board fabricators and assemblers. That is creating embedded components buried within the board structure rather than being soldered on to the surface of a completed but "unpopulated" board (that is, a board with no components attached). There are several approaches to both capacitors and resistors.

Capacitors typically involve normal processing to form capacitors inside the board structure. Capacitors require a "sandwich" of copper conductors on either side of a nonconductive material of known dielectric properties. These dielectrics can be designed through the use of filler material to produce the desired capacitance values.[43]

Resistors require the formation of sections of partially conductive material in contact with the conductor lines. They are typically embedded in the normal dielectric in line with the circuit lines, although some structures require a "resistance layer" located below or above the conductor lines. These layers are etched to form the resistors of the proper value.[44]

Another approach has been to utilize ceramic resistor materials that are modified for adhesion to copper foil. They are fired onto copper sheets that are later laminated to dielectric material with the ceramic resistors embedded in the dielectric. The copper foil is then processed normally to make circuit lines that connects the resistors to the other circuitry. More recent work has focused on improving the process, the reliability and operating values of the resistors.[45]

13.4 Regulatory and Nonregulatory Considerations

13.4.1 End-of-Life Considerations

While the preceding sections have given a brief overview of some options for reducing or eliminating pollution in plant process operations, there are additional considerations for reduction of waste. These include selling waste materials to recyclers for reuse or recycling rather than sending the waste to an incinerator or landfill, recycling in-plant as was noted for some process solutions such as etchants and plating baths, and, finally, working with materials suppliers on cooperatives efforts to reduce materials usage.

Recycling materials outside of the plant can include not only chemicals or materials recovered in recovery or closed-loop processes but also scrap materials. Even waste treatment sludge can often be recycled for the metals contained within it. Those plants that cannot regenerate their own process solutions can often work with suppliers that can regenerate them at their own facility and then ship them back to the manufacturer's plant. Raw materials can also be recycled—most printed wiring board shops send scrap solder and even solder dross back to suppliers for recovery and reuse of the solder. Even scrap circuit boards, while often recycled only for their metal content, have found a market as decorative items for clip boards, book covers, and other novelty items.

Even process solutions that contain materials that can't economically be recycled may be useful in a process that utilizes the characteristics of that solution, such as acidity. This often happens where a printed wiring board process solution is too contaminated for use in a board process but could be utilized by another industry that isn't affected by the contaminants.

As mentioned above, some plants are now working with their suppliers in cooperative agreements to supply not process chemistry, but completed parts. In other words, rather than selling manufacturers chemistry or materials, suppliers are charging based on the number of good parts processed. This allows both the supplier and the manufacturer to concentrate on using the minimum amount of material and splitting the savings generated. The result is usually a reduction in the amount of material utilized and requiring disposal or treatment.

13.4.2 Regulatory Considerations

The electronics industry is a large and highly visible industry, consuming significant energy, water, and resources to produce its products. In addition, its supply chain runs from chips made of highly purified silicon, connected by interconnection devices, to components soldered to these devices, to subunits created from several devices, to finished equipment in cabinets. These interconnection devices are made from ceramics, glasses, organic polymers, precious metals, and toxic heavy metals. They are applied in processes that use toxic chemicals, acidic and caustic solutions, molten metals, corrosive gases, and many other processes that rely on basic chemical reactions to add and subtract materials in precision patterning steps to create the circuits that make up these devices.

Because of its visibility both as a potential source of pollution and energy consumption, as well as one of the most significant impacts on daily life and industry, the electronics industry has been the focus of much environmental impact examination. But the electronics industry has also been a leader in its environmental proactivity. In this section, we discuss the regulatory impacts on the printed wiring board and assembly industry. Some of the proactive efforts, mentioned in earlier sections, have been undertaken by the industry to make itself into a model industry with significant advances in sustainable development.

13.4.2.1 Regulatory compliance issues. All industry is subject to a continually growing set of environmental laws promulgated during the 1980s and 1990s by the U.S. government and added to by state and local authorities. Noncompliance with such laws can lead to significant fines and/or imprisonment for responsible company officials as well as loss of permission to operate. We will touch on the more important aspects of these. We would refer the reader to the references for more information. While this chapter presents an overview of these laws, there are subtleties and complexities that require extensive training and knowledge to properly comply with them.

The applicability of many federal regulations is determined in part by the chemicals being used at a facility. Individual facilities, however, have their own chemical use patterns. As a result, each facility must identify the universe of rules that apply to it by examining the regulations themselves. Furthermore, implementation of many federal programs is delegated to states, with programs at least as stringent as the federal program. Thus, even where federal regulations apply, state laws may impose additional requirements that are not addressed in this document.

13.4.2.2 Clean Water Act. The Clean Water Act provides for development of publicly owned treatment works (POTWs), which are responsible for the virtual elimination of direct discharge of conventional sewage into the nation's waters. It also provides for setting limits on what materials and concentrations of materials may be discharged to these POTWs by various industrial categories.

There are general categories of materials that cannot be sent to either a municipal or regional sewage treatment system (POTW) defined by

- Pollutants that create a fire or explosion hazard
- Pollutants that cause corrosive structure damage
- Solid or viscous pollutants that will obstruct the flow in the POTW or cause interference

- Any pollutant, including biological oxygen demand, released at a flow rate or concentration that interferes with the POTW
- Materials hot enough to cause the POTW to exceed 104°F

In addition to general restrictions that limit the materials that may enter a POTW, the EPA publishes categorical standards that contain numerical limits for specific metals and organic compounds for the discharge of pollutants from specified industrial categories. Most PWB manufacturers are specified as "metal finishing," "electroplating," or both. At this writing, the EPA is in the process of collecting data to create a new category with lower allowable amounts of materials to be sent to sewer. It is known as the *metal products and machining effluent guidelines*. Additionally, through sewer use ordinances, local authorities may set tighter limits on and/or add other materials to those having limits, depending on locality. Local limits generally apply to the connection point with the community sewer system.

In areas were disposal to a POTW is not practical or prohibited, industrial companies sending waste to a sewage treatment system are permitted to send treated wastewaters to surface waters (lakes, rivers, streams, etc.) under an National Pollution Discharge Elimination System (NPDES) permit. These NPDES permits are site specific and generally issued by the state or EPA. They generally contain more limits and tests than a general sewer permit.

13.4.2.3 Resource Conservation and Recovery Act (RCRA)—hazardous wastes. The Resource Conservation and Recovery Act (RCRA) deals with federal laws that govern waste generation, disposal, storage, and minimization of "hazardous" waste that is removed by channels other than the sewer system or as a volatile air pollutant. These wastes are normally shipped outside the plant for recycling, land filling, or incineration.

Hazardous wastes are controlled by RCRA requirements from the time they are generated, through handling, "transportation to," and "treatment, storage or disposal at" a permitted site and beyond; i.e., paying someone to haul it away and do something with it. The rules are often complicated, require extensive record keeping and documentation, and can lead to significant fines if not followed.

There are also additional rules for facilities that qualify as "large generators." Most PWB facilities fall under the classification of "large generators" and as such must follow certain rules for accumulation, temporary storage (up to 90 days or, for some smaller shops under certain conditions, 180 days) and disposal of hazardous wastes. All wastes handled by large generators must be manifested using state or federal manifesting forms, which must be sent with the waste to the disposal or treatment facility. There must be documentation that the waste was treated as manifested with certain time restrictions. Violations of the time, documentation, or reporting rules are subject to fines.

13.4.2.4 Clean Air Act Amendments (CAAAs). The Clean Air Act and its amendments are designed to "protect and enhance the nation's air resources so as to promote the public health and welfare and the productive capacity of the population."[46] Under the CAA, there are six Titles that direct EPA to establish air quality standards and EPA and the states to enforce and maintain the standards. The new amendments for the CAA became law in Nov. 1990. Many of the provisions deal with specific chemicals known as *hazardous air pollutants (HAPs)*. Under federal regulations, these materials are restricted to emissions of less than 10 tons of any one HAP or 25 tons total of any combination of them. Installation of a high degree of control technology will be required, depending on the permit conditions and emission rates.

The Title I of these amendments deals with volatile organic compounds (VOCs). Almost all organic solvents and many chemicals that have been used by PWB manufacturers are included, e.g., 1,1,1-trichloroethane, MEK, and isopropanol. Many organic stripping solvents, liquid solder masks, and organic degreasers would be affected by control of these solvents. Standards have been established for these types of chemicals by industry categories, and, depending on a number of factors, emitting sources have to have permits specifying their emissions and control levels. They are also regulated and required to reduce emissions based on "lowest achievable levels" using "reasonably available technology." In addition, as states and even districts within states develop their own implementation plans (SIPs), they may lower the allowed levels of air pollutants

Title VI deals with stratospheric ozone protection, with provisions to limit the use and phase out of chlorofluorocarbons (CFCs), carbon tetrachloride, and methylchloroform (2002). There are also labeling and escalating tax requirements to provide negative incentives for use of these materials to speed phase out.

Titles V and VII indicate that permits and enforcement will again fall principally to the states. The federal ruling requires that state plans include permit fees, reporting requirements, and review of permits. This structure may be much like the current POTW or NPDES permitting system for water. The law also includes strengthening the power of the EPA to enforce regulations with severe penalties.

13.4.2.5 Miscellaneous reporting and record keeping regulations. The Comprehensive Environmental Response, Compensation, and Liability Act (CERCLA) also knows as the *Superfund,* authorizes EPA to respond to releases and threatened releases of hazardous substance representing a threat to the environment and public safety. The act also requires companies that have releases of hazardous substances above reportable quantities to report them to the National Response Center.

The Superfund Amendments and Reauthorization Act (SARA) of 1986 created the Emergency Planning and Community Right-to-Know Act (EPCRA, also known as SARA Title III–313). Its main provisions provide for emergency response plans for fires and potential hazardous releases through reporting and planning with local fire departments and local emergency response commissions (LEPCs). The act also requires manufacturers including PWB fabricators with 10 or more employees which use, process, or manufacturer specified amounts of certain chemicals to report them each year. EPA Form R covers "releases and transfers" of toxic chemicals to various facilities and environmental media. There continues to be a great deal of controversy around Form R and the Toxic Release Inventory (TRI), which compiles the results of Form R reporting. The definitions of *release* and *transfer* are subject of continuous discussion and clarification. TRI and EPCRA have recently been the focus of a number of environmental groups who believe that TRI represents a source of pollution rather than of use of materials.

13.4.2.6 Pollution Prevention Act of 1990. This act added additional reporting responsibilities to Form R filers, including information on "Total Waste Generation," "Emission Reduction Techniques," and "Anticipated Future Reductions."

13.4.3 Waste Disposal and Control Impacts

As you can see from the preceding section on regulations, all forms of waste emission and disposal are impacted by at least one federal regulation as well as any applicable state, regional, or local ones. These all generally apply to certain materials that are classed as hazardous to the environment or to human health. The regulations generally require significant cost by the PWB manufacturer or assembler in terms of disposal fees, transportation costs, insurance risk, documentation, waste treatment equipment, material and labor, etc.

Several companies have sought to eliminate or reduce the use of materials captured in the various regulations to the point that they are no longer subject to the regulations—they are no longer using regulated materials. This is also good for the environment and human health. It should also be pointed out that these approaches are not always available in all cases.

13.5 Summary

Environmentally conscious manufacturing of printed wiring boards and assemblies encompasses a broad spectrum of activities. From the simplest approaches to pollution prevention to the ultimate in design for the environment and eco-efficient technologies, all companies should have begun the journey to sustainable enterprise. These form the basis for an *environmentally conscious strategy (ECS)*. An ECS is the next step up from the simpler *environmental management systems* that became necessary in the later 1980s. An ECS must focus not only on processes and systems, but also products along their entire life cycle on a global basis. Companies that wish to build a sustainable enterprise must begin to develop such a strategy to reach that goal.

13.6 References

1. Inside Skinny Report, *Weekly News Report,* Electronic Industry Alliance, July 27, 2000.
2. The International Material Data System is an industry wide, web-based reporting system that will assist corporate efforts in meeting environmental, health, and recycling goals as well as regulatory requirements. The list of materials on the list administered through IMDS is being developed by the automakers.
3. (a) See for example, The Development of an Industry Standard Supply-Base Environmental Practices Questionnaire, J. Andersen and H. Choong, *Conference Record of the 1997 IEEE International Symposium on Electronics & the Environment (ISEE),* San Francisco, CA, 1997, pp. 276–281; and (b) Integrating Environmental Considerations Into Supplier Management Processes at Sun Microsystems, Inc., E. Craig, *Conference Record of the 1997 IEEE International Symposium on Electronics & the Environment (ISEE),* San Francisco, CA, 1997, pp. 282–289.
4. Product-related Environmental Attributes. Technical Report TR-70, ECMA Standardizing Information and Communication Systems—CDROM Version 4.0, February 1998, 114 Rue du Rhone, CH1204 Geneva, Switzerland, or download at www.ecma.ch.
5. A life cycle inventory is actually the first step in producing a life cycle assessment. This requires that the product be followed from its inception to its eventual disposal or recycling. The mass balance of what goes into the product, as well as the energy and other resources, is determined as well as the waste and product streams resulting. This is the inventory. The effect or impact of each of these inputs and outputs on the environment is the actual assessment.
6. (a) Framework for the Development of Metrics for Design for the Environment Assessment of Products, *Proceedings of the 1996 IEEE International Symposium on Electronics and the Environment (ISEE–1996),* A.D. Veroutis and J.A. Fava, 1996, pp. 13–18; and (b) Inclusion of Environmental Aspects and Parameters in Product Specification & Product Assessment: A Report on ECMA TC 38, *Proceedings of the 1996 IEEE International Symposium on Electronics and the Environment* (ISEE-1996), F. Herman, L. Scheidt, and H. Stadlbauer, 1996, pp. 305–306.

7. This process has been done for not only the printed wiring board industry, but also for the printing and dry cleaning industries. For an overview of the implementation of this process see EPA document: Design for the Environment—Building Partnerships for Environmental Improvement, EPA-600-K-93-002—September 1995.
8. For the Making Holes Conductive (direct metallization) process, see EPA-744-R-97-001 (February 1997), Implementing Cleaner Technologies in the Printed Wiring Board Industry: Making Holes Conductive.
9. *Environmental Management System—Demonstration Project—Final Report*, NSF International, December 1996, Ann Arbor, MI (www.nsf.org).
10. For example, Building an Environmental Management System: H-R Industries' Experience, EPA 744-F 97-010, Design for the Environment, Printed Wiring Board Case Study 8 and Case Study 9, available at http://www.epa.gov/dfe or from the Pollution Prevention Clearinghouse, U.S. EPA, 401 M Street, SW (7409), Washington, D.C. 20460.
11. (a) *The ISO 14000 Handbook*, Cascio, J., editor. CEEM Information Services with ASQC Quality Press, 1996; (b) Environmental Management Systems: An Implementation Guide for Small and Medium-Sized Organizations, NSF International, November 1966; and (c) *ISO 14000: A Guide to the New Environmental Management Standards,* Tibor, T., with Ira Feldman, Irwin Professional Publishing, 1996.
12. Inside Skinny Report, *Weekly News Report,* Electronic Industry Alliance, May 31, 2000 p. 1, July 27, 2001 p. 2, and other dates for the same report.
13. Environmental Performance and Shareholder Value, J. Blumberg, A. Korsvold and G. Blum, Publication of the World Business Council on Sustainable Development. Available from E & Y Direct, P.O. Box 934, Bournemouth, Dorset BH8 8YY, UK, or downloaded in PDF format from www.wbcsd.org.
14. See, for example, the Dow Jones Sustainability Group Index at www.sustainability-index.com and sites and organizations such as the Coalition for Environmentally Responsible Economies (CERES) at www.ceres.org and the Social Investment Organization found at www.socialinvest.com.
15. California has banned CRT monitors from municipal landfills. Several states either have or are in the process of banning mercury-containing LCDs from municipal landfills.
16. EPA 744-R-00-013A, available in draft form on the EPA DFE Surface Finishes web site: http://www.epa.gov/opptintr/dfe/pwb/ctsasurf/pwb-pub.htm.
17. (a) Examining the Environmental Impact of Lead-Free Soldering Alternatives, Laura Turbini et al., *Proceedings of the IEEE International Symposium on Electronic and the Environment, 2000,* pp. 46–53; and (b) Lead Free Solders–A Push in the Wrong Direction?, Ed Smith, *IPC Printed Circuits Expo '99 Proceedings,* 1999, pp. F10–1–6.
18. Dioxins have been supposedly linked to certain types of cancer and are considered highly hazardous. Their formation can be caused by the presence of halogens such as chlorine or bromine at certain temperatures in the presence of organic materials. The source of the halogens can be other than flame retardants, however, and can cause this reaction to occur.
19. Regulatory Status of the Flame Retardant Tetrabromobisphenol A, Marcia Hardy, *Proceedings of the IPC Expo 2000*, 2000, pp. S16–1–6.
20. The Impact of Non-brominated Flame-retardants on PWB Manufacturing, Jack Fisher, *IPC Review,* May 2000, p. 16.
21. However, some work has been done to show that at least some of the replacement materials are not without environmental and health risks. See, for example, the references cited Ref. 20.
22. See, for example, *Environmentally Benign Manufacturing*, WTEC Report, T. G. Gutowski et al., April 2001, International Technology Research Institute, pp. 87–88 and references cited therein, and Toxicology of Commercial PBDPOs and TBBPA, M.L. Hardy, presented at IPC Printed Circuits Expo, April 5, 2000.
23. Tertiary Recycling Process for Electronic Materials, R. Allred (Adherent Technologies), *ARPA Environmentally Conscious Electronics Manufacturing Workshop,* Raleigh, NC, September 19–21, 1995.

24. Recycling: A *PET* Project, *DuPont Magazine Online,* November 23, 1997.
25. Delphi press release, May 2, 2000; see also web site at www.delphiauto.com. Dupont also ran a polyester recycling business for several years under the trade name "PetroTec."
26. For example, Hewlett Packard already has a "take-back" program to recycle the plastics in its printer toner cartridges; Product Stewardship: Providing Customer Focused Solutions for Environmental Responsibility, J. Heusinkveld, Presentation at the 1998 International Symposium on Electronics and the Environment, Chicago, IL, May 4–6, 1998.
27. See, for example, Chapters 3 and 6 in *Water and Waste Control for the Plating Shop* (2nd edition), by J.B. Kushner & A.S. Kushner, Gardner Publications, Inc., 1981.
28. See, for example, Water Reuse Project, Jeff Erb, *IPC Expo 1998 Proceedings,* p. S07-2-3.
29. See, for example, *Pollution Prevention Work Practices,* EPA document EPA 774-F-95-004, July 1995 [Design for the Environment–Printed Wiring Board Case Study 1].
30. *A Continuous-Flow System for Reusing Microetchant,* EPA document EPA 774-F-96-024, December 1996 [Design for the Environment–Printed Wiring Board Case Study 5].
31. Water Reuse for Printed Circuit Boards–When Does It Make Sense?, *Proceedings of the Technical Conference—IPC EXPO '98,* April 1998, J.M. Hosea, pp. S07-1-1-4.
32. *Pollution Prevention and Control Technologies for Plating Operations,* by George Cushnie, 1994, National Center for Manufacturing Sciences, 3025 Boardwalk Drive, Ann Arbor, MI 48108.
33. *Printed Wiring Board Pollution Prevention and Control: Analysis of Survey Results,* EPA 744-R-95-006, September 1995, USEPA.
34. *On-Site Etchant Regeneration,* EPA 744-F-95-005, Printed Wiring Board Case Study 2, July 1995, USEPA.
35. *Opportunities for Acid Recovery and Management,* EPA 744-F-95-009, Printed Wiring Board Case Study 3, September 1996, USEPA.
36. Waste Minimization through Solution Recycling, *Circuitree,* December 1992, p. 10.
37. Innovative Uses For IMPRINTED, U-Shaped PWB Traces & Microvias, George D. Gregoire, *IPC Expo 1998 Proceedings,* pp. S01-1-5, April 26–28, 1998.
38. (a) Chapter 3, Conductor Via (Through-hole) Formation, *Dry Film Photoresist Processing Technology,* Karl Dietz, Electrochemical Publications, 2001, pp. 34–36 and (b) Chapter 2, Microvias, Built-up Multilayers, and High Density Circuit Boards, *High Performance Printed Circuit Boards,* Charles Harper, Ed., McGraw-Hill, 1999 pp. 2.1–2.16.
39. Microvias Using a Conductive Paste to Replace Electroless Plating, *Proceedings of the IPC Printed Circuits Expo 2001,* pp. S13-3-1-5.
40. For liquid, see Direct Imaging–Major Trends and Impact on PCB Manufacturers, *Proceedings of the IPC Printed Circuits Expo 2001,* Itzhak Taff, 2001, pp. S04-1-1-5. (b) For dry film, see LDI–The Real Benefits, *Proceedings of the IPC Printed Circuits Expo 2001,* Haring Fritz, 2001, pp. S17-4-1-4. See also (c) Laser Direct Imaging—Trends for the Next Generation of High Performance Resists for Electronic Packaging, *Proceedings of the IPC Printed Circuits Expo 2000,* Mark R. McKeever, 2000, pp. S12-1-1-5 and (d) sections on Laser Direct Imaging in Chapter 10—Exposure, in *Dry Film Photoresist Processing Technology,* Karl Dietz, Electrochemical Publications, Ltd., 2001.
41. UV Laser Direct Patterning (LDP) of HDI Applications, *Proceedings of the IPC Printed Circuits Expo 2001,* D.J. Meier, 2001, pp. S04-3-1-5.
42. Improved Yields and Etching Latitude with Wet Lamination of Thin Dry Film Photoresists, *Proceedings of the IPC Printed Circuits Expo 2001,* Thomas Foreman et al., 2001, pp. S17-3-1-8.
43. Development of Polyimide-Based Capacitance and Resistors for Integral Passives, Gary Min et al., *Proceedings of the IPC Printed Circuits Expo 2000,* pp. S08-4-1-6 and references cited therein.

44. Integral Planar Resistors in Low Cost High Density Substrates, *Proceedings of the IPC Printed Circuits Expo 2001*, Daniel Brandler, 2001, pp. S08–5–1.
45. Embedded Ceramic Passives in PWB: Process Development, *Proceedings of the IPC Printed Circuits Expo 2001*, John J. Felten and William J. Borland, 2001, pp. S08–2–1–4 and references cited therein.
46. Profile of the Electronics and Computer Industry, EPA Office of Compliance Sector Notebook Project, EPA/310-R-95-002, 1995, p. 91.

Index

A

Additive processes 241–242
Adhesives 14, 21, 39, 220, 458
   conductive 310
   inorganic 159
   organic 159
   silver-filled epoxy 160
Alpha particles 180
Alpha-particle absorption 182
AlSiC 521–522
Alumina, *see* Aluminum oxide
Aluminum nitride 508–511
Aluminum oxide 507–508
Anodes 57
Area array tape 172
Atmospheric pressure CVD 98
Atomic structure 52
Atomization 311

B

Back grinding 100
Backplanes 1
Backup material 232
Ball grid array 19
Ball grid arrays 3
Barrel cracks 241
Barrier layer 238
Bell, Alexander Graham 4
Belt speed 341–342
Bendix Corporation 18
Benzotriazole 354
Berry, Arthur 8, 213
Beryllia, *see* Beryllium oxide
Beryllium oxide 507–509
BGA 307, 339
Bipolar transistors 65
Bismaleimide triazine (BT) 144
Blind vias 260
Board warpage 336
Bond pads
   dual staggered row 165
   in-line 165
   triple stagger-row 166

Boron 60
   dopant 86
Boron nitride 513–514
Brazing pastes 310
Brominated flame retardants 637
B-stage 223
Build-up board 261

C

Capacitance 306
Capacitors, decoupling 69
Capillary tip 161
Carrier plate 249
Cathodes 57
Cavity UFP wire bond packages 290
Ceramic ball grid array rework 299
Ceramic flip chip modules 284
Ceramic materials 507–516
   aluminum nitride 508–511
   aluminum oxide 507–508
   beryllium oxide 507–509
   boron nitride 513–514
   diamond 512–513
   silicon carbide 515–516
Ceramic, bonding mechanisms 484–485
Ceramic, electrical properties 500–503
   breakdown voltage 500–501
   conductivity 500
   dielectric properties 501–503
   dissipation factor 502–503
   loss tangent 502–503
   resistivity 500
Ceramic, fabrication 503–506
   chemical vapor deposition, infiltration 505–506
   extrusion 505
   hot isostatic powder pressing 505
   powder pressing 505
   roll compaction 505
   sintering 506
   tape casting 504–505

660 Index

Ceramic, forming and shaping 524–526
Ceramic, mechanical properties 494–500
   coefficient of thermal endurance 499–500
   compressive strength 496–498
   hardness 498–499
   Hooke's law 495
   maximum stress 496
   modulus of elasticity 494–495
   modulus of rupture 496
   plain strain fracture toughness 497–498
   tensile strength 496–498
   thermal shock 498–499
Ceramic, surface properties 486–490
   camber 489–490
   surface roughness 486–489
   waviness 489–490
Ceramic, thermal properties 490–494
   specific heat 491–493
   temperature coefficient of expansion (TCE) 493–494
   thermal conductivity 490–492
Cermets 12
CFCs 358
Chemical recovery systems 645
Chemical vapor deposition 96
Chemicals in the workplace 415
Chip attach 160
Chip on board 19
Chip scale packaging 149
Chlorine atoms 55
Clean Air Act 416, 652
Clean rooms 90
Clean Water Act 416
Cleanability 338
Cleaner technology substitute assessment 633
   use clusters 632–633
Clock frequency 78
CMOSFET, see Transistors
Code marking
   ink printing 182
   laser scribing 182
Coefficient of thermal expansion 22, 313
Co-laminated printed circuit boards 262
Collet 101
Composite
   AlSiC 521–523
   aluminum nitride 520–522
   ceramic–ceramic 518
   ceramic–glass 518–522
   definition 483
   diamond 512–513
   Dymalloy® 523–524
   forming 524–526
   high-temperature cofired ceramic 519–520
   low-temperature cofired ceramic 520–521
   materials 515–524
   metal–ceramic 522–524
   organic–ceramic/glass 518
   organic–organic 518
Computer aided design 168
Conductive adhesives 310
Conductive inks 222
Conformal coatings 25, 429
   acrylic 431
   application methods 441
   drying and curing 447
   epoxy 431
   health and safety issues 450
   polyurethane 431
   prerequisites 437
   selection 433
   silicone 432
   specifications 430
   types 430
Controlled atmospheres 340
Corner cracks 241
Covalent bonds 55
Covalent solids 58
Creep 316
Cross-sectional analysis 240
Crystal orientation 87
Crystal structure 314
CSP 339
CTE mismatch 142
CTSA, see Cleaner technology substitute assessment
CVD, see Chemical vapor deposition

Index   661

Czochralski method 83

D
DCA 307
De Forest, Lee 5
Decoupling capacitors 69
Delamination 241
Depanelization 238
Depletion areas 69
Depletion region 62
Deposition layer 97
Design for the environment 631
    DFE 632
    DfE 632
Development 248
Dew point 342
Dielectric 1, 4, 91, 456
Dielectric constant 128, 199
Digital switches 68
Diodes 57
DIP 307
Dissipation factor 199
Dopants
    boron 86
    phosphorus 86
Doping 61
Double-clad laminates 228
Double-sided circuit process 245
Drain (FET) 67
Drill registration 241
Drilling 232, 250
Dry etching 96
Ducas, Charles 9, 213
Dymalloy® 523–524
Dynamic viscosity 311

E
Eco-efficiency 631
Ecolabels 636
Eco-Management and Audit Scheme (EMAS) 633
Edge beveling 240
Edison, Thomas 6, 212
E-glass 203, 205, 626
Eisler, Paul 17, 213
Electrical conductivity 313
Electrolytic deposition 311
Electromagnetic interference
    capacitance 106

crosstalk 106–107
electrostatic discharge 106, 186, 602
inductance 106
Electromigration 99
Electron beam patterning 94
Electron conductivity 313
Electron tube 57
Electron vacancies 61
Electronic Industries Alliance 108
Electronic Industries Association of Japan 115
Electronic Industries Association of Japan (EIAJ) 108
Electronic Industries Association, *see* Electronic Industries Alliance
Electroplating 350
    pattern plating 236
Electrostatic discharge 106, 186
EMI 310
Encapsulation 147
End-of-life considerations 648
Entry material 232
Environmental costs
    assigning correctly 641
    unrecognized downstream 641
Environmental management systems 642
    documented procedures 642
    procedures 642
    value of 634
Environmental noncompliance 650
Environmental regulations
    applicable documents 424
    Clean Air Act 416–417
    Clean Water Act 416
    European 421
    hazardous air pollutants 417
    heavy metals 421
    Maximum Achievable Control Technology standards 418
    National Ambient Air Quality Standards 417
    National Emission Standard for Hazardous Air Pollutant program 418
    noncompliance 650
    ozone-depleting substances 418

PWB fabrication and cleaning 402–427
Resource Conservation and Recovery Act 416, 422
sludge 421
small-quantity generators 423
solid wastes 420
state laws 650
volatile organic compounds 417
waste water 419
Environmentally conscious manufacturing 640
costs 643
Epoxy 119
difunctional resin 196
multifunctional resin 197
Epoxy underfill 149
Epoxy writing 160
Etchback 241, 250
Etching 92, 236, 248
Eutectic 317

F

Fatigue 316
FET transistors 67
Filament, in vacuum tube 56
First-time yield 308
Flexible circuits 254
Flexible laminates 219
polyester 220
polyethylene naphthalate 220
polyimide 220
Flip chip 105, 283
ceramic 284
organic modules 286
Flip chip interconnection 100
Flip chips 339
Fluxing 395
Flying wires 74
Foil cracks 241
Formica® 191
Forward bias 63

G

Gallium arsenide 58, 60
Gas flow rate 341
Gate (FET) 67
Gate count 78

Germanium 60
Glass cloths 218
Glass transition temperature 22, 197, 199, 204, 219, 604
Grain-boundary sliding 317
Greenhouse effect 414

H

Halogen-free laminate materials 639
Hanson, Albert 6
Hazardous substances, releases 653
Hazardous waste 651
HDI, *see* High-density interconnection
Heat spreaders 187
Hermeticity 118
High-density interconnection 257
substrate types 258
Hole cleaning 250
Hole preparation 232
Hole wall pull away 241
Hot air solder leveling 238, 347, 354, 356
Humidity 342
Hybrid circuit
definition 529
Hybrid circuits
copper metallization 581–589
active metal brazing 587–589
direct bond copper 581–585
plated copper technology 585–587
die bonding 602–607
hybrid circuit assembly 581–612
die bonding 602–607
wire bonding 607–612
hybrid circuit design 621–625
hybrid circuit packaging 612–621
hermetic 612–617
non-hermetic 617–621
HyperBGA package 282

I

Imaging 234
dry film 235
screen print 235
Inductance 306
Inner diameter saw 87

Index 663

Inner-lead bonding (ILB) 169
Integrated circuit packaging
  alpha-particle emission 125
  alumina ceramic 125
  area array packages 111
  ball grid arrays 21, 105, 114, 117
  ball placement system 183
  ball-wedge bonding 121
  bond pads 130, 158
  bond-pad-limited chips 155, 166
  bottom-brazed DIPs 111
  BQFP 115
  build-up process 153
  bump pitch 175
  CCGA 133
  ceramic ball grid array 132
  ceramic cap 124
  ceramic pin grid array 130
  CERQUAD 114
  chip attachment 157
  chip cavity 136
  chip encapsulation 121
  chip on board (COB) 134
  cofired laminated ceramic
    process flowchart 126
  column grid arrays 21
  compliant layer 150
  copper slugs 136, 154
  direct chip attachment 149
  dual in-line (DIP) 110
  elastomer layer 150
  embedded leadframe 123
  epoxy underfill 175
  eutectic bonding 132
  eutectic chip attach 160
  FC pads 174
  first-level assembly 156
  first-level interconnection, 157
  flame retardant additives 179
  flat packs 115
  flip chip 105
  flip chip attachment 152
  glass fillers 125
  glass sealing 124
  glob-top 136
  gull wing leads 115
  heat spreading internal planes 154
  hermetic 123
  hermetic packages 177
  hermeticity 118
  impedance 128
  inductance 128
  injection molded 121
  inorganic adhesives 159
  laminated ceramic technology 111
  laminated plastic packages 187
  lead coplanarity 114
  lead finishing 125
  lead pitch 114
  lead trim and form 125
  leaded 114
  leadframe 118
  leadframe paddle 158
  leadless 114
  leak testing 125
  lid sealing 124
  line capacitance 128
  low-alpha-radiation molding compounds 179
  low-temperature cofired ceramic 134
  MCM 134
  MCP 134
  moisture sensitivity 147
  molded plastic technology 110
  molding compounds 180
  molding process 178
  MQFP 115
  multilayer ceramic package 118
  multilayer plastic 118
  multilayer substrate 138
  nonhermetic 118
  organic adhesives 159
  organic sealants 177
  perimeter packages 114
  peripheral-lead-type packages 186
  pin grid arrays 21, 105
  plastic 118
  plastic ball grid array (PBGA) 135
  plastic leaded chip carrier 114
  plastic pin grid array (PPGA) 135
  plastic substrates 118
  popcorning 142
  pressed ceramic technology 110
  propagation delay 128
  quad flat pack 105, 115

resin dam 152
ribbon leads 150
room-temperature vulcanized rubber (RTV) systems 138
saw street separation 184
seal glass 125
seal ring 132
shrink DIP 110
side-brazed DIPs 111
silica fillers 178
single-in-line package (SIP) 112–113
singulation 151
skinny DIP 110
small-outline 114
soft error reliability 125
solder ball pads 151
solder ball pitch 114
solder balls 133, 151
solder bump interconnection 174
solder columns 133
solder joint reliability 147
solder mask 151
solder reflow 114
solderability 121
surface mount 107
tape automated bonding 105
thermal conductivity 153
thermal vias 136
thermoplastic polymer 121
thermosonic gold ball-wedge 136
through-hole mount 107
TQFPs 117
tungsten powder 126
ultrasonic aluminum wedge-wedge 136
vias 130
wedge-to-wedge wire bonding 124
window-frame package 123
wire bonding 105, 158
wire sweep 178
zigzag-in-line package 112–113
Integrated circuits 70, 105, 306
  chip size 83
  clock frequency 78
  DRAMs 78
  feature size 93
  gate count 78
  Kilby IC 73
  MOS 78
  packaging technology 307
  saw streets 83, 184
Interconnections
  first level
    C4 technology 270
    flip chip 270
    wire bond 269
  second level
    ceramic ball grid array 276
    ceramic column grid array 276
    double-sided double-pass 273
    pin-in-hole technology 272
    surface mount technology 273
  thermosonic 129
Intermetallic compounds 334
Interposers 4
Intrinsic semiconductor 60
Ionic conductivity 313
IPC-4103, "Specification for Base Materials for High-Speed/High-Frequency" 208
IPC-4104, "Specification for High Density Interconnect (HDI) and Microvia Materials." 207

J

JEDEC, *see* Joint Electron Device Engineering Council
JFET, *see* Transistors
Joint Electron Device Engineering Council 108, 115
Just-in-time manufacturing 107

K

Kapton® 24
Kilby, Jack 73
Known good die 150, 258

L

Laminate reinforcement 216
  aramid fibers 217
  paper 217
  quartz cloth 217
Laminate types
  BF 229
  BI 229
  CEM -1 229

Index 665

CEM -3 229
CEM-1 230
CEM-3 230
FR-1 192, 229
FR-2 194, 229
FR-3 194, 229
FR-4 229–230
FR-5 229–230
G-10 229–230
GC 229
GE 229
GF 229
GH 229
GI 229
GM 229
GP 229
GT 229
GX 229
GY 229
PX 229
SC 229
XXXP 229
XXXPC 229
Laminates
  double-clad 228
  single-clad 225
Lamination 223
  batch 224
  continuous 225
  lay-up for 249
  methods 249
  vacuum-assisted 225
  vacuum-assisted autoclave 225
Laser drilling 307, 646
Laser marking of wafers 90
Laser scribing 525
Lasers
  ablative laser beams 647
  exposure of photoresist 647
LCCC 307
Lead finishing 125
Lead trim and form 125
Leak testing 125
Life cycle assessment 632
Lifted lands 241
Liquidus temperature 313
Low-pressure CVD 98

M

Mass lamination 251
Material "black lists" 632
MCM, *see* Multichip module
Measling 216
Melting point 314
Metal deposition
  electroless 351
  electrolytic 351
  immersion 351
Metal foils 220
  electrodeposited 221
  rolled 222
  wrought 221
Metal-core PCBs 252
Metallization 98, 232
  molybdenum 132
Metal-oxide semiconductors 75
Microstructure of solder 336
Microvias 260, 646
  materials 205–207
Moisture preconditioning 108
Moisture sensitivity 147, 185
Molybdenum metallization 132
Monocrystalline silicon 86
Montreal Protocol 358
MOSFET, *see* Transistors
Multichip modules 625–630
  definition 530
  MCM-C 626–629
  MCM-D 629
  MCM-L 625–626
Multilayered ceramic packages 285
Mylar® 24

N

Nail heading 241
National Bureau of Standards 12
National Center for Manufacturing
  Sciences 301
National Electrical Manufacturers
  Association 193
National Electronics Manufacturing
  Initiative, Inc. (NEMI) 302
Negative-acting photoresist 92
Nodule 241
Nomex® 24
n-type semiconductor 61

O

Organic coatings 355
Organic flip chip modules 286
Organic resins 218
 bismalamide triazine 219
 cyanate esters 219
 epoxies 219
 phenolics 219
 polyimides 219
Organic solderability preservatives 26
Organic solderability protectants 238
Organic surface protectants 637
Organic wire bond modules 289
Outer-lead bonding (OLB) 169
Overetched circuits 237
Overmolding 147
Oxidation 91
Oxidation rate 92
Oxygen level 341, 343

P

Panasonic minidisk player 332–333
Panel plate 245
Parasitics 306
Parolini, Cesar 10
Passivation layer 103
Pasty range 313
Pattern plate 245
Peak temperature 334, 342
Peel strength 199
Periodic table 52–53
PGA 307
Phase transition temperature 313
Phosphorus 60
 dopant 86
Photoablation 30
Photoelectronic vacuum tubes 57
Photoimagable resists 234
 negative working resists 234
 positive working resists 234
Photolithography 8
 contact printing 93
 direct wafer stepping 93
 electron beam 93
 photomasking 93
 photoresist material 92
 proximity printing 93
 scanning projection printing 93
 x-ray 93
Photomasking 92
Photoresist 647
Photoresist materials 92
Plasma etching 307
Plasma-enhanced CVD 98
Plastic deformation 314–315
Plastic IC package cracking 336–337
Plastic range 313
PLCC 307
p-n junction 61
Poisson's ratio 155
Pollution prevention 640–641, 644
 chemical maintenance 645
 economic improvement 643
Polycrystalline structure 83
Polymer thick film 4
Popcorning 142
Positive-acting photoresist 92
Post separation 241
Precipitation hardening 315
Prepreg 223
Pressure-cooker test 108
Print-and-etch processing 243
Printed circuit boards
 3-D flex 11
 additive vs. subtractive 8, 26
 as chip carriers 40
 base materials
  introduction 191
 bismaleimide triazine 23
 ceramic 37
 classes 3–4
 conductor patterning 28
 conductors 24
 connections 30
 crossovers 3, 35
 cyanate ester 23
 definition 1
 difunctional epoxy resin 196
 double-access 34
 double-sided 3, 35–36
 electrical properties of resins 208
 electroless Au/electroless Ni 347
 electroplated Au/Ni 347
 electroplated Sn-Ni 347
 epoxies 23

Index

eyelets 214
flex packaging 41
flip chip strip 42
FR-4 22
global market 209
halogen-free movement 23
history 4–19, 212
immersion Au/electroless Ni 347
immersion Pd 347
immersion Pd/electroless Ni 347
jumpers 3, 35
laminate types
    bismaleimide triazines 205
    CEM-1 201, 203, 208
    CEM-3 202–203
    composite materials 199–203
    FR-2 201
    FR-3 201
    FR-4 208
    FR-4 materials 194–199
    high-performance materials 203–207
    high-speed/high-frequency materials 207–208
    paper-based materials 192–194
    polyimide 203
    polyimide vs. FR-4 204
    polytetrafluroethylene 207
laminate, definition 191
materials 21–26
microvia materials 205–206
    reinforced 206
    unreinforced 206
multifunctional epoxy resin 197
multilayer 4, 37
multilayer flex 38
photo-etching process 17
Placir process 21
polyimides 23
prepreg 27
prepreg, definition 191
processes 14–16, 20
PTH reliability 197
purpose 2
reinforcements 192
repair 204
resin systems 192, 208
rigid multilayer 37

semiadditive 29
single-layer 32
single-sided 3
singulation/depanelization 31
solder masks 31
surface finish 347
tape automated bonding 41
trends 45–47
woven E-glass 195
Printed wiring boards 363
  cleaning
    aqueous 378
    basic processes 363–364
    chemical 376, 386
    contaminants 367
    drying 378, 388
    final rinse 388
    handling after stripping 383
    mechanical 373, 383, 400
    metallic resist stripping 379
    no-clean process 365
    nonperoxide 380
    organic resists 381
    peroxide 379
    residuals 380
    solder mask over bare copper 379
    storage 393
    substrates 366
    surface preparation 383
    terms and definitions 367
    test methods 378
    testing 389
  fabrication
    environmental considerations 402–414
    environmental regulations (OSHA, EPA) 415–427
    flex and rigid flex 459–482
  fabrication and cleaning 368–379
  fabrication steps 368
  flex and rigid flex
    adhesive 458
    conductors 458
    covercoat (coverlay) 457
    dielectrics 456
    performance classes 455–456
    types 454–455

fluxing 395
handling 399
history 2
inspection and preclean 395
multilayer flex 469
rigid flex 475
single-sided flex 459
solder coating 396
solder leveling 394, 396
solder mask over bare copper 379
solder mask over tin-lead 397
Printed-through holes 245
PTH attach methods 292
p-type semiconductor 61

Q
QFP 307

R
Reaction chambers 97
Recovery 315
Recrystallization 315
Recycling 639, 649
Reflow soldering 184
Reflow temperature profile 335
Refractive optics 94
Refractory metals 99
Release films 249
Release tape 101
Rent's rule 269
Repair
  acrylic conformal coatings 431
  ceramic packages 616
  coatings 449
  epoxy 431
  metal packages 616
  polyimides 219
  printed circuit boards 204
  silicone 432
Residual stress 336
Residue 338
Resin smear 241
Resin-coated copper 220
Resist coating 232, 248
Resist stripping 236, 248
Resistance 306
Resistor trimming 590–601
Resource Conservation and Recovery
  Act 416

Reverse bias 63
Reverse osmosis 408
Rigid flex circuits 256
Room-temperature vulcanized rubber
  138
Rule of mixtures 516–517

S
SARA form R 653
Sb addition 315
Scanning electron microscope 169
Schoop, Max 9, 213
Seam sealer 129
Second-level board assembly 108
Semi-additive processes 242
Separator plate 249
Sequential laminated printed circuit
  board 263
S-glass 208
Silicides 99
Silicon 58
Silicon carbide 515–516
Silicon ingots 83
Silicon nitride 100
Silicon nuggets 83
Silicon wafers 83
Silver spot plating 119
Silver-filled epoxy 160
Single-clad laminates 225
Single-layer tape 169
SMT attach methods 294
Soft error reliability 125
Soft errors 182
SOIC 307
Solder 16, 637
  coatings 464
  lead-containing 309
  lead-free 301, 309
  lead-free alloys 344
  masks 458
  paste 310
  powder 311
  reflow 114, 475, 482
Solder alloys
  10Sn/90Pb 316
  42Sn/58Bi 317
  43Sn/43Pb/14Bi 317
  5Sn/85Pb/10Sb 316

Index   669

5Sn/95Pb 316
62Sn/36Pb/2Ag 317, 356
63Sn/37Pb 317, 322
85.2Sn/4.1Ag/2.2Bi/0.5Cu/8.0In 319
88.5Sn/3.0Ag/0.5Cu/8.0In 319
91.5Sn/3.5Ag/1.0Bi/4.0In 319
92.8Sn/0.7Cu/0,5Ga/6.0In 319
92Sn/3.3Ag/4.7Bi 319
93.3Sn/3.1Ag/3.1Bi/0.5Cu 319
95.4Sn/3.1Ag/1.5Cu 319
96.2Sn/2.5Ag/0.8Cu/0,5Sb 319
96.5Sn/3.5Ag 319–320
99.3Sn/0.7Cu 319–320, 330
Sn/0.5-0.7Cu/5.0-6.0In/0.4-0.6Ga 320
Sn/3.0-3.5Ag/0.5-1.5Cu 320
Sn/3.0-3.5Ag/0.5-1.5Cu/6.0-8.0In 320
Sn/3.0-3.5Ag/1.0-4.8Bi 320
Sn/3.0-3.5Ag/3.0-3.5Bi/0.5-0.7Cu 320
Sn/3.0-4.1Ag/2.2Bi/0.5Cu/8.0In 320
Sn/Ag 316
Sn/Ag/Bi 318, 321
Sn/Ag/Bi/Cu/In 318
Sn/Ag/Bi/In 318, 323–324, 328
Sn/Ag/Cu 318, 356
Sn/Ag/Cu/Bi 318, 325
Sn/Ag/Cu/In 318, 327–328
Sn/Cu 328
Sn/Cu eutectic 318
Sn/Cu/In/Ga 318, 329
Sn/Sb 316
Solder balling 336–337, 340
Solder balls 328
Solder beading 336–337
Solder bump interconnection 174
Solder bumps 148
Solder coatings 637
Solder joint reliability 147
Solder joint voids 336, 338
Solder masks 25, 228
    heat-curable resins 231
    photoimagable resins 231
    UV-curable resins 231
Solder void 241

Solderability 121, 340, 347, 350
Soldering 24–26, 31, 159, 211, 292–294, 301–303, 366, 570, 604–605, 607, 637
    cooling rate 335
    inert and reducing atmosphere 340–344
    peak temperature 335
    preheating temperature 335
    preheating time 335
    reflow 328–340
    Sn/Cu/In/Ga 330
    solder grapes 276
    spattering problem 328
    stress fracture solder joints 276
Solution hardening 315
Source (FET) 67
Sprague Electric Company 212
Sprague, Frank 212
Sputtering 99
Stack drilling 233
Standard multilayer circuit process 247
Stencil aperture design 345
Stencil making
    chemical etching 349
    electroforming 349
    electropolishing 349
    laser cut 349
Stencil materials
    alloy 42 348
    brass 348
    electroforming 348
    molybdenum 348
    performance 348
    stainless steel 348
Stencil thickness 344
Stencils 346
Stencils, selection 347
Stiffener ring 148
Strain hardening 315
Substrates 70
Subtractive processes 241–242
Superplastic deformation 315
Surface finishes
    Au/Ni 351, 356
    electroless Pd/Cu 352
    electroless Pd/Ni/Cu 352

electrolytic Au/electroless Ni 353
electrolytic Au/electrolytic Ni 353
HASL 356
HASL Sn/Cu eutectic 354
immersion Ag 354, 357
immersion Au/electroless Ni/Cu 353
immersion Au/electroless Pd/Cu 352
immersion Au/electroless Pd/electroless Ni/Cu 353
immersion Au/electrolytic Ni 353
immersion Bi 353
immersion Pd 357
immersion Pd/Cu 352
immersion Sn 354
Ni interlayer 352
Ni/Au 357
organic 356
OSP 357
Pd/Cu 356
Pd/Ni 356
Surface insulation resistance 430
Surface mount packages 107
Surface tension 313–314
Syringe dispensing 160

T

Technographic Printed Circuits, Ltd. 18
Temperature-humidity-bias test 108
Test probes 100
Thermal conductivity 128, 313
Thermal lag material 249
Thermionic effect 57
Thermoset 121
Thermosetting 119
Thermosonic interconnections 129
Thick film 530–571
  paste 530–538
  physical properties 535–538
Thick film materials 310, 555–571
Thin film 571–581
  electroplating 576
  evaporation 573–576
  photolithographic processes 577
  sputtering 571–573
Thin film materials 577–581

barriers 579
conductors 555–559, 579–580
dielectrics 567–569
overglazes 570–571
resistors 559–567, 578–580
Thin film processes 538–555
  drying 548–549
  firing 549–555
  screen printing 538–548
Three-layer tape 169
Through-hole mount packages 107
Transfer printing 160
Transistors 64
  bipolar 65
  complementary metal oxide semiconductor field-effect 70
  depletion areas 69
  FET 67
  junction field-effect 68
  metal oxide semiconductor field-effect 68, 70
  n-channel 68
  switching speeds 78
Triodes 57
Two-layer tape 169

U

Underetched circuits 237
Unpolymerized wafer areas 92
Unreinforced laminates, see Flexible laminates
UV light 94

V

Vacuum tube, history 56
Vacuum tubes 9
Valence electrons 54
Via fillers 646
Vias 30, 130
VOC, see volatile organic compounds
Voids 241
Volatile organic compounds 411, 416, 449

W

Wafers 87
Waste Electrical and Electronic Equipment (WEEE) Directive 301

Waste streams 416
Water assisted lamination 647
Water recycling 406
Water vapor pressure 342
Wet etching 96
Wettability 338
Wetting ability 334, 336
Wire bonding 25, 100, 158, 349, 607–612
   thermocompression 161
   thermosonic 161
   ultrasonic 161
Wire saw 90
Wire-bond interconnects 119
Wire-bondable tape 172
Working with suppliers 649
Woven E-glass 195

Y

Young's modulus 155

Z

Zero defect 107
z-wire 214

About the Editor

Charles A. Harper is President of Technology Seminars, Inc., an organization which provides educational training courses in electronic packaging and manufacturing to business and industry. He also had an earlier esteemed career with Westinghouse in this field. Widely recognized as one of the leaders in this industry, he has authored over a dozen highly respected books and is among the founders and past presidents of the International Microelectronics and Packaging Society (IMAPS). He is also Series Editor for the McGraw-Hill Electronic Packaging and Interconnection Series, a widely used book series in the electronics industry. Mr. Harper is a graduate of the Johns Hopkins University School of Engineering, where he has also served as Adjunct Professor.